THE ROYAL
HORTICULTURAL SOCIETY

WHAT PLANT WHERE

ENCYCLOPEDIA

LONDON, NEW YORK, MUNICH,
MELBOURNE, DELHI

Editor-in-Chief Zia Allaway

Senior Editors Chauney Dunford, Helen Fewster
Senior Art Editor Joanne Doran
Editors Jenny Hendy, Annelise Evans
Project Art Editors Vicky Read, Alison Shackleton
Senior Producer Alex Bell, Seyhan Esen
Pre-Production Producer Andy Hilliard
DK Images Romaine Werblow, Claire Bowers
Jacket Designer Nicola Powling
Managing Editor Penny Warren
Managing Art Editor Alison Donovan
Art Director Jane Bull
Publisher Mary Ling
RHS Editor Simon Maughan
RHS Publisher Rae Spencer-Jones

DK INDIA
Senior Editor Nidhilekha Mathur
Senior Art Editor Anchal Kaushal
Editors Divya Chandhok, Ekta Sharma
Art Editors Tarun Sharma, Vandna Sonkariya
DTP Designer Anurag Trivedi
Managing Editor Alicia Ingty
Managing Art Editor Navidita Thapa
Pre-Production Manager Sunil Sharma
Picture Researchers Nikhil Verma, Surya Sankash Sarangi

First published in 2013 in Great Britain in association with
The Royal Horticultural Society by Dorling Kindersley Limited,
80 Strand, London WC2R 0RL
Penguin Group (UK)

2 4 6 8 10 9 7 5 3 1 – 001 – 185918 – Oct/2013

A CIP catalogue record for this book is available from the British Library

ISBN 978-1-4093-8297-3

To find out more about the RHS membership, visit www.rhs.org.uk

Printed and bound in China

Discover more at
www.dk.com

Image right: *Lilium superbum*, American Turkscap lily (see p.162)

THE ROYAL HORTICULTURAL SOCIETY

WHAT PLANT WHERE ENCYCLOPEDIA

CONTENTS

PLANTS FOR SPECIAL EFFECTS

PLANTS FOR GARDEN STYLES

PLANTS FOR SEASONAL INTEREST

PLANTS FOR COLOUR AND SCENT

PLANTS FOR SHAPE AND TEXTURE

PLANTS FOR GARDEN PROBLEMS

About this book

The core of this book is divided into two sections. The first, Plant Locations, recommends plants for different growing conditions. Part two, Plants for Special Effects, suggests plants suitable for different uses, such as for scent, and for various garden styles, such as Asian-themed.

132 Plant focus: grasses and sedges

Elegant and versatile, the slender foliage and airy flowers of grasses and sedges contrast beautifully with broad-leaved plants.

ORNAMENTAL GRASSES AND SEDGES comprise a range of shapes and sizes, from tall and slender to short and squat, and form the key ingredient of many garden styles. Those with slim leaves and delicate flowerheads, such as *Calamagrostis*, blend perfectly into naturalistic schemes, while the blue spikes of *Festuca glauca* are ideal for container displays. Sedges, including those in the *Acorus* family, make elegant features in the boggy ground beside natural water features. Although sedges and grasses look similar, they are adapted to different environments. With a few exceptions, grasses are deciduous and prefer sun and free-draining soils, while the majority of sedges are evergreen and happy in sun or shade, and moist soils. Grasses also tend to have more dramatic flower- and seedheads, adding to their decorative value. Both grasses and sedges help to heighten the beauty of winter landscapes: the leaves of deciduous species tend to remain intact throughout the season, providing useful structure in herbaceous borders.

POPULAR GRASS AND SEDGE SPECIES

USING GRASSES AND SEDGES

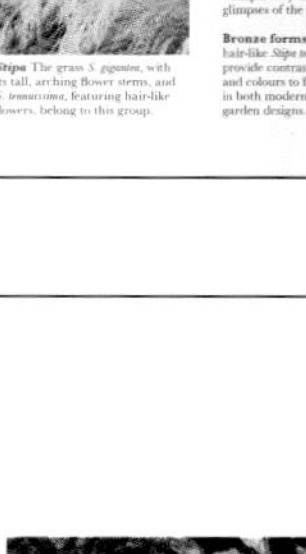

Plant focus pages Found within the Plant Locations section, these features give detailed advice on many popular groups of garden plants, such as grasses and sedges.

50 Plants for sandy soil

GARDENS IN SUN

SHRUB MEDIUM
Physocarpus opulifolius
This is a deciduous shrub, with rounded, lobed leaves. Dome-shaped, compact clusters of small, white flowers appear in early summer, followed by reddish brown fruits. Varieties include 'Diabolo' AGM (above), which sports deep purple foliage.
↕3m (10ft) ↔1.5m (5ft)

SHRUB LARGE
Pittosporum tenuifolium
An evergreen, columnar, later rounded, shrub or small tree, with purple shoots and wavy-edged, oval, glossy, mid-green leaves. In late spring, it produces tiny, honey-scented, purple flowers. The form 'Variegatum' has white-edged leaves.
↕10m (30ft) ↔5m (15ft)

PERENNIAL SMALL
***Potentilla* 'Gibson's Scarlet'**
CINQUEFOIL This clump-forming perennial has dark green leaves that resemble those of a strawberry plant. From early- to late summer, it produces saucer-shaped, bright scarlet flowers on branching stems. It also grows in part shade.
↕45cm (18in) ↔60cm (24in)

SHRUB SMALL
Prunus tenella
This is a deciduous, bushy, upright shrub, with narrowly oval, glossy, green leaves. From mid- to late spring, the stems are clothed in bright pink, saucer-shaped flowers. 'Fire Hill' is a popular red-flowered form.
↕↔1.5m (5ft)

TREE LARGE
Quercus coccinea
SCARLET OAK This deciduous tree forms a rounded head of glossy, dark green, lobed leaves. In autumn, the foliage turns bright red, producing a bold display that can last for several weeks. Small acorns form at the same time. Best in large gardens.
↕20m (70ft) ↔15m (50ft)

PERENNIAL LARGE
Romneya coulteri
TREE POPPY A vigorous, bushy, shrubby perennial, with deeply divided, grey-green leaves. It is grown for its large, fragrant, papery, golden-centred white flowers, which appear in late summer on stout stems. Mulch well with compost in autumn.
↕↔2m (6ft)

PERENNIAL MEDIUM
Rudbeckia hirta
This upright, short-lived perennial, often grown as an annual, has mid-green leaves. From summer to autumn, it produces daisy-like, golden or red flowers with brown centres. Varieties include 'Becky Mixed' (above).
↕up to 90cm (36in) ↔45cm (18in)

PERENNIAL MEDIUM
Salvia x sylvestris
WOOD SAGE This is a compact perennial, with small, dark green, aromatic leaves. From early- to midsummer, 'Mainacht' AGM (above) produces spikes of indigo-blue, two-lipped flowers that last for a long period. It is attractive to butterflies and bees.
↕80cm (32in) ↔30cm (12in)

PERENNIAL MEDIUM
Santolina chamaecyparissus
COTTON LAVENDER A rounded perennial, with soft, woolly, silvery white, finely divided foliage and yellow, pompon flowers in summer. Trim after flowering. Varieties include 'Lemon Queen' (above), which has grey foliage and cream blooms.
↕75cm (30in) ↔1m (3ft)

PERENNIAL MEDIUM
Scabiosa atropurpurea
PINCUSHION FLOWER A clump-forming perennial with lance-shaped, grey-green leaves. In summer, masses of slim stems are topped with small, domed, lilac-purple flowers, adorned with creamy anthers that look like pins. Deadhead regularly.
↕up to 1m (3ft) ↔30cm (12in)

Plant names All plants are listed under their Latin botanical names. Popular common names are also given, where they apply.

Key features The images illustrate the most important features of the plants, and will not necessarily show its overall appearance.

Plant hardiness This shows the plant's ability to survive low temperatures. However, plants may die of other causes as well.

Soil type This symbol shows the ideal soil type for the plant, although some plants will tolerate a range of growing conditions.

Sun or shade In the first section of the book, the ideal aspect for the plant is indicated by the sun or shade symbol located here.

Common names Some plants have more than one common name. Only the most widely used are given here.

PERENNIAL SMALL

Sedum spectabile ♀

ICE PLANT This sedum is a clump-forming perennial, with succulent stems of fleshy, grey-green leaves. From late summer to autumn, it produces flat heads of pink flowers, followed by winter seedheads. Varieties include 'Brilliant' AGM (above).

↕↔ 45cm (18in)

◇ ❄❄❄ (!)

Plant type and size These labels show what type the plant is, such as a tree, shrub, or perennial, and indicate its size. Use these for quick reference when choosing plants; the description provides detailed information.

Toxic plants Some plants have toxic sap and should only be handled when wearing protective gloves.

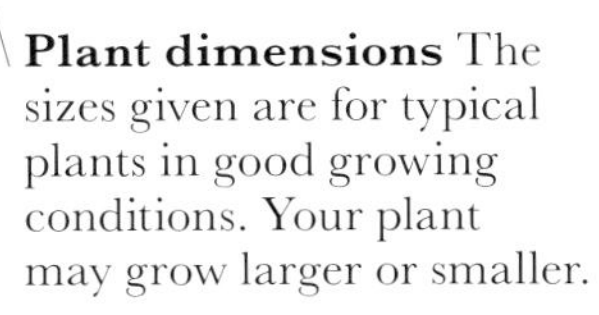

Plant dimensions The sizes given are for typical plants in good growing conditions. Your plant may grow larger or smaller.

Other suggestions Use these lists for further recommendations of plants suitable for an aspect, soil type, or special effect.

KEY TO SYMBOLS

- ☼ Prefers sun
- ◐ Prefers partial shade
- ● Tolerates full shade
- ◇ Prefers well-drained soil
- ◑ Prefers moist soil
- ◆ Prefers wet soil
- (!) Toxic
- pH Needs acid soil
- ❄❄❄ Fully hardy – withstands temperatures to -15°C (5°F)
- ❄❄ Frost hardy – withstands temperatures to -5°C (23°F)
- ❄ Half hardy – tolerates temperatures to 0°C (32°F)
- No symbol (Tender) - plant may be damaged by temperatures below 5°C (41°F)
- ♀ AGM - RHS Award of Garden Merit

SIZE CATEGORIES USED IN THIS BOOK

SMALL	MEDIUM	LARGE
TREES		
UP TO 9M (28FT)	10–14M (30–46FT)	15M (50FT) AND ABOVE
SHRUBS		
UP TO 1.4M (4½FT)	1.5–2.9M (5–10FT)	3M (10FT) AND ABOVE
CLIMBERS		
UP TO 1.9M (6FT)	2–3.9M (6–12FT)	4M (12FT) AND ABOVE
PERENNIALS; ANNUALS & BIENNIALS; WATER PLANTS		
UP TO 59CM (24IN)	60CM–1.1M (24IN–3½FT)	1.2M (4FT) AND ABOVE
BULBS		
UP TO 14CM (5½IN)	15–60CM (6–24IN)	61CM (24IN) AND ABOVE

INTRODUCTION

GARDENS IN SUN

GARDENS IN SHADE

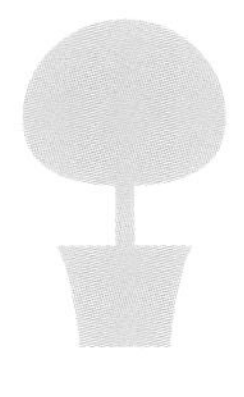
PLANTS FOR
GARDEN STYLES

PLANTS FOR
SEASONAL INTEREST

PLANTS FOR
COLOUR AND SCENT

PLANTS FOR
SHAPE AND TEXTURE

PLANTS FOR
GARDEN PROBLEMS

Identifying plant groups

Plants fall into various categories, defined by the length of their life cycle, size, shape, and habit, as well as physical structure. This guide explains how plants are grouped, and identifies how they will behave in your garden.

ANNUALS AND BIENNIALS

Plants that germinate, form stems and leaves, flower, set seeds, and die in one year are known as "annuals". Hardy annuals can withstand frost, while tender types require protection from the cold. Many bedding plants are annuals, and can either be sown from seed or bought as young plants. Confusingly, some bedding plants are actually perennials (*see below*), but are treated as annuals.

A biennial plant lives for two seasons, germinating, and forming stems and leaves during the first year, then, after overwintering, flowering, setting seed, and dying in the second.

Annuals This group of plants is ideal for short-lived summer colour.

Biennials These produce flowers in their second year, and then die.

True perennials These emerge and die back to the base each year.

Evergreen perennials This group gives year-round interest.

Woody perennials These retain a woody framework during winter.

Ferns Grown for their foliage, many are also evergreen perennials.

Grasses This is a large group of perennials and includes evergreens.

PERENNIALS

This group of plants lives for many years, emerging from underground roots in spring, then growing, flowering, and setting seeds, before dying back in autumn. Most become dormant in winter, although some perennials are evergreen and retain their foliage year-round, while others die back to a woody base, rather than their roots. Some perennials are tender, only re-emerging in spring if protected during winter or brought under cover, and are often treated as annuals for short-term colour.

As well as conventional flowering plants, this group also includes deciduous and evergreen grasses, as well as ferns, which produce spores instead of flowers.

Aquatics Water lilies float on the surface, but are rooted under water.

Marginals Many types of iris are ideal for pond side positions.

MARGINALS AND WATER PLANTS

Plants that thrive in water fall into two main categories: marginals, which grow in shallow water, usually 7–15cm (3–6in) deep; and aquatic plants, which include water lilies, and generally grow at depths of between 30cm (12in) and 1.9m (6ft). Ideally, a natural pond will include a range of marginals and deep water aquatics to create a balanced ecosystem, as well as a few "oxygenators". These are submerged, fast-growing plants that release oxygen into the pond, and compete for nutrients with pond weeds and algae, helping to keep the water clear. Hornwort (*Ceratophyllum demersum*) and *Ranunculus aquatilis* are examples of non-invasive oxygenators.

SHRUBS

These woody-stemmed plants produce a permanent network of branches that rise up from the ground or a central trunk to create a bushy form. They can be deciduous or evergreen, and provide the structural framework in most gardens. Diverse in size, shape, and habit, shrubs range from ground-huggers, such as thyme and heather, to tree-like mock orange, and include airy bamboos. Many give a long season of interest, bearing flowers, decorative fruits, and attractive autumn foliage. Other shrubs can be made into features by pruning and training, such as evergreen box, used in topiary.

Flowering Roses are highly popular summer-flowering shrubs.

Foliage and form Evergreen shrubs give texture and interest.

Bulbs Snowdrops are among the first bulbs to flower each year.

BULBS

This group includes all plants that form bulb-like structures, such as corms, tubers, rhizomes, and true bulbs. Popular bulbous plants include spring tulips and daffodils, which are planted in autumn, and summer-flowering dahlias, lilies, and gladioli, which are planted in spring.

Trees Highly varied, there are trees to suit all parts of the garden.

TREES

Providing the main architecture in the garden, trees are also grown for their flowers, bark, and foliage. Whether deciduous or evergreen, they can form a spreading, rounded, conical, weeping, or columnar canopy of leafy branches, from a central trunk, leading to trees of all shapes and sizes.

CLIMBERS

This group of woody-stemmed and perennial plants uses a variety of methods to climb over structures or through host plants. Twining climbers, such as clematis, use leaf stalks, tendrils, or stems to coil around a support, while the thorny stems of roses act as hooks to heave themselves up. Self-clinging climbers, including *Parthenocissus* and ivy, attach themselves to supports with adhesive pads or aerial roots. Climbers can be evergreen or deciduous, fast or slow growing, and while perennial forms die back each year, many shrubby climbers develop a substantial and permanent structure.

Woody climbers These require a sturdy support to hold their stems.

Perennial climbers Dying back each year, these suit smaller plots.

Understanding soil and site

Plants thrive in the conditions they enjoy most, so taking time to assess your soil type and the amount of sunlight your garden receives will greatly improve your chances of success.

UNDERSTANDING SOILS

Assessing the soil in your garden is the first important step to help you choose the most suitable plants for your conditions The structure and composition of soil determines how much water it can retain, which also affects its fertility, since plants absorb nutrients held in a water solution. Most soils contain a proportion of sand and clay particles, but one type usually dominates. Sand-rich soils tend to be free-draining and have low fertility, due to soluble plants nutrients being washed away. However, they are also light and easy to dig. Clay soils are heavier and more difficult to cultivate, and although they hold water and nutrients well, they are also prone to waterlogging and cracking when dry. The ideal soil type is loam, which is comprised of roughly equal parts of sand and clay, and is both free-draining and water-retentive. To test your soil type, take samples from around your garden and roll them between your fingers (*see below*).

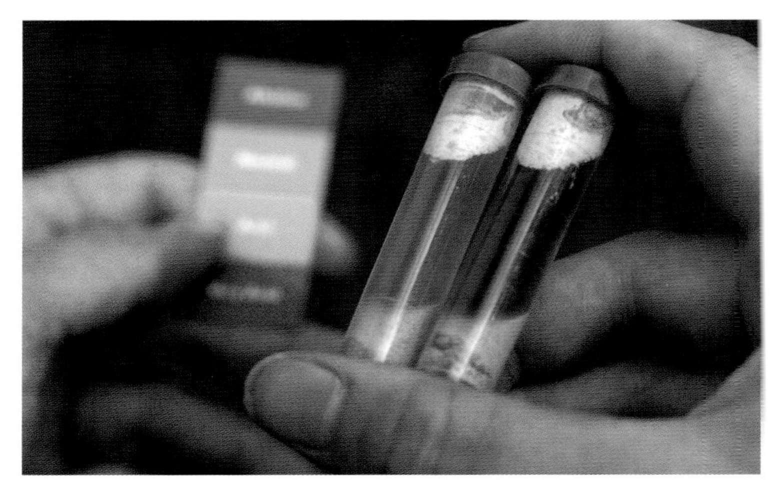

Acid or alkaline Soils range from acid to alkaline pH, which affects the plants you can grow. Most plants tolerate a range of pHs but some are particular. Test your soil with an easy-to-use pH testing kit.

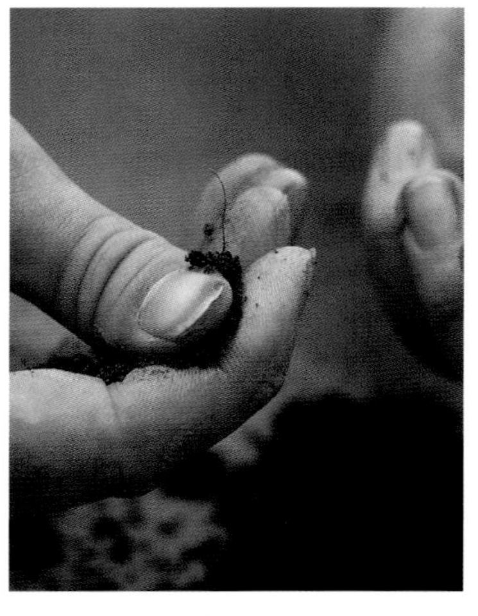

Sandy soil When rubbed between your fingers, sandy soil feels gritty, and can't easily be rolled into a ball.

Clay soil Soils rich in clay feel smooth and sticky, and can easily be moulded into a ball or sausage.

IMPROVING SOILS

All soils benefit from the regular addition of organic matter, such as well-rotted manure or garden compost. This improves soil structure, helping sandy soils retain water and nutrients, and improving drainage in clay soils. Dig organic matter into the top 30cm (12in) of soil, or use it as a mulch by applying a thick layer over the surface.

On very heavy clay soils, horticultural grit can also be added to increase drainage.

Adding organic matter Dig organic matter into sandy soils in spring and into clay soils in autumn.

Mulching Add a 5cm (2in) layer of well-rotted organic matter over the soil surface each spring.

Adding grit Before planting, add horticultural grit to very heavy clay soils to improve drainage.

WHICH WAY DOES YOUR GARDEN FACE?

When selecting plants, take time to assess the direction in which your garden faces and the amount of sunlight it receives every day. Knowing its aspect will help you make the right plant choices and also plan where to place trees and shrubs that may cast shade over your plot. With plants suitable for areas in both sun and shade, there is a great choice for every garden.

SOUTH-FACING GARDENS

Sunny for most of the day in summer, south-facing gardens provide optimum growing conditions for many flowering plants. However, soils dry out quickly here, and additional watering may be needed to support some species.

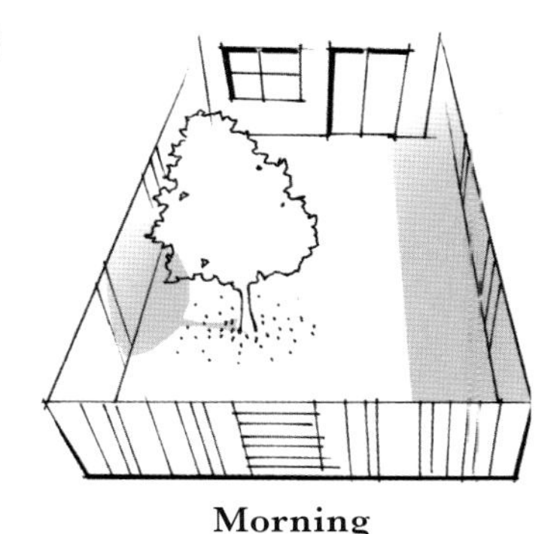

Morning

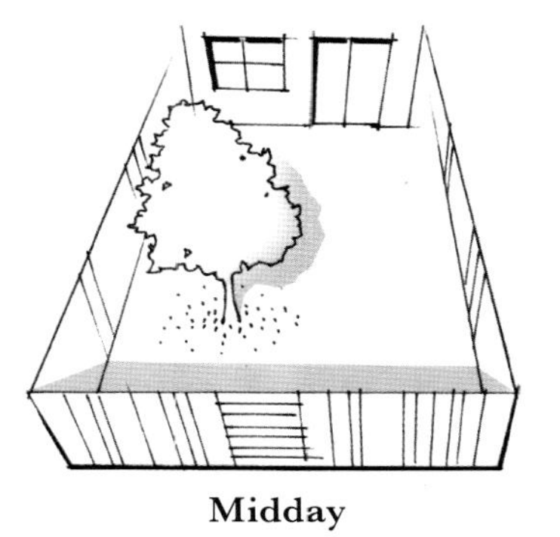

Midday

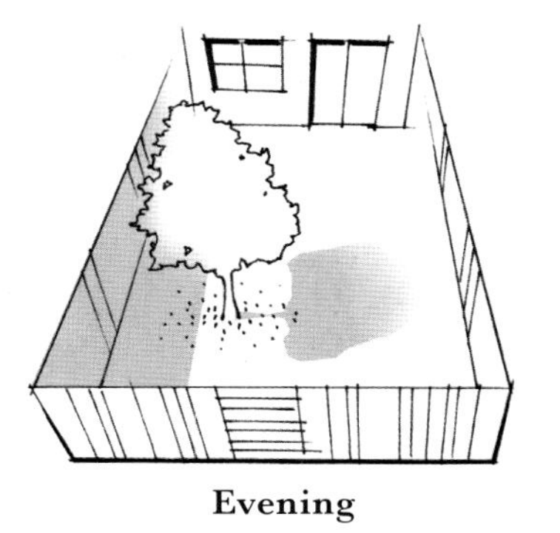

Evening

NORTH-FACING GARDENS

Many architectural foliage plants and a selection of flowers enjoy the shady, cool conditions of north-facing gardens. When planting, position trees and large shrubs at the end of your plot to prevent them casting even more shade.

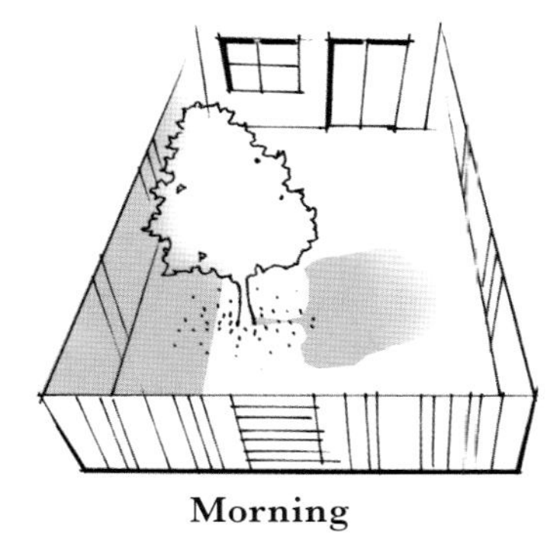

Morning

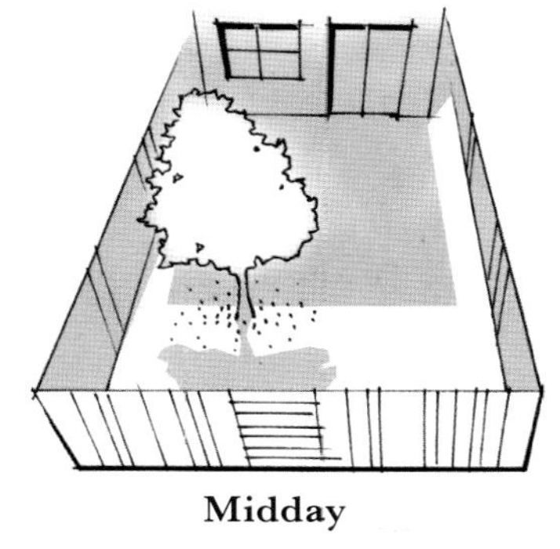

Midday

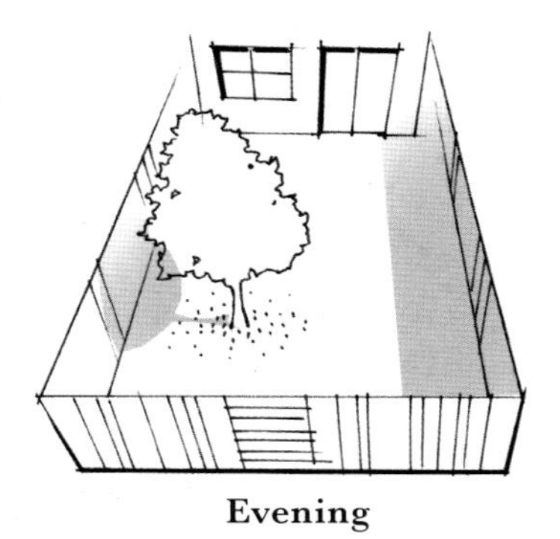

Evening

EAST-FACING GARDENS

These gardens are sunny for many hours until late afternoon, and will support a large group of flowering and foliage plants that thrive in part or dappled shade. Grow sun-loving plants in areas of the garden that are bright for the longest periods.

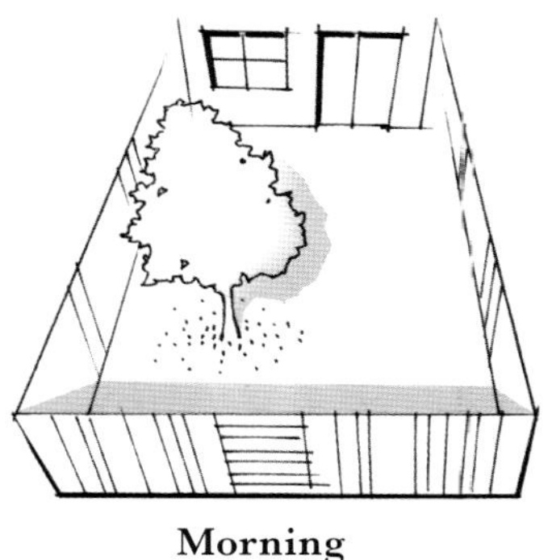

Morning

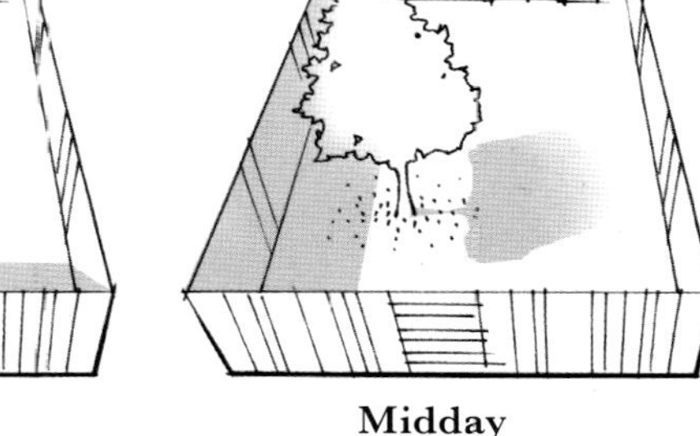

Midday

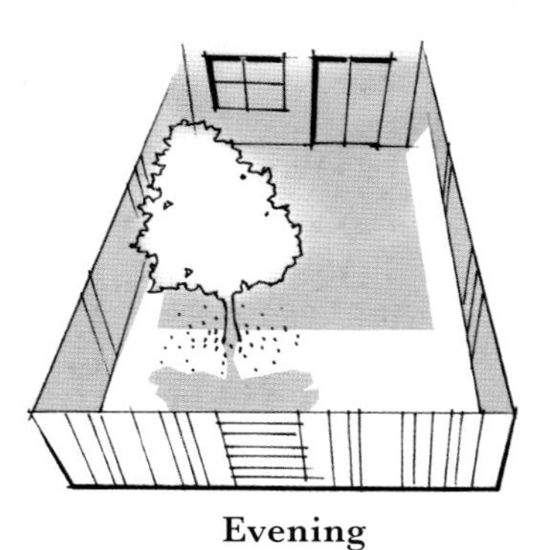

Evening

WEST-FACING GARDENS

Sunny from late morning to evening in summer, west-facing gardens provide ideal conditions for plants that are happy in full sun, or partial and dappled shade. Sun-loving plants should thrive in the brightest areas of a west-facing garden.

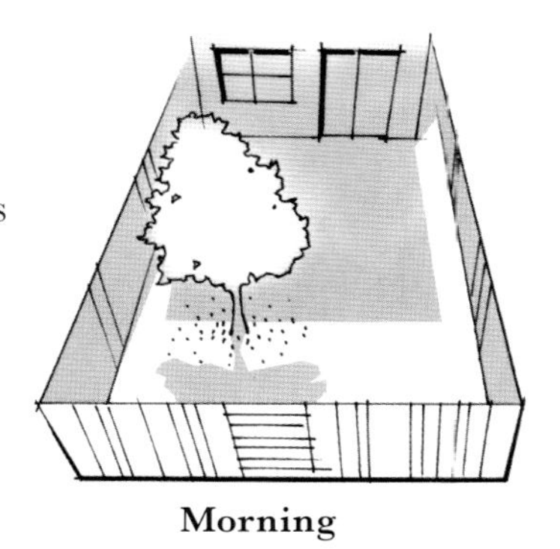

Morning

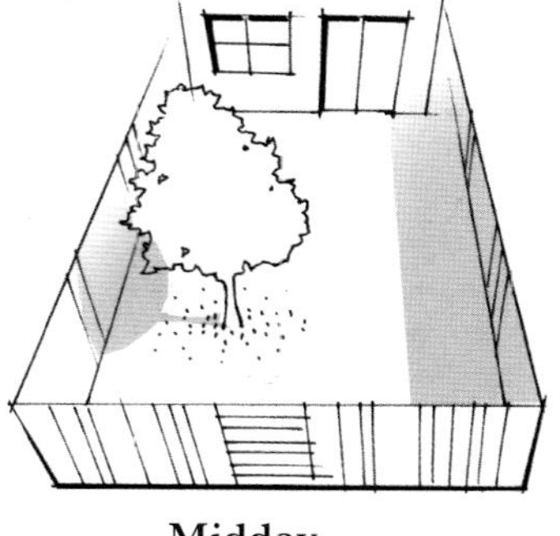

Midday

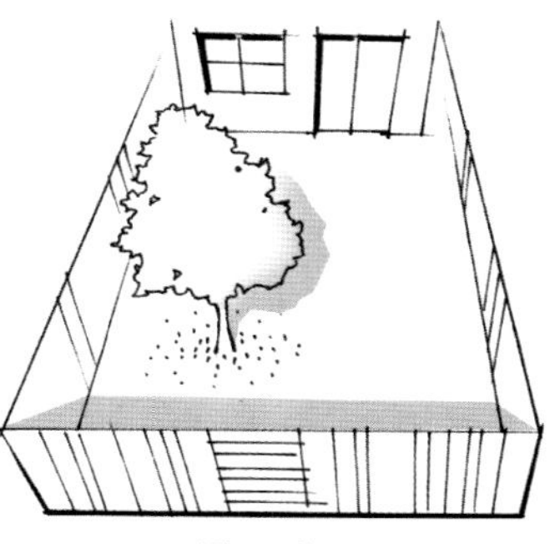

Evening

Planting styles

Your selection of plants and the way they are grouped convey the style or theme of a garden, be it formal, naturalistic, or traditional. Follow these guidelines to create your chosen design.

NATURAL SWATHES

Contemporary perennial, wildlife, and informal design styles reflect nature by replicating the way plants grow in the wild. Mark out beds in organic shapes with sweeping, curved edges, avoiding fussy, wiggly lines, and fill them with a restricted number of species, set out in bold, flowing swathes. Designers also include smaller pockets of plants within these drifts to mimic the effects of self-seeding. Grasses and perennials, such as *Achillea* and *Stipa* (*see below*) are the mainstay of naturalistic schemes, with contrasting flower forms heightening interest.

Natural groups To create a naturalistic design, use a restricted palette of plants, set them out in organic-shaped swathes, and mix perennials and grasses with different shapes and forms to create exciting visual patterns.

Modern blocks Chic and sophisticated, modernist schemes are highly structured, with plants grouped in blocks in geometric-shaped beds, laid out in an asymmetric pattern. Clipped hedging is popular in these designs

Formal symmetry Strict symmetry and geometric beds are key to a formal design. Planting styles can be modern or traditional.

Traditional layers To achieve deep, layered schemes, create a bed that slopes slightly, and group plants in descending height order.

MODERN vs TRADITIONAL

Both formal and modernist planting schemes focus on visual effects, rather than attempting to mirror nature. Formal garden styles are rooted in the classical architecture of Greece and Italy, and French Renaissance gardens, such as Versailles. The designs create elegant patterns with plants, and present them in symmetrical beds. Parterres and knot gardens made from clipped box and topiary shapes lend a sense of formality to schemes, while more informal perennial planting is often included, but confined to geometric beds to maintain the style.

Designers who follow modernist principles adhere to the maxim, "less is more", and use a limited palette of plants, set out in geometric beds, usually rectangles. Unlike formal garden designs, modernist schemes employ asymmetry, which creates a more dynamic effect. Upright grasses and hedging are often used, and perennials are limited to just one or two varieties to create uniform blocks of colour and shape.

Traditional borders, as epitomized by Gertrude Jekyll, take elements from both formal and informal styles. While contemporary perennial schemes mix plants of different heights and forms on one plane, traditional schemes are often laid out in beds on a slight incline, with the tallest plants at the back and shortest in front, allowing all to be seen clearly. However, plants are not designed in formal lines, but grouped in interlocking lozenge-shaped swathes that reflect natural landscapes and create a sense of movement.

USING PLANT TYPES TO CREATE A DESIGN

When creating a design, look at the structure and habit of the plants you have selected, as well as their flowers, and consider the role each will play in your scheme. Start by planning the position of permanent structural plants, such as trees and shrubs, which will form the main framework of your design and lend a sense of solidity. Then, plan in seasonal plants to create a sequence of colour and interest throughout the year, and mid-range perennials to fill gaps between the structural elements. Focal plants with eye-catching shapes or colours, such as spiky phormiums or shrubs with dramatic leaves, provide highlights in a border, while those with spreading stems cover bare soil with a carpet of foliage and flowers.

Balancing shape and form This summer border shows how different plant forms can be used to create a harmonious display.

Focal plant Spiky grasses and plants, such as *Libertia*, draw the eye and can be used to punctuate a border.

Mid-range plants Use perennials and small shrubs to knit together seasonal flowers and larger plants.

Structural plants Create a permanent framework with well-placed evergreen and deciduous trees and shrubs.

Seasonal plants Sustain interest throughout the year by including a range of seasonal plants and flowers.

Ground cover Plants with spreading stems help to fill gaps in a border, and can suppress weed growth too.

PLANTING YEAR

Gardens are like stage sets, with a change of scenery every few weeks helping to maintain the audience's interest. To put on this show, remember that while evergreens offer year-round colour, they can't compete with the fleeting beauty of a tree in blossom or the unfurling petals of a rose in summer.

To create a garden that evolves month by month, you need to plan ahead. Punctuate boundaries and borders with trees and shrubs that produce different effects as the seasons turn, with flowers in spring and summer, followed by fiery foliage and berries in autumn. Dress up these structural plants with climbers, such as clematis, that prolong their interest, and weave into your scheme a variety of early, mid- and late-flowering perennials for a kaleidoscope of colour. In autumn, plant a range of spring bulbs in any gaps in your borders, to provide a bright burst of flowers as winter fades.

Early to mid-spring Autumn-planted bulbs, such as crocuses, grape hyacinths (*Muscari*), and daffodils, together with early-flowering shrubs, including camellias, provide a blaze of colour from early- to mid-spring.

Late spring Bridge the gap between spring and summer with late-flowering tulips and alliums; clematis and shrubs, including *Ceanothus* and lilacs; and perennials, such as *Lamprocapnos* and *Aquilegia*.

Summer The choice of summer flowers is legion and includes every colour, size, and shape imaginable. Select plants that perform early, such as roses, alongside late bloomers, like dahlias, to sustain the performance.

Autumn As summer flowers start to fade, autumn brings its own treasure trove of colour and excitement. Flaming foliage is the main attraction, but also find a place for berried shrubs, grasses, and late flowers, such as asters.

Winter Choices dwindle as temperatures drop, but winter is still full of surprises. Flowers and bulbs, including snowdrops, brave the cold, and shrubs with evergreen leaves or bright stems, such as *Cornus*, also add colour.

How to plant

Following tried and tested methods when planting will pay dividends, helping to get plants off to a good start, ensuring they thrive, and improving the long-term survival rate of permanent plants.

PREPARING THE SITE

The best times to plant are in the autumn or spring, although container-grown plants can be planted at any time, unless the soil is frozen, waterlogged, or exceptionally dry. Borderline hardy plants are best planted in spring, which allows them to establish a mature root system before winter sets in.

A few weeks before you plan to plant, prepare your soil. Start by removing large stones and rubble, and then tackle the weeds. Dig out by hand perennial weeds, such as dandelions, dock, and bindweed, ensuring you remove the roots. Then hoe off the annual weeds. Leave the bed for a week or two and remove any new weeds that have appeared. If your soil needs improving, dig in well-rotted organic matter and/or horticultural grit (*see p.12*).

SHRUBS AND PERENNIALS

Most container-grown shrubs and perennials are planted in the same way. About an hour before planting, give your plant a good soak, either with a watering can, or by placing it in a bucket of water, submerging the pot and waiting until no bubbles appear. Then lift it out and leave it to drain. Meanwhile, dig a hole twice as wide as the pot and a little deeper. Place the plant in the hole and check that it will be at the same depth when planted as it is in its pot. Add or remove soil accordingly, and fork over the bottom of the hole. Mix some slow-release granular fertilizer with the excavated soil, as directed on the pack. Remove the plant from its pot, place it in position, and refill around the root ball. Firm the soil with your hands to remove any air pockets, and water the plant well.

Setting plants out Before planting, set out your plants on the ground to check that you are happy with their arrangement, facing their best sides forward for an engaging display.

Planting out Most plants are planted at the same depth as they were in their pots. After planting, firm the soil well with your hands to remove air pockets. Water well.

TREES AND LARGE SHRUBS

Use the following method to plant container-grown trees and large shrubs. Bare-root plants, available through autumn and winter, are planted in the same way, but refill around the roots in stages to ensure you eliminate all the air gaps. Do not add extra fertilizer or organic matter when planting, but prepare the soil well in advance.

Water your tree or shrub, using the method described for perennials.

Remove the plant from its container or wrapping. Tease out the roots carefully and scrape away the top layer of soil if the plant is in a pot. Dig a planting hole of the same depth, and up to three times as wide as the root ball, and loosen the sides of the hole with a fork. Place the tree into the hole, ensuring that the area where the roots start to flare out is near the surface. Refill around the root ball, and firm the soil to remove air pockets. Water well.

Planting a tree Plant trees slightly proud of the soil surface, so the area where the roots flare out from the trunk is just visible. Use a cane set over the pot to judge the right depth, and adjust the soil accordingly.

Providing support For large trees, drive a stake into the ground at 45 degrees, so that it reaches a third of the way up the trunk. Insert the stake vertically for smaller trees. Attach the tree using a tree tie.

Aftercare After planting, apply a mulch, leaving a space of 10cm (4in) around the trunk. Water consistently for the first two years, and check and loosen tree ties regularly. Remove the stake after four years once established.

BULBS

Spring-flowering plants that grow from bulbs, such as daffodils and tulips, are planted in autumn, while those that bloom in summer, including lilies and gladioli, are planted in spring. Most prefer free-draining soil, so if you have heavy clay, add sharp grit to the area where you plan to plant. Alternatively, plant them in pots filled with a mixture of soil-based compost and grit.

Although bulbs can be planted individually, they tend to look better and are easier to plant in a group. First, dig a wide hole to the appropriate depth. Most bulbs are planted at a depth of three times their height, except for tulips, which are planted at four times their height. For example, a 5cm (2in) daffodil is planted at 15cm (6in), while a 5cm (2in) tulip should be planted at a depth of 20cm (8in). Water well after planting.

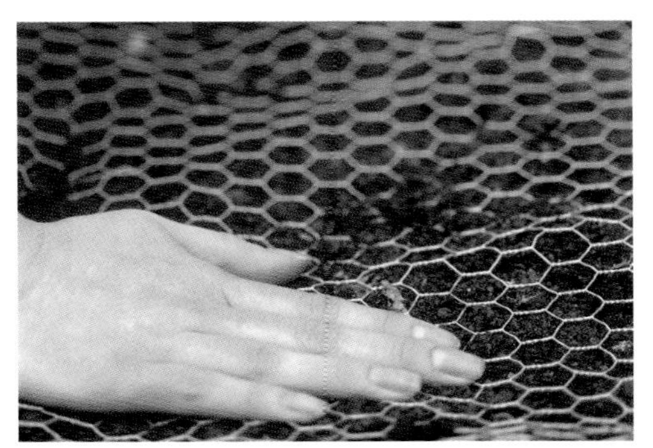

Planting dormant bulbs Plant bulbs with the pointed tip facing upwards. To prevent animals digging them up, cover the soil after planting with wire mesh. Secure the mesh with stakes.

Planting snowdrops Unlike other bulbs, snowdrops are best planted when in leaf, just after they have flowered in spring. This is known as planting "in the green".

CLIMBERS

When planting climbers, use canes to guide the stems towards a permanent support, such as a fence. Prepare the soil and planting hole as for perennials (*see p.18*), at least 30cm (12in) from the support; dig a deeper hole for clematis, which should be planted with the root ball 8cm (3in) below the soil. Place the canes in the hole before positioning the plant with its root ball angled slightly towards the supporting structure.

Providing initial support Dig a planting hole and insert three or four evenly spaced canes in to the base. Tie the top of the canes to the wires to hold them in place.

Attaching horizontal wires Screw in vine eyes 45cm (18in) apart along a fence, or use a drill and Rawlplugs to secure them into walls. Fix horizontal wires between the eyes.

Planting the climber After planting, backfill with soil and firm to remove air pockets. Then, spread out the stems and tie them loosely to the canes with soft twine.

Aftercare After planting, water well, and add a mulch of organic matter or chipped bark to supress weeds. Keep the mulch clear of the stems. Water regularly in the first year.

AQUATICS

Marginals and deep water aquatics are planted using one simple method. Plant them into sterilized topsoil or aquatic compost in pond baskets or planting bags. These hold the soil in place while allowing water to pass through, aerating the roots and keeping the plants healthy.

Deep water aquatics, such as waterlilies, are best set on bricks on the bottom of the pond after planting, so that their leaves float on the surface. Remove the bricks as the plant grows. Alternatively, deep water plants can be suspended in the water using string, and gradually lowered to the correct depth.

Set marginal plants on shelves in shallow water at the edge of the pond. Oxygenators, such as hornwort, do not require planting; simply place them in water at least 30cm (12in) deep, and they will spread out just beneath the surface.

Filling the pond basket First, line the bottom of the pond basket with compost or sterilized topsoil. Tip the plant out of its original container and place it in the basket.

Planting the aquatic Fill around the plant with more compost and firm gently. Apply a layer of pea shingle over the surface to prevent the compost floating out.

Aftercare Divide mature aquatics and marginals in spring. Remove plants from their baskets, cut the root balls up with a knife, and repot healthy sections. Top-dress with grit.

CONTAINERS

When planting in containers, choose the ones large enough for your plants, and ensure they have drainage holes in the base. Remember, too, that large pots need watering less frequently than small ones, since they hold greater volumes of compost and water. Also consider the material the container is made of. Choose frost-proof terracotta and lightweight materials, such as plastic or resin, for balconies and roof terraces.

Filling with compost Add a layer of compost over the broken pots, selecting an appropriate type for your particular plant. Also add some slow-release fertilizer.

Ensuring good drainage Place pieces of broken terracotta pot or polystyrene over the drainage holes to prevent the compost falling through or blocking them.

Planting the plants Having first watered the plants, tip them out of their pots and place on the compost. Fill in around the plants, firming them in gently.

Aftercare Leave a 5cm (2in) gap between the top of the compost and the pot rim, to make watering easier. Water regularly, especially in summer. Top-dress annually.

SEEDS

Growing annuals and vegetables from seeds is both rewarding and cheaper than buying mature plants. Hardy seeds can be sown directly outdoors from late spring, where they are to grow; half hardy seeds must be started under cover.

To sow indoors, fill a clean seed tray or pot with seed compost, and firm it down gently. Water the compost using a can fitted with a rose, and allow to drain. Sprinkle seeds thinly on the compost surface. Check the seed packet for the required planting depth, and cover the seeds with sifted compost to that depth. Label the tray or pot, and cover with a lid or clear plastic bag. Place in a lit, warm area, such as a windowsill, and wait for the seeds to germinate. Keep the compost moist, and remove the lid or bag as soon as the seedlings emerge. Grow the seedlings on under cover and pot them up as they grow. Plant them out when the risk of frost has passed.

Using pots for large seeds Fill a small pot with compost, firm gently, and make holes to the depth noted on the seed packet.

Sowing the seeds Drop a seed into each hole, cover with compost, and water in. Cover with a clear plastic bag.

Growing on the seedlings Once your seedlings have a few leaves, pot them on individually into modular trays or small pots.

Potting up the seedlings Plant one seedling per module or pot, firm them in gently, and water well. Grow on under cover.

PLANT LOCATIONS

GARDENS IN SUN
GARDENS IN SHADE

Choosing plants for sun or shade

Some plants like it hot while others shy away from the sun, but whatever the conditions found in your garden, there's a wealth of beautiful plants to choose from to create a visual feast. The key is to match the plants to your conditions.

SUN-LOVING PLANTS

Dry, dusty soil and blazing hot sun are challenging conditions for plant growth, and species unsuited to such sites will soon start to suffer. However, some of the most beautiful plants and flowers have adapted to thrive in these areas, including a whole host of colourful blooms, and sculptural shrubs and trees, which you can use even if your soil is sandy and your garden faces south.

A plant's label will usually tell you the conditions it prefers, although you can often tell when plants are suited to hot, dry sites, as many share key characteristics that help them survive. These include small leaves to reduce moisture loss, silvery foliage to reflect light, and hairs that trap water droplets, protecting the surface from the sun and reducing evaporation. Succulents have fleshy stems and leaves that store water, while other sun-lovers have colourful flowers that shine in strong light, attracting pollinating insects. Use these visual clues, alongwith the plant labels, to identify plants best suited to your garden.

Soaking up the rays These sun-loving specimens show the range of beautiful plants that can cope with strong light.

KEY CHARACTERISTICS OF SUN-LOVING PLANTS

Silver foliage Most plants with reflective, silver leaves enjoy hot, sites and shimmer in strong light.

Small leaves Foliage with a small surface area helps to reduce moisture-loss during transpiration.

Furry foliage Hairy leaves trap moisture, helping to sustain plants during drought.

Fleshy stems and leaves Some plants from dry regions trap moisture in their leaves and stems.

SHADE-LOVING PLANTS

Cool woodland floors and other shady areas make perfect homes for a whole host of large-leaved, architectural plants. Unlike sun-lovers, these shade-dwellers must mop up as much of the available light as possible, and as a consequence their foliage is generally large and thin. Some also sport shiny leaves, with a waxy outer surface that helps to conserve moisture, as the soil beneath trees may also be quite dry. Pastel colours stand out in the gloom, which is why many shade-loving plants bear pale flowers that will catch the attention of pollinating insects. Other plants bloom in winter, making the most of the increased light levels when any neighbouring deciduous trees are not in leaf.

There are many degrees of shade, from deep shade, such as that experienced at the back of north-facing borders, to dappled and partial shade, where the sun may shine for a few hours of the day. Plant choices are restricted in areas of deep shade and the best option is to introduce as much light as possible. Prune overhead trees and shrubs, and consider removing shade-causing structures. Moist shade suits more plants than dry shade, so try increasing soil moisture by improving it with well-rotted organic matter, such as garden compost or manure, or create a plastic-lined bog garden.

To transform your gloomy garden into a cool, calm oasis, combine foliage with different textures, sizes, and shapes, and include a range of seasonal flowers to highlight your scheme. As you will see, shade is a bonus, never a problem.

Cooling off in the shade Large, dramatic leaves and delicate flowers can scorch under hot sun.

KEY CHARACTERISTICS OF SHADE-LOVING PLANTS

Large, thin leaves Plants adapted to shade have large leaves that can absorb low light.

Glossy foliage Shiny foliage reflects the available light onto adjacent leaves and adds sparkle.

Winter flowers Some plants flower only in winter when deciduous trees are bare.

Evergreen foliage Permanent leaf cover enables many shade-lovers to grow all year.

Recipes for sunny sites

SUMMER FOCAL POINT FOR A PATIO

A dramatic *Abutilon* creates the central feature in this summer display, lending height and stature to the mix – you may need to support its stems with a discreet cane or two – while purple verbenas, white petunias, and the trailing silver *Helichrysum* form a skirt around the edge. Plant this combination in multi-purpose compost, incorporating a few slow-release fertilizer granules as you do, and keep it well watered throughout summer. If the flowers start to flag towards the end of the season, apply a liquid tomato fertilizer to perk them up.

Plant list
1 *Petunia* 'Trailing White'
2 *Verbena* - purple form
3 *Abutilon* 'Nabob'
4 *Helichrysum petiolare*

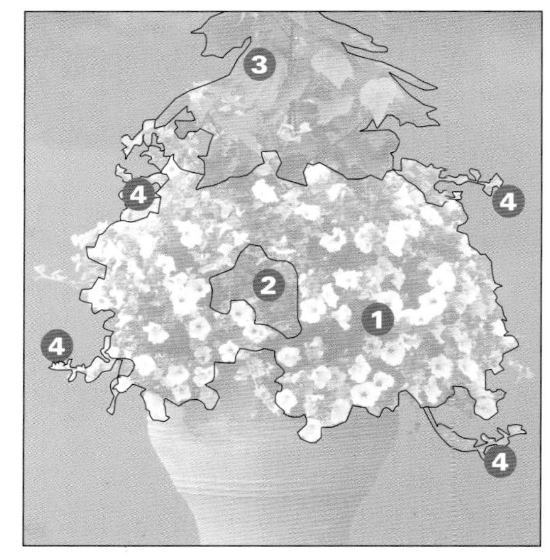

HERB FUSION

Herbs slot easily into a sunny, mixed border on free-draining soil. *Origanum* makes a colourful edge, and bronze and sweet fennel make decorative feathery towers in midsummer, shading plants like *Brunnera* that prefer cooler conditions – remember to snip off the fennel seeds frequently before they self-seed. Also try a bright yellow-flowered broom (*Cytisus*) at the back of the bed to inject a splash of colour in spring.

Plant list
1 *Origanum vulgare* 'Aureum'
2 *Brunnera macrophylla*
3 *Foeniculum vulgare* 'Purpureum' (purple fennel)
4 *Foeniculum vulgare* (sweet fennel)
5 *Potentilla recta* 'Warrenii'

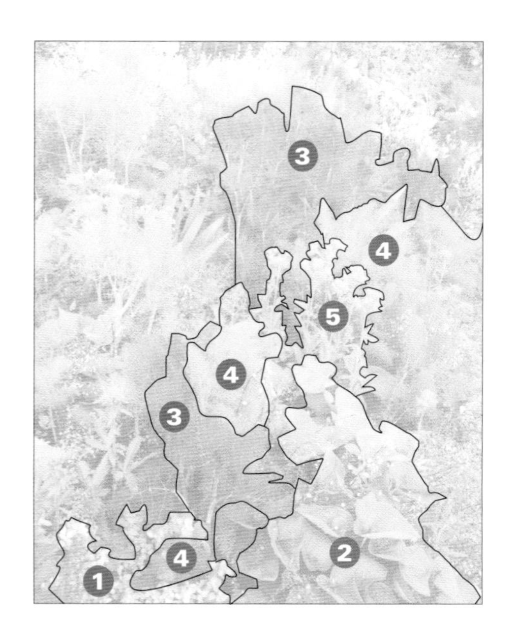

FIERY BORDER FOR SANDY SOIL

This confection of fiery shades will light up a sunny area of the garden throughout summer. Combine perennials, such as *Achillea*, *Helenium*, and *Kniphofia* in rich shades of gold and crimson, and tone them down with the wispy grass, *Stipa tenuissima*, which thrives at the front of sunny borders. Tall decorative dahlias inject late season colour into the back of this display, while their dark foliage also provides a foil for the green leaves. The dark painted fence sharpens the effect of the hot hues.

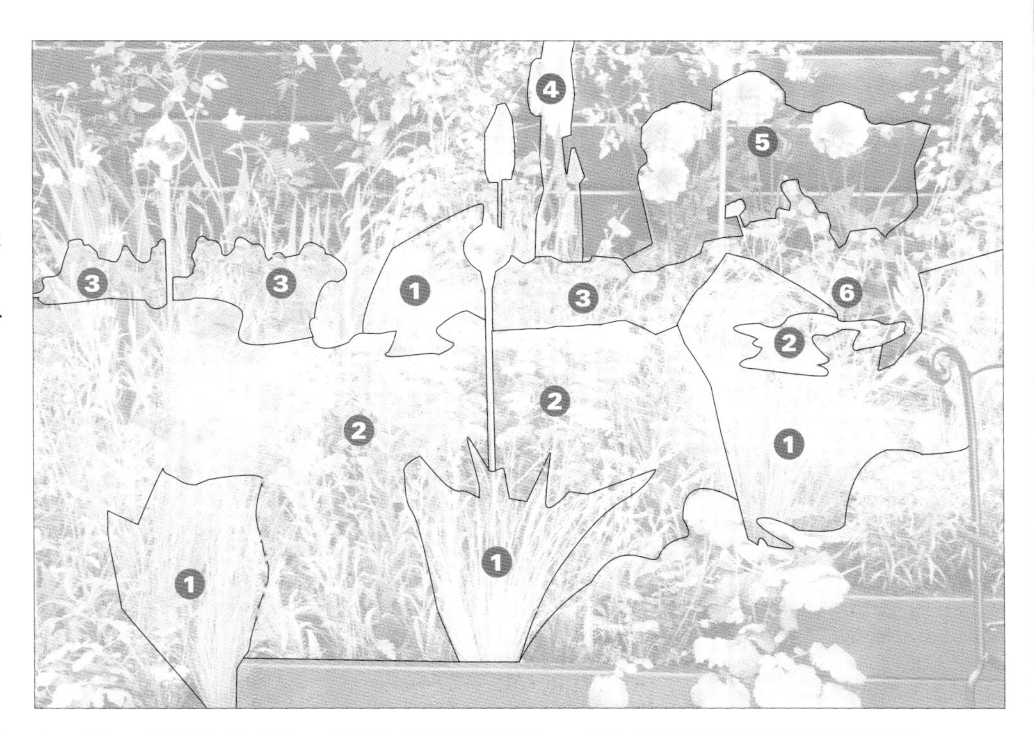

Plant list

1 *Stipa tenuissima*
2 *Achillea* 'Terracotta'
3 *Helenium* varieties
4 *Kniphofia* - orange form
5 Decorative dahlia
6 *Geum* 'Prinses Juliana'

PROFUSE POND PERIMETER

A small artificial pond creates a beautiful oasis, reflecting light into the garden and attracting a wide range of birds, small animals, and aquatic wildlife. Decorate the water surface with a small water lily, such as *Nymphaea pygmaea*, and create a boggy area around your pond for a range of flowers and foliage plants. This pond edge features *Gunnera*, together with colourful *Astilbe*, and gold-flowered *Ligularia dentata*, which also sports beautiful leaves that hold the interest before the blooms appear. *Cornus alba* will provide structure throughout the year, with its colourful winter stems, and decorative variegated leaves in summer. In smaller spaces, replace the giant-sized *Gunnera* with a more manageable *Rodgersia*.

Plant list

1 *Nymphaea pygmaea* (dwarf water lily)
2 *Astilbe* x *arendsii*
3 *Ligularia dentata*
4 *Cornus alba* 'Elegantissima' (variegated dogwood)
5 *Gunnera manicata* (giant rhubarb)

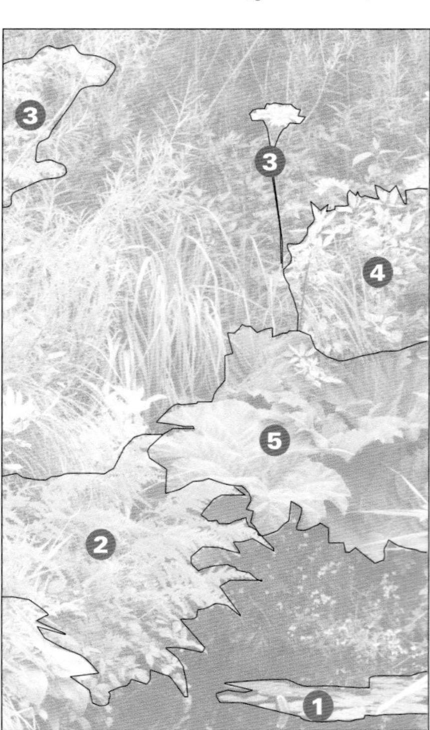

MONOCHROME BORDER FOR CLAY

The rose and perennials here will thrive in clay soils in a sunny site, although the box, geranium, and dark purple *Actaea* are equally happy in part shade. Prune the rose annually to keep it in check, and site it where you can enjoy its sweet fragrance.

Plant list

1 *Geranium sanguineum* 'Album'
2 *Rosa* 'Winchester Cathedral'
3 *Actaea simplex* 'Brunette'
4 *Veronica spicata* 'Alba'
5 *Buxus sempervirens* 'Suffruticosa' (box)

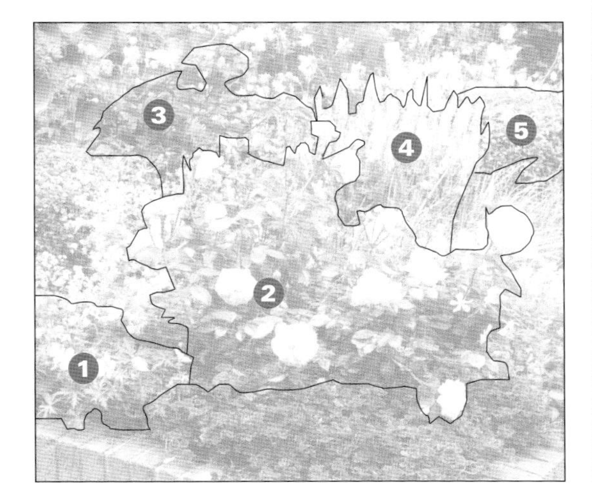

SUMMER SELECTION FOR A GRAVEL GARDEN

Gravel beds provide the free-draining conditions that salvias and spiky *Sisyrinchium* enjoy, and they both flourish at the front of sunny beds and borders. The bronze sedge will also be happy in gravel and sun or part shade, but delphiniums need more moisture and an annual application of all-purpose granular fertilizer each spring, so leave them out if you cannot provide these conditions.

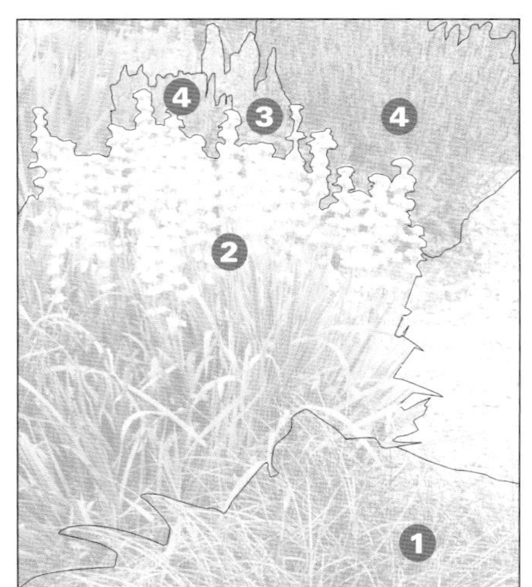

Plant list

1 *Carex comans* - bronze form
2 *Sisyrinchium striatum*
3 *Delphinium elatum*
4 *Salvia nemerosa*

Recipes for shady sites

LEAFY EDIBLES FOR PART SHADE

Lettuces tend to bolt quickly in bright sun, while thriving in a cool spot in part shade. Swiss chard will also adapt to some shade, and varieties with colourful stems make pretty partners for green and purple lettuces. Dot pot marigolds between your leafy crops for spots of orange or yellow to create a border that looks as good as it tastes. Guard against slugs, especially when the lettuces are young, and keep the border moist at all times.

Plant list
1 Swiss chard
2 *Calendula officinalis* (Pot marigold)
3 Purple lettuce
4 Purple frilly-leaved lettuce ('Lollo rossa')
5 Green lettuce (cut and-come-again variety)

FOLIAGE TEXTURES FOR CLAY SOIL

This rich mix of foliage plants would be perfect for a shady site just beyond a tree canopy, and needs few flowers to create a spectacular effect. Combine large-leaved hostas and bergenias to smother the soil, and the majestic ostrich fern, *Matteuccia*, for additional height and texture towards the back of the border. Flowers will appear from spring to summer, starting with the bergenias and the acid-yellow blooms of the *Euphorbia*, and continuing with pale lilac hosta blooms as summer progresses, and tall stems of yellow *Ligularia* flowers, which appear later in the season. Take precautions against slug damage to ensure the hostas are not shredded.

Plant list
1 *Bergenia purpurascens*
2 *Hosta elegans*
3 *Polygonatum* x *hybridum*
4 *Euphorbia palustris*
5 *Ligularia dentata*
6 *Ligularia stenocephala*
7 *Matteuccia struthiopteris* (ostrich fern)
8 *Darmera peltata*

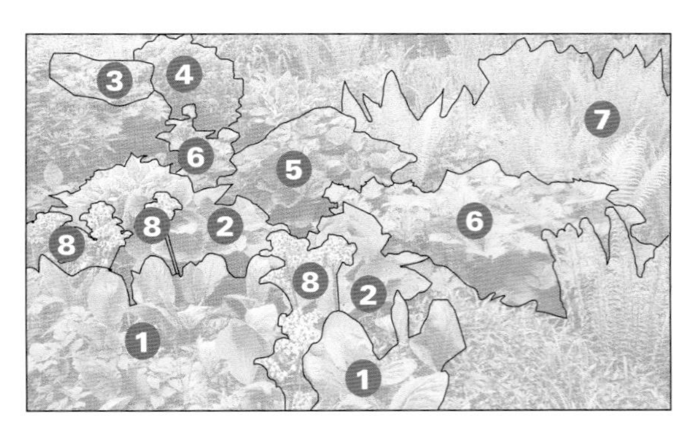

WOODLAND UNDERSTOREY

This simple combination of ferns, purple *Astilbe,* and the stark white stems of a Himalayan birch (*Betula*) form an eye-catching display when the understorey plants are repeated and set out in bold swathes. Ensure your soil is damp enough to please the *Astilbe*, which prefers constant moisture. Adding a thick mulch of organic matter over clay soils in spring will help, or substitute them with *Anemone* x *hybrida* on free-draining, sandy soils.

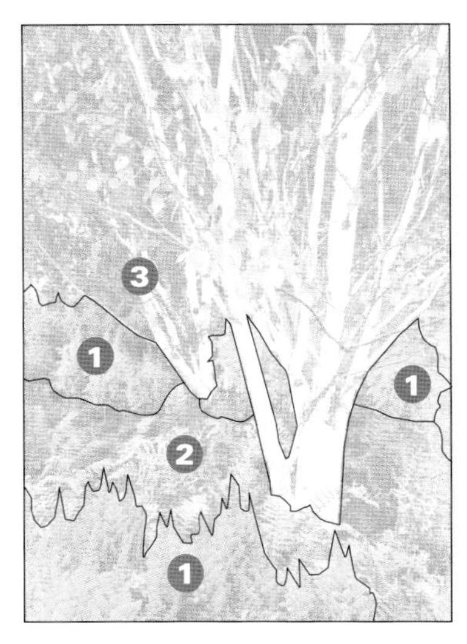

Plant list
1 *Astilbe* x *arendsii*
2 *Dryopteris affinis* 'Cristata'
3 *Betula utilis* var. *jacquemontii*

Recipes for shady sites

LEAFY GLADE

The focus of this planting group is the purple-leaved hazel (*Corylus*), which will colour up well in light, dappled shade. *Trollius*, *Euphorbia*, and geraniums are also best in part shade, but you can tuck the *Alchemilla mollis* in a darker corner, as it will grow almost anywhere. If you have space, squeeze in a few ferns, such as *Polystichum* or *Dryopteris*, both of which tolerate deep shade. The hazel has a number of benefits, offering beautiful leaf colour, catkins from late winter to early spring, and edible nuts in autumn, which also provide food for wildlife. It will grow to a substantial size, but can be kept in check in small gardens by pruning the stems back in late winter or early spring.

Plant list

1 *Alchemilla mollis*
2 *Euphorbia cyparissias*
3 *Trollius* x *cultorum*
4 *Corylus avellana* 'Fuscorubra'
5 *Geranium sylvaticum* 'Alba'

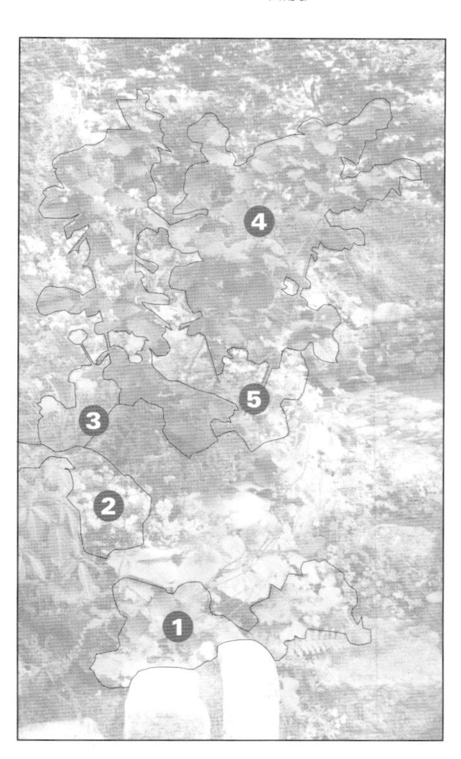

FOLIAGE BORDER FOR A CITY SPACE

This sophisticated combination will offer lasting colour in a cool, shady, urban garden. The ferns (*Polystichum*) and *Epimedium* are evergreen, creating permanent cushions of foliage, while the columbines produce dainty late spring flowers in pastel shades of pink, blue, purple, and white. They tend to self-seed prolifically, so deadhead the blooms promptly to contain them. *Deschampsia* adds textural contrast to the mix and is one of the few grasses that performs best in shade.

Plant list

1 *Epimedium perralderianum*
2 *Polystichum* 'Braunlaub'
3 *Deschampsia cespitosa*
4 *Aquilegia vulgaris* (granny's bonnet)

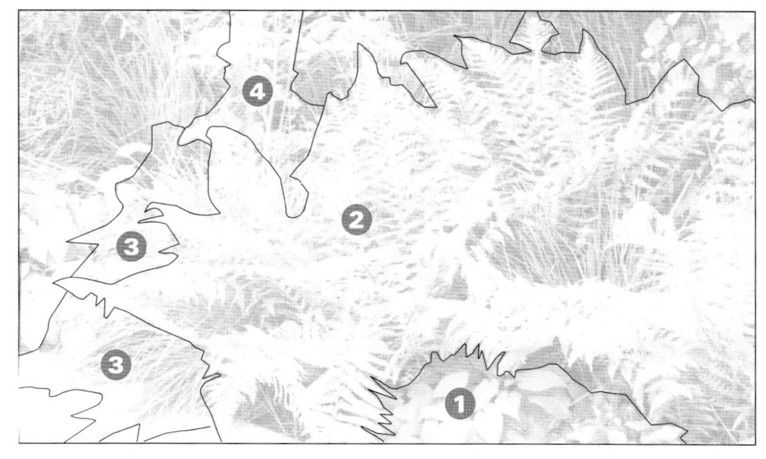

COLOURFUL MIX FOR A CONTEMPORARY POT

While most summer bedding plants prefer full sun, a few tolerate part shade, and are particularly useful in town gardens that are shaded by buildings. The best include fuchsias, lobelia, *Plectranthus*, and busy Lizzies – opt for disease-resistant New Guinea types. Upright fuchsias make strong statement features for the centre of a display, while trailing lobelia and *Plectranthus* add a decorative frill. Squeeze the busy Lizzies in between to add bright highlights.

Plant list

1 *Lobelia erinus* - trailing form
2 Tender fuchsia - bedding variety
3 *Impatiens* - white form
4 *Plectranthus madagascariensis* 'Variegated Mintleaf'

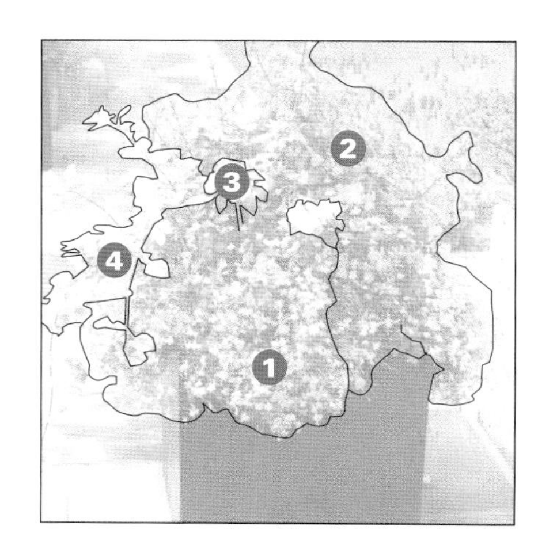

GARDENS in SUN

Sun-drenched gardens are perfect for a wealth of flowering plants, as well as those with sparkling silver foliage and richly coloured leaves. Brightly lit patios are also a boon, providing ideal conditions for many spring bulbs, summer bedding, and edible plants, including soft fruits, Mediterranean vegetables, and tender green beans. And if you have a pond, indulge in a sumptuous water lily or two, and fringe your feature with marginals that will mirror their shapes and colours in the glassy surface.

Plants for clay soil

Moisture retentive and rich in nutrients, clay soils provide the perfect home for a whole host of garden plants, including roses, peonies, and clematis.

Cottage borders filled with delphiniums, asters, geums, and monarda, or traditional shrub and perennial schemes will all thrive in clay soils in the sun. If you have modern designs in mind, you can opt for bamboos, cannas, and *Veronicastrum*. Although clay soil can be a little difficult to manage, if treated the right way it can be extremely fertile. Clay is prone to waterlogging and cracking during dry spells, but these problems can be remedied by applying a thick mulch of well-rotted manure every year in spring. This will help to improve the structure of clay and increase drainage.

PERENNIAL MEDIUM

Aconitum 'Stainless Steel'

MONKSHOOD A perennial with deeply divided, dark green leaves and dense, upright spikes of hooded, silvery blue flowers, which appear from mid- to late summer. All parts of the plant are poisonous.

↕1m (3ft) ↔60cm (24in)

ANNUAL/BIENNIAL LARGE

Amaranthus caudatus

LOVE-LIES-BLEEDING This is a bushy annual, with oval, pale green leaves. From summer to autumn, it produces long, pendulous, tassel-like, red flowers, which can be used for indoor displays.

↕1.2m (4ft) ↔45cm (18in)

PERENNIAL MEDIUM

Amsonia tabernaemontana

This is an upright perennial, with slim, tapering leaves that develop attractive tints before falling in autumn. It produces clusters of star-shaped, blue flowers from late spring to summer. Grow in front of or in the centre of a border.

↕1m (3ft) ↔30cm (12in)

PERENNIAL LARGE

Anemone hupehensis ♈

A perennial with dark green, divided leaves. Tall, upright stems of purple, red, pink, or white flowers with yellow eyes appear over many weeks from late summer to early autumn. 'Hadspen Abundance' AGM (above) has pale pink blooms.

↕1.2m (4ft) ↔45cm (18in)

PERENNIAL MEDIUM

Aquilegia vulgaris var. *stellata* ♈

An upright perennial, with round leaves and tall stems bearing pompon-like flowers from late spring to early summer. Deadhead after flowering. 'Nora Barlow' AGM (above) has white-tipped greenish pink blooms.

↕75cm (30in) ↔50cm (20in)

PERENNIAL LARGE

Aster novae-angliae

A perennial, with slim, hairy leaves and daisy-like, purple, pink, red, or white flowers on branched stems from late summer to autumn. Best in neutral to alkaline soils; it has mildew resistance. 'Purple Dome' (above) is a popular variety.

↕1.5m (5ft) ↔60cm (24in)

PERENNIAL MEDIUM

Astilbe 'Venus'

This leafy perennial produces divided leaves, topped with feathery, tapering plumes of tiny, pale pink flowers in midsummer. The seedheads become dry and provide interest well into winter. Divide every 3 or 4 years to maintain vigour.

↕↔1m (3ft)

PERENNIAL MEDIUM

Astrantia major

MASTERWORT A clump-forming perennial, with divided, mid-green leaves. It produces sprays of small, greenish white, pink, or red flowers from midsummer to early autumn. Deadhead regularly to prolong the display. Remove spent growth in autumn.

↕60cm (24in) ↔45cm (18in)

SHRUB LARGE

Berberis darwinii ♈

DARWIN'S BARBERRY This vigorous, evergreen, arching shrub has small, glossy, dark leaves. Numerous rounded, deep orange-yellow flowers appear from mid- to late spring, followed by bluish berries. Its vicious spines are good for security hedges.

↕↔3m (10ft)

PERENNIAL LARGE

Campanula lactiflora

An upright, branching perennial, with narrowly oval, green leaves and slender stems of large, nodding, bell-shaped, blue, occasionally pink or white, flowers in midsummer. Needs reliably moist and fertile soil. May need staking on windy sites.
↕up to 1.2m (4ft) ↔60cm (24in)

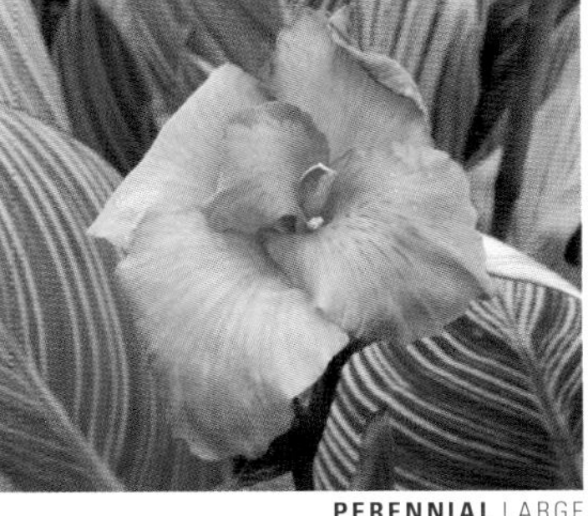

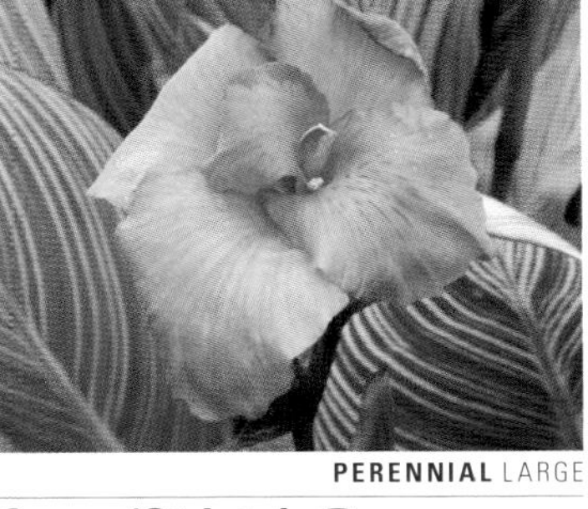

PERENNIAL LARGE

Canna 'Striata' 🏆

This upright perennial is grown for its green- and yellow-striped foliage and bright orange flowers, which appear from midsummer to early autumn. Grow it in fertile soil, keep well watered, and protect the rhizomes from frost.
↕1.5m (5ft) ↔50cm (20in)

TREE MEDIUM

Cercis canadensis

This deciduous, spreading tree or shrub has heart-shaped, dark green leaves that turn yellow in autumn. Pale pink, pea-like flowers emerge from bare stems in mid-spring. *C. canadensis* var. *alba* (above) has white flowers.
↕↔10m (30ft)

SHRUB MEDIUM

Chaenomeles speciosa 🏆

This is a vigorous, deciduous, bushy shrub, with thorny stems and oval, dark green leaves. Red, white, or pink flowers are borne from early- to mid-spring, followed by fragrant, yellow fruits. 'Moerloosei' AGM (above) is a popular variety.
↕2.5m (8ft) ↔5m (15ft)

PERENNIAL SMALL

Chrysanthemum 'Grandchild'

This upright perennial has dark green, lobed leaves, and from late summer to early autumn, it produces branching heads of bright mauve, double flowers. It is good for cutting; deadhead regularly to prolong the display.
↕45cm (18in) ↔40cm (16in)

PERENNIAL SMALL

Dodecatheon meadia 🏆

AMERICAN COWSLIP This is a clump-forming perennial, with pale green leaves. In spring, clusters of small, purple flowers with reflexed petals appear, but it then becomes dormant in summer. *D. meadia* f. *album* (above) has white flowers.
↕20cm (8in) ↔15cm (6in)

CLIMBER MEDIUM

Clematis JOSEPHINE 🏆

This deciduous climber has green, lance-shaped leaves. From summer to early autumn, a succession of large, double flowers appear; the pinkish purple petals are cream beneath, giving a layered effect. Best in rich soil; keep the roots shaded.
↕2.5m (8ft)

SHRUB MEDIUM

Corylopsis pauciflora 🏆

BUTTERCUP WITCH HAZEL A deciduous, bushy shrub, with oval, bright green leaves, similar to a beech tree's, which are bronze when young. It bears fragrant, tubular to bell-shaped, pale yellow flowers from early- to mid-spring.
↕1.5m (5ft) ↔2.5m (8ft)

SHRUB SMALL

Deutzia x *elegantissima* 🏆

This deciduous shrub has a flaky bark and dark green leaves. It bears clusters of sweet-smelling, star- or cup-shaped flowers from late spring to early summer. Varieties include the pink-tinged 'Rosealind' AGM (above).
↕1.2m (4ft) ↔1.5m (5ft)

SHRUB LARGE

Enkianthus deflexus

A large, deciduous, spreading shrub, with red shoots and tufts of dull green leaves that turn bright red in autumn. Small, bell-shaped, creamy yellow flowers appear in late spring or early summer. It prefers acid soil.
↕4m (12ft) ↔3m (10ft)

PERENNIAL MEDIUM

Euphorbia griffithii 'Fireglow'

This bushy, upright perennial has decorative, lance-shaped, mid-green leaves, with pale red midribs. In early summer, it produces rounded clusters of small, orange-red flowers. The sap is a skin irritant.

↕75cm (30in) ↔90cm (36in)

PERENNIAL LARGE

Filipendula rubra

QUEEN OF THE PRAIRIE This is an upright perennial that grows in damp soils. It has large, green, deeply cut, fragrant leaves. From early- to midsummer, it produces fluffy, pink flowerheads on tall, slender stems. Best in boggy, wildlife areas.

↕up to 2.5m (8ft) ↔1.2m (4ft)

PERENNIAL MEDIUM

Filipendula ulmaria

MEADOWSWEET This deciduous perennial bog plant has leafy stems that bear divided, mid-green leaves. Plume-like spikes of creamy white flowers appear in midsummer. Self-seeds quite vigorously. Best placed next to garden ponds, or in wild areas.

↕90cm (36in) ↔60cm (24in)

SHRUB LARGE

Forsythia x *intermedia*

This vigorous, upright, deciduous shrub has long stems covered with small, star-shaped, yellow flowers from late winter to mid-spring, before its small, bright green leaves appear. Varieties include 'Lynwood Variety' AGM (above).

↕↔3m (10ft)

SHRUB SMALL

Fothergilla gardenii

This is a deciduous, bushy shrub, with oval, dark blue-green leaves that turn red, orange, and yellow in autumn. Clusters of fragrant, bottlebrush, white flowers appear from mid- to late spring. Best in fertile soil.

↕↔1m (3ft)

PERENNIAL SMALL

Geranium 'Ann Folkard'

CRANESBILL A spreading perennial, forming mats of ivy-shaped, deeply cut, bright yellowish green foliage. It produces an abundance of small, rich magenta blooms with black centres and veins in midsummer. It may also flower in autumn.

↕50cm (20in) ↔1m (3ft)

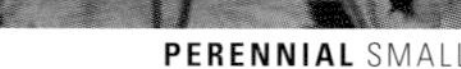

Geum coccineum

This clump-forming perennial has irregularly lobed leaves, above which rise slender, branching, hairy stems bearing single, orange flowers with prominent yellow stamens in summer. Best in damp soils. 'Cooky' (above) is a popular variety.

↕↔30cm (12in)

PERENNIAL MEDIUM

Helenium 'Moerheim Beauty'

An upright perennial with dark green foliage below strong, branching stems. Sprays of daisy-like, rich coppery red flowers appear from mid- to late summer. A valuable late season border plant.

↕1m (3ft) ↔60cm (24in)

PERENNIAL MEDIUM

Helenium 'Wyndley'

An upright perennial with branching stems bearing sprays of daisy-like, orange-yellow flowerheads for a long period in late summer and autumn. The foliage is dark green. It needs regular division in spring or autumn.

↕80cm (30in) ↔50cm (20in)

PERENNIAL LARGE

Helianthus 'Lemon Queen'

This sunflower is a tall, vigorous, upright perennial, with stout, branched stems, bearing coarse, lance-shaped, dark green leaves and large, daisy-like, pale yellow flowers from late summer to autumn. Needs fertile soil; not drought-tolerant.

↕1.5m (5ft) ↔60cm (24in) or more

PERENNIAL MEDIUM

Heliopsis helianthoides

NORTH AMERICAN OX-EYE This is an upright perennial, with large, lance-shaped leaves and rich yellow or orange, daisy-like flowers, which appear from late summer to early autumn. 'Summer Sun' (above) has large, yellow flowers.
↕90cm (36in) ↔60cm (24in)

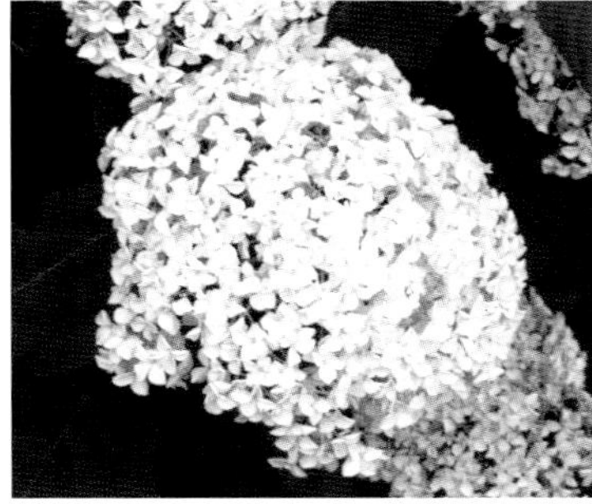

SHRUB MEDIUM

Hydrangea arborescens ♈

A rounded shrub, with mid-green, tapered, oval leaves. In midsummer, flattened, dome-shaped clusters of long-lasting white flowers appear; papery flower skeletons feature all winter. 'Annabelle' (above) has extra large flowerheads.
↕↔1.5m (5ft)

PERENNIAL MEDIUM

Leucanthemum x *superbum*

SHASTA DAISY This upright perennial has lance-shaped, dark green leaves and large, white flowers on long stems from midsummer to autumn. Hidden staking helps keep it tidy. 'Wirral Pride' (above) is a double-flowered form.
↕1m (3ft) ↔60cm (24in)

TREE LARGE

Liquidambar styraciflua

SWEET GUM A deciduous, conical to spreading tree, its shoots develop corky ridges and the large, lobed, glossy, dark green leaves turn brilliant orange, red, and purple in autumn. Weeping and slender, compact cultivars are available.
↕25m (80ft) ↔12m (40ft)

PERENNIAL MEDIUM

Lysimachia punctata

GARDEN LOOSESTRIFE This is an upright, very vigorous, tall, clump-forming perennial, with mid-green leaves. In summer, it produces spikes of bright yellow flowers. Plant in moist soils for the best results.
↕1m (3ft) ↔60cm (24in)

PERENNIAL LARGE

Monarda 'Prärienacht'

An upright, clump-forming perennial, with mid-green, scented leaves. It produces tufted, two-lipped, tubular, pinky violet blooms from mid- to late summer. Provide shade from hot sun; protect from winter wet.
↕1.2m (4ft) ↔60cm (24in)

BULB MEDIUM

Narcissus 'Scarlet Gem'

A mid-spring flowering daffodil that produces strap-shaped foliage and scented flowers, with golden petals and deep red-orange cups. It spreads to form clumps over time. Needs good drainage or the bulbs will rot.
↕35cm (14in)

PERENNIAL MEDIUM

Paeonia mlokosewitschii ♈

CAUCASIAN PEONY This is an erect, clump-forming perennial, with pinkish leaf buds that unfurl as oval, bluish green leaves, which turn orange-brown in autumn. Large, bowl-shaped, lemon-yellow flowers appear from late spring to early summer.
↕↔75cm (30in)

PERENNIAL MEDIUM

Persicaria bistorta 'Superba' ♈

A vigorous, clump-forming perennial, with oval, tapered leaves, which spread to form a dense mat. From early summer to autumn, spikes of soft pink flowers appear above the foliage.
↕75cm (30in) ↔60cm (24in)

Plants for clay soil

SHRUB LARGE

Phyllostachys nigra

BLACK BAMBOO This is a large, evergreen shrub that produces clumps of grooved, greenish brown stems or "culms" that turn black in their second season. The leaves are narrow and rustle in the wind. Protect from cold, drying winds.

↕8m (25ft) ↔ indefinite

PERENNIAL SMALL

Primula 'Wanda'

Primroses are evergreen or semi-evergreen perennials, with lance-shaped to oval, bronze to dark green foliage. From winter to mid-spring, they bear flowers in shades of blue, yellow, purple, red, and pink.

↕10cm (4in) ↔ 15cm (6in)

PERENNIAL MEDIUM

Ranunculus aconitifolius

This upright, clump-forming perennial has deeply divided, green leaves. Yellow-centred white flowers are produced on branched stems from late spring to summer. 'Flore Pleno' AGM (above) has double, white blooms.

↕75cm (30in) ↔ 50cm (20in)

PERENNIAL MEDIUM

Ratibida columnifera

This perennial, grown as an annual, has pale green, slender leaves. It produces unusual flowers, with reflexed yellow petals that surround a prominent brown "cone", in summer. Deadhead frequently to prolong the display; it is drought-resistant.

↕80cm (32in) ↔ 30cm (12in)

SHRUB LARGE

Rhododendron luteum

This is a deciduous shrub, with oblong- to lance-shaped leaves that develop orange and red tints in autumn. In spring, it produces very fragrant, funnel-shaped, bright yellow flowers. Prefers a cool spot.

↕↔ 4m (12ft)

SHRUB SMALL

Ribes sanguineum

This deciduous, spreading shrub has aromatic, green leaves and pendent clusters of small, pink or red spring flowers, followed by white-bloomed, blue-black fruits. 'Brocklebankii' (above) has pale green leaves; needs a cool spot.

↕↔ 1.2m (4ft)

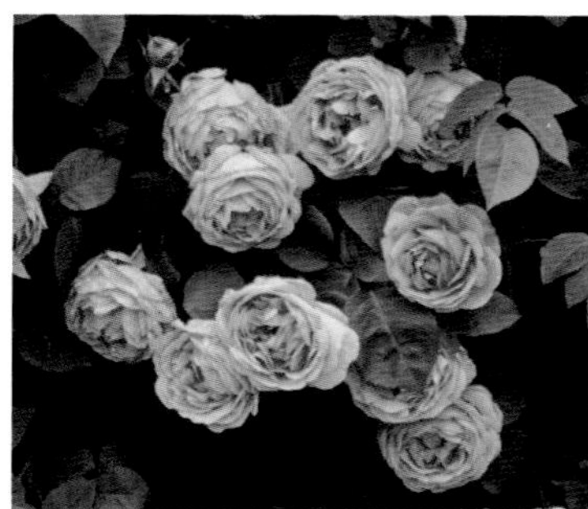

CLIMBER MEDIUM

Rosa CONSTANCE SPRY

This climbing rose produces greyish green leaves and an abundance of large, bowl-shaped, clear pink, double flowers with a myrrh fragrance. The blooms appear in summer and last for about a month. Mulch annually with organic matter.

↕2m (6ft)

SHRUB SMALL

Rosa DOUBLE DELIGHT

This repeat-flowering hybrid tea rose has dark green foliage, and, throughout summer and autumn, it produces a succession of highly fragrant, fully double, creamy white flowers with dark red edges. Mulch annually with organic matter.

↕1m (3ft) ↔ 60cm (24in)

SHRUB SMALL

Rosa KNOCK OUT

A repeat-flowering, modern shrub rose, with disease-resistant, dark green foliage. The cherry red, single flowers are produced throughout summer and early autumn, and have a light fragrance. Mulch annually with organic matter.

↕↔ 1.2m (4ft)

PERENNIAL MEDIUM

Rudbeckia fulgida var. *sullivantii* 'Goldsturm'

A compact, upright perennial with narrowly oval, mid-green leaves. From late summer to autumn, daisy-like, golden flowers with brown centres appear on sturdy stems. Leave faded flowerheads for winter interest.

↕75cm (30in) ↔ 30cm (12in) or more

PERENNIAL MEDIUM

Sanguisorba menziesii

An upright, clump-forming perennial, with attractive blue-grey, feathery foliage. The fluffy, bottlebrush-like, red flowers are held on tall stems and appear from early summer to early autumn. The seedheads retain their colour through winter.
↕75cm (30in) ↔60cm (24in)

PERENNIAL LARGE

Thalictrum aquilegiifolium

This is a clump-forming perennial with finely divided, grey-green leaves, resembling those of a fern or aquilegia. Bunched heads of fluffy, lilac-purple flowers appear from early- to midsummer, followed by attractive seedheads.
↕1.2m (4ft) ↔45cm (18in)

PERENNIAL LARGE

Rudbeckia laciniata

A tall, upright perennial, with lance-shaped foliage and branched stems of daisy-like, single, yellow flowers with green centres from late summer to autumn. Leave faded flowerheads for winter interest. Varieties include 'Herbstsonne' AGM (above).
↕2.3m (7ft) ↔75cm (30in)

PERENNIAL LARGE

Veronicastrum virginicum

CULVER'S ROOT An upright perennial, with tall stems of lance-shaped, dark green leaves. From late summer to early autumn, it produces spikes of small, star-shaped, white, blue, or pink blooms. 'Album' (above) has white flowers.
↕1.2m (4ft) ↔45cm (18in)

PERENNIAL LARGE

Sanguisorba canadensis

CANADIAN BURNET A clump-forming, upright perennial with mid-green, divided foliage that turns red when hit by autumn frosts. It bears spikes of bottlebrush-like, white flowers on stout stems from late summer to early autumn.
↕2m (6ft) ↔60cm (24in)

PERENNIAL LARGE

Solidago 'Goldkind'

This is an upright, clump-forming perennial, with narrow, pointed, green leaves. From midsummer to early autumn, it produces flattened heads of small, golden-yellow flowers. Deadhead to prolong the display.
↕1.2m (4ft) ↔1m (3ft)

SHRUB MEDIUM

Viburnum acerifolium

MAPLE LEAF VIBURNUM An upright, deciduous shrub with bright green leaves that turn orange, red, and purple in autumn. Decorative red fruits, which turn purple-black, follow creamy white flowerheads in early summer. The berries are poisonous.
↕2m (6ft) ↔1.2m (4ft)

SHRUB LARGE

Syringa vulgaris

COMMON LILAC This vigorous, deciduous shrub has mid-green, heart-shaped foliage. In late spring, it produces large, conical clusters of highly fragrant flowers in shades of purple, pink, and white. 'Paul Thirion' (above) has rose-pink flowers.
↕↔7m (22ft)

OTHER SUGGESTIONS

Annuals

Adonis annua

Perennials

Bergenia cordifolia • *Camassia leichtlinii* • *Carex elata* 'Aurea' • *Carex oshimensis* • *Hemerocallis* 'Red Precious' • *Lythrum salicaria* 'Feuerkerze' • *Prunella vulgaris* • *Ratibida pinnata* • *Vernonia noveboracensis*

Shrubs

Abies concolor 'Compacta' • *Arbutus unedo* f. *rubra* • *Berberis* x *stenophylla* • *Berberis thunbergii* f. *atropurpurea* 'Atropurpurea Nana' • *Calluna vulgaris* • *Cryptomeria japonica* 'Vilmoriniana' • *Erica carnea* • *Hypericum* 'Hidcote' • *Photinia* x *fraseri* 'Red Robin' • *Rhododendron* 'Gristede' • *Salix lanata*

Trees

Crataegus laevigata 'Paul's Scarlet' • *Crataegus viridis* 'Winter King'

Plant focus: roses

Reigning over the flower garden, roses are coveted for their exquisite blooms, tantalizing perfume, and long flowering season.

Roses have been grown for thousands of years, and are celebrated for their beautiful, scented blooms and diverse habits. Ranging from ground-hugging shrubs to lofty climbers, they never fall out of fashion and suit almost every design style, including contemporary, wildlife, and cottage. The flowers comprise many different forms, spanning the single flowers of the species to spherical, fully double blooms with 40 petals or more. Most of them are also fragrant, and while some roses only flower once a year, modern bushes and shrubs often perform for many months from summer to autumn, especially when deadheaded regularly. In addition, species and old garden roses produce large, brightly-coloured hips to decorate designs through autumn and winter. Roses suffer from a number of diseases, most notably black spot, which causes the foliage to discolour and fall. To prevent it, buy modern resistant forms or tough species that tend to not succumb to the disease.

USING ROSES

Shrub roses can be trained as standards; you can use them in pots to flank a doorway or as a centrepiece in formal designs. The elegant flowers of bush roses enhance the colour and fragrance of mixed herbaceous borders in informal gardens, and climbing roses can be trained to create curtains of colour.

Species roses make excellent hedges, with their tough, disease-resistant foliage and early-summer flowers, followed by decorative hips. Their thorns also help to deter intruders.

Climbers trained on wires fixed to a pergola or series of arches provide a tunnel of flowers, foliage, and scent in a traditional garden design.

TYPES OF ROSE

Bush roses This group includes large-flowered hybrid teas, which bear one bloom per stem, and cluster-flowered floribundas.

Species roses These large shrubs bear simple, often fragrant flowers, in a single flush in early summer, followed by decorative hips

Old garden roses This group produces a single flush of semi- or fully double, scented flowers in early summer, followed by red hips.

Modern shrub roses Usually larger than bushes and more compact than species roses, modern shrubs flower from summer to autumn.

Patio and miniature roses Single to double flowers are freely produced on compact plants in flushes from summer to autumn.

Climbing roses These tall shrubs produce flowers over many months from summer to autumn on long stems that need to be supported.

Rambling roses These flexible shrubs weave through trees and shrubs, and produce a single flush of small flowers in early summer.

Ground-cover roses These spreading or trailing roses produce single- to fully double flowers on short shoots throughout summer.

Plants for sandy soil

Water and nutrients drain through sandy soils, but many plants live happily on such meagre rations and may rot or flower poorly if given a richer diet.

Choose plants that tolerate drought and dislike winter wet for this free-draining soil. Lavender, agapanthus, santolina, and other plants from Mediterranean climes thrive in sandy conditions, as do most bulbs. Fashionable prairie plants, including sedums, achilleas, salvias, and *Echinacea purpurea*, are also at home in sandy, sunny sites, while wildflower meadows will put their best performance on here. If you are looking for larger, more permanent plants, choose from the many shrubs and trees that like drier conditions, such as *Spiraea*, *Choisya*, and *Cistus*.

SHRUB MEDIUM

Abelia 'Edward Goucher'

Deciduous or semi-evergreen, this arching shrub has oval, tapered, green leaves that are bronze when young. From midsummer to autumn, it produces masses of trumpet-shaped, pink, slightly fragrant blooms. Prune in spring.

↕1.5m (5ft) ↔2m (6ft)

PERENNIAL MEDIUM

Achillea 'Coronation Gold'

YARROW An upright perennial, with fern-like, grey-green foliage, and flat heads of golden-yellow flowers in summer. The stems may need discrete support, ideally put into place in spring so the plants can grow up into them.

↕1m (3ft) ↔60cm (24in)

PERENNIAL MEDIUM

Achillea 'Paprika'

This is a semi-evergreen, upright perennial, with feathery, silvery green leaves. In summer, it produces flat heads of tiny, scarlet flowers that fade with age. The dried flowerheads can be left to provide winter interest.

↕80cm (32in) or more ↔40cm (16in)

PERENNIAL MEDIUM

Agapanthus africanus

AFRICAN LILY An evergreen perennial, with strap-shaped, mid-green leaves and spherical heads of deep blue, trumpet-shaped flowers, which appear from summer to early autumn on tall, sturdy stems. Protect from frost in winter.

↕1m (3ft) ↔50cm (20in)

PERENNIAL SMALL

Agastache aurantiaca

GIANT HYSSOP An upright perennial, with grey-green, scented leaves and, in summer, long-lasting spikes of small, orange, two-lipped, tubular flowers. Varieties include 'Apricot Sprite' (above), with pale orange blooms.

↕50cm (20in) ↔1m (3ft)

BULB MEDIUM

Allium cristophii

This short-stemmed, summer-flowering bulb has narrow, grey-green leaves that fade as the flowers appear. Large balls of star-shaped, violet flowers are borne on glaucous heads in early summer. Attractive seedheads follow the blooms.

↕40cm (16in)

BULB MEDIUM

Alstroemeria ligtu hybrids

PERUVIAN LILY These summer-flowering tubers produce narrow, twisted leaves and funnel-shaped flowerheads in shades of pink, yellow, or orange, often spotted or streaked with contrasting colours. In cold areas, apply a deep mulch in winter.

↕60cm (24in)

PERENNIAL MEDIUM

Anthemis tinctoria

DYER'S CHAMOMILE A clump-forming perennial, with finely divided, aromatic leaves. It bears a mass of daisy-like, white or yellow flowers on slim stems over a long period in summer. 'E.C. Buxton' (above) has lemon-yellow flowers.

↕↔1m (3ft)

SHRUB SMALL

Artemisia 'Powis Castle'

WORMWOOD This evergreen, rounded, dwarf shrub is grown for its finely divided, silver leaves. The small, yellow late summer flowers are often removed to maintain the plant's foliage effect. May need protection in cold winters.

↕↔1m (3ft)

PERENNIAL LARGE

Asphodeline lutea

YELLOW ASPHODEL This is a tall, upright, clump-forming perennial that produces narrow, textured, grassy leaves and dense spikes of star-shaped, bright yellow, fragrant flowers in late spring. The blooms are followed by decorative seed pods.

↕1.2m (4ft) ↔1m (3ft)

SHRUB SMALL

Ballota pseudodictamnus

This low-growing, compact, evergreen subshrub is prized for its small, round, grey-green leaves, which are covered with woolly, white hairs. Small, pink flowers appear in early summer at the end of the leaf stems. Clip after flowering.
↕60cm (24in) ↔90cm (36in)

PERENNIAL MEDIUM

Baptisia australis

FALSE INDIGO A clump-forming perennial, with grey-green, divided foliage and spikes of small, violet-blue flowers, which appear in summer. Decorative, dark grey seed pods form in autumn. Provide support in spring.
↕75cm (30in) ↔60cm (24in)

PERENNIAL SMALL

Campanula carpatica

TUSSOCK BELLFLOWER A low-growing, clump-forming perennial, with round, open bell-shaped, violet-blue or white flowers, which appear over a mound of heart-shaped leaves in summer. Varieties include 'White Clips', with pure white blooms.
↕10cm (4in) ↔30cm (12in)

PERENNIAL MEDIUM

Centranthus ruber

RED VALERIAN This upright perennial produces fleshy, grey-green leaves and branching heads of small, star-shaped, deep reddish pink or white flowers from late spring to late summer. It thrives in poor soil and may self-seed widely.
↕90cm (36in) ↔60cm (24in) or more

SHRUB SMALL

Ceratostigma willmottianum

BLUE-FLOWERED LEADWORT A deciduous shrub, with small, green leaves that turn red in autumn. Clusters of small, sky-blue flowers appear from late summer to autumn. May die back in winter, reshooting in spring.
↕1m (3ft) ↔1.5m (5ft)

SHRUB LARGE

Cercis chinensis

This is a large, branching shrub or small tree, with pea-like, pink flowers that appear on bare stems in late spring, before the foliage emerges. The large, decorative, green foliage is heart shaped. Mulch with compost in spring.
↕6m (20ft) ↔5m (15ft)

TREE MEDIUM

Cercis siliquastrum

JUDAS TREE This deciduous, spreading, bushy tree has attractive green, heart-shaped leaves. In mid-spring, before the leaves emerge, it produces bright pink, pea-like flowers on bare stems, followed by long, purple pods in late summer.
↕↔10m (30ft)

SHRUB MEDIUM

Choisya ternata

MEXICAN ORANGE BLOSSOM An evergreen, rounded shrub, with aromatic, glossy, bright green leaves, divided into three leaflets. Clusters of fragrant, white blooms open in late spring and autumn. The popular 'Aztec Pearl' has slender, green leaves.
↕↔2.5m (8ft)

SHRUB SMALL

Cistus x dansereaui

This evergreen shrub has grey-green leaves and white, pink, or purple, papery, bowl-shaped summer flowers. 'Decumbens' AGM (above) is low growing and compact, and produces white blooms with crimson blotches in the centre.
↕60cm (24in) ↔1m (3ft)

ANNUAL/BIENNIAL LARGE

Cleome hassleriana

SPIDER FLOWER This is an erect annual, with spiny stems of divided, palm-shaped leaves. In summer, it produces rounded clusters of small, spidery, white, pink, and purple flowers, with long stamens and a light fragrance.
↕1.2m (4ft) ↔45cm (18in)

PERENNIAL SMALL

Coreopsis verticillata

This is a compact perennial, with feathery, dark green foliage. It produces a profusion of flowers, held on wiry stems, over a long period in late summer. 'Moonbeam' AGM (above) has lemon-yellow flowers.
↕50cm (20in) ↔45cm (18in)

SHRUB LARGE

Cornus alba

DOGWOOD This deciduous, upright shrub is grown for its winter display of bright scarlet young shoots and dark green foliage. In early summer, creamy white flowers appear, followed by white fruits. 'Sibrica' AGM (above) is a popular variety.

↕↔3m (10ft)

ANNUAL/BIENNIAL LARGE

Cosmos bipinnatus

A tall, bushy, erect, half-hardy annual, with feathery leaves and an abundance of large, daisy-like, red, pink, or white flowers, which appear from early summer to autumn. Deadhead to encourage a long display of blooms.

↕up to 1.2m (4ft) ↔45cm (18in)

PERENNIAL LARGE

Crambe cordifolia ♈

GREATER SEA KALE This is a tall, robust perennial that produces a mound of large, crinkled and lobed, dark green foliage. In summer, branched stems that support clouds of small, fragrant, white flowers appear above the leaves.

↕2m (6ft) ↔1.2m (4ft)

BULB LARGE

Dahlia 'Café au Lait'

DECORATIVE DAHLIA This clump-forming perennial has toothed-edged, dark green leaves. From midsummer to autumn, it produces large, double, peach-flushed cream flowers. Deadhead to encourage more blooms. Protect tubers from frost.

↕90cm (36in)

SHRUB LARGE

Daphne bholua

NEPALESE PAPER PLANT This is an evergreen, occasionally deciduous, upright shrub, with leathery, dark green foliage, and highly fragrant, deep pink winter flowers. 'Jacqueline Fost II' AGM (above) has purplish pink and white blooms.

↕3m (10ft) ↔1.5m (5ft)

PERENNIAL LARGE

Delphinium elatum

This delphinium is an upright perennial, with deeply cut, green leaves. In midsummer, it produces tall spikes of bowl-shaped flowers in a wide range of colours, with many varieties to choose from. The stems require sturdy stakes.

↕2m (6ft) ↔90cm (36in)

PERENNIAL SMALL

Dianthus 'Dad's Favourite'

GARDEN PINK This dwarf, evergreen perennial forms a mound of grey-green, grass-like foliage. Throughout summer, it produces clove-scented, semi-double, white flowers with maroon markings. Pink-flowered 'Doris' AGM is also popular.

↕up to 45cm (18in) ↔30cm (12in)

BULB LARGE

Dierama pulcherrimum

ANGEL'S FISHING ROD This is an upright, summer-flowering corm with narrow, strap-shaped, evergreen leaves. In summer, it bears long, arching stems from which dangle funnel-shaped, deep pink flowers. Enrich the soil with organic matter.

↕1.5m (5ft)

SHRUB LARGE

Dipelta floribunda ♈

This is a vigorous, deciduous, upright, tree-like shrub, with peeling, pale brown bark and pointed, mid-green leaves. In late spring and early summer, it produces fragrant, pale pink blooms that are marked yellow inside

↕↔4m (12ft)

PERENNIAL LARGE

Echinacea purpurea

An erect, clump-forming perennial, with slim, tapering, green leaves. The daisy-like flowers have reflexed petals and a central brown disk. Varieties include the compact 'Kim's Knee High' (above), with orange-centred bright pink flowers.

↕1.5m (5ft) ↔45cm (18in)

PERENNIAL LARGE

Echinops bannaticus

This is an upright perennial, with narrow, deeply cut leaves and pale to mid-blue, spiky, spherical flowerheads, borne in late summer. The seedheads look good through winter. Varieties include 'Taplow Blue' (above), with steel-blue flowers.

↕1.5m (5ft) ↔75cm (30in)

SHRUB MEDIUM

Edgeworthia chrysantha

This is a deciduous, rounded, open shrub, with oval, dark green leaves. From late winter to early spring, it produces rounded heads of fragrant, tubular, yellow flowers covered in silky white hairs. It needs a warm, sheltered site.

↕↔1.5m (5ft)

BULB LARGE

Eremurus stenophyllus

This is a tall, upright, summer-flowering bulb that forms a clump of strap-shaped leaves. In summer, it produces long stems topped with torpedo-shaped heads of small, yellow flowers. May need staking in windy sites; apply fertilizer in spring.

↕1m (3ft)

PERENNIAL MEDIUM

Eryngium alpinum

This is an upright perennial, with heart-shaped, divided, green leaves. In summer, it produces tall, blue stems topped with cone-shaped, purple flowers that are surrounded by spiny bracts. 'Superbum' (above) has dark blue blooms.

↕70cm (28in)

BULB MEDIUM

Eucomis bicolor

PINEAPPLE LILY A late-summer flowering bulb, with wavy-edged leaves and spotted stems topped with clusters of greenish white flowers with purple-edged petals. With "hats" of leaf-like bracts, the flowers resemble pineapples. Protect during winter.

↕50cm (20in)

PERENNIAL LARGE

Euphorbia characias subsp. *wulfenii*

MEDITERRANEAN SPURGE An evergreen, upright subshrub, with finger-like, grey-green leaf clusters forming a frill around the stems and large, yellow-green spring flowerheads. Wear gloves to remove spent flower stems.

↕↔1.2m (4ft)

SHRUB LARGE

Fatsia japonica

CASTOR OIL PLANT An evergreen, rounded, dense shrub, with large, deeply lobed, hand-shaped, dark green leaves. Dense clusters of tiny, white flowers in mid-autumn are followed by black fruits. Forms include 'Variegata', with white-edged leaves.

↕↔4m (12ft)

PERENNIAL SMALL

Geranium ROZANNE

This sprawling, deciduous perennial produces rounded, deeply divided, mid-green leaves, with marbled, pale green markings. From summer to autumn, it bears masses of shallow, cup-shaped, blue flowers for many weeks.

↕↔50cm (20in) or more

ANNUAL/BIENNIAL MEDIUM

Gypsophila elegans

This fast-growing, erect, bushy annual has lance-shaped, greyish green leaves. From summer to early autumn, it bears clouds of tiny, white flowers on branching stems. Varieties include 'Covent Garden', with single, clear white flowers. Good for cutting.

↕60cm (24in) ↔30cm (12in) or more

SHRUB SMALL

Hebe 'Great Orme'

An evergreen, rounded, open shrub, with deep purple shoots and glossy, dark green foliage. Slender spikes of pink flowers that fade to white are produced from midsummer to mid-autumn. Trim in early spring to maintain a compact habit.

↕↔1.2m (4ft)

Plants for sandy soil

PERENNIAL MEDIUM

Helictotrichon sempervirens ♀

BLUE OAT GRASS An evergreen, perennial grass, with stiff, silvery blue leaves, forming a large, airy mound and long, arching stems of straw-coloured flowers in early summer, followed by long-lasting seedheads.

↕1m (3ft) ↔60cm (24in)

SHRUB LARGE

Hibiscus syriacus

A deciduous, upright shrub, with dark green, lobed leaves. From late summer to mid-autumn, it produces an abundance of large, saucer-shaped blooms in blue, pink, or white. 'Red Heart' AGM (above) has white flowers with pink centres.

↕3m (10ft) ↔2m (6ft)

SHRUB SMALL

Juniperus communis 'Compressa' ♀

This dwarf, slow-growing juniper is an evergreen, coniferous shrub. It forms a slim, dense, tapering column of blue-grey foliage, and may need protection from cold winds in winter.

↕80cm (32in) ↔45cm (18in)

SHRUB LARGE

Juniperus scopulorum

ROCKY MOUNTAIN JUNIPER This conifer has a neat, narrow, columnar habit, and does not require pruning. It has bright, steel-blue, scale-like foliage and tolerates hot, dry sites. Plant it at the back of borders or as a "focal" plant.

↕6m (20ft) ↔75cm (30in)

PERENNIAL LARGE

Kniphofia uvaria

RED-HOT POKER An upright perennial, with strap-shaped, evergreen leaves, above which rise tall spikes of bicoloured, cone-shaped flowerheads in late summer. The small, tubular, red and yellow flowers open from the base of the cluster.

↕1.2m (4ft) ↔60cm (24in)

SHRUB SMALL

Lavandula 'Willow Vale' ♀

A hybrid form of the traditional French lavender, this is an evergreen, compact shrub, with narrow, aromatic, grey-green foliage. From early- to midsummer, it produces purple blooms, topped with wavy flower bracts.

↕↔70cm (28in)

BULB LARGE

Lilium 'Gran Paradiso'

ASIATIC LILY This bulb produces upright stems with slender, red-flushed green leaves that are topped with heads of large, open, unscented, deep orange flowers. It will gradually form a clump if left in the soil over winter. Stems may require support.

↕1m (3ft)

BULB LARGE

Lilium 'Red Hot'

ORIENTAL LILY This large bulb produces long, arching stems of slender green leaves, topped with clusters of large, richly scented, white-edged rose-pink flowers in summer. The stems may need support in exposed gardens.

↕1m (3ft)

PERENNIAL SMALL

Limonium sinuatum

This is a bushy, upright perennial, grown as an annual, with lance-shaped, lobed, dark green leaves and, from summer to early autumn, clusters of tiny, blue, pink, or white flowers that are ideal for drying. Varieties include the Fortress Series (above).

↕40cm (16in) ↔30cm (12in)

ANNUAL/BIENNIAL SMALL

Linaria maroccana

TOADFLAX A fast-growing, bushy annual, with lance-shaped, pale green leaves and tiny, snapdragon-like flowers in shades of red, pink, purple, yellow, or white during summer. Varieties include 'Fairy Lights' (above), with flowers of many colours.

↕20cm (8in) ↔15cm (6in)

PERENNIAL MEDIUM

Linum narbonense

A clump-forming, short-lived perennial, with lance-shaped, greyish green leaves and clusters of saucer-shaped, pale to deep blue flowers in summer. Trim after flowering to encourage further blooms. 'Heavenly Blue' is a popular choice.

↕60cm (24in) ↔30cm (12in)

PERENNIAL MEDIUM

Lupinus 'The Page'

LUPIN This short-lived perennial forms bushy clumps of dark green leaves, divided into "fingers". In early summer, it bears tall stems of pink-red, pea-like blooms that may need staking in exposed sites.

↕1m (3ft) ↔75cm (30in)

PERENNIAL MEDIUM

Lychnis coronaria

A clump-forming, upright perennial, often grown as a biennial, with oval, grey, slightly downy leaves. From mid- to late summer, small, round, bright pink flowers appear over a long period on branched, grey stems. Deadhead to prevent self-seeding.

↕60cm (24in) ↔45cm (18in)

PERENNIAL LARGE

Melianthus major

HONEYBUSH Grown for its foliage, this perennial has blue-grey leaves, divided into toothed-edged leaflets. Small, tubular, brownish red flowers appear in late spring. Shelter from cold winds and provide a dry mulch in winter.

↕↔3m (10ft)

PERENNIAL SMALL

Oenothera speciosa

PINK EVENING PRIMROSE A short-lived, clump-forming perennial, with fragrant, saucer-shaped, pure white summer flowers that age to pink and spoon-shaped, deeply cut leaves. The blooms open in the evening. Varieties include 'Rosea' (above).

↕↔ up to 30cm (12in)

SHRUB MEDIUM

Osmanthus delavayi

An evergreen, rounded, bushy shrub, with arching branches and small, glossy, dark green leaves. From mid- to late spring, it produces a profusion of highly fragrant, tubular white flowers that are followed by small, blue-black fruits.

↕2m (6ft) ↔4m (12ft) or more

PERENNIAL MEDIUM

Penstemon 'Sour Grapes'

A semi-evergreen, upright perennial, with slender, lance-shaped, green leaves. The tubular, bell-shaped, purple-blue flowers are suffused with violet and white inside. They appear over many weeks from midsummer to autumn. Prefers rich soil.

↕60cm (24in) ↔45cm (18in)

SHRUB SMALL

Perovskia 'Blue Spire'

This deciduous, upright, subshrub has aromatic, silvery grey foliage. From late summer to autumn, it produces upright, branching spikes of tiny, violet-blue flowers that last for several weeks. Taller plants may need support.

↕1.2m (4ft) ↔1m (3ft)

SHRUB SMALL

Philadelphus 'Belle Etoile'

MOCK ORANGE This deciduous, arching shrub has small, mid-green leaves and sweetly fragrant, white flowers from late spring to early summer. Plant at the back of a border or along a boundary wall or fence. The scent is strongest on sunny days.

↕1.2m (4ft) ↔2.5m (8ft)

PERENNIAL MEDIUM

Phlomis russeliana

An evergreen perennial, with large, rough-textured, heart-shaped leaves. In summer, it produces unusual hooded, butter-yellow flowers, set at intervals up the tall, stout stems. The seedheads provide interest over winter.

↕1m (3ft) ↔60cm (24in) or more

SHRUB SMALL

Phygelius x *rectus*

CAPE FUCHSIA A semi-evergreen, upright subshrub, which may die back in winter when young. It has lance-shaped, green leaves and clusters of long, tubular flowers in summer. 'African Queen' AGM (above) has orange-red blooms.

↕1m (3ft) ↔1.2m (4ft)

Plants for sandy soil

SHRUB LARGE

Physocarpus opulifolius

This is a deciduous shrub, with rounded, lobed leaves. Dome-shaped, compact clusters of small, white flowers appear in early summer, followed by reddish brown fruits. Varieties include 'Diabolo' AGM (above), which sports deep purple foliage.
↕3m (10ft) ↔1.5m (5ft)

SHRUB LARGE

Pittosporum tenuifolium

An evergreen, columnar, later rounded, shrub or small tree, with purple shoots and wavy-edged, oval, glossy, mid-green leaves. In late spring, it produces tiny, honey-scented, purple flowers. The form 'Variegatum' has white-edged leaves.
↕10m (30ft) ↔5m (15ft)

PERENNIAL SMALL

Potentilla 'Gibson's Scarlet'

CINQUEFOIL This clump-forming perennial has dark green leaves that resemble those of a strawberry plant. From early- to late summer, it produces saucer-shaped, bright scarlet flowers on branching stems. It also grows in part shade.
↕45cm (18in) ↔60cm (24in)

SHRUB MEDIUM

Prunus tenella

This is a deciduous, bushy, upright shrub, with narrowly oval, glossy, green leaves. From mid- to late spring, the stems are clothed in bright pink, saucer-shaped flowers. 'Fire Hill' is a popular red-flowered form.
↕↔1.5m (5ft)

PERENNIAL MEDIUM

Rudbeckia hirta

This upright, short-lived perennial, often grown as an annual, has mid-green leaves. From summer to autumn, it produces daisy-like, golden or red flowers with brown centres. Varieties include 'Becky Mixed' (above).
↕up to 90cm (36in) ↔45cm (18in)

TREE LARGE

Quercus coccinea

SCARLET OAK This deciduous tree forms a rounded head of glossy, dark green, lobed leaves. In autumn, the foliage turns bright red, producing a bold display that can last for several weeks. Small acorns form at the same time. Best in large gardens.
↕20m (70ft) ↔15m (50ft)

PERENNIAL LARGE

Romneya coulteri

TREE POPPY A vigorous, bushy, shrubby perennial, with deeply divided, grey-green leaves. It is grown for its large, fragrant, papery, golden-centred white flowers, which appear in late summer on stout stems. Mulch well with compost in autumn.
↕↔2m (6ft)

PERENNIAL MEDIUM

Salvia x *sylvestris*

WOOD SAGE This is a compact perennial, with small, dark green, aromatic leaves. From early- to midsummer, 'Mainacht' AGM (above) produces spikes of indigo-blue, two-lipped flowers that last for a long period. It is attractive to butterflies and bees.
↕80cm (32in) ↔30cm (12in)

PERENNIAL MEDIUM

Santolina chamaecyparissus

COTTON LAVENDER A rounded perennial, with soft, woolly, silvery white, finely divided foliage and yellow, pompon flowers in summer. Trim after flowering. Varieties include 'Lemon Queen' (above), which has grey foliage and cream blooms.
↕75cm (30in) ↔1m (3ft)

PERENNIAL MEDIUM

Scabiosa atropurpurea

PINCUSHION FLOWER A clump-forming perennial, with lance-shaped, grey-green leaves. In summer, masses of slim stems are topped with small, domed, lilac-purple flowers, adorned with creamy anthers that look like pins. Deadhead regularly.
↕up to 1m (3ft) ↔30cm (12in)

PERENNIAL SMALL

Sedum spectabile ♡

ICE PLANT This sedum is a clump-forming perennial, with succulent stems of fleshy, grey-green leaves. From late summer to autumn, it produces flat heads of pink flowers, followed by winter seedheads. Varieties include 'Brilliant' AGM (above).

↕↔ 45cm (18in)

PERENNIAL LARGE

Solidago 'Goldkind'

This perennial forms an upright clump of stems with narrow, pointed green leaves. From midsummer to early autumn, it produces branching stems of small, golden-yellow blooms. Deadhead to prolong the flowering display.

↕ 1.2m (4ft) ↔ 1m (3ft)

SHRUB MEDIUM

Spiraea nipponica

This is a spreading shrub, with arching stems of narrow, dark green leaves and dense clusters of white or pink flowers, which appear in early summer. Varieties include the white-flowered 'Snowmound' AGM (above).

↕↔ 2.5m (8ft)

ANNUAL/BIENNIAL LARGE

Verbascum olympicum

OLYMPIAN MULLEIN A semi-evergreen biennial or short-lived perennial, with rosettes of grey, felted leaves and tall, branching stems of saucer-shaped, bright golden flowers, which appear from mid- to late summer. It may need staking.

↕ 2m (6ft) ↔ 1m (3ft)

SHRUB LARGE

Stachyurus praecox ♡

A deciduous, spreading, open shrub, with purplish red shoots and pale greenish yellow, bell-shaped flowers that appear from late winter to early spring, before the slim, tapering, dark green leaves emerge. Can be grown as a wall shrub.

↕ 4m (12ft) ↔ 3m (10ft)

PERENNIAL MEDIUM

Veronica austriaca subsp. *teucrium*

SAW-LEAVED SPEEDWELL This upright perennial has dark green, toothed-edged leaves and spikes of rich blue flowers in early summer. Deadhead to promote more flowers the following year.

↕↔ 60cm (24in)

PERENNIAL MEDIUM

Stipa tenuissima ♡

MEXICAN FEATHER GRASS A deciduous perennial that forms a tuft of fine, green leaves. From early summer, it produces panicles of silvery green flowers that turn beige as seeds form, giving the plant a hair-like appearance.

↕ 60cm (24in) ↔ 40cm (16in)

BULB MEDIUM

Zephyranthes candida

This is a tender bulb, with narrow, erect, grassy leaves. In summer, leafless stems carry crocus-like, white flowers. Plants often burst into bloom following heavy rain. Add plenty of organic matter to the soil. Bulbs can be lifted for winter.

↕ 25cm (10in)

PERENNIAL SMALL

Stokesia laevis

STOKES ASTER An evergreen perennial, with narrow, mid-green leaves. It produces large, cornflower-like, lavender- or purple-blue flowers on short stems from summer to mid-autumn. Varieties include 'Purple Parasols' (above).

↕↔ 45cm (18in)

OTHER SUGGESTIONS

Perennials

Acanthus hirsutus • *Achillea filipendulina* 'Gold Plate' ♡ • *Alstroemeria* 'Oriana' ♡ • *Anchusa azurea* 'Loddon Royalist' • *Anemone hupehensis* 'Bowles's Pink' ♡ • *Artemisia lactiflora* ♡ • *Aster cordifolius* 'Chieftain' ♡ • *Aster ericoides* 'Brimstone' ♡ • *Campanula latiloba* 'Percy Piper' ♡ • *Cephalaria gigantea* • *Coreopsis verticillata* 'Grandiflora' ♡ • *Delphinium* 'Conspicuous' ♡ • *Echinops ritro* • *Salvia nemorosa* 'Caradonna' ♡ • *Verbascum* 'Caribbean Crush'

Shrubs

Clerodendrum bungei • *Euonymus fortunei* 'Emerald Gaiety' ♡ • *Hebe* 'Margret' ♡ • *Hebe ochracea* 'James Stirling' ♡ • *Indigofera heterantha* • *Rhododendron* 'Praecox' ♡ • *Spiraea japonica* 'Goldflame'

Trees

Ginkgo biloba 'Troll' ♡

Plant focus: irises

Celebrated for their intricate and widely coloured flowers, the iris family includes plants for borders, ponds, and pots.

Irises comprise a range of different species adapted to widely different conditions. They grow from rhizomes or bulbs, and while some thrive in shallow water, others require dry, well-drained soils. Irises that grow from rhizomes are divided into two main groups: bearded and beardless forms. Bearded types are ideal for sunny, free-draining borders, with shorter species suitable for rock gardens, alpine beds, or troughs. Prone to rotting, their rhizomes must sit at or above soil level where they are exposed to full sun. Rhizomatic beardless irises have slimmer petals, and most prefer well-drained borders in sun or partial shade. However, Louisiana and Laevigatae irises, which thrive in or near ponds, are also part of this group; some are invasive and should be restrained in pond baskets to prevent them spreading. Bulbous irises include tall Dutch forms suitable for borders and cutting, colourful Juno irises, and tiny Reticulata types, ideal for planting in rock gardens and spring containers.

POPULAR IRIS FORMS

Bearded irises These early summer-flowering irises are characterized by tufted "beards" of soft hairs on the lower petals.

Beardless irises This group has slim, hairless petals, and includes water irises and Siberian forms that thrive in sun or part shade.

Water irises Beardless *I. laevigata*, *I. versicolor*, *I. ensata*, and the invasive flag iris, *I. pseudacorus*, grow well in shallow water and full sun.

Crested irises With a ridge on their lower petals, these early summer-flowering irises include *I. cristata* and *I. gracilipes*.

Dutch irises Grown from bulbs, these early summer-flowering forms have slim petals and grassy foliage; plant them in sun or part shade.

Reticulata irises These spring-flowering irises bear often scented, small blooms on short stems, and are dormant from summer to winter.

USING IRISES

Irises flower from late winter to early summer, and provide elegant forms and colour in many parts of the garden. Plant the bulbs in autumn and rhizome-forming plants in autumn or spring.

Bold drifts of bearded irises create a naturalistic effect in a border teamed with late-flowering Oriental poppies.

Water irises make perfect pond-side plants. The white-striped foliage of *I. laevigata* 'Variegata' (*above*) prolongs the season of interest well after the summer flowers have faded.

Bearded irises add glamour to large containers filled with gritty, free-draining compost, and will flower every year if fed annually in spring.

Plants for pond perimeters

Plants that enjoy cool, damp soil at the water's edge are known as bog plants and include flowers such as irises, lobelias, and many forms of primula.

Bog plants not only add interest to a garden, they also help develop and maintain wildlife habitat. If you have an artificial pond the water is contained, so the surrounding soil will be the same as elsewhere in the garden. However, you can replicate naturally occurring boggy conditions by digging out an area around your pond, lining it with pond liner punctured with a few drainage holes, and replacing the excavated soil, enriched with some well-rotted manure, on top of the liner. Also remember to water the area well during dry spells.

PERENNIAL MEDIUM

Acorus calamus

SWEET FLAG An upright, deciduous or semi-evergreen perennial marginal, with grass-like, green leaves and spikes of small, insignificant, brown flowers in summer. Plant it in shallow water or in a bog garden.

↕90cm (36in) ↔60cm (24in)

💧 ❄❄❄

PERENNIAL LARGE

Aruncus dioicus 🏆

GOAT'S BEARD This form of goat's beard is a tall, clump-forming perennial, with large, light green leaves, divided into smaller leaflets. In midsummer, it produces arching stems that carry plumes of tiny, creamy white flowers.

↕2m (6ft) ↔1.2m (4ft)

💧 ❄❄❄

PERENNIAL MEDIUM

Acorus calamus 'Argenteostriatus'

An upright, deciduous or semi-evergreen perennial marginal, with sword-like, cream-variegated leaves that are flushed rose-pink in spring. It forms a large clump that can be divided in spring.

↕75cm (30in) ↔60cm (24in)

💧 ❄❄❄

PERENNIAL SMALL

Acorus gramineus

SLENDER SWEET FLAG This is a compact, semi-evergreen perennial marginal for bogs or ponds, with grass-like, dark green leaves. Varieties include 'Ogon' (above), which has attractive creamy-yellow-striped foliage.

↕25cm (10in) ↔15cm (6in)

💧 ❄❄❄

PERENNIAL MEDIUM

Astilbe chinensis

This is a clump-forming perennial with deeply cut, dark green leaves. In late summer, it produces upright plumes of tiny, pink-white flowers. *A. chinensis* var. *pumila* AGM has deep pink-red blooms. Keep well watered.

↕60cm (24in) ↔20cm (8in)

💧 ❄❄❄

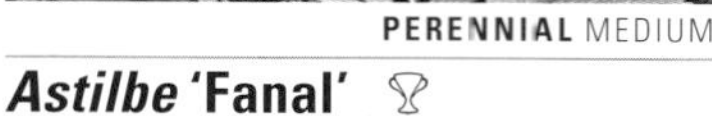

PERENNIAL MEDIUM

Astilbe 'Fanal' 🏆

This is a clump-forming perennial, with deeply divided, fern-like, dark green foliage. In midsummer, feathery, dark pink flowerheads appear on slim stems. Prefers rich soil; also grows well in shade.

↕60cm (24in) ↔45cm (18in)

💧 ❄❄❄

BULB LARGE

Camassia leichtlinii

An upright, deciduous bulb, with long, narrow, erect, lower leaves, and tall spikes of starry, creamy white blooms from late spring to early summer. It flowers later than most camassias, extending their season. 'Semiplena' (above) has double flowers.

↕1.5m (5ft)

💧 ❄❄

PERENNIAL LARGE

Canna 'Wyoming' 🏆

An erect perennial, with decorative foliage and flowers. It produces large, oval, purple-bronze leaves, with darker purple veins and gladiolus-like, pale orange flowers from midsummer to early autumn. Protect it from frost in winter.

↕1.8m (6ft) ↔50cm (20in)

💧 ❄

PERENNIAL LARGE

Darmera peltata ♈

UMBRELLA PLANT A spreading perennial, grown for its large, round, deeply veined leaves that can grow larger than dinner plates, and turn red in autumn. In spring, it bears white or pale pink flower clusters on hairy stems, before the foliage appears.

↕1.2m (4ft) ↔60cm (24in)

PERENNIAL SMALL

Carex elata 'Aurea' ♈

BOWLES' GOLDEN SEDGE An evergreen, perennial sedge, with arching, golden-yellow leaves. Triangular stems bearing blackish brown flower spikes appear in summer. Plant at the edge of a bed or pond for a grassy effect.

↕40cm (16in) ↔15cm (6in)

PERENNIAL LARGE

Eupatorium cannabinum

HEMP AGRIMONY This is a tall, upright perennial, with large, divided leaves held on red stems. From late summer to early autumn, it produces clusters of fluffy, light pink or purple flowers, which attract butterflies.

↕1.5m (5ft) ↔1.2m (4ft)

PERENNIAL MEDIUM

Carex muskingumensis

PALM BRANCH SEDGE This sedge is a deciduous perennial, with light green, grassy foliage, which turns yellow before dying back. In spring, it produces small, insignificant flowers, followed by attractive brown seedheads. Cut back in autumn.

↕↔75cm (30in)

PERENNIAL MEDIUM

Carex pendula

PENDULOUS SEDGE An evergreen, perennial sedge, with long, green, grassy leaves. In summer, greenish brown, catkin-like flowers dangle from arching, triangular stems. Best in wildlife gardens or wild areas, where it will self-seed freely.

↕1m (3ft) ↔30cm (12in)

PERENNIAL LARGE

Eupatorium purpureum

JOE PYE WEED This is a stately, upright perennial, with coarse, oval leaves held on purplish green stems. From late summer to early autumn, it produces fluffy, pinkish purple flowerheads on tall, sturdy stems. Ideal for wildlife gardens.

↕2.2m (7ft) ↔1m (3ft)

PERENNIAL LARGE

Filipendula purpurea

PURPLE MEADOWSWEET This is an upright perennial, with deeply cut, green leaves. In late summer, it produces large clusters of tiny, rich reddish purple flowers, held on tall, leafy stems.

↕1.2m (4ft) ↔60cm (24in)

PERENNIAL LARGE

Filipendula rubra

QUEEN OF THE PRAIRIE An upright perennial, with large, aromatic leaves, divided into jagged-edged, green leaflets and, in midsummer, feathery plumes of tiny, soft, rose-pink flowers, which pale as they age. Use it at the back of a boggy site.

↕up to 2.5m (8ft) ↔1.2m (4ft)

PERENNIAL MEDIUM

Geum rivale

WATER AVENS This perennial forms neat rosettes of rounded, green leaves. From late spring to summer, nodding, bell-shaped, pink or dark orange flowers appear on top of slender stems. Plant it at the front of a bed; it may self-seed and spread slowly.

↕↔60cm (24in)

PERENNIAL SMALL

Houttuynia cordata 'Chameleon'

A vigorous, deciduous perennial, with aromatic, heart-shaped, green leaves, splashed with yellow and red. In summer, it produces small sprays of greenish white flowers. Can be invasive.

↕10cm (4in) ↔indefinite

PERENNIAL MEDIUM

Iris ensata

JAPANESE FLAG An upright, clump-forming perennial, with strap-like, drooping, green foliage and beardless, purple or red-purple flowers, with a yellow blaze on the lower petals, from early- to midsummer. 'Rose Queen' AGM (above) has lilac-pink flowers.

↕90cm (36in) ↔60cm (24in)

PERENNIAL MEDIUM

Iris sibirica

SIBERIAN IRIS This clump-forming perennial has upright, sword-like, blue-green leaves and large, beardless flowers in blue, pink, white, and yellow from late spring to early summer. 'Butter and Sugar' AGM (above) has yellow and white blooms.

↕1m (3ft) ↔indefinite

PERENNIAL MEDIUM

Lobelia cardinalis

CARDINAL FLOWER This is a deciduous, upright perennial, with narrow, lance-shaped, glossy, green leaves. In summer, it produces spires of striking scarlet, two-lipped flowers on tall stems. The plant is toxic; wear gloves when handling it.

↕75cm (30in) ↔23in (9in)

PERENNIAL MEDIUM

Iris versicolor

AMERICAN BLUE FLAG IRIS This upright, clump-forming perennial, with arching, strap-shaped foliage and blue-purple flowers in early summer can be grown in a bog or shallow water. Varieties include the purple-pink 'Kermesina' (above).

↕80cm (32in) ↔indefinite

TREE MEDIUM

Liquidambar styraciflua

A deciduous, conical to spreading tree with large, lobed, glossy, dark green leaves that turn brilliant orange, red, and purple in autumn. 'Variegata' (above) has gold-splashed foliage. Site it some distance from the pond edge.

↕15m (50ft) ↔8m (25ft)

PERENNIAL MEDIUM

Lobelia siphilitica

BLUE CARDINAL FLOWER This is an upright perennial, with narrow, lance-shaped, light green leaves. From mid- to late summer, it produces tall spikes of long-lasting, tubular, two-lipped, blue flowers. Plant it towards the back of a bed.

↕1m (3ft) ↔23cm (9in)

PERENNIAL MEDIUM

Lysimachia clethroides

GOOSENECK LOOSESTRIFE This is a vigorous, clump-forming, spreading perennial, with narrow, lance-shaped, grey-green foliage. In late summer, it produces long, tapering flowerheads comprising small, white blooms.

↕↔1m (3ft)

PERENNIAL MEDIUM

Persicaria microcephala 'Red Dragon'

A spreading perennial, bearing heart-shaped, reddish green leaves, with silver and bronze markings. Tiny, white flowers appear in midsummer. Cut back spent flower stems in autumn.

↕70cm (28in) ↔1m (3ft)

PERENNIAL LARGE

Physostegia virginiana var. *speciosa*

OBEDIENT PLANT An erect perennial, with lance-shaped, toothed, green leaves and spikes of hooded, two-lipped, rose-purple flowers in late summer. 'Variegata' (above) has white-edged leaves.

↕1.2m (4ft) ↔60cm (24in)

PERENNIAL LARGE

Primula florindae 🏆

GIANT COWSLIP A clump-forming perennial, with oval, green leaves and clusters of fragrant, bell-shaped, nodding flowers on slim stems in summer. Deadhead to prevent self-seeding and to encourage repeat-flowering.

↕1.2m (4ft) ↔1m (3ft)

PERENNIAL SMALL

Primula japonica 🏆

CANDELABRA PRIMULA This deciduous perennial produces rosettes of oval, toothed-edged, pale green leaves. In early summer, clusters of tubular, deep red flowers appear on stout stems. 'Miller's Crimson' (above) has deep crimson flowers.

↕↔45cm (18in)

PERENNIAL MEDIUM

Primula vialii 🏆

ORCHID PRIMROSE This clump-forming perennial produces rosettes of green, oval leaves. In late spring, tapering cones of small, tubular, bluish purple and red flowers appear above the leaves on slim stems.

↕60cm (24in) ↔30cm (12in)

PERENNIAL LARGE

Salvia uliginosa 🏆

BOG SAGE This tall, upright, branching perennial has oblong to lance-shaped, deeply toothed, slightly sticky, mid-green leaves. From late summer to autumn, two-lipped, clear blue flowers appear. The stems require staking; mulch well in autumn.

↕2m (6ft) ↔90cm (36in)

PERENNIAL LARGE

Sanguisorba canadensis

CANADIAN BURNET This perennial has a clump-forming habit, and produces mid-green, divided foliage that turns red in autumn. From late summer to early autumn, it produces upright spikes of bottlebrush-like, white flowers on sturdy stems.

↕2m (6ft) ↔60cm (24in)

TREE LARGE

Taxodium distichum

SWAMP CYPRESS This is a deciduous, broadly conical conifer, with feathery, green foliage that turns yellow-brown in late autumn before falling. Small, oval cones appear in summer. It is a large tree, best suited to big gardens.

↕40m (130ft) ↔9m (28ft)

PERENNIAL MEDIUM

Trollius x *cultorum*

GLOBEFLOWER An upright, clump-forming perennial, with lobed, toothed, dark green leaves. From late spring to early summer, it produces buttercup-like, bowl-shaped, semi-double flowers. 'Lemon Queen' (above) has pale yellow blooms.

↕60cm (24in) ↔45cm (18in)

OTHER SUGGESTIONS

Perennials

Anagallis tenella • *Aruncus* 'Misty Lace' • *Cardamine raphanifolia* • *Carex testacea* • *Cephalanthus occidentalis* • *Eupatorium fortunei* 'Pink Frost' • *Filipendula ulmaria* • *Gunnera magellanica* • *Hibiscus coccineus* • *Hymenocallis caroliniana* • *Iris chrysographes* 🏆 • *Iris virginica* • *Lobelia* 'Russian Princess' • *Lythrum salicaria* • *Persicaria microcephala* • *Primula beesiana* 🏆 • *Primula* x *bulleesiana* • *Primula bulleyana* 🏆 • *Ranunculus acris* 'Flore Pleno' 🏆 • *Vernonia noveboracensis*

Grasses

Arundo donax • *Miscanthus sinensis* 'Gracillimus' • *Miscanthus sinensis* 'Variegatus'

Shrubs

Salix integra 'Hakuro-nishiki' 🏆

Plants for ponds

Reflections of colourful aquatic plants in pools of sparkling water create a magical scene in gardens large and small, in an open, sunny site.

Plant up a tiny, formal water feature in an urban garden, or wildlife pond in a rambling plot to create a textural backdrop. Use plants with different foliage forms, such as strappy irises alongside spear-shaped *Pontederia* leaves, and add a splash of colour with bright yellow *Caltha* species and scarlet lobelias for hot schemes or cool-hued water lilies, forget-me-nots, and veronicas for a pastel palette. Check plant labels for the correct planting depths, since some like their roots completely submerged, while others are happier languishing in the shallows.

WATER PLANT MEDIUM

Alisma plantago-aquatica

WATER PLANTAIN This deciduous perennial marginal has large, oval, bright green leaves that are held well above the water. It produces slim, branching stems of small, pale pink to white flowers in summer. May spread rapidly.

↕75cm (30in) ↔45cm (18in)

❄❄❄

PERENNIAL SMALL

Calla palustris

BOG ARUM This deciduous or semi-evergreen, spreading perennial marginal, with heart-shaped, glossy, dark green leaves. In late spring, it bears large, white, petal-like "spathes", usually followed by red or orange berries in autumn.

↕25cm (10in) ↔30cm (12in)

PERENNIAL SMALL

Caltha palustris

MARSH MARIGOLD This is a deciduous perennial marginal, with rounded, dark green leaves and clusters of cup-shaped, bright golden-yellow flowers, which appear on upright stems above the foliage in spring. Also suitable for pond perimeters.

↕40cm (16in) ↔45cm (18in)

❄❄❄

PERENNIAL SMALL

Caltha palustris 'Flore Pleno' 🏆

MARSH MARIGOLD This perennial has dark green, rounded leaves, and clusters of bright golden, double spring flowers. *C. palustris* var. *alba* is compact, with single, white flowers. Also suitable for pond perimeters.

↕↔25cm (10in)

❄❄❄

WATER PLANT MEDIUM

Ceratophyllum demersum

HORNWORT This is a deep water, deciduous, aquatic plant, which helps to keep pond water clear and provides cover for wildlife. It produces slim, spreading stems of feathery, dark green foliage, and tiny, white flowers in summer.

↔indefinite

❄❄❄

PERENNIAL SMALL

Eriophorum angustifolium

COTTON GRASS This is a vigorous, spreading perennial marginal, with grassy, green leaves. Downy, white flowers that resemble cotton wool appear on slim stems in summer. Restrict its spread in a pond basket.

↕30cm (18in) ↔indefinite

❄❄❄ pH

PERENNIAL MEDIUM

Iris laevigata

JAPANESE WATER IRIS This is an upright, clump-forming perennial marginal, with sword-shaped, mid-green leaves. In early summer, rich purple flowers with gold marks on the lower petals are borne on tall, slim stems.

↕90cm (36in) or more ↔indefinite

❄❄❄ (!)

PERENNIAL MEDIUM

Iris laevigata 'Variegata' 🏆

JAPANESE WATER IRIS This is an upright, clump-forming perennial marginal, with sword-shaped, white- and-green-striped foliage. The purple-blue flowers, which appear in summer, sometimes bloom a second time in early autumn.

↕1m (3ft) ↔50cm (20in)

❄❄❄

PERENNIAL MEDIUM

Iris versicolor

AMERICAN BLUE FLAG This is an upright, clump-forming perennial marginal, with sword-shaped leaves. In early summer, it produces blue-purple flowers on branched stems. 'Whodunit' (above) has purple-edged white flowers, with purple veining.

↕80cm (32in) ↔indefinite

❄❄❄

PERENNIAL SMALL

Juncus effusus f. *spiralis*

CORKSCREW RUSH This is an evergreen perennial marginal or bog plant, with leafless, tubular stems that twist and curl, and are often prostrate. In summer, it produces small, greenish brown flowers that seed easily; deadhead to keep in check.

↕30cm (12in) ↔60cm (2ft)

❄❄❄

WATER PLANT SMALL

Menyanthes trifoliata

BOG BEAN This perennial marginal has unusual, upright, green leaves, split into three leaflets, held at the end of floating, spreading stems. Clusters of fringed, small, white flowers open from cerise buds in early spring.

↕23cm (9in) ↔indefinite

❄❄❄

PERENNIAL MEDIUM

Mimulus cardinalis 🏆

SCARLET MONKEY FLOWER This is a spreading perennial marginal, often grown as an annual, with small, toothed-edged, downy, green leaves. Masses of tomato-red flowers cover the plant from summer to early autumn.

↕90cm (36in) ↔60cm (24in)

💧 ❄❄❄

WATER PLANT LARGE

Nuphar lutea

YELLOW WATER LILY This is a vigorous, spreading, deep water aquatic perennial, with round, floating, green leaves, similar to those of a water lily. In summer, buttercup-like flowers that smell of alcohol appear on stalks above the foliage.

↔1.5m (5ft)

❄❄❄

WATER PLANT LARGE

Nymphaea 'Escarboucle' 🏆

WATER LILY This deep water aquatic perennial has deciduous, floating, dark green leaves. In summer, it bears cup-shaped, deep crimson flowers, with bright golden-yellow centres. It spreads and is best for larger ponds with calm water.

↔3m (10ft)

WATER PLANT SMALL

Nymphaea pygmaea 'Rubra'

WATER LILY This aquatic perennial has round, green, floating leaves, with red undersides and purple blotches. Rose-coloured, cup-shaped flowers appear from mid- to late summer, maturing to garnet-red. Ideal for smaller ponds; requires still water.

↔40cm (16in)

WATER PLANT SMALL

Orontium aquaticum

GOLDEN CLUB A deciduous, deep water perennial that helps to keep pond water clear. It produces blue-grey, floating leaves and, in spring, pencil-like, gold and white flower spikes. Restrict its spread by planting in a large pond basket.
↕45cm (18in) ↔75cm (30in)

❄❄❄

PERENNIAL SMALL

Persicaria amphibia

AMPHIBIOUS BISTORT A vigorous perennial marginal, with lance-shaped, dark green, floating leaves. From midsummer to autumn, it produces conical clusters of pink flowers on stout stems above the foliage, followed by glossy, brown fruits.
↕30cm (12in) ↔2m (6ft)

❄❄❄

PERENNIAL MEDIUM

Pontederia cordata

PICKEREL WEED This is an upright perennial marginal, with large, spear-shaped, glossy, mid-green leaves. In late summer, dense spikes of blue flowers emerge on stout stems. Varieties include 'Alba', with white flowers.
↕75cm (30in) ↔45cm (18in)

❄❄❄

WATER PLANT SMALL

Ranunculus aquatilis

WATER CROWFOOT This perennial aquatic is one of the best flowering oxygenators, helping to keep pond water clear. It produces mats of ivy-shaped, toothed leaves and yellow-centred white, buttercup-like, floating flowers in summer.
↕1cm ($\frac{1}{2}$in) ↔ indefinite

❄❄❄

PERENNIAL MEDIUM

Ranunculus flammula

LESSER SPEARWORT This is a spreading perennial marginal, with narrow, spear-shaped leaves. In summer, it produces masses of yellow, buttercup-like, single flowers on wiry stems. The plant is toxic, and the sap can cause skin irritation.
↕70cm (28in) ↔75cm (30in)

❄❄❄ (!)

PERENNIAL LARGE

Sagittaria latifolia

AMERICAN ARROWHEAD This vigorous perennial marginal is grown for its large, linear, arrow-shaped, green leaves. In summer, it produces small, white, three-petalled flowers, with yellow centres, on leafless stems.
↕1.5m (5ft) ↔60cm (24in)

❄❄❄

PERENNIAL SMALL

Sagittaria sagittifolia

COMMON ARROWHEAD This perennial marginal is grown mainly for its upright, arrow-shaped, mid-green leaves. In summer, it produces clusters of three-petalled, white flowers, with dark purple centres.
↕45cm (18in) ↔30cm (12in)

❄❄❄

PERENNIAL MEDIUM

Saururus cernuus

SWAMP LILY This is a perennial marginal, which produces clumps of heart-shaped, mid-green leaves and long, slim stems of creamy white flowers in summer. It can be invasive; restrict its spread by planting in a large pond basket.
↕1m (3ft) ↔ indefinite

❄❄❄

PERENNIAL LARGE

Schoenoplectus lacustris subsp. *tabernaemontani*

GREY CLUB RUSH A tall, upright, spreading perennial aquatic sedge, with leafless, grassy stems. Brown flowerheads appear in summer. 'Zebrinus' (above) has creamy yellow leaves with horizontal, green stripes.

↕1.5m (5ft) ↔ indefinite

❄❄❄

PERENNIAL LARGE

Thalia dealbata

POWDERY ALLIGATOR FLAG An evergreen perennial marginal, with large, narrowly oval, grey-green leaves, which are covered with white powder. In summer, it bears small, purple flowers on top of long, slender stems.

↕1.5m (5ft) ↔ 60cm (24in)

❄

PERENNIAL LARGE

Typha latifolia

BULRUSH A large, vigorous perennial marginal, with slender, grass-like foliage. It produces decorative, sausage-shaped, dark brown flowerheads in summer. It is invasive; best in large ponds, confine to a planting basket to prevent it spreading.

↕2.5m (8ft) ↔ 60cm (24in)

❄❄❄

PERENNIAL LARGE

Typha angustifolia

LESSER REED MACE A tall, vigorous, clump-forming perennial marginal, with arching, grass-like leaves. Sausage-shaped, dark brown flowerheads appear on slender stems in summer. Restrict its spread by planting in a large pond basket .

↕1.5m (5ft) ↔ indefinite

❄❄❄

PERENNIAL MEDIUM

Typha minima

This miniature reed mace is a perennial marginal, with grass-like, mid-green leaves. Spikes of short, sausage-shaped, blackish brown flowerheads are produced in profusion in late summer. Suitable for smaller ponds and patio water features.

↕60cm (24in) ↔ 30cm (12in)

❄❄❄

PERENNIAL MEDIUM

Zantedeschia aethiopica

ARUM LILY This is a perennial marginal or bog plant, with large, arrow-shaped, dark green leaves. In summer, it produces white, petal-like spathes, each with a yellow spike in the centre. 'Crowborough' AGM is a popular variety.

↕↔ 90cm (36in)

❄❄

PERENNIAL MEDIUM

Zantedeschia aethiopica 'Green Goddess' 🏆

This semi-evergreen perennial marginal has large, dark green leaves and dark green, hood-like "spathes", with splashes of white in the throat and yellow, central spikes in summer. Also suitable for bog gardens.

↕1m (3ft) ↔ 60cm (2ft)

❄❄ ⓘ

OTHER SUGGESTIONS

Marginals

Butomus umbellatus • *Cardamine pratensis* • *Cotula coronopifolia* • *Cyperus papyrus* 'King Tut' • *Eleocharis palustris* • *Equisetum hyemale* • *Iris pseudacorus* • *Iris pseudacorus* var. *bastardii* • *Iris versicolor* 'Rowden Concerto' • *Juncus ensifolius* • *Lysichiton americanum* • *Lysichiton camtschatcensis* 🏆 • *Mentha aquatica* • *Mimulus guttatus* • *Myosotis palustris* 'Alba' • *Myosotis scorpioides* • *Oenanthe fistulosa* • *Phragmites australis* • *Ruellia brittoniana* 'Katie' • *Sagittaria graminea* • *Veronica beccabunga*

Plant focus: water lilies

The most flamboyant of all aquatic plants, water lilies are prized for their large, colourful summer flowers and decorative foliage.

WATER LILIES, KNOWN BOTANICALLY AS *NYMPHAEA* , make beautiful features in ponds and pools, where their dramatic blooms and rounded leaves create waves of colour throughout the summer months. As well as providing ornamental value, water lilies play a useful role in aquatic ecosystems, their large leaves helping to regulate the water temperature and shield resident flora and fauna from pests and predators. When choosing a water lily, check the plant label for its final spread to ensure it will suit the size and depth of your pond. Also note that the tropical blue and purple forms are not hardy and must be grown in a warm, frost-free environment, such as a conservatory. Most water lilies grow at depths of between 45cm (18in) and 1.2m (4ft), although dwarf hybrids will thrive in shallow water less than 30cm (12in) deep. All forms prefer full sun and still water, and do not grow well in streams or too close to fountains or waterfalls.

WATER LILY FORMS

Small water lilies Best suited to smaller ponds and features, this group prefers shallow water and a planting depth of 30–45cm (12–18in). Varieties include the pink-red *N.* 'Ellisiana' (*top*) and the pale yellow *N.* 'Pygmaea Helvola' (*above*), which both spread to 90cm (36in).

Medium water lilies This group is ideal if you have a deeper, mid-sized pond, or plan to grow several plants in a larger pond. Requiring a planting depth of 60–100cm (24–39in), *N.* 'Gonnère' (*top*) and *N.* 'Escarboucle' (*above*) both spread to 1.5m (5ft).

Large water lilies Only suitable for large ponds and lakes, these water lilies require a planting depth of 1–2m (3–6ft) and spread to over 1.8m (6ft) across. *N.* 'Attraction' (*top*) produces white or pink blooms, while *N. alba* (*above*) has large, pure white flowers in summer.

USING WATER LILIES

Although water lilies will tolerate light shade, they flower best in full sun. New plants may take a year or two to reach full flowering size, and should be split every 4–5 years to maintain vigour.

Miniature ponds Small-sized pygmy water lily varieties are ideal for growing in watertight patio containers, such as a half-barrel fitted with pond liner.

Wildlife ponds The leaves of water lilies help protect pond wildlife by cooling the water during summer. They also help to reduce algal growth, which keeps the water clear, healthy, and well oxygenated.

Formal ponds One or two lilies grown as specimens in a raised pool add an elegant note to a formal garden design.

Plants for boundaries, hedges, and windbreaks

Plants make beautiful screens, carving up the space within a garden, or marking out boundaries with foliage and flowers.

Choose from smooth, green walls made from clipped yew or box to set off a formal scheme, or colourful, textured screens using photinia or berberis for an informal design. Hedges that bear flowers include hypericums, lavender, and fuschias, which offer a long season of interest and, if wildlife watching is your hobby, make a home for birds, animals, and insects with a medley of malus, roses, and hornbeam. Some shrubs can also provide protection, their thorns helping to keep intruders at bay, so if security is an issue, select the fiercely armoured berberis or a spiny holly.

SHRUB LARGE

Abelia x *grandiflora*

This is a deciduous or semi-evergreen, arching shrub, with oval, green leaves that are bronze when young. From midsummer to autumn, it produces masses of trumpet-shaped, pink-tinged white blooms. Clip in spring.

↕3m (10ft) ↔4m (12ft)

SHRUB MEDIUM

Berberis julianae

WINTERGREEN BARBERRY An evergreen, bushy shrub, with glossy, oval, dark green leaves, spiny stems, and yellow flowers in late spring, followed by blue-black fruits. Trim lightly in summer, after the flowers have faded. Use as a security hedge.

↕2.5m (8ft) ↔3m (10ft)

SHRUB LARGE

Berberis x *stenophylla*

BARBERRY An evergreen, arching, prickly shrub, with narrow, spine-tipped, dark green leaves, blue-grey beneath. Small, yellow flowers appear in spring, followed by small, blue-black fruits. Trim lightly in summer, after the flowers have faded.

↕3m (10ft) ↔5m (15ft)

SHRUB MEDIUM

Buxus sempervirens 'Elegantissima'

BOX A compact, rounded, evergreen shrub, bearing oval, green leaves, with white margins. It makes a dense, waist-high, evergreen formal hedge; clip twice a year in early and late summer.

↕↔1.5m (5ft)

SHRUB SMALL

Buxus sempervirens 'Suffruticosa'

BOX This compact, slow-growing, evergreen shrub produces woody stems of small, oval, green leaves. Ideal for topiary and for low, formal hedges or border edging. Clip twice a year in early and late summer.

↕1m (3ft) ↔1.5m (5ft)

TREE LARGE

Carpinus betulus

COMMON HORNBEAM A deciduous tree, with oval, veined, dark green leaves that turn yellow and orange in autumn; young stems retain the dried foliage in winter. Bears green catkins in late spring. Use for formal or wildlife hedging; prune in midsummer.

↕25m (80ft) ↔20m (70ft)

SHRUB LARGE

Cornus mas

CORNELIAN CHERRY A deciduous shrub, with oval, green leaves that turn purple in autumn. From winter to early spring, small, yellow flowers appear on bare shoots, followed by edible fruits. Use as a screen; prune in early summer.

↕↔5m (15ft)

SHRUB LARGE

Corylus avellana

HAZEL A large, spreading shrub or small tree, with rounded, deeply veined, green leaves that turn yellow in autumn. Yellow catkins in spring are followed by edible nuts in autumn. Ideal for a wildlife or naturalistic garden; prune in late winter.

↕↔5m (15ft)

TREE LARGE

x *Cuprocyparis* 'Castlewellan'

LEYLAND CYPRESS An upright, vigorous conifer, slightly slower growing than the species, with golden foliage tinged with bronze. Makes an excellent hedge or windbreak in large gardens. Cut annually in early autumn to keep in check.

↕25m (80ft) ↔4m (12ft)

SHRUB LARGE

Elaeagnus x *ebbingei*

This vigorous, evergreen shrub has broadly oval, dark green leaves covered with a silvery dusting. Clip it in late summer. Colourful variegated forms, such as 'Gilt Edge', are also available, which form attractive informal screens.
↕↔5m (15ft)

SHRUB MEDIUM

Escallonia 'Apple Blossom'

A compact, evergreen shrub, with small, glossy, leathery, dark green leaves. From early- to midsummer, it produces a profusion of pink flowers on leafy stems. Suitable for coastal areas. Clip hedges in late summer after flowering.
↕↔2.5m (8ft)

TREE LARGE

Fagus sylvatica

COMMON BEECH The oval, wavy-edged, textured, green leaves of this large, deciduous tree turn orange-brown in autumn and persist through winter, if pruned regularly. Trim hedges in winter to keep dense and compact.
↕25m (80ft) ↔15m (50ft)

SHRUB LARGE

Forsythia x *intermedia*

This deciduous shrub bears masses of pale yellow flowers on bare stems from late winter to early spring, before the small, green leaves appear. Makes a dense screen when pruned in spring after flowering. 'Lynwood Variety' AGM (above) is popular.
↕↔3m (10ft)

TREE LARGE

Ilex aquifolium

ENGLISH HOLLY A slow-growing, evergreen tree, with dark green, glossy leaves armed with long, pointed spines and red berries in autumn. Suitable for formal or wildlife hedges, and intruder-proof boundaries. Prune in late summer with secateurs.
↕up to 20m (70ft) ↔6m (20ft)

SHRUB MEDIUM

Fuchsia magellanica var. *molinae*

This hardy fuchsia is an upright, deciduous shrub, with small, lance-shaped, green leaves. It bears pendent, pale pink flowers from summer to early autumn on arching stems. Prune hedges in spring.
↕↔2m (6ft)

SHRUB LARGE

Griselinia littoralis

NEW ZEALAND BROADLEAF A fast-growing evergreen, this shrub has oval, leathery, mid-green leaves. Suitable for coastal areas; prune hedges lightly in summer with secateurs to avoid damaging the leaves.
↕8m (25ft) ↔5m (15ft)

SHRUB SMALL

Hypericum 'Hidcote'

ST JOHN'S WORT An evergreen or semi-evergreen shrub, with oval, dark green leaves and large, golden-yellow flowers that appear from midsummer to early autumn. Use it as an informal hedge; prune in spring with secateurs.
↕1.2m (4ft) ↔1.5m (5ft)

SHRUB SMALL

Lavandula angustifolia

ENGLISH LAVENDER An evergreen, bushy, dwarf shrub, with slim, aromatic, silvery grey leaves. In summer, it bears spikes of small, fragrant, violet-blue flowers. Creates a colourful low screen. Trim in late winter to encourage strong, compact growth.
↕80cm (32in) ↔60cm (24in)

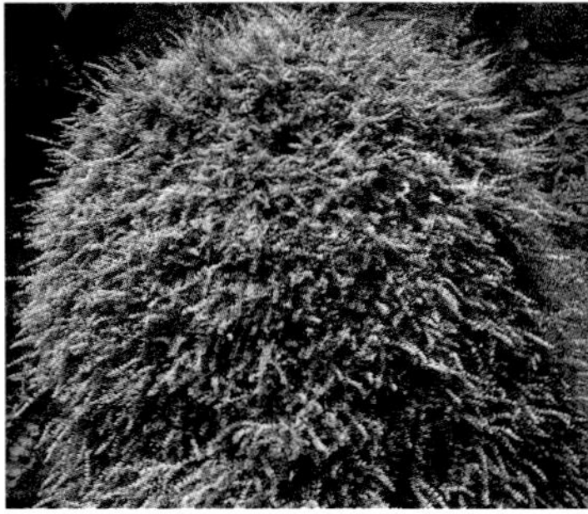

SHRUB MEDIUM

Lonicera nitida

SHRUBBY HONEYSUCKLE This evergreen, bushy shrub has arching shoots covered with tiny, dark green leaves. Suitable for formal hedges; trim regularly between spring and autumn. 'Baggesen's Gold' (above) has golden leaves.
↕2m (6ft) ↔3m (10ft)

TREE LARGE

Magnolia grandiflora

This dense, evergreen tree has large, glossy leaves and white, cup-shaped, fragrant flowers in summer. Use it to create a tall, informal boundary rather than clipped hedge. Prune lightly in spring using secateurs.

↕18m (60ft) ↔15m (50ft)

SHRUB LARGE

Malus sargentii

A deciduous, spreading shrub with arching branches, it bears white or pink flowers in mid-spring, followed by small, red or yellow fruits, and autumn tints. Use it to create an informal boundary with year-round interest. Prune in late winter.

↕4m (12ft) ↔5m (15ft)

SHRUB LARGE

Osmanthus x *burkwoodii* 🏆

An evergreen, dense shrub, with glossy, dark green foliage and a profusion of small, highly fragrant, white flowers from mid- to late spring. Use it as a tall, formal or informal hedge; prune after flowering.

↕↔3m (10ft)

SHRUB LARGE

Photinia x *fraseri* 'Red Robin' 🏆

An evergreen, dense shrub, with glossy, dark reddish green leaves, which are bright red when young. It makes a decorative, informal hedge or screen; prune with secateurs in spring and again in summer.

↕↔5m (15ft)

SHRUB LARGE

Phyllostachys viridiglaucescens

This evergreen, clump-forming bamboo has greenish brown canes that mature to yellow-green, and bright green, linear leaves. Use it as an informal screen; cut out dead, weak, and old canes in spring.

↕8m (25ft) ↔indefinite

SHRUB LARGE

Pyracantha 'Mohave'

A dense, evergreen shrub, with small, green leaves and spiny stems. It bears white flowers in early summer, followed by clusters of red berries. Cut formal hedges in spring and late summer, or leave to grow as a wildlife or security hedge.

↕↔4m (12ft)

TREE LARGE

Pinus strobus

EASTERN WHITE PINE This conifer has grey-green foliage and cylindrical cones. The bark becomes fissured with age. Prune regularly in spring and use as an informal boundary. Only suitable for large gardens.

↕35m (120ft) ↔8m (25ft)

SHRUB LARGE

Prunus laurocerasus

CHERRY LAUREL This evergreen, dense shrub has large, glossy, dark green leaves. In spring, it bears spikes of small, white flowers, followed by red fruits that turn black. Makes a dense, informal hedge; prune in spring.

↕8m (25ft) ↔10m (30ft)

SHRUB LARGE

Prunus lusitanica 🏆

PORTUGAL LAUREL This is a dense, evergreen shrub, with large, glossy leaves. Slender spikes of small, fragrant, white summer flowers are followed by purple fruits. Use as a formal hedge or informal boundary; prune in spring.

↕↔10m (30ft)

SHRUB MEDIUM

Ribes sanguineum 🏆

FLOWERING CURRANT This deciduous shrub has lobed, ivy-shaped leaves and drooping clusters of crimson flowers in spring, followed by blue-black fruits with a white bloom. Use it as an informal hedge; prune in spring after flowering.

↕2m (6ft) ↔2.5m (8ft)

SHRUB MEDIUM

Rosa glauca 🏆

BLUE-LEAVED ROSE Arching red stems of this species rose bear greyish purple leaves and single, rose-pink summer flowers with pale centres, followed by red hips in autumn. Use as an informal hedge; prune in late winter.

↕2m (6ft) ↔1.5m (5ft)

SHRUB MEDIUM

Symphoricarpos x *doorenbosii*

This vigorous, deciduous, dense shrub has small, round, dark green leaves and greenish white summer flowers, which are followed by round, white fruits. Use it as an informal hedge; trim annually in early spring.

↕2m (6ft) ↔indefinite

SHRUB LARGE

Syringa vulgaris

COMMON LILAC This upright, spreading, deciduous shrub has large, green, glossy leaves and spikes of blue, pink, or white, scented blooms in summer. Use as an informal boundary; clip lightly after flowering each year.

↕↔7m (22ft)

TREE LARGE

Taxus baccata ♀

YEW This bushy, evergreen tree has needle-like, dark green leaves, and red berries in autumn. It makes a superb hedge for boundaries or dividing up a garden. Clip it in spring and summer; tolerates hard pruning. All parts are highly toxic.

↕15m (50ft) ↔10m (30ft)

TREE SMALL

Taxus x *media* 'Hicksii'

HICKS YEW This evergreen conifer has flattened, needle-like, dark green leaves and tiny, white summer blooms, followed on female plants by red berries. It is faster growing than *T. baccata*; trim in spring and summer. All parts are toxic.

↕6m (20ft) ↔4m (12ft)

TREE LARGE

Thuja occidentalis

WHITE CEDAR This evergreen conifer produces flat sprays of scale-like, yellowish green leaves that are pale or greyish green beneath, and yellow-green cones that ripen to brown. It makes a dense, formal hedge; trim in early autumn.

↕15m (50ft) ↔5m (15ft)

TREE LARGE

Thuja plicata

WESTERN RED CEDAR An evergreen conifer, with scale-like, glossy, dark green leaves, which have a pineapple-like aroma when crushed, and green cones that ripen to brown. Use it as a formal hedge; trim in early autumn.

↕30m (100ft) ↔8m (25ft)

TREE LARGE

Tilia cordata

SMALL-LEAVED LIME This deciduous tree has heart-shaped, glossy, dark green leaves that turn yellow in autumn. Small, yellowish white flowers appear in summer. Ideal for pleaching or as a formal hedge; trim in summer.

↕30m (100ft) ↔12m (40ft)

TREE LARGE

Tsuga heterophylla ♀

WESTERN HEMLOCK An evergreen conifer, with drooping stems bearing needle-like, flattened, dark green leaves that have silvery bands beneath. The green cones ripen to dark brown. Makes a tall, dense formal hedge; trim in early autumn.

↕30m (100ft) ↔10m (30ft)

SHRUB LARGE

Viburnum farreri ♀

This deciduous, upright shrub has dark green foliage, bronze-tinted when young. From late autumn to early spring, it bears fragrant, white or pale pink blooms. Use as an informal screen; trim after flowering.

↕3m (10ft) ↔2.5m (8ft)

SHRUB LARGE

Viburnum tinus

LAURUSTINUS This is a bushy, evergreen shrub, with oval, dark green leaves. Abundant flat heads of small, white blooms open from pink buds during late winter and spring. Use as an informal screen; trim after flowering.

↕↔3m (10ft)

OTHER SUGGESTIONS

Shrubs

Berberis x *ottawensis* 'Auricoma' • *Cotinus coggygria* GOLDEN SPIRIT ♀ • *Cotoneaster simonsii* ♀ • *Euonymus fortunei* 'Emerald 'n' Gold' ♀ • *Poncirus trifoliata* • *Rosa arvensis* • *Rosa canina* • *Rosa pimpinellifolia* • *Rosa rugosa* 'Alba'

Trees

Alnus glutinosa • *Carpinus caroliniana* • *Chamaecyparis lawsoniana* 'Aurea Densa' ♀ • *Crataegus monogyna* • *Fagus sylvatica* Atropurpurea Group • *Ilex* x *altaclerensis* 'Golden King' • *Ilex aquifolium* 'Ferox Argentea' ♀ • *Ilex aquifolium* 'Golden Queen' ♀ • *Ligustrum lucidium* ♀ • *Ligustrum ovalifolium* 'Aureum' ♀ • *Prunus cerasifera* • *Prunus spinosa* • *Quercus ilex* • *Sorbus acuparia*

Plants for beside hedges, walls, and fences

Plants grown beside walls and fences soften the hard surfaces with flowers and foliage, while hedges provide a leafy backdrop to colourful blooms.

The soil close to a man-made vertical surface may be in a rain shadow, and will remain dry, even after a heavy downpour. Likewise, hedging plants soak up large volumes of water, drying out the soil next to them. The drought-tolerant plants that cope with these conditions are the same as those that enjoy sandy soil, and include species such as *Artemisia* and *Buddleja* from rocky areas, and those like *Euphorbia characias* and rosemary from Mediterranean regions. A sunny, south-facing wall is also the perfect location for many tender plants, providing the heat and shelter they need to survive outside all year in cooler climes.

PERENNIAL LARGE

Agastache foeniculum

ANISEED HYSSOP This perennial has an upright habit and lance-shaped, mid-green leaves that smell and taste of liquorice. In late summer, it produces spikes of fluffy, lavender-blue flowers that are attractive to bees and butterflies.
↕1.2m (4ft) ↔30cm (12in)

PERENNIAL LARGE

Anemone x *hybrida*

JAPANESE ANEMONE This upright, branching perennial has divided, dark green leaves. It bears white or pink flowers from late summer to early autumn, and is useful for late-season colour. 'Königin Charlotte' AGM (above) has pale pink flowers.
↕1.2m (4ft) ↔ indefinite

PERENNIAL SMALL

Antennaria dioica 'Rubra'

MOUNTAIN EVERLASTING A semi-evergreen, spreading, mat-forming perennial, with tiny, oval, woolly leaves, and small clusters of fluffy, rose-pink flowerheads from late spring to early summer. Ideal for planting at the base of hedges.
↕15cm (6in) ↔30cm (12in)

SHRUB LARGE

Argyrocytisus battandieri

MOROCCAN BROOM A rounded, deciduous shrub, with grey-green leaves, divided into leaflets that are silky when they unfurl. Spikes of yellow, pineapple-scented flowers appear in late summer. Prefers acid soil and the protection of a south-facing screen.
↕↔4m (12ft)

PERENNIAL LARGE

Artemisia ludoviciana

WESTERN MUGWORT This bushy perennial is grown for its aromatic, lance-shaped, toothed-edged, silvery grey, woolly leaves. In summer, slender plumes of tiny, brownish yellow flowers appear. Forms include 'Silver Queen', with tall stems of silver leaves.
↕1.2m (4ft) ↔60cm (24in)

PERENNIAL LARGE

Boltonia asteroides

FALSE ASTER This is a tall, branching perennial, with linear, grey-green leaves and sprays of tiny, white, daisy-like flowers, which appear from late summer. Reduce stems by one-third in spring to prevent the stems collapsing.
↕2m (6ft) ↔1m (3ft)

SHRUB LARGE

Buddleja alternifolia

BUTTERFLY BUSH This vigorous, deciduous shrub produces arching stems bearing narrow, grey-green leaves. Long clusters of fragrant, lilac-purple flowers appear in early summer. The stems can be trained to form a small, weeping tree.
↕↔4m (12ft)

BULB MEDIUM

Calochortus venustus

FAIRY LANTERN This early-summer-flowering bulb has grey-green, grassy foliage and white, yellow, purple, or red, cup-shaped flowers, with a yellow-margined dark red blotch on each petal. Grow in dry soil beside a sunny wall.
↕60cm (24in)

PERENNIAL LARGE

Campanula latifolia

GIANT BELLFLOWER A tall, upright perennial, with oval, toothed-edged, mid-green foliage, and long clusters of tubular, violet-purple flowers throughout summer. Deadhead regularly; best on alkaline soils. Varieties include 'Brantwood' (above).

↕1.2m (4ft) ↔60cm (24in)

SHRUB MEDIUM

Carpenteria californica

TREE ANEMONE An evergreen, bushy shrub, with glossy, dark green, slim leaves and fragrant, yellow-centred white flowers from early- to midsummer. Plant it against a sheltered south- or west-facing wall in cold areas.

↕2m (6ft) or more ↔2m (6ft)

SHRUB MEDIUM

Ceanothus 'Dark Star'

A spreading, evergreen shrub, with oval, dark green leaves. It produces round clusters of deep blue-purple flowers that are held on arching stems in late spring. Train the stems against a sunny fence or wall.

↕2m (6ft) ↔3m (10ft)

SHRUB SMALL

Chaenomeles x *superba*

JAPANESE QUINCE This deciduous, spiny shrub produces cup-shaped flowers in shades of red, pink, orange, and white in spring before the leaves appear. It also bears yellow, edible fruits. 'Crimson and Gold' AGM (above) has orange-red blooms.

↕1m (3ft) ↔2m (6ft)

PERENNIAL MEDIUM

Gaillardia x *grandiflora*

A bushy, short-lived perennial, with slim, green leaves, and a profusion of large, red, daisy-like flowers having yellow-tipped petals from midsummer to early autumn. Deadhead regularly. Varieties include the dwarf form 'Kobold' (above).

↕90cm (36in) ↔45cm (18in)

PERENNIAL SMALL

Coreopsis grandiflora

TICKSEED A compact perennial, often grown as an annual, with serrated, green foliage. Masses of daisy-like, bright yellow, single or double flowerheads appear throughout summer. 'Sunray' (above) has double flowers.

↕↔45cm (18in)

PERENNIAL LARGE

Euphorbia characias subsp. *wulfenii*

MEDITERRANEAN SPURGE This evergreen, upright subshrub has tall stems covered with slender, grey-green leaves the first year, then large, vivid yellow-green flower spikes in spring. The sap is a skin irritant.

↕↔1.2m (4ft)

SHRUB LARGE

Fremontodendron 'California Glory'

FLANNEL BUSH A vigorous, evergreen or semi-evergreen, tree-like shrub, with rounded, lobed, dark green leaves. Large, saucer-shaped, bright yellow flowers appear from late spring to mid-autumn.

↕6m (20ft) ↔4m (12ft)

SHRUB LARGE

Garrya elliptica

SILK-TASSEL BUSH This dense, evergreen shrub has leathery, wavy-edged, grey-green leaves. From midwinter to early spring, it is covered with grey-green catkins. Plant it against a wall or fence, and prune when the catkins fade.

↕↔4m (12ft)

PERENNIAL LARGE

Gaura lindheimeri

WAND FLOWER An upright, clump-forming perennial, with lance-shaped, green leaves and tall, slim stems dotted with star-shaped, white flowers, which appear from midsummer to early autumn. 'Siskiyou Pink' has pink flowers.

↕1.5m (5ft) ↔60cm (24in)

PERENNIAL SMALL

Geranium macrorrhizum

CRANESBILL A semi-evergreen, carpeting perennial, with aromatic, rounded, deeply lobed leaves, and a profusion of small, magenta flowers in early summer. Cut back foliage in midsummer to restrict spread and promote new growth.

↕38cm (15in) ↔60cm (24in)

Plants for beside hedges, walls, and fences

PERENNIAL MEDIUM

Iris 'Edith Wolford'

BEARDED IRIS This tall iris is an upright, semi-evergreen perennial, with sword-shaped, grey-green foliage. The blooms appear in late spring and are pale yellow with blue-violet lower petals. Plant the rhizomes just above the soil surface.

↕90cm (36in) ↔60cm (24in)

PERENNIAL LARGE

Helianthus 'Lemon Queen'

This vigorous, upright perennial has sturdy, branched stems of oval, dark green leaves. From late summer to early autumn, it produces an abundance of large, daisy-like, pale yellow flowers, with dark yellow centres. Prefers fertile soil.

↕1.5m (5ft) ↔60cm (24in) or more

SHRUB LARGE

Hibiscus syriacus

ROSE MALLOW This deciduous, upright shrub has lobed, dark green foliage, and single, blue, violet, or white flowers, with dark centres from late summer to early autumn. 'Oiseau Blue' (above) is a blue-flowered variety.

↕3m (10ft) ↔2m (6ft)

PERENNIAL LARGE

Iris pallida 'Argentea Variegata'

DALMATIAN IRIS A semi-evergreen, upright perennial, with sword-shaped, grey-green and white striped foliage and pale blue-purple, early-summer flowers. Plant the rhizomes just above the soil surface.

↕1.2m (4ft) ↔indefinite

SHRUB LARGE

Juniperus scopulorum

ROCKY MOUNTAIN JUNIPER This is a narrow, upright conifer, with a columnar habit. It produces bright, steel-blue, scale-like foliage and tolerates hot, dry sites in front of walls or hedges. No pruning is required.

↕6m (20ft) ↔75cm (30ft)

SHRUB MEDIUM

Lupinus arboreus

TREE LUPIN This is a fast-growing, semi-evergreen, sprawling shrub that, in early summer, bears short spikes of fragrant, clear yellow flowers above hairy, feathery pale green leaves. Best in sandy soil; it can be short-lived.

↕↔2m (6ft)

TREE LARGE

Magnolia grandiflora

An evergreen, rounded tree, this magnolia has large, glossy, dark green leaves and large, fragrant, white flowers from midsummer to early autumn. It prefers the shelter provided by south-facing walls. Prune in spring to keep it in check.

↕18m (60ft) ↔15m (50ft)

SHRUB LARGE

Magnolia stellata

STAR MAGNOLIA A deciduous, rounded shrub, with fragrant, star-shaped, white flowers that open from silky buds in early spring before the slim, green leaves appear. Plant close to a south-facing wall to protect the blooms from frost.

↕↔10m (30ft)

BULB LARGE

Nectaroscordum siculum

HONEY GARLIC An upright bulb, with narrow, green, garlic-scented leaves that fade when fountain-shaped clusters of pendent, bell-shaped, cream and purple summer flowers appear on top of tall stems. Known for its shuttlecock-like seedheads.

↕1.2m (4ft)

PERENNIAL MEDIUM

Nepeta 'Six Hills Giant'

CATMINT A vigorous, clump-forming perennial, with narrow, oval, toothed, aromatic, grey-green leaves. In summer, it is covered with spikes of tubular, lavender-blue flowers. Cut back in late summer to keep it tidy.

↕1m (3ft) ↔1.2m (4ft)

💧 ❄❄❄

SHRUB MEDIUM

Ribes speciosum 🏆

FUCHSIA-FLOWERED GOOSEBERRY A deciduous, bushy shrub, with spiny stems of glossy, lobed, green foliage. Pendent, tubular, fuchsia-like, red flowers appear in mid-spring, followed by red fruits. Best planted against a sunny wall in cold climes.

↕↔2m (6ft)

💧 ❄❄

SHRUB MEDIUM

Rosmarinus officinalis

ROSEMARY An evergreen shrub, with aromatic, needle-like, dark green foliage and small, blue flowers in spring that appear on leafy stems. Varieties include 'Miss Jessopp's Upright', which has a slim, erect habit.

↕↔1.5m (5ft)

💧 ❄❄

TREE SMALL

Sophora SUN KING 🏆

This bushy, evergreen tree produces long, leafy stems comprised of small, oval leaflets. From late winter to early spring, golden-yellow, bell-shaped flowers appear in large clusters. Thrives in the shelter of a south-facing wall.

↕↔up to 3m (10ft)

💧 ❄❄❄

SHRUB SMALL

Rosmarinus officinalis Prostratus Group

ROSEMARY An evergreen, spreading shrub, with aromatic, narrow foliage and clusters of small, blue flowers from spring to early summer. It has a prostrate habit, ideal for a border edge. Suitable for culinary use.

↕15cm (6in) ↔1.5m (5ft)

💧 ❄❄

PERENNIAL MEDIUM

Rumex sanguineus

BLOODY DOCK A clump-forming perennial, grown for its decorative, oval, green leaves, with distinctive red veins, which are edible when young. Spikes of small, cream flowers appear in summer. Clip back after flowering to encourage new leaves.

↕90cm (36in) ↔30cm (12in)

💧 ❄❄❄

SHRUB MEDIUM

Weigela florida 'Variegata' 🏆

This deciduous, bushy shrub bears oval, cream-edged green leaves and a profusion of funnel-shaped, pink flowers from late spring to summer. An easy-going plant, it tolerates the dry soil close to a wall or fence.

↕↔2.5m (8ft)

💧 ❄❄❄

PERENNIAL SMALL

Stachys byzantina

LAMB'S EARS An evergreen, mat-forming perennial, with soft, silvery grey foliage, and spikes of small, mauve-pink flowers in summer. Drought-tolerant, it makes an excellent ground-cover plant for the front of a south-facing border.

↕38cm (15in) ↔60cm (24in)

💧 ❄❄❄

BULB MEDIUM

Tigridia pavonia

TIGER FLOWER This tender bulb has sword-shaped, pleated, green leaves, and unusual, three-petalled summer flowers in red, pink, orange, yellow, and white, with contrasting markings. The blooms often open just for a day. Protect bulbs from frost.

↕45cm (18in)

💧 ❄

OTHER SUGGESTIONS

Annuals

Cosmos bipinnatus Sensation Series

Perennials

Achillea 'Paprika' • *Alcea rosea* • *Cirsium rivulare* • *Echinops bannaticus* • *Echinops ritro* 'Veitch's Blue' • *Lychnis chalcedonica* 🏆 • *Penstemon pinifolius* 🏆 • *Penstemon* 'White Bedder' 🏆 • *Phlomis russeliana* 🏆

Grasses

Schizachyrium scoparium

Bulbs

Agapanthus inapertus • *Allium* 'Purple Sensation'

Shrubs

Buddleja davidii 'Black Knight' 🏆 • *Camellia sasanqua* 'Crimson King' 🏆 • *Cotoneaster horizontalis* • *Euonymus fortunei* 'Emerald Gaiety' 🏆 • *Genista hispanica* • *Indigofera heterantha* 🏆

Plants for walls, fences, and vertical surfaces

Climbers and wall shrubs are perfect for decorating walls and fences, with flowers in all colours and interesting leaf textures and plant shapes.

To extend your flower display, grow climbers and wall shrubs that bloom at different times. Try teaming rambling roses that peak in early summer with late-flowering clematis, but avoid partnering two clematis that require different pruning methods; you risk cutting off the blooms of an early-flowering variety when pruning a later one to the ground in spring. While self-clinging climbers, such as *Parthenocissus*, stick to surfaces with adhesive pads or roots, other climbers and wall shrubs, such as climbing roses and abutilon, need support – use trellis or galvanized wires secured to a vertical surface or structure.

SHRUB MEDIUM

Abutilon megapotamicum ♈

TRAILING ABUTILON An evergreen shrub, with long, flexible stems of slim, heart-shaped, dark green leaves. It bears yellow and red, pendent, bell-shaped flowers all summer. Train against a sunny wall in summer; protect against frost in winter.

↕↔2m (6ft)

💧 ❄

CLIMBER LARGE

Actinidia kolomikta ♈

A vigorous, deciduous, woody-stemmed, twining climber, grown for its oval, green leaves, splashed with creamy white and pink markings on mature plants. Small, cup-shaped, white flowers appear in summer. Grow it on a large wall or fence.

↕4m (12ft)

💧 ❄❄❄

CLIMBER MEDIUM

Clematis alpina

ALPINE CLEMATIS This deciduous clematis has divided, mid-green leaves and lantern-shaped, blue, pink, or white flowers from early- to late spring, followed by fluffy, silvery seedheads. It is ideal for exposed sites; keep the roots shaded.

↕3m (10ft)

💧 ❄❄❄

CLIMBER MEDIUM

Billardiera longiflora ♈

CLIMBING BLUEBERRY A woody-stemmed, evergreen, twining climber, with narrow, green leaves. Decorative, purple-blue fruits in autumn follow small, bell-shaped, green-yellow flowers. Grow it against a south-facing wall; protect in winter in cold climes.

↕2m (6ft)

💧 ❄❄

CLIMBER LARGE

Campsis x *tagliabuana*

TRUMPET VINE A vigorous, deciduous, self-clinging climber, with large leaves, divided into oval leaflets. Trumpet-shaped, orange flowers appear in summer, followed by bean-like seed pods. Provide frost protection in cold winters.

↕12m (40ft)

💧 ❄❄

CLIMBER SMALL

Clematis ANGELIQUE

A compact, mid- to late-season, deciduous clematis, with mid-green leaves. It produces an abundance of lilac-blue flowers for a long period from early summer to mid-autumn. Use it to decorate a rose arch, pillar, or tripod.

↕1.2m (4ft)

💧 ❄❄❄

CLIMBER LARGE

Clematis armandii

This vigorous, evergreen, early-flowering clematis has lance-shaped, dark green foliage and scented, single, white flowers in early spring. Best grown up a tree or wall in a sheltered, south- or south-west-facing site; keep the roots shaded.

↕5m (15ft)

💧 ❄❄

CLIMBER LARGE

Clematis 'Bill MacKenzie' 🏆

This vigorous, long-flowering, deciduous clematis has dark green leaves. Yellow, bell-shaped flowers with thick, almost waxy petals from midsummer to late autumn are followed by fluffy seedheads. Provide a large support; keep the roots shaded.

↕7m (22ft)

CLIMBER SMALL

Clematis CHANTILLY

A large-flowered clematis, with mid-green leaves. From summer to autumn, it produces single, scented, pale pink flowers with a deeper pink central bar on each petal. Plant it against a wall, fence, or up a rose arch, with the roots shaded.

↕1.2m (4ft)

💧 ❄❄❄

CLIMBER MEDIUM

Clematis 'Doctor Ruppel'

This early-flowering, deciduous clematis has mid-green leaves and large, rich pink flowers, with dark pink stripes in the centre of each petal, which appear throughout summer. Grow it up an arch, fence, or wall, keeping the roots shaded.

↕2.5m (8ft)

💧 ❄❄❄

CLIMBER MEDIUM

Clematis 'Duchess of Albany'

This deciduous clematis has mid-green leaves and small, tulip-shaped, soft pink flowers, with a deeper pink stripe inside each petal from summer to early autumn. Grow it on a fence, wall, or over an arch, keeping the roots shaded.

↕2.5m (8ft)

💧 ❄❄

CLIMBER LARGE

Clematis 'Etoile Violette' 🏆

A late-flowering, deciduous clematis, with mid-green leaves. It produces an abundance of flattish, single, violet-purple flowers, with cream stamens, from midsummer to autumn. Grow through a tree, or up a fence, wall, or pergola. Keep the roots shaded.

↕5m (15ft)

💧 ❄❄❄

CLIMBER LARGE

Clematis 'Fireworks'

This deciduous clematis has mid-green leaves and, from late spring to early summer, large, blue-mauve flowers, with a central magenta stripe on each petal. Another flush of smaller blooms appears in late summer. Keep the roots shaded.

↕4m (12ft)

💧 ❄❄❄

CLIMBER LARGE

Clematis flammula

VIRGIN'S BOWER A vigorous clematis, with dark green, divided, deciduous foliage and masses of small, almond-scented, starry, white flowers from summer to early autumn, followed by fluffy seedheads. Requires a large support; keep the roots shaded.

↕5m (15ft)

💧 ❄❄

CLIMBER SMALL

Clematis 'Fleuri'

A compact, deciduous clematis, with mid-green leaves. From late spring to summer, it produces deep purple flowers, with a red stripe down the centre of each petal. Grow it up a rose pillar or tripod, with the roots in the shade.

↕1.2m (4ft)

💧 ❄❄❄

CLIMBER MEDIUM

Clematis florida var. *florida* 'Sieboldiana'

This deciduous or semi-evergreen clematis has mid-green foliage and creamy white flowers, each with a central boss of petal-like, purple stamens. Fluffy seedheads form in autumn. Train on a sunny, sheltered wall.

↕2.5m (8ft)

💧 ❄❄

CLIMBER MEDIUM

Clematis 'Frances Rivis' 🏆

A deciduous, early-flowering clematis, with mid-green, divided foliage. In spring, it produces nodding, bell-shaped, violet-blue flowers with white centres. Makes a good specimen for a tripod, arch, wall, or fence. Keep the roots shaded.

↕3m (10ft)

💧 ❄❄❄

CLIMBER LARGE

Clematis montana var. _grandiflora_ ♡

A large, early-flowering, deciduous clematis, with divided, mid-green foliage and small, yellow-centred white flowers from late spring to early summer. Ideal for a tree, large wall, or pergola; shade the roots.

↕10m (30ft)

CLIMBER LARGE

Clematis 'Huldine' ♡

A vigorous, late-flowering, deciduous clematis with mid-green, divided foliage. From mid- to late summer, it produces a profusion of small, white flowers, mauve beneath. Ideal for a pergola, arch, wall, or fence; keep the roots shaded.

↕4m (12ft)

CLIMBER LARGE

Clematis montana var. _rubens_ 'Tetrarose' ♡

A vigorous, early-flowering, deciduous clematis, with purple-green leaves and small, cream-centred pink flowers in late spring. Good subject for a tree, house wall, or large pergola; shade the roots.

↕10m (30ft)

CLIMBER MEDIUM

Clematis JOSEPHINE

This deciduous climber has green, broadly lance-shaped leaves. From summer to early autumn, a succession of large, double flowers appears; the pinkish purple petals are cream beneath and produce a layered effect.

↕2.5m (8ft)

CLIMBER MEDIUM

Clematis 'Perle d'Azur'

This late-flowering deciduous clematis has mid-green leaves. From midsummer to autumn, it produces large, azure blue flowers, with creamy green centres and pink-purple stripes on each petal. Plant the roots in shade.

↕3m (10ft)

CLIMBER LARGE

Clematis 'Purpurea Plena Elegans' ♡

This late-flowering, deciduous clematis has pale green leaves and, from midsummer to late autumn, frilly, double, purple-pink flowers, with pale pink centres. Grow it on a large wall or fence; shade the roots.

↕4m (12ft)

CLIMBER LARGE

Clematis rehderiana ♡

Vigorous and late-flowering, this deciduous clematis has coarse, green leaves. From midsummer to autumn, it bears masses of fragrant, tubular, nodding, yellow flowers. Ideal for growing through a tree, or on a large wall or pergola. Shade the roots.

↕7m (22ft)

CLIMBER LARGE

Clematis tangutica

This vigorous, late-flowering clematis has mid-green leaves and lantern-shaped, yellow flowers throughout summer and early autumn, followed by fluffy, silvery seedheads for winter interest. Grow it on a large wall or fence; shade the roots.

↕6m (20ft)

CLIMBER LARGE

Cobaea scandens ♡

CUP-AND-SAUCER VINE An evergreen, perennial climber, often grown as an annual, with dark green leaves. From midsummer to autumn, it produces scented, cup-shaped, creamy green flowers that age to purple. Will quickly cover a wall, arch, or tripod.

↕5m (15ft)

CLIMBER LARGE

Distictis buccinatoria

MEXICAN BLOOD FLOWER This tender, evergreen, twining climber has dark green leaves and trumpet-shaped, crimson-red flowers in summer. It requires warm growing conditions and is best planted in a frost-free conservatory.

↕5m (15ft) or more

CLIMBER MEDIUM

Eccremocarpus scaber

CHILEAN GLORY FLOWER This evergreen, perennial tendril climber, often grown as an annual, produces small, green leaves and tubular, orange-red flowers from summer to autumn. Grow it in a sheltered spot on an arch, tripod, or fence.

↕3m (10ft)

SHRUB LARGE

Garrya elliptica

SILK-TASSEL BUSH An evergreen shrub, with leathery, wavy-edged, grey-green leaves. From midwinter to early spring, it bears silvery grey catkins that tremble in the wind. Train it on wires against a wall or fence, and prune when the catkins fade.

↕↔4m (12ft)

CLIMBER LARGE

Gelsemium sempervirens

YELLOW JESSAMINE An evergreen, twining climber, with slim, lance-shaped, dark green leaves. In summer, it produces clusters of trumpet-shaped, bright yellow flowers. Grow against a sunny, sheltered wall or fence. All parts are toxic.

↕6m (20ft)

CLIMBER LARGE

Humulus lupulus 'Aureus'

GOLDEN HOP A perennial, twining climber, grown for its rough, toothed, lobed, yellow-green foliage. In autumn, it bears straw-coloured, pendent flowerheads that resemble cones. Grow it over a pergola, or on a large fence or wall.

↕6m (20ft)

CLIMBER MEDIUM

Ipomoea coccinea

RED MORNING GLORY A tender, annual, twining climber, with heart-shaped, mid-green leaves and small, fragrant, tubular, scarlet flowers, with yellow throats, from late summer to autumn. Ideal for a tripod, rose arch, or pillar.

↕3m (10ft)

CLIMBER MEDIUM

Ipomoea tricolor 'Heavenly Blue'

MORNING GLORY A fast-growing annual twining climber, with heart-shaped leaves and, from summer to early autumn, large, funnel-shaped, sky-blue morning flowers. Quickly covers a fence, wall, or rose arch.

↕3m (10ft)

CLIMBER MEDIUM

Jasminum humile 'Revolutum'

ITALIAN JASMINE An evergreen, shrubby climber, with glossy, bright green leaves, divided into leaflets, and clusters of small, fragrant, tubular, yellow flowers from early spring to early summer.

↕2.5m (8ft)

CLIMBER LARGE

Jasminum officinale

COMMON JASMINE A semi-evergreen or deciduous, woody-stemmed, twining climber, with green leaves, divided into leaflets. From midsummer to autumn, clusters of highly fragrant, white flowers appear. Cut back in late winter.

↕12m (40ft)

CLIMBER LARGE

Lablab purpureus

AUSTRALIAN PEA This deciduous, twining climber, often grown as an annual, has small, dark green leaves and pink, pea-like summer flowers, followed by decorative long, purple pods with edible seeds. Grow it on a fence, arch, or pillar.

↕6m (20ft)

CLIMBER LARGE

Lonicera x *brownii* 'Dropmore Scarlet'

SCARLET TRUMPET HONEYSUCKLE A semi-evergreen or deciduous, twining climber, with rounded, blue-green leaves and tubular, scarlet, unscented blooms from summer to early autumn. May need protection in winter.

↕4m (12ft)

CLIMBER LARGE

Lonicera x *heckrottii*

This deciduous or semi-evergreen, twining climber has oval, dark green leaves that are blue-green beneath. In summer, it produces fragrant, pink flowers, with orange throats, sometimes followed by red berries. Provide a large support.

↕5m (15ft)

CLIMBER MEDIUM

Mandevilla x *amabilis* 'Alice du Pont'

A tender, evergreen climber, with dark green foliage and funnel-shaped, pink flowers. Grow it indoors in cold climes, or treat as an annual, and train up a tripod or arch. Shade from the midday sun in hot summers.

↕3m (10ft)

CLIMBER LARGE

Lonicera periclymenum

COMMON HONEYSUCKLE This deciduous, twining climber produces oval, dark green leaves, whitish green below, and richly scented, tubular, creamy white, yellow, or red flowers in summer. Use it to cover pergolas, walls, fences, or trees.

↕7m (22ft)

CLIMBER LARGE

Lonicera x *tellmanniana*

A deciduous, twining climber, with oval, blue-green leaves. Bright coppery orange, unscented flowers are carried in clusters from late spring to summer. Ideal for growing through a tree, or on a house wall or large fence.

↕5m (15ft)

CLIMBER LARGE

Parthenocissus henryana

This vigorous, deciduous, self-clinging climber is grown for its palm-shaped, dark green leaves with cream veining, which turns crimson in autumn. Blue-black berries are produced in autumn. Leaf colour is best in light shade.

↕10m (30ft) or more

CLIMBER LARGE

Parthenocissus quinquefolia

VIRGINIA CREEPER A deciduous, vigorous, large, self-clinging climber, with rounded, green leaves, divided into oval, toothed-edged leaflets, which turn bright red and orange in autumn. Provide a large support.

↕15m (50ft) or more

CLIMBER LARGE

Parthenocissus tricuspidata

BOSTON IVY This vigorous, deciduous, self-clinging climber is grown for its large, lobed, green leaves that turn spectacular shades of crimson in autumn. Use it to cover large expanses of a wall or boundary fence.

↕20m (70ft)

CLIMBER LARGE

Passiflora caerulea ♡

BLUE PASSION FLOWER This evergreen or semi-evergreen, tendril climber has glossy, lobed, dark green leaves. From summer to autumn, it bears white flowers with purple filaments, followed by egg-shaped, orange fruits. Requires a large support.
↕10m (30ft) or more

CLIMBER LARGE

Plumbago auriculata ♡

CAPE LEADWORT An evergreen, scrambling climber, with small, dark green leaves. Large clusters of sky-blue flowers appear from summer to early autumn. Grow it against a sunny, sheltered wall outside, or under cover in regions with cold winters.
↕6m (20ft)

CLIMBER MEDIUM

Rhodochiton atrosanguineus

PURPLE BELL VINE An annual climber, with heart-shaped, mid-green leaves. From late spring to late autumn, it bears dangling, umbrella-shaped flowers, with pinky red "hats" and dark maroon lower tubes. Ideal for a tripod.
↕3m (10ft)

CLIMBER LARGE

Rosa 'Albertine' ♡

This vigorous rambler rose has arching, thorny, reddish stems and glossy, dark green foliage. In summer, it produces clusters of scented, fully double, salmon-pink flowers in a single flush. Grow it up a tree or on a large pergola.
↕5m (15ft)

CLIMBER MEDIUM

Rosa 'Aloha' ♡

A climbing hybrid tea rose with leathery, disease-resistant, dark green leaves. Fragrant, fully double, rose- and salmon-pink flowers appear in summer and again in autumn. Train the stems over an arch or on a wall. Mulch annually in spring.
↕2.5m (8ft)

CLIMBER LARGE

Rosa banksiae 'Lutea' ♡

YELLOW BANKSIAN A vigorous, thornless, climbing rose, with small leaves. In late spring, it bears clusters of small, unscented, fully double, rosette-shaped, yellow flowers in one flush. Needs a sunny, sheltered wall or fence and protection from drying winds.
↕10m (30ft)

CLIMBER LARGE

Rosa 'Climbing Cécile Brünner' ♡

This vigorous climber has disease-resistant, dark green leaves and, from summer to late autumn, masses of small, sweetly-scented, fully double, blush pink flowers, fading to pearl. Ideal for a pergola or wall.
↕5m (15ft)

CLIMBER LARGE

Rosa 'Climbing Lady Hillingdon' ♡

This vigorous, climbing, hybrid tea rose has dark green foliage, coppery mahogany when young. Large, fragrant, apricot-yellow flowers are produced all summer. Train on a house wall, or over a pergola.
↕5m (15ft)

CLIMBER MEDIUM

Rosa 'Compassion' ♡

A climbing hybrid tea rose with disease-resistant, glossy, dark green leaves. From summer to autumn, it produces fragrant, double, salmon-apricot flowers, which are tinted pink, on thorny stems. Grow it on a wall or over a pergola.
↕3m (10ft)

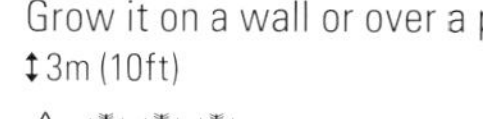

CLIMBER MEDIUM

Rosa 'Dortmund'

A climbing rose, with disease-resistant, glossy, dark green foliage. Clusters of flat, single, red flowers, with white eyes and a slight scent, are produced from summer to autumn. Ideal for an arch or pergola. Mulch in spring.

↕3m (10ft)

CLIMBER MEDIUM

Rosa DUBLIN BAY

This climbing rose has disease-resistant, glossy, dark green foliage and clusters of double, bright crimson flowers in summer and again in autumn. Grow it against a wall, or on an arch or post. Mulch in spring with garden compost.

↕2.2m (7ft)

CLIMBER LARGE

Rosa 'Félicité Perpétue'

A vigorous, rambling rose, with semi-evergreen, glossy, green leaves. Clusters of scented, fully double, blush pink to white flowers are produced in a single flush in summer. Grow it up a tree or large wall, or over a pergola.

↕5m (15ft)

CLIMBER MEDIUM

Rosa LAURA FORD

This stiffly branching, climbing rose has small, dark, glossy leaves with good disease resistance. Sprays of scented, urn-shaped to flat, double, yellow flowers appear in summer and again in autumn. Grow it on arches or pillars.

↕2.2m (7ft)

CLIMBER LARGE

Rosa 'Mermaid'

This climbing rose has glossy, green, disease-resistant foliage, and flowers repeatedly from summer to autumn, producing single, primrose-yellow blooms. The thorny stems can be used to deter intruders.

↕6m (20ft)

CLIMBER LARGE

Rosa 'Seagull'

A vigorous rambler rose, with disease-resistant, glossy, light green leaves. In late spring, it produces clusters of fragrant, single, white flowers with golden centres in a single flush. Grow it up a tree, large wall, or fence. Best for larger gardens.

↕6m (20ft)

CLIMBER LARGE

Rosa 'New Dawn'

A climbing rose with disease-resistant, glossy, dark green leaves and fragrant, double, pale pearl-pink flowers, borne in clusters from summer to autumn. Perfect for sunny walls, fences, or pergolas, it also tolerates north-facing sites. Mulch in spring.

↕5m (15ft)

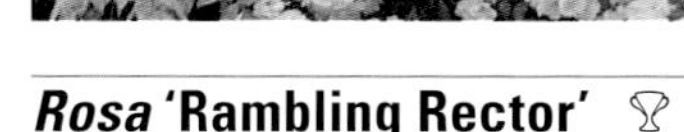

CLIMBER LARGE

Rosa 'Rambling Rector'

This vigorous rambler rose has disease-resistant, greyish green foliage. Clusters of scented, semi-double, creamy white flowers, with golden centres, appear in a single flush in summer. Grow it up a tree, large wall, or fence. Best for larger gardens.

↕6m (20ft)

CLIMBER MEDIUM

Rosa SUMMER WINE

This climbing rose has disease-resistant, dark green leaves and small clusters of fragrant, flat-faced, semi-double, coral-pink flowers from summer to autumn. Ideal for a wall, fence, arch, or pillar. Mulch in spring.

↕3m (10ft)

CLIMBER LARGE

Schizophragma integrifolium ♀

CHINESE HYDRANGEA VINE A deciduous, self-clinging climber, with heart-shaped green leaves. Clusters of tiny flowers with petal-like bracts, resembling lacecap hydrangea blooms, appear in summer.

↕12m (40ft)

CLIMBER LARGE

Solanum crispum 'Glasnevin' ♀

CHILEAN POTATO TREE A semi-evergreen, scrambling climber, with slim, oval, green leaves and clusters of fragrant, violet-blue flowers, with yellow eyes, from summer to autumn. Grow in a sheltered, sunny site.

↕6m (20ft)

CLIMBER LARGE

Stauntonia hexaphylla

An evergreen, twining climber, with rounded leaves, divided into oval leaflets. Clusters of fragrant, violet-tinged white flowers appear in spring, followed by egg-shaped, purple fruits on female plants. Grow it in a sheltered site.

↕10m (30ft)

CLIMBER MEDIUM

Thunbergia alata

BLACK-EYED SUSAN This moderately fast-growing, annual twining climber has toothed, oval to heart-shaped leaves and small, rounded, flat, orange or golden-yellow flowers, with dark brown centres, from early summer to early autumn.

↕3m (10ft)

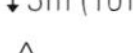

CLIMBER LARGE

Trachelospermum asiaticum ♀

ASIATIC JASMINE An evergreen, twining climber, with small, glossy, dark green leaves. Small, scented, starry, cream flowers that age to yellow appear in summer. Grow it on a sunny, sheltered wall.

↕6m (20ft)

CLIMBER LARGE

Trachelospermum jasminoides ♀

STAR JASMINE An evergreen, twining climber, with small, glossy, dark green leaves that turn bronze in winter. Clusters of fragrant, starry, white flowers appear in summer. Grow it on a sunny, sheltered wall.

↕9m (28ft)

CLIMBER MEDIUM

Tropaeolum speciosum ♀

FLAME CREEPER A perennial, twining climber, with rounded, blue-green leaves, divided into oval leaflets. Scarlet flowers appear in summer, followed by spherical, bright blue fruits. Grow it through a tree, with the roots in shade. Best in fertile soil.

↕3m (10ft)

CLIMBER LARGE

Vitis coignetiae ♀

CRIMSON GLORY VINE A vigorous, deciduous tendril climber, with large heart-shaped, textured leaves, which turn red and purple in autumn. Small, black berries follow insignificant green, summer flowers. Grow it on a large wall or pergola.

↕15m (50ft)

CLIMBER LARGE

Vitis vinifera 'Purpurea' ♀

GRAPE VINE This deciduous, tendril climber is grown for its lobed, maple-like, purplish leaves that turn bright crimson in autumn. Tiny, pale green, summer flowers are followed by unpalatable purple berries. Grow it on a large wall or pergola.

↕7m (22ft)

OTHER SUGGESTIONS

Annual climbers

Ipomoea lobata • *Ipomoea quamoclit*

Perennial climbers

Akebia quinata • *Akebia quinata* 'Alba' • *Aristolochia macrophylla* • *Bignonia capreolata* • *Bougainvillea* x *buttiana* 'Poulton's Special' ♀ • *Campsis radicans* 'Indian Summer' • *Celastrus scandens* • *Clematis* x *triternata* 'Rubromarginata' • *Codonopsis convolvulacea* • *Hardenbergia comptoniana* 'Rosea' • *Jasminum* x *stephanense* • *Mutisia decurrens* • *Passiflora edulis* • *Passiflora incarnata* • *Schisandra rubriflora* ♀ • *Solanum laxum* 'Album' ♀ • *Vitis* 'Brant' ♀ • *Wisteria floribunda* • *Wisteria floribunda* 'Alba' ♀ • *Wisteria* x *formosa* • *Wisteria frutescens* 'Amethyst Falls' • *Wisteria macrostachya* 'Blue Moon' • *Wisteria sinensis*

Plant focus: clematis

The vast range of flower forms and plant sizes make clematis invaluable additions to gardens, large and small.

VALUED FOR THEIR VERSATILITY, clematis can be used to dress up arches and pergolas, scramble through trees, and cover walls and fences with climbing stems of colourful flowers. As well as beautiful blooms, many species also sport decorative seedheads, and with careful selection, clematis can decorate the garden almost all year round. Clematis comprise a wide variety of species and cultivars. Many summer-flowering forms produce large, showy blooms, while plants that perform in spring, late summer, and autumn often bear smaller flowers. Most clematis, barring a few shrubby types, are hardy climbers and require support for their twining leaf stems to cling to. Use horizontal wires on walls and fences, or slim canes to coax them onto trees and shrubs. To produce the best displays, clematis should be pruned annually, and the method you use depends on when they flower. Those that bloom in spring and early summer generally require a light trim, while later flowering forms are cut back hard.

USING CLEMATIS

Cover walls and fences with the flowers and foliage of clematis by fixing horizontal wires or trellis securely to the surface. Roses and clematis are a classic planting combination for arches and pergolas. Select plants that bloom at the same time for the most colourful display.

Compact clematis have been specially bred to grow in pots and are the best choice for patio displays. Plant them in a large container, and provide a tripod to support their stems.

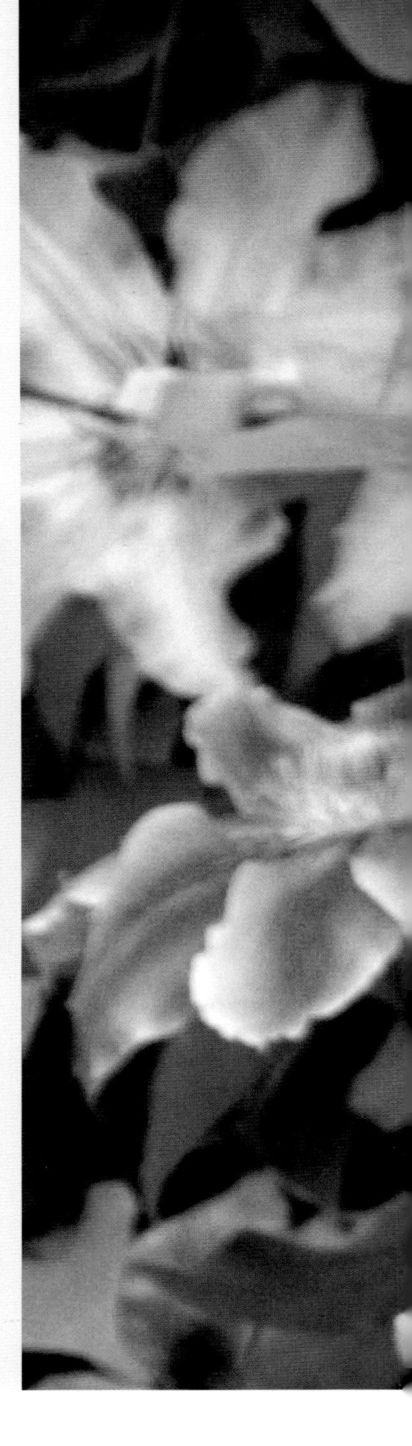

Large early summer-flowering clematis can be combined to decorate walls and fences with blooms, which form richly coloured backdrops for beds and borders.

POPULAR CLEMATIS SPECIES

Evergreens The early spring-flowering *C. cirrhosa* and *C. armandii* have evergreen leaves and are not as hardy as deciduous types. Plant them in a sheltered site, away from cold winds.

Spring-flowering forms *C. macropetala* and *C. alpina* fall into this group, with their nodding, bell-shaped flowers, as do forms of the rambling *C. montana*, with their star-shaped blooms.

Early-season large-flowered forms These include many popular forms that bear large, single and double flowers in a wide range of colours from late spring to midsummer.

Late summer large-flowered forms These bear large, colourful flowers from midsummer to early autumn on the current season's stems; prune them hard in spring.

Late-flowering species This group includes *C. tangutica* and *C. orientalis*, with their dainty, nodding flowers, and the vigorous small-flowered *C. rehderiana* and *C. flammula* species.

Winter interest Many late-flowering clematis produce attractive fluffy seedheads that may persist into winter. These include *C. tangutica*, *C. orientalis*, *C. rehderiana*, and *C. flammula*.

Plants for cracks in walls and paving

Stone or old brick walls punctuated with small, spreading plants nestling in the cracks transform a practical landscaping feature into a focal point.

You can also soften the sharp outlines of paths and patios with colourful, low-growing plants squeezed between the slabs and bricks. Designs are most effective where the plants look like they have self-seeded, and this is one option, using plants such as *Lobularia* and *Erigeron* that grow easily from seed, given a little soil in which to germinate. Alternatively, remove the pointing between the bricks or pavers, or remove one altogether, and wedge young plants or plugs wrapped in sticky clay-rich soil into the gaps. Dribble water into the gaps until the plants have established.

BULB MEDIUM

Allium schoenoprasum

CHIVES A clump-forming, upright, deciduous bulb, with narrow, hollow, dark green, edible leaves. In summer, it produces fluffy, pompon, pale purple flowerheads on slim stems. Ideal for paving cracks and tight spaces.

↕60cm (24in)

PERENNIAL SMALL

Arabis alpina* subsp. *caucasica

A mat-forming, evergreen perennial, with small, hairy, grey-green leaves. From early spring to early summer, it produces masses of small, white flowers. Grow it in wall or paving cracks. Cut back faded flowers.

↕↔15cm (6in)

PERENNIAL SMALL

Armeria maritima

THRIFT An evergreen, clump-forming perennial, with grassy, dark green leaves. Round heads of small, white to pink flowers appear on slim stems in summer. Use it to fill gaps in paving; suitable for coastal areas.

↕10cm (4in) ↔15cm (6in)

PERENNIAL SMALL

***Aubrieta* 'Argenteovariegata'**

This evergreen, trailing perennial has mats of oval, green leaves with cream edges. Small, pinkish lavender flowers cover the plant throughout spring. Ideal for cracks in walls and paving. Cut back after flowering.

↕5cm (2in) ↔15cm (6in)

PERENNIAL SMALL

Aubrieta deltoidea

AUBRETIA A ground-hugging, evergreen perennial, with spreading stems of rounded, grey-green leaves and masses of pale mauve flowers in spring. Varieties include 'Red Carpet', which has dark maroon flowers.

↕5cm (2in) ↔60cm (24in)

PERENNIAL SMALL

Aurinia saxatilis

GOLD DUST An evergreen, clump-forming perennial, with oval, hairy, grey-green leaves. From spring to summer, it produces a mass of tiny, chrome-yellow flowers on slim stems. Use it to brighten up walls and paving.

↕23cm (9in) ↔30cm (12in)

SHRUB SMALL

Calluna vulgaris

This spreading, evergreen shrub has upright stems covered with tiny, bright green leaves and pink, red, or white flowers from midsummer to late autumn. 'Spring Cream' AGM (above) has white flowers and cream-coloured new shoots in early summer.

↕60cm (24in) ↔75cm (30in)

PERENNIAL SMALL

Campanula carpatica

TUSSOCK BELLFLOWER This clump-forming perennial produces a carpet of heart-shaped, green leaves and open bell-shaped, violet-blue or white flowers throughout summer. Use it to fill gaps in walls and paving.

↕10cm (4in) ↔30cm (12in)

PERENNIAL SMALL

Campanula poscharskyana

TRAILING BELLFLOWER This vigorous, spreading perennial has round, green leaves, with serrated edges. In summer, it is covered with star-shaped, violet flowers. The sprawling leaf and flower stems make a colourful addition to a wall.

↕15cm (6in) ↔ indefinite

PERENNIAL SMALL

Chamaemelum nobile

CHAMOMILE A mat-forming, evergreen perennial, with aromatic, ferny foliage and pompon, white flowers held on erect stems in summer. Use it to fill cracks in paving. 'Flore Pleno' (above) is a double-flowered variety.

↕30cm (12in) ↔ 45cm (18in)

SHRUB SMALL

Daboecia cantabrica

IRISH HEATH An evergreen, spreading shrub, with small, oval, dark green leaves, silver-grey beneath. From early summer to autumn, it produces urn-shaped, single or double, white, purple, or mauve flowers. 'Bicolor' AGM (above) is a popular variety.

↕45cm (18in) ↔ 60cm (24in)

PERENNIAL SMALL

Delosperma nubigenum

YELLOW ICE PLANT A ground-hugging, evergreen perennial that produces mats of succulent, triangular, green leaves and daisy-like, lemon-yellow flowers in summer. Plant it on top of a wall or gravel beside paving.

↕5cm (2in) ↔ 50cm (20in)

PERENNIAL SMALL

Dianthus alpinus

A mat-forming, evergreen perennial, with dark grey-green foliage and white, pink, or crimson flowers, held on short stems in summer. 'Joan's Blood' AGM (above) is a popular variety, with deep crimson, dark-centred flowers.

↕8cm (3in) ↔ 10cm (4in)

PERENNIAL SMALL

Dryas octopetala

MOUNTAIN AVENS An evergreen, mat-forming perennial, with oval, lobed, leathery, dark green leaves. Small, cup-shaped, creamy white flowers appear from late spring to early summer, followed by feathery seedheads. Use it in paving cracks.

↕6cm (2½in) ↔ indefinite

SHRUB SMALL

Erica carnea

WINTER HEATH An evergreen, spreading shrub, with needle-like, dark green leaves. It produces tiny, tubular, pink, red, or white flowers, which appear from early winter to late spring. Grow it in paving cracks.

↕30cm (12in) ↔ 45cm (18in) or more

PERENNIAL SMALL

Erigeron karvinskianus

MEXICAN FLEABANE This spreading perennial has lance-shaped, hairy, green leaves and, from summer to early autumn, daisy-like flowers that open white, turn pink, and fade to purple. A perfect addition to a wall or paving.

↕15cm (6in) ↔ indefinite

PERENNIAL SMALL

Erinus alpinus

FAIRY FOXGLOVE A semi-evergreen perennial, with soft, mid-green foliage. From late spring to summer, small, purple, pink, or white flowers appear on leafy stems. It will self-seed in crevices in walls and paving.

↕↔ 8cm (3in)

PERENNIAL SMALL

Erodium manescavii

HERON'S BILL This clump-forming perennial has divided, ferny, green leaves. It produces loose clusters of magenta-purple flowers throughout summer, and will self-seed freely in the cracks in walls and paving.

↕45cm (18in) ↔ 60cm (24in)

PERENNIAL SMALL

Festuca glauca

BLUE FESCUE An evergreen, tuft-forming, perennial grass, with steel-blue foliage. The leaves are joined by short flower spikes in summer. It will self-seed in cracks in paving. The cultivar 'Blaufuchs' has the brightest foliage.

↕↔50cm (20in)

PERENNIAL SMALL

Gentiana septemfida

CRESTED GENTIAN This evergreen perennial produces trailing stems of small, oval leaves and trumpet-shaped, mid-blue flowers from summer to autumn. It will trail from cracks in walls, or spread out when planted in paving.

↕20cm (8in) ↔30cm (12in)

PERENNIAL SMALL

Geranium asphodeloides

ASPHODEL CRANE'S BILL A spreading, deciduous perennial, with small, rounded, deeply dissected leaves and masses of starry, white or light pink, magenta-veined flowers in early summer. Grow it between paving stones or flowing over walls.

↕↔30cm (12in)

PERENNIAL SMALL

Geranium sanguineum

BLOODY CRANESBILL This spreading perennial forms a neat mound of deeply dissected, dark green leaves. It flowers freely in summer, bearing an abundance of round, magenta-pink blooms. Grow it in paving cracks; water during dry spells.

↕25cm (10in) ↔30cm (12in)

SHRUB SMALL

Helianthemum 'Rhodanthe Carneum'

ROCK ROSE An evergreen shrub, with small, slim, grey-green leaves and saucer-shaped, orange-centred pale pink flowers, held on lax stems throughout summer. It looks beautiful trailing from a wall.

↕↔30cm (12in) or more

SHRUB SMALL

Helichrysum italicum

CURRY PLANT An evergreen, bushy subshrub, grown for its narrow, curry-scented, silvery grey leaves. Domed clusters of small, bright yellow flowers are produced on long, upright, white shoots in summer. Grow it in cracks in paving.

↕60cm (24in) ↔1m (3ft)

SHRUB SMALL

Lavandula angustifolia 'Hidcote'

A bushy, evergreen shrub, with narrow, aromatic, silver-grey leaves and dense spikes of fragrant, deep purple flowers that appear from mid- to late summer. Remove a paving stone and plant it in the gap.

↕60cm (24in) ↔75cm (30in)

ANNUAL/BIENNIAL SMALL

Limnanthes douglasii

POACHED-EGG FLOWER A spreading annual, easy to grow from seed, with feathery green foliage and cup-shaped, yellow-centred white flowers throughout summer. Sow seeds in the cracks in walls and paving. It may then self-seed.

↕15cm (6in) ↔10cm (4in)

PERENNIAL SMALL

Linaria alpina

ALPINE TOADFLAX A short-lived perennial, which produces trailing stems of lance-shaped, fleshy, grey-green leaves and a succession of snapdragon-like, orange-centred, purple-violet flowers in summer. Use it to decorate walls and paving.

↕↔15cm (6in)

SHRUB SMALL

Lithodora diffusa 'Heavenly Blue'

This evergreen shrub forms a carpet of small, hairy, green leaves on trailing stems. Funnel-shaped, deep blue flowers cover it throughout summer. Grow in walls or paving cracks; trim stems after flowering.

↕30cm (12in) ↔45cm (18in)

ANNUAL/BIENNIAL SMALL

Lobularia maritima

SWEET ALYSSUM An easy-to-grow annual with lance-shaped, green leaves and masses of rounded, sweetly fragrant, white or pink flowerheads throughout summer. Sow seeds in wall and paving cracks. 'Easter Bonnet' AGM (above) is popular.

↕10cm (4in) ↔30cm (12in)

BULB MEDIUM

Muscari armeniacum

GRAPE HYACINTH This spring-flowering bulb produces grassy, green leaves and short spikes of small, fragrant, bell-shaped, deep blue flowers, held in cone-shaped clusters. Plant it in cracks or tight spaces.

↕20cm (8in)

PERENNIAL SMALL

Oenothera speciosa

PINK EVENING PRIMROSE A spreading perennial, with small, divided green leaves and cup-shaped, fragrant, white flowers suffused with pink from late spring to early summer. Use it to fill gaps in paving or walls.

↕↔30cm (12in)

PERENNIAL SMALL

Origanum vulgare 'Aureum'

GOLDEN MARJORAM This clump-forming perennial forms a dense mat of aromatic, golden-yellow, rounded leaves that turn pale yellow-green in midsummer. It also produces tiny, mauve flowers in summer. Grow it in cracks in walls or paving.

↕8cm (3in) ↔ indefinite

PERENNIAL SMALL

Sedum 'Ruby Glow'

STONECROP This short, clump-forming, and deciduous perennial has oval, fleshy, purplish green leaves. In late summer, it produces loose heads of star-shaped, pink and ruby red flowers on dark red stems. Use it in cracks in paving.

↕20cm (8in) ↔40cm (16in)

PERENNIAL SMALL

Sempervivum tectorum

COMMON HOUSELEEK A mat-forming evergreen perennial, with rosettes of red-flushed, green, fleshy leaves. In summer, it bears clusters of star-shaped, reddish purple flowers. Plant it in gaps in walls and paving, but not where it will be stepped on.

↕15cm (6in) ↔20cm (8in)

PERENNIAL SMALL

Silene schafta

AUTUMN CATCHFLY This low-growing perennial has a spreading habit, forming mats of slender, dark green leaves. From late summer to autumn, it produces dainty, starry, magenta flowers. It will self-seed in cracks, but is easy to control.

↕25cm (10in) ↔30cm (12in)

SHRUB SMALL

Thymus serpyllum

THYME Evergreen and mat-forming, this subshrub produces trailing stems of small, hairy, aromatic, dark green leaves. In summer, it is covered with masses of tiny, purple or pink flowers. Plant in crevices in walls and paving. Suitable for culinary use.

↕25cm (10in) ↔45cm (18in)

PERENNIAL SMALL

Veronica prostrata

PROSTRATE SPEEDWELL A dense, mat-forming, evergreen perennial, with narrow, toothed, green leaves and upright spikes of small, saucer-shaped, bright blue or lilac flowers in early summer. Use it in crevices in walls or paving.

↕30cm (12in) ↔ indefinite

OTHER SUGGESTIONS

Annuals

Crepis rubra • *Iberis umbellata* Fairy Series

Perennials

Arabis caucasica • *Arabis procurrens* 'Variegata' • *Arenaria balearica* • *Arenaria purpurascens* • *Aubretia* 'Doctor Mules' • *Calamintha nepeta* • *Callirhoe involucrata* • *Campanula garganica* • *Cerastium tomentosum* • *Ceratostigma plumbaginoides* • *Crepis incana* • *Dianthus deltoides* • *Erodium* x *kolbianum* • *Geranium cinereum* 'Ballerina' • *Mentha requienii* • *Phlox subulata* 'McDaniel's Cushion' • *Saponaria officinalis* • *Silene acaulis*

Shrubs

Aethionema 'Warley Rose' • *Cassiope* 'Edinburgh' • *Cytisus* x *kewensis* • *Frankenia thymifolia* • *Iberis sempervirens*

Plants for patios, balconies, and windowsills

Decorative plants in pots will brighten up the smallest of spaces, while containers of different sizes help to create a dynamic effect.

The choice of container plants is extensive, from pretty annuals for baskets and windowboxes, to perennials to make up permanent displays, to shrubs for dramatic, sculptural statements. If your container is large enough and you are able to supply sufficient fertilizer and water, almost any plant will survive, but if time and space are limited, opt for those that will cope with some drought, such as hebes, lavender, and houseleeks. Trees, shrubs, and perennials are best housed in big pots that hold plenty of compost, which means they also hold more water and fertilizer, and plant them *in situ* as they will be very heavy when full.

SHRUB LARGE

Abutilon 'Nabob'

FLOWERING MAPLE This evergreen shrub produces large, maple-like, green foliage and bowl-shaped, deep crimson flowers in summer. Grow it in a large pot of soil-based compost, with cane stem supports; protect from frost.

↕↔ 3m (10ft)

TREE SMALL

Acer palmatum 'Bloodgood'

JAPANESE MAPLE A deciduous tree, with dark reddish purple, maple-like leaves that turn bright red in autumn. Small, purple flowers in spring are followed by winged, red fruits. Grow in a large pot of soil-based compost, shaded from full sun. Water well.

↕↔ 5m (15ft)

PERENNIAL LARGE

Agapanthus Headbourne hybrids

This perennial has strap-shaped foliage and spherical clusters of funnel-shaped, blue flowers from late summer to early autumn. Grow in a sheltered site in a large pot of soil-based compost mixed with grit.

↕ 1.2m (4ft) ↔ 60cm (24in)

PERENNIAL LARGE

Agave americana 'Variegata'

This succulent has lance-shaped, sharply pointed, grey-green leaves, with cream edges. In hot summers, tall stems of cream flowers may appear. Use a mix of soil-based compost and grit; protect from frost.

↕↔ up to 1.5m (5ft)

PERENNIAL MEDIUM

Anthemis tinctoria

DYER'S CHAMOMILE This clump-forming perennial has lacy, aromatic leaves and daisy-like, white or yellow flowers in summer. Plant it in large pots of soil-based compost. Varieties include the lemon-yellow-flowered 'E.C. Buxton' (above).

↕↔ 1m (3ft)

ANNUAL/BIENNIAL SMALL

Antirrhinum majus

SNAP DRAGON This upright annual has lance-shaped, green leaves. From late spring to autumn, it produces spikes of two-lipped flowers in pink, red, crimson, burgundy, white, and yellow. Grow it in groups in pots of multi-purpose compost.

↕↔ 45cm (18in)

ANNUAL/BIENNIAL MEDIUM

Arctotis fastuosa

CAPE DAISY This upright annual has lobed, silvery green leaves and masses of daisy-like, bright orange flowers, with dark maroon and black eyes, throughout summer. Plant it in groups in pots of multi-purpose compost.

↕ 60cm (24in) ↔ 30cm (12in)

SHRUB SMALL

Argyranthemum foeniculaceum

MARGUERITE An evergreen subshrub, with ferny, grey-green foliage and yellow-centred white flowers throughout summer and early autumn. Grow in large pots of multi-purpose compost; protect from frost.

↕↔ 80cm (32in)

PERENNIAL SMALL

Begonia Cocktail Series

This bushy, evergreen perennial, grown as an annual, has rounded, bronze-green leaves and pink, red, or white flowers from summer to autumn. Use it to edge pots, windowboxes, and baskets; plant in multi-purpose compost.

↕↔30cm (12in)

PERENNIAL MEDIUM

Begonia sutherlandii 🏆

This tuberous perennial has trailing stems and lobed, green leaves and clusters of single, orange flowers in summer. Plants die back in late autumn. Treat as an annual, or lift and protect tuber from frost. Plant in baskets of multi-purpose compost.

↕1m (3ft) ↔indefinite

PERENNIAL SMALL

Bellis perennis

DAISY A mound-forming perennial, grown as an annual, with oval, mid-green leaves and pompon, pink, white, or red spring flowers. Plant it with spring bulbs in pots, baskets, and windowboxes in multi-purpose compost.

↕↔20cm (8in)

ANNUAL/BIENNIAL SMALL

Brachyscome iberidifolia

SWAN RIVER DAISY A bushy annual, with feathery, green leaves and daisy-like blue, pink, purple, or white flowers that appear from summer to early autumn. Perfect for baskets and containers; plant in multi-purpose compost.

↕25cm (10in) ↔45cm (18in)

SHRUB SMALL

Buxus sempervirens 'Suffruticosa'

BOX An evergreen shrub that forms a dense mass of oval, bright green leaves. Plant it in a large container filled with soil-based compost, and clip to form decorative topiary shapes.

↕1m (3ft) ↔1.5m (5ft)

ANNUAL/BIENNIAL MEDIUM

Calendula officinalis

POT MARIGOLD A bushy annual, with lance-shaped, aromatic, pale green leaves and, from spring to autumn, daisy-like, single or double flowers in shades of yellow and orange. Grow it in containers or windowboxes in multi-purpose compost.

↕↔60cm (24in)

PERENNIAL SMALL

Calibrachoa Million Bells Series

A semi-trailing perennial, grown as an annual, with slender, dark green leaves and, from summer to early autumn, small, colourful, trumpet-shaped flowers. Trail from containers, windowboxes, and baskets.

↕30cm (12in) ↔1m (3ft)

ANNUAL/BIENNIAL MEDIUM

Callistephus chinensis

CHINA ASTER This bushy annual has oval, lobed, green leaves and, from late summer to autumn, bears flowers in shades of pink, purple, white, and yellow. The Pom Pom Series (above) produces small, rounded flowers.

↕60cm (24in) ↔45cm (18in)

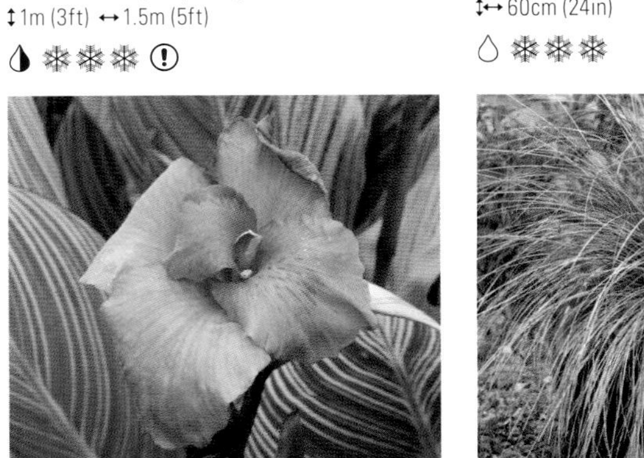

PERENNIAL LARGE

Canna 'Striata' 🏆

This upright perennial is grown for its green- and yellow-striped foliage and bright orange flowers, which appear from midsummer to early autumn. Grow it in soil-based compost; keep well watered, and protect the rhizomes from frost.

↕1.5m (5ft) ↔50cm (20in)

PERENNIAL SMALL

Carex comans

NEW ZEALAND SEDGE An evergreen perennial sedge with dense tufts of fine, grassy, bronze-coloured leaves. In summer, it also bears small, brown flower spikes. Grow it in a pot of soil-based compost. Remove dead growth in spring.

↕35cm (14in) ↔75cm (30in)

PERENNIAL MEDIUM

Carex 'Ice Dance'

JAPANESE SEDGE This evergreen perennial sedge bears mounds of grassy, green leaves, with creamy white margins, and small, white flowers in spring. Plant in windowboxes and pots of soil-based compost; keep well watered. May spread if planted in borders.
↕60cm (24in) ↔75cm (30in)

PERENNIAL SMALL

Carex testacea

ORANGE NEW ZEALAND SEDGE This evergreen perennial sedge produces mounds of olive-green to orange-brown, grassy leaves and dark brown flower spikes in summer. It looks good in tall pots of soil-based compost.
↕45cm (18in) ↔60cm (24in)

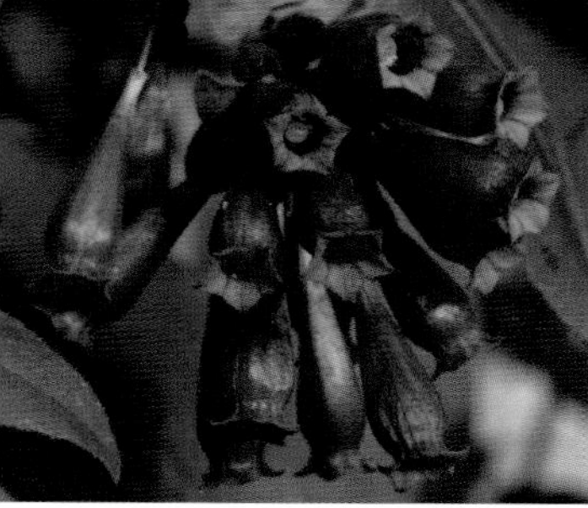

SHRUB LARGE

Cestrum elegans

A tender, evergreen shrub, with large, spear-shaped, mid-green foliage and dense clusters of scented, tubular, bright pink-red flowers during summer. Grow it in a large pot of soil-based compost; protect from severe frost.
↕↔3m (10ft)

ANNUAL/BIENNIAL LARGE

Cleome hassleriana

SPIDER FLOWER A tall, erect annual, with spiny stems of divided, spear-shaped leaves. In summer, it produces rounded heads of spidery, pink, white, and purple flowers, with long stamens. 'Rose Queen' (above) has pink flowers.
↕1.5m (5ft) ↔45cm (18in)

SHRUB MEDIUM

Choisya ternata 🏆

MEXICAN ORANGE BLOSSOM This is an evergreen, rounded shrub, with aromatic, glossy, bright green leaves, divided into oval leaflets. Clusters of fragrant, white blooms open in late spring and autumn. SUNDANCE AGM (above) has golden foliage.
↕↔ up to 2.5m (8ft)

TREE SMALL

Cordyline australis Purpurea Group

CABBAGE PALM This small, evergreen tree is grown for its fountain of spiky, purple foliage. Use as a specimen plant in a tall or large container of soil-based compost, and protect from frost in winter.
↕up to 3m (10ft) ↔1m (3ft)

ANNUAL/BIENNIAL MEDIUM

Coreopsis tinctoria

TICKSEED An erect annual, with finely divided, ferny green leaves. Daisy-like, lemon-yellow flowers appear on wiry stems throughout summer. Grow it in groups in a pot of soil-based compost.
↕60cm (24in) ↔30cm (12in)

CLIMBER SMALL

Clematis SHIMMER

A compact, deciduous, long-flowering climber, with mid-green leaves and large, deep lilac summer-long blooms, which fade to mid-blue and have a paler central bar. Plant it in a large pot of soil-based compost, with a tripod support.
↕1.8m (6ft)

ANNUAL/BIENNIAL LARGE

Cosmos bipinnatus

This compact annual, easy to grow from seed, produces feathery, green foliage and large, daisy-like flowers in shades of pink, red, and white from summer to early autumn. Plant it in groups in pots of multi-purpose compost. Deadhead regularly.
↕up to 1.2m (4ft) ↔45cm (18in)

PERENNIAL LARGE

Ensete ventricosum

ABYSSINIAN BANANA An evergreen, palm-like perennial, with large, paddle-shaped leaves, featuring cream midribs and red undersides. Plant it in a large pot of soil-based compost and grit; protect from frost during winter.

↕↔3m (10ft) in a pot

BULB MEDIUM

Dahlia Gallery Series 'Gallery Art Deco'

This bulb bears lobed, dark green leaves and double, burgundy-edged pale orange flowers from midsummer to autumn. Grow in pots of multi-purpose compost. It dies back in autumn; protect tubers from frost.

↕45cm (18in)

BULB MEDIUM

Eucomis bicolor

PINEAPPLE LILY A late-summer flowering bulb, with wavy-edged leaves and spotted stems that hold clusters of greenish white flowers with purple-edged petals. With "hats" of leaf-like bracts, the flowers look like pineapples. Protect from hard frost.

↕50cm (20in)

ANNUAL/BIENNIAL MEDIUM

Dianthus barbatus

SWEET WILLIAM An upright biennial, with lance-shaped, green leaves and domed heads of sweetly scented, pink, red, white, or burgundy flowers, which appear in early summer. Grow it in groups in pots of multi-purpose compost.

↕70cm (28in) ↔30cm (12in)

SHRUB MEDIUM

Fatsia japonica

CASTOR OIL PLANT An evergreen shrub, with large, hand-shaped, glossy, dark green leaves and small, spherical, white flowerheads in autumn, followed by black fruits. Ideal for winter interest, plant in large pots of soil-based compost.

↕↔2m (6ft) in a pot

SHRUB SMALL

Felicia amelloides

BLUE MARGUERITE Grown as an annual, this subshrub forms a bushy clump of small, dark green foliage and a mass of yellow-eyed blue flowers from summer to autumn. Grow in pots, windowboxes, and baskets in multi-purpose compost. Deadhead often.

↕↔25cm (10in)

SHRUB SMALL

Fuchsia 'Mrs Popple'

This deciduous, upright shrub bears small, dark green leaves and single, red and purple flowers in summer. It makes a statement in a large pot of soil-based compost. Overwinter outside in a sheltered site.

↕↔1.1m (3.5ft)

SHRUB SMALL

Fuchsia 'Thalia'

A deciduous, upright shrub, with dark green leaves that are maroon beneath. Clusters of pendent, tubular, red flowers appear from summer to early autumn. Plant it in a large pot of soil-based compost; protect from frost.

↕↔60cm (24in)

ANNUAL/BIENNIAL SMALL

Gazania Talent Series

TREASURE FLOWER A dwarf annual, with narrow, grey-felted leaves and daisy-like yellow, orange, pink, or maroon summer flowers. Plant in pots or windowboxes with other annuals in multi-purpose compost. Varieties include 'Talent Yellow' (above).

↕↔25cm (10in)

PERENN AL SMALL

Gerbera 'Mount Rushmore'

FLORIST GERBERA This perennial has oval, lobed, green leaves and tall stems of daisy-like, black-eyed pink, yel ow, or orange flowers, which appear from summer to autumn. Plant in containers of multi-purpose compost; protect from frost.

↕↔40cm (16in)

SHRUB SMALL

Hebe 'Red Edge'

A small, evergreen shrub that produces oval, blue-green leaves with red margins, and clusters of pale mauve to white flowers in summer. Plant it in a pot of soil-based compost, and place in a sheltered site. Trim after flowering to keep it compact.

↕45cm (18in) ↔60cm (24in)

ANNUAL/BIENNIAL LARGE

Helianthus debilis

WHITE SUNFLOWER This erect annual has slim, mid-green leaves and large, chocolate-centred creamy white summer flowers. Grow it in groups in a large pot of multi-purpose compost in a sheltered site. Stake the tall flower stems.

↕1.2m (4ft) ↔60cm (24in)

SHRUB MEDIUM

Helichrysum italicum

CURRY PLANT A bushy, evergreen shrub grown for its slender, curry-scented, silvery grey leaves. It bears domed clusters of small, bright yellow flowers on long, upright stems in summer. Grow it in pots of soil-based compost and grit.

↕60cm (24in) ↔1m (3ft)

PERENNIAL SMALL

Heuchera 'Amber Waves'

This evergreen, clump-forming perennial has lobed, ruffled, orange-yellow leaves, pale burgundy underneath, and sprays of small, bell-shaped, pink summer flowers. Plant in soil-based compost. Use it to edge pots, windowboxes, and baskets.

↕30cm (12in) ↔50cm (20in)

PERENNIAL SMALL

Ipomoea batatas 'Blackie'

SWEET POTATO VINE This trailing evergreen perennial, grown as an annual, produces lobed, ivy-shaped, almost black leaves. A beautiful edging plant for tropical-style arrangements, grow it in a tall container of multi-purpose compost.

↕25cm (10in) ↔60cm (24in)

ANNUAL/BIENNIAL MEDIUM

Ismelia carinata

CORN MARIGOLD An erect, branching annual, with feathery, grey-green leaves. From summer to autumn, daisy-like, brown-eyed flowers in various colour combinations appear. 'Court Jesters Mixed' (above) is a popular variety.

↕60cm (24in) ↔30cm (12in)

PERENNIAL SMALL

Isotoma axillaris

STAR FLOWER This mound-forming perennial, grown as an annual, has feathery, green foliage and masses of star-shaped, lilac or blue flowers from summer to early autumn. Grow it with other annuals in pots of multi-purpose compost.

↕↔30cm (12in)

CLIMBER MEDIUM

Lathyrus odoratus

SWEET PEA An annual climber that bears oval, green leaves and scented, white, pink, purple, and red summer flowers. 'Knee High' (above) is a dwarf form, suitable for hanging baskets or pots with a tripod support. Deadhead regularly.

↕3m (10ft)

TREE SMALL

Laurus nobilis 𝕍

BAY LAUREL This evergreen, conical tree is grown primarily for its leathery, aromatic, glossy, dark green leaves, used for cooking. Grow it in a large container of soil-based compost, and clip it to form topiary shapes. Prune regularly to restrict its size.

↕3m (10ft) ↔1m (3ft) in a pot

SHRUB SMALL

Lavandula stoechas

FRENCH LAVENDER This evergreen shrub produces grey-green, aromatic foliage and, from late spring to summer, scented, blue, white, purple, or pink flowers with upright "ears". 'Kew Red' (above) has deep pink flowers. Plant in soil-based compost.

↕↔45cm (18in)

PERENNIAL SMALL

Lotus berthelotii 𝕍

PARROT'S BILL This trailing perennial, grown as an annual has feathery, silver-green foliage and clusters of summer flowers, with pointed petals that resemble birds' beaks. Grow it in windowboxes or baskets of multi-purpose compost.

↕20cm (8in) ↔indefinite

BULB LARGE

Lilium Golden Splendor Group 𝕍

YELLOW TRUMPET LILY This upright perennial bulb has slim, green leaves and tall stems topped with scented, trumpet-shaped, golden summer flowers. Plant in gritty, soil-based compost; stake the stems.

↕2m (6ft)

ANNUAL/BIENNIAL SMALL

Lobelia erinus

A bushy annual, with tiny, green leaves. It is covered with masses of small, white, blue, pink, or mauve flowers in summer. Use it to edge pots, windowboxes, or hanging baskets, and plant in multi-purpose compost. 'Snowball' (above) has white flowers.

↕23cm (9in) ↔15cm (6in)

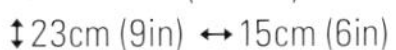

BULB MEDIUM

Muscari latifolium 𝕍

GRAPE HYACINTH A dwarf bulb, with grey-green, strap-shaped leaves and tiny, bell-shaped, two-tone spring blooms dark blue at the base and pale blue on top. Plant the bulbs in autumn in pots of soil-based compost mixed with grit.

↕25cm (10in)

SHRUB SMALL

Mahonia aquifolium

OREGON GRAPE This spreading evergreen shrub produces glossy, spiny, dark green leaves and clusters of yellow flowers in spring, followed by blue-black berries. Ideal for year-round colour; plant it in a large pot of soil-based compost.

↕1m (3ft) ↔1m (3ft) in a pot

ANNUAL/BIENNIAL SMALL

Matthiola incana

STOCK An upright annual, with oval, green leaves and spikes of clove-scented, pink, purple, and white flowers throughout summer. Plant in pots of multi-purpose compost. The Cinderella Series (above) is a popular range.

↕↔25cm (10in)

BULB MEDIUM

Narcissus 'Canaliculatus'

DAFFODIL This bulb has linear, green leaves and slim stems bearing clusters of small, fragrant spring flowers, with reflexed white petals and yellow cups. Plant bulbs in autumn in pots of soil-based compost mixed with grit.

↕23cm (9in)

BULB MEDIUM

Narcissus 'Fortune'

DAFFODIL An upright bulb that produces linear, green leaves. From early- to mid-spring, single flowers with lemon-yellow petals and orange cups appear. Plant the bulbs in autumn in pots of soil-based compost mixed with grit.

↕40cm (16in)

PERENNIAL SMALL

Nemesia caerulea

A compact, mat-forming perennial, grown as an annual, with narrow, dark green leaves and fragrant, blue-mauve flowers from summer to early autumn. Use it to edge pots, windowboxes, and hanging baskets; plant in multi-purpose compost.

↕↔30cm (12in)

PERENNIAL SMALL

Nemesia 'KLM'

This compact, mat-forming perennial, grown as an annual, has narrow, green leaves and produces small, blue and white flowers throughout summer. Useful for edging pots, windowboxes, and baskets; plant in multi-purpose compost.

↕30cm (12in) ↔16cm (6in)

ANNUAL/BIENNIAL SMALL

Nemophila menziesii

BABY BLUE EYES An easy-to-grow, spreading annual, with serrated, grey-green leaves. Small, saucer-shaped, blue flowers with white centres appear in summer. Grow it in pots, windowboxes, and baskets in multi-purpose compost.

↕20cm (8in) ↔15cm (6in)

PERENNIAL MEDIUM

Nicotiana alata

TOBACCO PLANT An upright perennial, grown as an annual, with oval, mid-green leaves. Clusters of trumpet-shaped, white, pink, or green flowers, which are fragrant at night, appear from summer to early autumn. Plant it in pots of multi-purpose compost.

↕75cm (30in) ↔30cm (12in)

PERENNIAL MEDIUM

Osteospermum 'Buttermilk'

AFRICAN DAISY An evergreen, clump-forming perennial, with narrow, grey-green leaves and daisy-like, pale yellow flowers from summer to early autumn. Plant in containers of multi-purpose compost. Protect during winter in cold areas.

↕60cm (24in) ↔30cm (12in)

PERENNIAL SMALL

Pelargonium 'Lord Bute' 🏆

An upright, evergreen perennial, often grown as an annual. It has rounded, hairy leaves and produces deep purple-red, ruffled flowers throughout summer. Plant it in pots and windowboxes in multi-purpose compost.

↕45cm (18in) ↔30cm (12in)

PERENNIAL SMALL

Pelargonium Mini Cascade Series

GERANIUM An evergreen, trailing perennial, grown as an annual, with ivy-shaped, glossy, green foliage and masses of single, red or lilac summer to autumn flowers. Ideal for a hanging basket or pot.

↕↔45cm (18in)

PERENNIAL SMALL

Pelargonium Multibloom Series

GERANIUM An erect, bushy, evergreen perennial, grown as an annual, with rounded leaves and white, red, pink, and purple flowers from summer to autumn. Grow in pots or baskets in multi-purpose compost.

↕↔30cm (12in)

ANNUAL/BIENNIAL SMALL

Petunia Prism Series

A dwarf, spreading annual, with dark green leaves and large, trumpet-shaped, yellow flowers from summer to early autumn. Grow it in pots, baskets, and windowboxes in multi-purpose compost. 'Prism Sunshine' (above) is a popular variety.

↕35cm (14in) ↔50cm (20in)

PERENNIAL SMALL

Phormium 'Bronze Baby'

NEW ZEALAND FLAX An evergreen, upright perennial, grown for its fountain of arching, sword-shaped, purple-bronze leaves. Use it to make a statement in a large container of soil-based compost.

↕↔ 60cm (24in)

PERENNIAL MEDIUM

Rehmannia elata

CHINESE FOXGLOVE An erect perennial, with slim, toothed-edged, green leaves and foxglove-like, yellow-throated, rose-purple flowers in summer. Plant it in a sheltered site in a large pot of soil-based compost.

↕ 1m (3ft) ↔ 45cm (18in)

SHRUB SMALL

Picea abies 'Ohlendorffii'

A compact, bushy conifer, with tightly packed, short green needles. It is slow growing and develops a rounded shape as it matures. Plant in large containers of soil-based compost for evergreen interest; keep it well watered.

↕↔ 1m (3ft)

SHRUB MEDIUM

Pieris japonica

LILY OF THE VALLEY BUSH A compact, evergreen shrub, with leathery, dark green leaves, bright red when young. In spring, it produces clusters of urn-shaped, white, pink, or red flowers. Grow it in a large pot of ericaceous compost.

↕ up to 2m (6ft) in a pot ↔ 1.5m (5ft)

SHRUB SMALL

Pinus mugo 'Mops'

DWARF MOUNTAIN PINE This compact, evergreen conifer forms a neat mound of dense, needle-like, dark green leaves, providing year-round interest. Grow it in a large container of soil-based compost, underplanted with small spring bulbs.

↕ 1m (3ft) ↔ 2m (6ft)

PERENNIAL SMALL

Rhodanthemum hosmariense

MORROCAN DAISY An evergreen, shrubby perennial, with silvery green leaves and white, daisy-like flowers from late spring to early autumn. Grow in soil-based compost mixed with grit. Deadhead regularly.

↕ 15cm (6in) or more ↔ 30cm (12in)

CLIMBER MEDIUM

Rhodochiton atrosanguineus

PURPLE BELL VINE This deciduous, annual climber has heart-shaped, mid-green leaves and dangling, umbrella-shaped flowers, with pinky red "hats" and dark maroon tubes, from summer to autumn. Grow in pots of soil-based compost with a tripod support.

↕ 3m (10ft)

SHRUB MEDIUM

Rhododendron yakushimanum

YAKUSHIMA RHODODENDRON A dome-shaped, evergreen shrub, with dark green leaves and funnel-shaped, pink to white, late spring flowers. Grow in ericaceous compost, shaded from full sun.

↕↔ 2m (6ft)

SHRUB MEDIUM

Rosa GERTRUDE JEKYLL

An upright shrub rose, it has disease-resistant, greyish green leaves. It produces highly fragrant, fully double, rose-pink flowers that bloom throughout summer. Grow it in a large pot or tub of soil-based compost.

↕ 2m (6ft) ↔ 1.2m (4ft)

PERENNIAL MEDIUM

Salvia farinacea

MEALY SAGE An upright perennial, often grown as an annual, with lance-shaped, green foliage and spikes of small, purple-blue flowers from midsummer to autumn. Grow it in mixed displays in pots of multi-purpose compost; deadhead regularly.

↕60cm (24in) ↔30cm (12in)

PERENNIAL SMALL

Salvia splendens

SCARLET SAGE This upright perennial, grown as an annual, has spear-shaped, dark green leaves. Spikes of tubular, scarlet, pink, purple, or white flowers appear through summer. Plant in pots, baskets, and windowboxes in multi-purpose compost.

↕25cm (10in) ↔35cm (14in)

ANNUAL/BIENNIAL SMALL

Salvia viridis

ANNUAL CLARY An upright annual, with grey-green leaves and, in summer, small, tubular flowers, concealed beneath leaf-like bracts, in shades of white, pink, and purple. Plant in multi-purpose compost. The stems can be cut and dried.

↕45cm (18in) ↔20cm (8in)

ANNUAL/BIENNIAL SMALL

Sanvitalia procumbens

CREEPING ZINNIA This spreading annual has slim, oval, green leaves. A profusion of small, daisy-like, yellow flowers with black centres appear in summer. Grow it in baskets, windowboxes, and pots of multi-purpose compost.

↕15cm (6in) ↔30cm (12in)

PERENNIAL SMALL

Scaevola aemula

FAIRY FAN-FLOWER An evergreen perennial, grown as an annual, with toothed, green leaves on trailing stems and fan-shaped, blue, lilac, or white flowers in summer. Ideal for windowboxes and baskets; plant it in multi-purpose compost.

↕↔50cm (20in)

PERENNIAL SMALL

Solenostemon 'Black Prince'

FLAME NETTLE This bushy perennial, grown as an annual, produces spear-shaped, dark purple foliage, with a bright pink midrib. Plant it with brightly coloured flowering annuals in pots and baskets of multi-purpose compost.

↕↔50cm (20in)

PERENNIAL SMALL

Solenostemon scutellarioides

FLAME NETTLE A bushy perennial, grown as an annual, with spear-shaped foliage in a variety of colours, including pink, red, green, and yellow. Plant it in pots, baskets, and windowboxes in multi-purpose compost.

↕45cm (18in) ↔30cm (12in) or more

ANNUAL/BIENNIAL SMALL

Tagetes patula

FRENCH MARIGOLD This annual has deeply divided, aromatic, dark green leaves and single or double flowers in shades of yellow, orange, red, or mahogany from summer to early autumn. Plant it in pots of multi-purpose compost. Deadhead regularly.

↕↔30cm (12in)

PERENNIAL SMALL

Tanacetum parthenium 'Aureum'

GOLDEN FEVERFEW This bushy perennial, grown as an annual, has golden, aromatic foliage and white, daisy-like, summer to early autumn flowers. Grow in pots or windowboxes in multi-purpose compost.

↕↔45cm (18in)

SHRUB SMALL

Teucrium chamaedrys

WALL GERMANDER A bushy, evergreen subshrub, with oval, aromatic, dark green leaves. From late summer to early autumn, spikes of purple-pink flowers appear. Grow it in soil-based compost mixed with grit.

↕60cm (24in) ↔30cm (12in)

SHRUB SMALL

Thymus vulgaris

THYME This ground-cover subshrub has tiny, round, aromatic, grey-green leaves. In summer, it produces a mass of purple, white, or pink flowers. Combine it with other herbs in pots of soil-based compost mixed with grit. Attractive to bees.

↕30cm (12in) ↔40cm (16in)

BULB MEDIUM

Tulipa 'Ballerina'

TULIP This late spring, lily-flowered tulip produces grey-green leaves and single, goblet-shaped, orange flowers, with an orange-red central section on the petals. Plant bulbs in groups in autumn, in pots of soil-based compost mixed with grit.

↕60cm (24in)

BULB MEDIUM

Tulipa 'Cape Cod'

TULIP A mid- to late spring-flowering bulb, with grey-green leaves that have maroon mottling and single, bowl-shaped, yellow flowers with a red strip on the petals. Plant bulbs in groups during autumn, in pots of soil-based compost mixed with grit.

↕45cm (18in)

BULB MEDIUM

Tulipa 'Purissima'

TULIP An early- to mid-spring-flowering bulb, with purple-marked, grey-green leaves and single, bowl-shaped, creamy white flowers. Plant bulbs in groups in autumn, in pots of soil-based compost mixed with grit.

↕40cm (16in)

ANNUAL/BIENNIAL SMALL

Tropaeolum majus

NASTURTIUM This bushy annual has round, green leaves, and red, yellow, or orange, trumpet-shaped flowers from summer to autumn. Grow in tall pots, windowboxes, or baskets in multi-purpose compost. Alaska Series (above) has cream-splashed foliage.

↕30cm (12in) ↔45cm (18in)

ANNUAL/BIENNIAL SMALL

Verbena 'Peaches and Cream'

This bushy annual, with narrow, toothed-edged leaves, bears clusters of pink, peach, and cream flowers throughout summer and autumn, if deadheaded regularly. Grow it in containers and baskets of multi-purpose compost.

↕↔50cm (20in)

PERENNIAL SMALL

Viola x *wittrockiana*

PANSY A perennial, grown as an annual, with lobed, green leaves and large, ruffled flowers in shades of red, purple, yellow, and white from early spring to summer. Plant it in groups in pots, baskets, and windowboxes in multi-purpose compost.

↕23cm (9in) ↔30cm (12in)

ANNUAL/BIENNIAL SMALL

Zinnia 'Thumbelina Mix'

A bushy, dwarf annual, with oval, dark green leaves. From summer to early autumn, it bears semi-double flowers in shades of red, yellow, maroon, and pink. Grow it in mixed containers, baskets, and windowboxes in multi-purpose compost.

↕15cm (6in) ↔21cm (8in)

OTHER SUGGESTIONS

Annuals

Anagallis monellii 'Skylover' • *Chrysanthemum carinatum* syn. *Ismelia carinata* • *Cuphea llavea* 'Tiny Mice' • *Leonotis leonurus*

Perennials

Eustoma grandiflorum Yodel Series • *Oenothera macrocarpa* • *Oxalis tetraphylla* 'Iron Cross'

Bulbs

Begonia 'Illumination Rose'

Shrubs and climbers

Camellia japonica 'Konronkoku' • *Hebe* 'Silver Queen' • *Hypericum* 'Magical Red Star' • *Iberis sempervirens* 'Snowflake' • *Punica granatum* • *Rhododendron* 'Dora Amateis' • *Skimmia japonica* 'Rubella'

Trees

Olea europaea • *Salix integra* 'Hakuro-nishiki'

Plant focus: dahlias

Bold and beautiful, dahlias decorate the garden from summer to early autumn with vividly coloured and varied flowers.

COMPRISING TEN DIFFERENT FLOWER FORMS, and with plant sizes ranging from compact and floriferous to towering giants, dahlias can be used in a wide variety of garden situations. Use ball, pompon, and rounded decorative dahlias to create spherical accents in a border of plant spires, such as verbascum and salvias, and try spiky cactus forms to add drama and texture to a scheme. Single and collarette dahlias complement naturalistic designs, such as prairie-style borders, and many single varieties attract bees and other beneficial insects. Diminutive bedding dahlias are perfect for patio containers, while taller varieties suit large pots, but their bloom-laden stems may need staking. Dahlias are easy to grow from tubers or young plants bought in spring. Set them outside in a sunny position and free-draining soil after all risk of frost has passed. Also guard them against slugs, which can decimate plants. Bring tubers inside over winter or in mild areas, cover them with a thick mulch.

USING DAHLIAS

Short-stemmed dahlias are perfect for patio planters and bedding schemes, and give a long display if deadheaded regularly. Taller varieties are best suited to borders, where they can be given additional support. Dahlias with long stems are ideal to use as cut flowers, and can be picked regularly from midsummer.

Dwarf dahlias that don't fit any of the eight main groups are often sold as bedding for pots and baskets. Overwinter the tubers, or discard them at the end of the season.

Dahlias such as 'Bishop of Llandaff' (*right*) and 'Yellow Hammer' have dark foliage, providing a dramatic contrast to brightly flowered plants.

TYPES OF DAHLIA

Decorative The fully double, ruffled flowers are composed of broad petals that may have gently twisted or curved edges.

Anemone The fully double flowers comprise one or more rings of flattened petals overlaid with shorter, tubular petals.

Water lily The fully double blooms are comprised of flat or slightly curved petals and resemble those of water lilies.

Collarette The flowers are formed of oval petals with an inner ruffled "collar" made up of smaller petals surrounding a central disc.

Single The flowers have a daisy-like appearance, with a central disc surrounded by oval petals. They are attractive to bees.

Ball The fully double, spherical flowers, sometimes flattened on top, are formed of small, densely packed, tubular petals.

Pompon These are miniature forms of ball dahlia, but more spherical, with fully double, tubular petals.

Cactus The fully double blooms are formed of narrow, pointed petals that can be straight or curved inwards, giving a shaggy effect.

Plants for productive patios

The beauty of a sunny patio is that you can combine flowers with ornamental crops to produce delicious displays that taste as good as they look.

Many crops are happy to live in pots, especially the tender types, such as chillies, tomatoes, and aubergines, which need a sheltered, hot spot for their fruits to ripen. Dwarf and miniature fruit trees can also thrive here, given large containers and plenty of water; consider installing an automatic watering system to ensure a constant supply. Fruits in pots also benefit from regular feeding. Pots provide the perfect habitat for leafy salads, too – container cultivation makes it easier to protect their leaves against marauding molluscs – while potatoes, beans, and peas will crop quite well in big tubs.

BULB MEDIUM

Onions

Allium cepa This onion crop can be planted as small bulbs or "sets" during spring, and harvested from mid- to late summer. Lift and dry the bulbs once the tops bend over. Plant in large pots and growing bags in a sunny position.

↕60cm (24in)

BULB MEDIUM

Spring onions

Allium cepa This small onion is sown from seed from spring to autumn, and is ready to harvest in 6–8 weeks. It is ideal for containers and windowboxes placed in a sunny position. White- and red-stemmed varieties (above) are available.

↕25cm (10in) when harvested

BULB MEDIUM

Chives

Allium schoenoprasum A perennial bulb with hollow, dark green, grassy, mild onion-flavoured leaves, which can be used in salads and savoury dishes. It bears pale purple or pink, pompon flowers in summer. Suitable for containers and windowboxes.

↕30cm (12in)

SHRUB LARGE

Lemon verbena

Aloysia triphylla This shrub has spear-shaped leaves that smell and taste of lemon, and can be used to make tea or to flavour sweet dishes. In summer, it bears spikes of small, white flowers. Grow it in a large container; protect from hard frosts.

↕↔3m (10ft)

ANNUAL/BIENNIAL LARGE

Amaranth

Amaranthus tricolor This upright plant is raised from seed and grown for its leaves, which can be used in salads, or left to flower and harvested for its edible seeds, which are used as grain. Add this colourful plant to decorative displays.

↕1.3m (4½ft) ↔45cm (18in)

ANNUAL/BIENNIAL MEDIUM

Celeriac

Apium graveolens Sown from seed in spring, this crop produces large, swollen roots, with a nutty, celery-like flavour. Harvest it in autumn or in summer as baby roots. Best in large containers; keep the plant well watered.

↕90cm (36in) ↔45cm (18in)

ANNUAL/BIENNIAL SMALL

Beetroot

Beta vulgaris Sown from seed in spring and summer, this earthy, sweet-tasting root can be harvested as soon as it reaches a good size. It has attractive stems and leaves, and suits well in decorative container displays.

↕23cm (9in) ↔45cm (18in)

ANNUAL/BIENNIAL SMALL

Swiss chard

Beta vulgaris Grown from seed in spring, this crop produces large, glossy, green leaves, which can be used raw in salads or steamed, and fleshy stems that are best cooked. Harvest plants lightly throughout summer, and water regularly.

↕↔45cm (18in)

ANNUAL/BIENNIAL SMALL

Mustard greens

Brassica juncea This crop is grown for its peppery-tasting leaves. Sow seeds in spring and summer, and harvest the plant young as "cut-and-come-again" leaves, or allow them to mature. Grow in pots and windowboxes. Water well.

↕↔30cm (12in)

ANNUAL/BIENNIAL SMALL

Kohl rabi

Brassica oleracea* var. *gongylodes This crop is grown for its nutty-tasting, swollen stems that mature in as little as six weeks. Sow seeds in repeated batches in growing bags and containers in spring and summer, and harvest when large enough.

↕↔45cm (18in)

ANNUAL/BIENNIAL MEDIUM

Cayenne chillies

Capsicum annuum This chilli produces long, tapering, fiery fruits that are packed with seeds. The fruits ripen through green and yellow to red. It is ideal for containers and windowboxes in full sun. Harvest regularly for the best crop.

↕90cm (36in) ↔30cm (12in)

ANNUAL/BIENNIAL MEDIUM

Jalapeño chillies

Capsicum annuum Producing short, pointed, fleshy-skinned fruits, this chilli ripens through green and yellow to red. It is ideal for containers and windowboxes in full sun. Harvest regularly for the best crop.

↕90cm (36in) ↔30cm (12in)

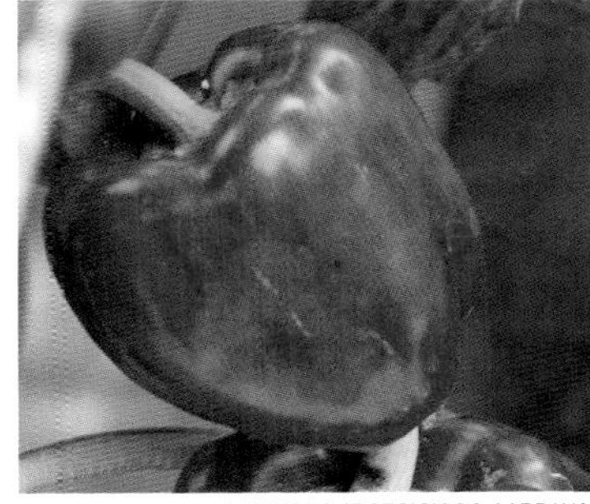

ANNUAL/BIENNIAL MEDIUM

Sweet peppers

Capsicum annuum This plant produces large, fleshy, yellow, orange, red, and black summer fruits, with a mild, sweet flavour. Plant in large containers in full sun. Water and feed regularly; tall varieties may need staking.

↕90cm (36in) ↔45cm (18in)

ANNUAL/BIENNIAL MEDIUM

Aji chillies

Capsicum baccatum With a mild, fruity flavour, this chilli ripens through various colours in summer, although it can be picked at any stage once large enough. Grow in full sun in large pots. Water regularly; stake tall plants.

↕90cm (36in) ↔45cm (18in)

ANNUAL/BIENNIAL MEDIUM

Scotch bonnet chillies

Capsicum chinense This extremely hot chilli bears flattened, rounded fruits in shades of green, yellow, and red in summer. Best grown in warmer areas with long summers. Plant it in containers in full sun. Water regularly.

↕90cm (36in) ↔30cm (12in)

TREE SMALL

Lemons

Citrus x limon This evergreen tree has glossy, green leaves. In spring, it produces fragrant, white blooms, followed by fruits, which take up to a year to ripen. Feed the plant well while in flower and fruit, and protect from frost in winter.

↕3m (10ft) ↔2m (6ft)

ANNUAL/BIENNIAL SMALL

Coriander

Coriandrum sativum A quick-growing annual herb, with rounded, green, aromatic leaves. Grown from seed, the whole plant can be used in cooking, and is a common feature in Asian cuisine. Grow in containers and windowboxes; keep it well watered.

↕50cm (20in) ↔20cm (8in)

ANNUAL/BIENNIAL MEDIUM

Courgettes

Cucurbita pepo This plant produces a mound of leaves, below which the bright yellow flowers and green fruits develop in summer. Harvest the fruits while young and tender. Ideal for growing bags and pots, water and feed the plant well.

↕↔60cm (24in)

ANNUAL/BIENNIAL MEDIUM

Summer squashes

Cucurbita pepo This vigorous plant flowers and fruits in summer, bearing squashes in a range of colours and shapes. Harvest squashes while small and tender. Ideal for large pots; train plants vertically or leave to trail. Feed and water regularly.

↕↔90cm (36in)

ANNUAL/BIENNIAL MEDIUM

Maincrop carrots

Daucus carota Sow seeds for this type of carrot throughout spring and summer. It can be harvested young and tender a few weeks after sowing, or left longer to produce larger roots. Best planted in deep containers; water well.

↕60cm (24in) ↔25cm (10in)

ANNUAL/BIENNIAL SMALL

Salad carrots

Daucus carota This type of carrot bears short, round, sweet-tasting roots. Sow seeds directly in containers and windowboxes in spring and summer, and harvest once large enough, then sow new batches. Water regularly.

↕30cm (12in) ↔15cm (6in)

Plants for productive patios

TREE SMALL

Figs

Ficus carica This tree has large, glossy, lobed foliage and, in summer, small, pear-shaped fruits. In cooler areas, the fruits ripen the following summer and need winter protection. Plant it in a large pot of soil-based compost.
↕3m (10ft) ↔4m (12ft)

💧 ❄❄❄

PERENNIAL LARGE

Sweet fennel

Foeniculum vulgare A tall, upright perennial, with ferny, aromatic leaves that taste and smell of aniseed. It bears clusters of tiny, yellow summer flowers, followed by edible seeds. Best planted in a large container; use it as a foil for other plants.
↕1.8m (6ft) ↔45cm (18in)

💧 ❄❄❄

PERENNIAL SMALL

Perpetual strawberries

Fragaria* x *ananassa This strawberry plant produces fruits with a mild flavour in summer, followed by a second, smaller crop into autumn. Ideal for smaller plots, plant it in containers, growing bags, or windowboxes; water and feed regularly.
↕30cm (12in) ↔indefinite

💧 ❄❄❄

PERENNIAL SMALL

Summer srawberries

Fragaria* x *ananassa This variety of strawberry gives a single crop of fruits, which have a rich, sweet flavour, from early- to midsummer. Ideal for containers that can be moved from view after cropping. Water and feed regularly.
↕30cm (12in) ↔50cm (20in)

💧 ❄❄❄

PERENNIAL SMALL

Alpine strawberries

Fragaria vesca This strawberry plant has a compact habit, and throughout summer, it bears tiny, bright red fruits, which have a sweet, intense flavour. It is ideal for containers and windowboxes. Water and feed regularly.
↕30cm (12in) ↔indefinite

💧 ❄❄❄

ANNUAL/BIENNIAL SMALL

Lettuces

Lactuca sativa Sown or planted in spring and summer, this leafy crop can be picked young as "cut-and-come-again" leaves or left to mature into full heads. Many colours and leaf textures are available; water plants regularly.
↕↔30cm (12in)

💧 ❄

TREE MEDIUM

Bay laurel 🏆

Laurus nobilis A large, evergreen tree, with aromatic, dark green leaves that can be clipped into shapes to create topiary. Grow it in a sheltered area. Use the leaves to make bouquet garni to flavour savoury dishes.
↕up to 12m (40ft) ↔up to 10m (30ft)

💧 ❄❄

TREE SMALL

Cooking apples

Malus domestica This small tree bears large, sour fruits from summer to autumn. Choose trees grafted onto dwarfing rootstocks to limit their size. Plant in a large container of soil-based compost; feed and water regularly from spring to summer.
↕8m (25ft) ↔2m (6ft) in pot

💧 ❄❄❄

TREE SMALL

Dessert apples

Malus domestica This tree bears fruits from late summer to autumn. Choose trees grafted onto dwarfing rootstocks. Plant in a large container of soil-based compost; feed and water regularly. You may need a pollinator variety.
↕6m (20ft) ↔2m (6ft) in pot

💧 ❄❄❄

PERENNIAL SMALL

Lemon balm

Melissa officinalis This herb has toothed and textured, green leaves that emit a rich lemon scent when crushed. Remove the summer flowers to encourage leaf growth. 'Aurea' (above) has bright gold-marked leaves.
↕↔45cm (18in)

💧 ❄❄❄

ANNUAL/BIENNIAL MEDIUM

Sweet basil

Ocimum basilicum A bushy annual or short-lived perennial, with fragrant, bright green leaves and pink summer flower spikes. Add the leaves to salads or use to season savoury dishes. Grow it in pots; water and feed well, and harvest regularly.
↕60cm (24in) ↔30cm (12in)

💧

ANNUAL/BIENNIAL MEDIUM

Purple basil

Ocimum basilicum* var. *purpurascens A bushy annual or short-lived perennial, with aromatic, purple leaves, which can be used in savoury dishes and salads. 'Dark Opal' (above) has pink-purple flowers in summer. Water and feed well, and pick regularly.
↕60cm (24in) ↔30cm (12in)

💧

ANNUAL/BIENNIAL LARGE

Borlotto beans

Phaseolus vulgaris This climbing crop is grown for its beans, which are borne in long, green pods, often flushed red. Harvest the beans while young and tender to eat fresh, or leave to mature fully for drying. Provide support; water and feed well.

↕1.8m (6ft)

PERENNIAL SMALL

Oregano

***Origanum* 'Kent Beauty'** A spreading perennial herb, with trailing stems of small, aromatic, oval leaves and unusual, pale pink summer blooms. Use the leaves in tea or to flavour savoury dishes and salads. Ideal for baskets and windowboxes.

↕20cm (8in) ↔30cm (12in)

ANNUAL/BIENNIAL MEDIUM

Parsley

Petroselinum crispum This clump-forming biennial, best grown as an annual, has divided, aromatic, green leaves, which are used as an edible garnish and to flavour savoury dishes. Grow in pots and windowboxes; water well in summer.

↕80cm (32in) ↔60cm (24in)

CLIMBER SMALL

Runner beans

Phaseolus coccineus Suitable for large containers, this climbing bean bears long, flattened, green pods throughout summer. For the best crop, harvest pods regularly, before the beans inside are fully formed. It needs tall supports; water and feed well.

↕1.8m (6ft)

ANNUAL/BIENNIAL LARGE

French beans

Phaseolus vulgaris This climber bears long, rounded, green or purple pods all summer. For the best crop, harvest before the seeds swell. Dwarf and tall varieties are available. Grow in large pots, provide support, and water and feed well.

↕1.8m (6ft)

PERENNIAL SMALL

Tomatillos

Physalis philadelphica This tender perennial has large, green leaves, yellow flowers, and tangy-tasting fruits that are concealed within papery cases from mid- to late summer. Ideal for large pots, it needs full sun and a long summer to crop well.

↕45cm (18in) ↔60cm (24in)

ANNUAL/BIENNIAL MEDIUM

Garden peas

Pisum sativum This form of pea is grown for traditional, podded peas, which appear from late spring to autumn. Climbing or dwarf varieties are available. Plant or sow in large containers from spring to early summer. Water well and pick regularly.

↕70cm (28in) ↔50cm (20in)

ANNUAL/BIENNIAL MEDIUM

Mangetout peas

Pisum sativum This form of pea is sown or planted in spring or early summer. From late spring to autumn, it bears flattened, crisp pods, which are eaten whole. Grow in large containers, water regularly, and harvest often for the best crop.

↕90cm (36in) ↔10cm (4in)

ANNUAL/BIENNIAL MEDIUM

Sugar snap peas

Pisum sativum Sown from seed in spring or early summer, this short, climbing plant produces crisp pods from late spring to autumn. Pick regularly, before the seeds inside are fully formed, and eat whole. Ideal for large pots; water regularly.

↕60cm (24in) ↔50cm (20in)

TREE SMALL

Acid cherries

Prunus cerasus This tree bears white or pink flowers in spring, followed by sharp-tasting, red fruits in summer. Choose self-fertile varieties on dwarfing rootstocks. Plant in large pots of soil-based compost. Water and feed regularly.

↕↔4m (12ft)

TREE SMALL

Sweet cherries

Prunus cerasus This tree bears pink or white blossom in spring, followed by sweet-tasting fruits from summer to late autumn. Select a self-fertile variety on a dwarfing rootstock. Plant in a large pot of soil-based compost; water and feed well.

↕↔4m (12ft)

TREE SMALL

Peaches

Prunus persica This crop produces large, juicy fruits, which ripen from midsummer to early autumn. Choose a modern, compact variety for the best harvest. Plant in a pot of soil-based compost; water and feed well, and protect from rain and frost in spring.

↕↔2m (6ft) in a pot

ANNUAL/BIENNIAL SMALL

Summer radishes

Raphanus sativus This crop can be sown from spring to early autumn, and forms succulent, peppery roots in just five weeks. Pull the roots once large enough, then re-sow a new batch. Ideal for windowboxes and pots. There are many varieties to grow.

↕↔20cm (8in)

SHRUB MEDIUM

Blackcurrants

Ribes nigrum This upright, branching shrub flowers in spring, followed by strings of sharp-tasting, black fruits. Plant it in a large container of soil-based compost. Mulch with compost in spring, and water regularly during summer.

↕↔1.5m (5ft)

SHRUB MEDIUM

Redcurrants

Ribes rubrum A branching shrub that flowers in spring and bears trailing strings of sharp-tasting, bright red fruits in summer. It is best planted in a large pot of soil-based compost. Mulch it in spring, and water regularly during summer.

↕↔1.5m (5ft)

SHRUB MEDIUM

Whitecurrants

Ribes rubrum This upright, branching shrub flowers in spring, then forms strings of sharp-tasting, almost transparent, white berries in summer. Plant in a large pot of soil-based compost. Mulch with compost in spring; water regularly all summer.

↕↔1.5m (5ft)

SHRUB MEDIUM

Culinary gooseberries

Ribes uva-crispa This branching, often thorny, shrub, produces large crops of sharp-tasting, round berries in summer that are best cooked before eating. Plant in a container of soil-based compost; water regularly and mulch in spring.

↕↔1.5m (5ft)

SHRUB MEDIUM

Dessert gooseberries

Ribes uva-crispa This branching, often thorny, shrub flowers in spring. It produces round, sweet-tasting, red or green fruits in summer, which can be eaten raw or cooked. Plant in a container of soil-based compost and water well. Mulch in spring.

↕↔1.5m (5ft)

SHRUB MEDIUM

Rosemary

Rosmarinus officinalis A tall, evergreen, bushy shrub, with aromatic, needle-like, dark green leaves and purplish blue blooms from spring to summer. Use the leaves with roast potatoes and to flavour meat dishes. Grow in a large pot of soil-based compost.

↕↔1.5m (5ft)

CLIMBER MEDIUM

Thornless blackberries

Rubus fruticosus This climbing shrub has dissected leaves and thornless stems. From late summer to autumn, it produces clusters of large, glossy, black berries. Plant in a large pot of soil-based compost, and provide support for the stems.

↕2.5m (8ft)

PERENNIAL MEDIUM

Common sorrel

Rumex acetosa This upright perennial has oval, aromatic, green leaves that are patterned with red veins in some forms. Pick the young, tangy leaves throughout summer to add to salads, or cook them like spinach to make soups and purées.

↕↔60cm (24in)

SHRUB SMALL

Sage

Salvia officinalis This evergreen, shrubby perennial has textured, grey-green, aromatic leaves, which are used to flavour savoury dishes. It bears spikes of lilac-pink flowers in summer. Plant in soil-based compost; it can also be used in winter displays.

↕up to 80cm (32in) ↔1m (3ft)

SHRUB SMALL

Purple sage

***Salvia officinalis* 'Purpurascens'** 🏆 This evergreen, shrubby perennial has aromatic, oval, purple leaves, which fade with age. The lilac-pink summer flower spikes can be removed to promote leaf growth. Use the leaves to flavour savoury dishes.

↕up to 80cm (32in) ↔1m (3ft)

PERENNIAL MEDIUM

Cotton lavender

Santolina chamaecyparissus An evergreen perennial with aromatic, finely cut, silvery foliage that can be dried for potpourri or used as a moth repellant. The yellow pompon blooms last many weeks in summer. Plant in soil-based compost.
↕75cm (30in) ↔1m (3ft)

ANNUAL/BIENNIAL SMALL

Summer savoury

Satureja hortensis This upright annual has slender, bronze-green leaves, which can be used to flavour salads, or meat and fish dishes. It bears small, pale pink, tubular blooms in late summer. Grow it among other plants in patio containers.
↕25cm (10in) ↔30cm (12in)

ANNUAL/BIENNIAL SMALL

Bush tomatoes

Solanum lycopersicum This tomato plant has a bushy habit, and is suitable for growing outdoors. Varieties with fruits in a range of colours, shapes, and flavours are available. Ideal for pots and growing bags; feed, water, and harvest regularly.
↕↔45cm (18in)

ANNUAL/BIENNIAL MEDIUM

Tumbling tomatoes

Solanum lycopersicum This type of tomato has a trailing habit, and is ideal for hanging baskets and raised containers. Several varieties are available, and all produce small fruits. Water and feed well in summer, and harvest regularly.
↕60cm (24in) ↔45cm (18in)

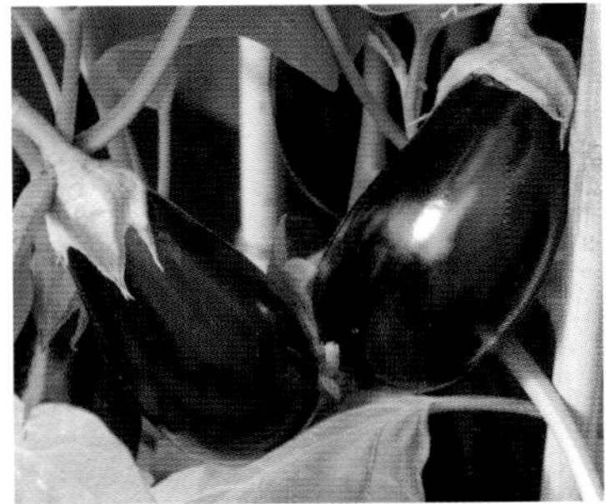

ANNUAL/BIENNIAL LARGE

Aubergines

Solanum melongena An upright, branching plant, with downy, green leaves and large, glossy, purple fruits from late summer to early autumn. It is suitable for pots and growing bags. Feed and water regularly. Best in areas with long summers.
↕1.2m (4ft) ↔50cm (20in)

BULB LARGE

Maincrop potatoes

Solanum tuberosum Plant these potatoes in spring in large pots, barrels, and raised beds, and allow to grow until autumn to produce full-sized tubers. There are many varieties to choose from, with either red or white skins.
↕1m (3ft)

ANNUAL/BIENNIAL LARGE

Sweetcorn

Zea mays This upright, annual crop is grown for its large, sweet-tasting "cobs" that mature in late summer. It can be grown in clusters in large containers to help pollinate the flowers. Provide support for stems and harvest regularly.
↕2m (6ft) ↔60cm (24in)

BULB LARGE

Salad potatoes

Solanum tuberosum The tubers of this type of potato are small and thin-skinned. Plant them in spring in large pots or barrels, and turn the potatoes out when large enough to harvest in summer. There are many varieties to choose from.
↕1m (3ft)

SHRUB SMALL

Thyme

Thymus vulgaris A low-growing subshrub, with tiny, aromatic, grey-green leaves that spread to form a low carpet. It bears a mass of purple to white flowers in summer. Ideal for windowboxes, containers, and baskets; it can also be used for winter interest.
↕30cm (12in) ↔40cm (16in)

SHRUB MEDIUM

Blueberries

Vaccinium corymbosum This deciduous shrub bears white flowers in spring, followed by clusters of sweet-tasting, blue-black berries that ripen in late summer and autumn. Plant in pots of ericaceous compost; water and feed well.
↕↔up to 1.5m (5ft)

OTHER SUGGESTIONS

Fruit trees

Apple 'Ashmead's Kernel' 🏆 • Apple 'Egremont Russet' 🏆 • Apple 'Red Falstaff'• Apple 'Royal Gala' 🏆 • Lemon 'Meyer' • Pear 'Doyenné de Comice' 🏆 • Pear 'Williams' Bon Chrétien' 🏆

Soft fruits

Blackberry 'Loch Ness' 🏆 • Blackcurrant 'Ben Sarek' • Blueberry 'Bluetta' • Blueberry 'Spartan' 🏆 • Redcurrant 'Red Lake' • Strawberry 'Lucy'

Vegetables

Aubergine 'Rosa Bianca' • Beetroot 'Red Ace' 🏆 • Carrot 'Chantenay Red Cored' • Chilli 'Numex Twilight' • Chilli 'Sweet Banana' • Courgette 'Eight Ball' • French bean 'Purple Queen' • Potato 'Pink Fir Apple' 🏆 • Potato 'Charlotte' 🏆 • Runner bean 'Painted Lady' • Runner bean 'Hiesta' • Spring onion 'White Lisbon' 🏆 • Tomato 'Sungold' 🏆 • Tomato 'Tigerella' 🏆

Plant focus: tomatoes

Delicious and versatile, tomatoes are an ideal crop for small spaces, producing bumper yields on relatively compact plants.

BEARING BEAUTIFUL FRUITS that are both tasty and decorative, tomatoes come in a wide variety of shapes and sizes. Choose from tiny cherries, mid-sized fruits, or large beefsteaks, and look out for reliable disease-resistant and grafted forms that guarantee a good crop. Tomatoes are divided into two main groups: cordons that fruit on a single stem, and bushes, which produce fruit on spreading side-stems. Both groups can be grown outside in a sunny, sheltered area, although some larger fruiting varieties require the extra heat of a greenhouse. Pot up young plants in growing bags or large containers of multi-purpose compost, and provide a long stake for cordons. Dwarf bushes can be grown in smaller pots or hanging baskets. As cordons grow, tie in the main stem to the stake, pinch out any sideshoots that form, and remove the top shoot when the stem has reached about 1.5m (5ft) height. Water tomato plants regularly in summer, and feed them weekly using a tomato feed once the fruits begin to show.

FORMS OF TOMATO

Bush tomatoes These plants fruit on spreading sideshoots that are allowed to grow. Insert several stakes to support the stems.

Cordon tomatoes Fruits are produced on a single, tall, upright stem that requires support. There are many cordon varieties to grow.

Patio crops Many compact bush varieties are ideal for patio containers and require no pinching out or staking. The fruits are small.

Coloured fruits Although most varieties produce red fruits, tomatoes range in colour from yellow to blackish-red when ripe.

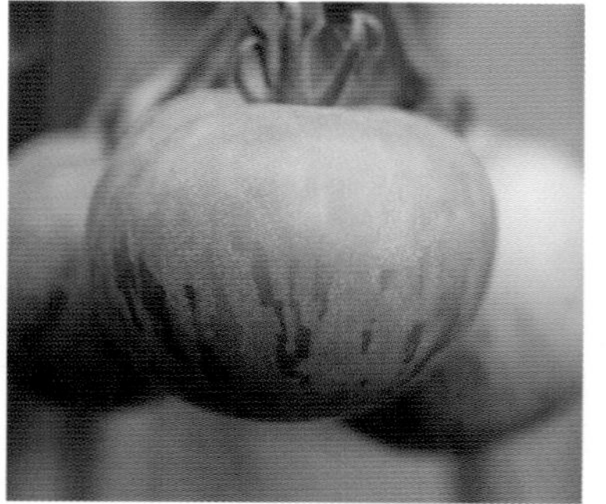

Heritage varieties Many older varieties have distinctive fruits that vary in size, shape, colour, and flavour from modern cultivars.

USING TOMATOES

Tomatoes are easy to grow and need less frequent watering if grown directly in the soil or in a raised bed, as the volume of soil retains more water and nutrients than compost in a container.

Many tomatoes grow happily on sunny patios and make decorative displays when matched with other sun-lovers.

Growing bags are ideal for bush and cordon tomatoes, and can be used in a greenhouse or outside. Most bags are large enough to grow three plants.

Compact dwarf bushes, such as the trailing, cherry-fruited Tumbler types, are perfect for large hanging baskets in sheltered sites.

Plants for herb gardens

Culinary and medicinal herbs have been grown for centuries, and many modern herb gardens still reflect medieval and Renaissance designs.

You can opt for a traditional style, and dedicate small, rectangular beds to individual plants, or try combining herbs with ornamentals in a mixed bed. Low-growing herbs, such as thyme and marjoram, also make a great edging for a border. Many compact herbs are suitable for containers, too, but restrict rampant growers, such as mint and lemon balm, to a pot of their own. Shrubby herbs, including rosemary and sage, will develop into substantial plants and are best grown in the ground in the long term, and pruned in spring to encourage new growth.

BULB MEDIUM

Chives

Allium schoenoprasum This summer-flowering perennial bulb has hollow, dark green, grassy leaves and pompon, pale purple or pink flowers in summer. Use the mild onion-flavoured leaves in salads and savoury dishes.

↕30cm (12in)

💧 ❄❄❄

SHRUB LARGE

Lemon verbena

Aloysia triphylla This spreading shrub has lance-shaped leaves that smell and taste of lemon. Spikes of small, white flowers appear in summer. Infuse the leaves in water to make a soothing tea, and use them to flavour sweet dishes.

↕↔3m (10ft)

💧 ❄❄

ANNUAL/BIENNIAL LARGE

Dill

Anethum graveolens A tall, upright annual, with fern-like, aromatic, grey-green leaves and rounded clusters of gold-green summer flowers, followed by edible seeds. Use the leaves in fish and poultry dishes; add dried seeds to salads.

↕1.2m (4ft) ↔60cm (24in)

💧 ❄❄❄

ANNUAL/BIENNIAL LARGE

Angelica

Angelica archangelica A tall biennial that produces decorative, deeply cut, green leaves and domed, tiny, green flowerheads in summer. The leaves are used for tea infusions, and tender young stems can be candied. It self-seeds freely.

↕2m (6ft) ↔1m (3ft)

💧 ❄❄❄

ANNUAL/BIENNIAL SMALL

Chervil

Anthriscus cerefolium A biennial member of the parsley family, it has fern-like, green foliage, with an aniseed flavour, and flat clusters of tiny, white flowers in summer. Use the leaves in salads and savoury dishes.

↕50cm (20in) ↔24cm (10in)

💧 ❄❄❄

PERENNIAL LARGE

Tarragon

Artemisia dracunculus An upright perennial, with linear, green, aromatic leaves on wiry stems. Grow in a sheltered area or in a container; it can be short-lived. Use its peppery leaves to flavour chicken and egg dishes, or in salads.

↕1.2m (4ft) ↔30cm (12in)

💧 ❄❄❄

ANNUAL/BIENNIAL MEDIUM

Borage

Borago officinalis This upright annual has oval, green, aromatic leaves and small, star-shaped, blue summer flowers. Add the cucumber-flavoured leaves to salads or summer drinks. Use the edible flowers as a garnish. Deadhead to prevent self-seeding.

↕90cm (36in) ↔30cm (12in)

💧 ❄❄❄

ANNUAL/BIENNIAL MEDIUM

Pot marigold

Calendula officinalis An annual with aromatic, green leaves. From spring to autumn, daisy-like flowers in shades of yellow and orange appear. The young leaves can be added to salads, and the flowers used as a garnish.

↕↔60cm (24in)

💧 ❄❄❄

PERENNIAL SMALL

Chamomile

Chamaemelum nobile An evergreen, mat-forming perennial, with finely divided, aromatic leaves and daisy-l ke, white flowers from late spring to summer. Infuse the flowers in boiling water to make a soothing tea.

↕10cm (4in) ↔45cm (18in)

PERENNIAL LARGE

Chicory

Cichorium intybus The wild form of the salad vegetable, this upright perennial produces small, green leaves and blue summer blooms. Use the leaves and flowers in salads, and the ground roots to make a coffee substitute.

↕1.2m (4ft) ↔45cm (18in)

PERENNIAL MEDIUM

Hyssop

Hyssopus officinalis This upright perennial has aromatic leaves that smell and taste of aniseed, and spikes of fluffy, lavender-blue summer flowers, attractive to bees and butterflies. Use the leaves to make tea and to flavour savoury dishes.

↕60cm (24in) ↔1m (3ft)

ANNUAL/BIENNIAL SMALL

Coriander

Coriandrum sativum An annual herb, easily grown from seed, with rounded, green, aromatic leaves. The leaves, stems, and seeds are all used in cooking, and commonly feature in Asian cuisine. Keep it well watered to prevent bolting.

↕50cm (20in) ↔20cm (8in)

BULB SMALL

Saffron crocus

Crocus sativus An autumn-flowering perennial bulb, with saucer-shaped, dark-veined, purple flowers, with bright rec stigmas that yield saffron. The flowers appear just before the grass-like leaves emerge. Plant the bulbs in late summer.

↕5cm (2in)

PERENNIAL LARGE

Fennel

Foeniculum vulgare This tall, upright perennial produces fern-like, aromatic leaves that taste of aniseed. It bears clusters of tiny, yellow flowers, which are followed by edible seeds. Add the seeds to salads and savoury dishes.

↕1.8m (6ft) ↔45cm (18in)

TREE MEDIUM

Bay laurel

Laurus nobilis A large, evergreen tree, with aromatic, dark green leaves that can be clipped into shapes to create topiary. Grow it in a sheltered site, and use the leaves to make bouquet garni to flavour savoury dishes.

↕up to 12m (40ft) ↔up to 10m (30ft)

SHRUB SMALL

Lavender

Lavandula angustifolia An evergreen, bushy, dwarf shrub, with slim, aromat c, silvery grey leaves. In summer, it prod Jces spikes of small, fragrant, violet-blue flowers. Use the flowers in potpourri, and to flavour ice cream and sweet dishes.

↕80cm (32in) ↔60cm (24in)

PERENNIAL LARGE

Lovage

Levisticum officinale A tall, upright perennial, with divided, celery-flavoured leaves and flat heads of tiny, yellow flowers in summer. The leaves and stems can be used to flavour savoury dishes, such as soups and stews.

↕1.2m (4ft) ↔1m (3ft)

PERENNIAL SMALL

Spearmint

Mentha spicata A spreading perennial, with textured, bright green, aromatic foliage and short spikes of pink summer flowers. Infuse the leaves to make tea, or use to flavour savoury dishes. Plant it in a container to prevent the roots spreading.

↕50cm (20in) ↔indefinite

PERENNIAL MEDIUM

Sweet bergamot

Monarda didyma A tall, clump-forming perennial, with lance-shaped, green, aromatic leaves and hooded, red, pink, white, or lilac summer flowers. Infuse the leaves either alone or with black tea to make an Earl Grey-style tea.

↕90cm (36in) ↔45cm (18in)

ANNUAL/BIENNIAL MEDIUM

Cinnamon basil

***Ocimum basilicum* 'Cinnamon'** A bushy annual or short-lived perennial, with lance-shaped, aromatic, green leaves, with a cinnamon-basil taste, and purple summer flowers. Use the leaves to flavour hot drinks and salads. Plant in pots; protect from frost.

↕60cm (24in) ↔30cm (12in)

ANNUAL/BIENNIAL MEDIUM

Purple basil

Ocimum basilicum* var. *purpurascens A bushy annual or short-lived perennial, with aromatic, purple, lance-shaped leaves, used in savoury dishes and salads. 'Dark Opal' (above) has fragrant, pink-purple summer flowers. Grow in pots; protect from frost.

↕60cm (24in) ↔30cm (12in)

PERENNIAL MEDIUM

Sweet cicely

Myrrhis odorata This tall perennial resembles cow parsley, and bears aromatic, fern-like, mid-green foliage and fragrant, creamy white flowers in early summer. The leaves have an aniseed flavour and are used in many savoury dishes.

↕90cm (36in) ↔60cm (24in)

PERENNIAL SMALL

Oregano

***Origanum* 'Kent Beauty'** 🏆 This spreading perennial produces trailing stems clothed in small, aromatic, oval leaves. Pale pink flowers appear in summer. The leaves can be used as an infusion to make tea, or in savoury dishes and salads.

↕20cm (8in) ↔30cm (12in)

ANNUAL/BIENNIAL MEDIUM

Thai basil

***Ocimum basilicum* 'Horapha'** A bushy annual or short-lived perennial, with red stems and pink-white summer flowers. Its dark green leaves have a liquorice scent and basil taste. Use them in curries and savoury dishes. Plant it in pots; protect from frost.

↕60cm (24in) ↔30cm (12in)

ANNUAL/BIENNIAL MEDIUM

Parsley

Petroselinum crispum A clump-forming biennial, best grown as an annual, which produces deeply divided, aromatic, green leaves, which are used to flavour a wide range of savoury dishes, and make a decorative, edible garnish.

↕80cm (32in) ↔60cm (24in)

ANNUAL/BIENNIAL MEDIUM

Sweet basil

Ocimum basilicum A bushy annual or short-lived perennial, with lance-shaped, aromatic, bright green leaves and fragrant, pink summer flowers. Use the leaves to season savoury dishes and add to salads. Grow it in pots; protect from frost.

↕60cm (24in) ↔30cm (12in)

SHRUB MEDIUM

Rosemary

Rosmarinus officinalis This tall, evergreen, bushy shrub has aromatic, needle-like, dark green leaves. Small, purplish blue flowers appear from spring to summer. Use the leaves with roast potatoes, and to flavour meat dishes.

↕↔1.5m (5ft)

PERENNIAL MEDIUM

Common sorrel

Rumex acetosa An upright perennial, with oval, aromatic, green leaves, patterned with red veins in some forms. Pick the young, tangy leaves throughout summer to add to salads, or cook them like spinach to make soups and purées.
↕↔ 60cm (24in)

SHRUB SMALL

Variegated sage

***Salvia officinalis* 'Tricolor'** This evergreen perennial subshrub has aromatic, cream- and pink-edged grey-green foliage and spikes of lilac-pink summer flowers. Use the leaves to flavour meat dishes, and as an infusion to make a soothing tea.
↕ 80cm (32in) ↔ 1m (3ft)

PERENNIAL LARGE

Common comfrey

Symphytum officinale A tall, upright perennial, with large, coarse, green leaves. From late spring to summer, it bears pink, purple, or white flowers. Steep the nutrient-rich leaves in water to make organic fertilizer or compost to improve soil.
↕ 1.5m (5ft) ↔ 2m (6ft)

PERENNIAL MEDIUM

Cotton lavender

Santolina chamaecyparissus
An evergreen, mound-forming perennial, with finely cut, aromatic, silvery grey foliage and pompon-like, yellow summer flowers. Use the leaves in potpourri or in sachets to repel moths.
↕ 75cm (30in) ↔ 1m (3ft)

SHRUB SMALL

Lemon thyme

Thymus citriodorus A bushy, evergreen shrub, with small, oval, lemon-scented foliage and pale lavender-pink flowers in summer. The citrus-flavoured leaves are used in fish dishes. 'Variegata' has cream-edged leaves.
↕ 30cm (12in) ↔ 25cm (10in)

SHRUB SMALL

Sage

Salvia officinalis This evergreen, perennial subshrub produces oval, aromatic, grey-green foliage and spikes of lilac-pink flowers in summer. The leaves are used to flavour meat dishes, and to make stuffing.
↕↔ 80cm (32in)

SHRUB SMALL

Thyme

Thymus vulgaris An evergreen, low-growing shrub, with small, oval, aromatic, grey-green leaves and purple to white flowers in early summer. Use the leaves to make a bouquet garni, and add to stews and soups. Trim annually to encourage growth.
↕ 30cm (12in) ↔ 40cm (16in)

SHRUB SMALL

Purple sage

***Salvia officinalis* 'Purpurascens'** 🏆
An evergreen perennial subshrub, with oval, aromatic, purple leaves, red-purple when young. Lilac-pink flowers appear in summer. Use the leaves to flavour meat dishes, and to make stuffing.
↕↔ 30cm (12in)

ANNUAL/BIENNIAL SMALL

Summer savoury

Satureja hortensis This upright annual produces slim, linear, bronze-green leaves and small, rose-white, tubular flowers in late summer. Use the leaves to add flavour to salads, or in meat and fish dishes.
↕ 25cm (10in) ↔ 30cm (12in)

OTHER SUGGESTIONS

Annuals and biennials

Wild celery (*Apium graveolens*) • White borage (*Borago officinalis* 'Alba') • Golden mustard (*Brassica* 'Golden Streaks') • Caraway (*Carum carvi*) • Lime basil (*Ocimum americanum*) • Purslane (*Portulacea oleracea*)

Perennials

Anise hyssop (*Agastache foeniculum*) • Korean mint (*Agastache rugosa*) • Sea kale (*Crambe maritima*) • Coneflower (*Echinacea angustifolia*) • Giant fennel (*Ferula communis*) • Wild strawberry (*Fragaria vesca*) • Sweet rocket (*Hesperis matronalis*) • Sea lovage (*Ligusticum scoticum*) • Lemon balm (*Melissa officinalis*) • Salad burnet (*Sanguisorba minor*)

Bulbs

Welsh onion (*Allium fistulosum*) • Garlic chives (*Allium tuberosum*)

Plants for gravel gardens

Gravel adds a decorative finish to designs, and is especially suited to gardens on sandy soil and drought-tolerant plants.

Take inspiration from the designs of celebrated plantswoman Beth Chatto who transformed a dry and dusty car park into a world famous garden, using plants that need no watering and take care of themselves once established. To recreate this on a smaller scale, combine plants with different forms and textures, such as the blue grass *Helictotrichon*, flat-topped achilleas, and daisy-flowered *Anthemis*. Hardy annuals, including California poppies, will self-seed in the gravel, creating a naturalistic effect, while bulbs, such as tulips, alliums, and nerines, add seasonal highlights from spring to autumn.

PERENNIAL MEDIUM

Achillea 'Coronation Gold'

YARROW This upright perennial has fern-like, grey-green leaves and, in summer, it bears flattened heads of golden-yellow flowers that are highly attractive to bees. Can be short-lived; the stems may need support.

↕1m (3ft) ↔60cm (24in)

PERENNIAL MEDIUM

Achillea 'Summerwine'

YARROW This upright perennial has feathery, dark green leaves. In summer, it produces flat heads of white-eyed crimson flowers that fade to pink. Plant it towards the front of a border; it may also require staking.

↕↔80cm (32in)

PERENNIAL MEDIUM

Aeonium 'Zwartkop'

This tender, evergreen perennial is grown for its rosettes of fleshy, glossy, dark purple-black leaves that are produced on branching stems. Protect it from frost during winter by bringing it under cover.

↕60cm (24in) ↔1m (3ft)

PERENNIAL MEDIUM

Agastache 'Black Adder'

HYSSOP An upright perennial, with lance-shaped, bright green leaves and spikes of fluffy, violet-purple flowers, which appear from late summer to mid-autumn. Plant in a sheltered spot with grasses and round-headed flowers, such as dahlias.

↕60cm (24in) ↔45cm (18in)

PERENNIAL SMALL

Agave parryi

An evergreen, succulent perennial, grown for its large rosette of spiky, grey-green to blue-green leaves. It is not hardy so protect from hard frost during winter by covering or bringing it under cover. Use it as a statement, plant in a gravel bed.

↕50cm (20in) ↔1m (3ft)

BULB LARGE

Allium cernuum

This upright, perennial bulb produces strap-shaped, grey-green leaves that fade before tall stems holding clusters of pendent, purplish pink flowers appear in summer. Plant the bulbs in bold groups and drifts in autumn.

↕70cm (28in)

BULB LARGE

Allium 'Purple Sensation'

An upright, perennial bulb, with strap-shaped, grey-green foliage, which fades as sturdy stems of spherical, deep violet flowerheads appear in early summer. Plant the bulbs in autumn between other plants that will hide its dying leaves.

↕80cm (32in)

PERENNIAL SMALL

Anaphalis triplinervis

PEARL EVERLASTING This low-growing, clump-forming perennial produces lance-shaped, pale grey-green leaves and clusters of small, white flowers in summer. 'Sommerschnee' AGM (above) is a popular variety.

↕50cm (20in) ↔60cm (24in)

PERENNIAL SMALL

Androsace carnea

ROCK JASMINE An evergreen, mound-forming, dwarf perennial, with linear, green leaves and rounded clusters of yellow-eyed pink flowers, which appear in late spring. Use it between paving stones set into a gravel path.

↕↔5cm (2in)

PERENNIAL MEDIUM

Anemanthele lessoniana

NEW ZEALAND WIND GRASS A clump-forming, semi-evergreen grass, with arching, green leaves, tinged with red and orange in summer and autumn. Sprays of red-brown flowers appear in late summer. It self-seeds prolifically; weed out unwanted seedlings.
↕↔ 1m (3ft)

PERENNIAL MEDIUM

Anthemis tinctoria

DYER'S CHAMOMILE A clump-forming perennial that produces finely dissected, green leaves and an abundance of daisy-like, white or yellow flowers on branching stems throughout summer. 'E.C. Buxton' (above) is a popular variety.
↕↔ 1m (3ft)

PERENNIAL MEDIUM

Aquilegia canadensis

ROCK BELLS This upright perennial produces fern-like, green leaves and yellow-centred red flowers that hang gracefully from wiry stems from mid-spring to summer. Grow it in groups with late-flowering spring bulbs.
↕ 60cm (24in) ↔ 30cm (12in)

PERENNIAL LARGE

Calamagrostis x acutiflora

FEATHER REED GRASS This perennial grass produces a clump of arching, green leaves, above which appear stems of upright, architectural summer flowerheads, which can be left for autumn interest. 'Overdam' (above) has variegated leaves.
↕↔ 1.2m (4ft)

ANNUAL/BIENNIAL MEDIUM

Arctotis fastuosa

CAPE DAISY This is an upright annual, with lobed, silvery green leaves. Throughout summer, it bears yellow, orange, or white, daisy-like blooms, with contrasting eyes. The flowers commonly close on cloudy days.
↕ 60cm (24in) ↔ 30cm (12in)

ANNUAL/BIENNIAL MEDIUM

Argemone mexicana

DEVIL'S FIG An upright annual, with prickly, deeply divided, dark green leaves and saucer-shaped, pale lemon-yellow flowers from late summer to autumn. Sow seeds in spring directly into the gravel; it will self-seed readily in following years.
↕ 60cm (24in) ↔ 30cm (12in)

PERENNIAL MEDIUM

Asclepias tuberosa

BUTTERFLY WEED An upright perennial, with lance-shaped, green leaves. From late summer to early autumn, it produces clusters of small, bright orange-red flowers that are magnets for bees and butterflies, and are followed by decorative seed pods.
↕ 75cm (30in) ↔ 45cm (18in)

PERENNIAL LARGE

Calamagrostis brachytricha

KOREAN FEATHER REED GRASS A deciduous clump-forming, grass, with arching, grey-green leaves and upright, feather-shaped, tall, pink-tinted silver summer flowerheads, followed by decorative, straw-coloured seedheads that persist through winter.
↕ 1.4m (4½ft) ↔ 50cm (20in)

SHRUB LARGE

Ceanothus 'Concha'

CALIFORNIAN LILAC This evergreen shrub produces arching stems of small, oval, dark green leaves. In late spring, rounded clusters of red-purple buds open to reveal dark blue flowers. Plant it close to a fence or wall; prune lightly after flowering.
↕↔ 3m (10ft)

Plants for gravel gardens

PERENNIAL MEDIUM

Centranthus ruber

RED VALERIAN A perennial that forms a clump of fleshy leaves. From late spring to autumn, it bears heads of small, star-shaped, deep reddish pink or white flowers. Thrives in poor soil and self-seeds readily; stems may need staking.

↕90cm (36in) ↔60cm (24in) or more

ANNUAL/BIENNIAL MEDIUM

Cerinthe major

HONEYWORT An upright annual, with erect stems that bear oval, grey-green leaves, topped with tubular, purple and yellow summer flowers. Stems may need some support; the plant may self-seed in gravel. 'Purpurascens' (above) is a popular variety.

↕60cm (24in) ↔30cm (12in)

BULB MEDIUM

Chionodoxa forbesii

GLORY OF THE SNOW This perennial bulb produces linear, mid-green foliage. In early spring, star-shaped, blue flowers with white centres appear on slender stems. Plant the bulbs in groups in autumn.

↕15cm (6in)

SHRUB SMALL

Cistus x *lenis* 'Grayswood Pink'

ROCK ROSE An evergreen shrub, with oval, grey-green leaves and saucer-shaped, pale pink flowers, each lasting a day, which appear in succession throughout summer. Plant it in a sheltered spot.

↕60cm (24in) ↔1m (3ft)

TREE SMALL

Cordyline australis

CABBAGE PALM This evergreen, palm-like tree produces a dramatic fountain of sword-shaped, green leaves. Use it as a focal plant, and protect young specimens from frost. Other colourful variegated forms are available.

↕3m (10ft) ↔1m (3ft)

PERENNIAL MEDIUM

Coreopsis grandiflora

TICKSEED A bushy perennial, often grown as an annual, with lance-shaped, green leaves, and single or double, golden-yellow blooms throughout summer. Sow seeds in early spring, or set out young plants in spring in bold groups.

↕75cm (30in) ↔60cm (24in)

PERENNIAL SMALL

Coreopsis 'Mango Punch'

A bushy perennial, grown as an annual, with feathery, green foliage and masses of saucer-shaped, orange-yellow flowers, which bloom throughout summer. Use it to edge a pathway or at the front of a bed.

↕45cm (18in) ↔30cm (12in)

PERENNIAL MEDIUM

Crocosmia x *crocosmiiflora*

MONTBRETIA An upright perennial, with arching, sword-shaped, green leaves and clusters of small, trumpet-shaped flowers from summer to autumn. A useful plant for the middle of a gravel bed. 'Star of the East' AGM (above) is a popular variety.

↕70cm (28in) ↔8cm (3in)

BULB SMALL

Crocus 'Gipsy Girl'

This dwarf *C. chyrsanthus* hybrid bulb has grassy, green foliage. In early spring, it bears fragrant, bowl-shaped, yellow flowers with purple stripes on the outer petals. Plant the bulbs in autumn in bold groups.

↕7cm (3in)

PERENNIAL LARGE

Cynara cardunculus

CARDOON A tall, upright perennial, with spiny, silver-grey foliage and large, thistle-like, purple flowers, which appear on long stems from summer to autumn. Plant it at the back of a gravel bed; provide support in exposed sites.

↕2m (6ft) ↔1m (3ft)

BULB MEDIUM

Dahlia 'HS First Love'

An upright perennial bulb, often sold as annual bedding, with divided, dark maroon-green foliage and, from summer to early autumn, contrasting bright peach flowers, with a red-ringed eye. Plant it in groups at the front of a bed.

↕60cm (24in)

BULB MEDIUM

Dahlia 'Roxy'

This upright perennial bulb has divided, dark purple-green leaves and contrasting magenta-pink flowers, with yellow-edged dark centres, from summer to autumn. Pot up tubers indoors in spring, or plant directly outside after the frosts.

↕45cm (18in)

PERENNIAL SMALL

Dianthus 'Dad's Favourite'

GARDEN PINK A low-growing, evergreen perennial that forms a neat mound of grey-green, grassy foliage. During summer, it bears fragrant, semi-double, white flowers with maroon markings. Deadhead regularly to prolong the display.

↕up to 45cm (18in) ↔30cm (12in)

PERENNIAL LARGE

Echinacea purpurea

An upright, clump-forming perennial, with narrow, tapering, green leaves. From mid- to late summer, it bears simple, daisy-like flowers with brown, prickly centres. Best planted in drifts. 'Kim's Knee High' (above) is a compact variety.

↕1.5m (5ft) ↔45cm (18in)

PERENNIAL SMALL

Erigeron glaucus

BEACH ASTER A compact, clump-forming perennial, with slim, spoon-shaped, green leaves and daisy-like, yellow-centred mauve-pink flowers from late spring to midsummer. Use to edge a pathway or at the front of a bed. Ideal for coastal areas.

↕30cm (12in) ↔45cm (18in)

PERENNIAL LARGE

Eryngium giganteum 🏆

SEA HOLLY An upright, short-lived perennial, with marbled, heart-shaped, grey-green foliage and tall stems that are topped with cone-like, silvery grey summer flowers, surrounded by spiny bracts. It self-seeds freely in gravel.

↕90cm (36in) ↔30cm (12in)

PERENNIAL MEDIUM

Erysimum 'Bowles's Mauve' 🏆

PERENNIAL WALLFLOWER This evergreen perennial has narrow, dark grey-green foliage and small, mauve flower spikes from early spring to late autumn. Deadhead regularly. Can be short-lived.

↕60cm (24in) ↔40cm (16in)

ANNUAL/BIENNIAL SMALL

Eschscholzia californica

CALIFORNIA POPPY A bushy, clump-forming annual, with feathery, blue-green leaves and a profusion of small, bowl-shaped, orange, red, and yellow flowers in summer. Sow seeds in beds in spring; it may self-seed after its first season.

↕30cm (12in) ↔15cm (6in)

PERENNIAL LARGE

Gaura lindheimeri 🏆

WAND FLOWER This upright perennial produces spoon-shaped, green leaves and pink-tinged buds that open to reveal small, white flowers from midsummer to autumn. 'Karalee White' (above) is a popular modern variety.

↕1.5m (5ft) ↔90cm (36in)

PERENNIAL MEDIUM

Gentiana asclepiadea

WILLOW GENTIAN This clump-forming perennial produces arching stems of lance-shaped, mid-green leaves and trumpet-shaped, dark- to light blue flower clusters from late summer to early autumn. Plant in acid soil in the middle of a bed.

↕90cm (36in) ↔60cm (24in)

BULB LARGE

Gladiolus callianthus

ABYSSINIAN GLADIOLUS An upright perennial bulb, with arching, grassy, green leaves and fragrant, hooded, maroon-centred white flowers in summer. Plant the bulbs in groups in spring. Protect from frost by lifting in autumn after flowering.

↕1m (3ft)

SHRUB SMALL

Hebe ochracea 'James Stirling'

A dwarf, evergreen, dome-forming shrub, with yellow-green, scale-like leaves, resembling a conifer. In late spring, it bears small, white flowers. Best planted at the front of a bed. Trim after flowering.

↕45cm (18in) ↔60cm (24in)

PERENNIAL SMALL

Helianthemum 'Fire Dragon'

ROCK ROSE This spreading, evergreen perennial has trailing stems of soft, grey-green leaves and yellow-eyed orange-red flowers in summer. Use as a ground cover at the front of a bed. Deadhead regularly.

↕30cm (12in) ↔45cm (18in)

PERENNIAL MEDIUM

Helictotrichon sempervirens

BLUE OAT GRASS An evergreen grass, with upright, blue-grey leaves and straw-coloured flowerheads, held on long, arching stems in summer. Use it as a focal plant with perennials that have rounded foliage.

↕1m (3ft) ↔60cm (24in)

PERENNIAL MEDIUM

Hemerocallis 'Golden Chimes'

DAYLILY A tall, semi-evergreen perennial, with arching, grassy, green leaves and trumpet-shaped, golden-yellow flowers that appear in succession in summer and each last a day. Plant towards the back of a bed.

↕1.1m (3½ft) ↔60cm (24in)

PERENNIAL MEDIUM

Hordeum jubatum

FOXTAIL BARLEY A short-lived perennial grass, with green leaves and silvery pink, feathery flower plumes in summer. Sow seeds directly in the gravel in spring, and combine with brightly coloured plants, such as dahlias. It may self-seed.

↕60cm (24in) ↔30cm (12in)

PERENNIAL MEDIUM

Iris 'Harriette Halloway'

BEARDED IRIS An upright, rhizome-forming perennial, with grey-green, sword-like leaves. From early- to midsummer, it bears large, scented, pale blue, yellow-centred, bearded flowers. Plant the rhizomes just above the soil surface.

↕70cm (28in) ↔30cm (12in)

BULB MEDIUM

Iris reticulata

DWARF IRIS A perennial bulb, with grassy, green leaves and small, blue early spring flowers with yellow and white marks on the lower petals. In autumn, plant the bulbs in pots or at the front of a bed, where they can be easily seen.

↕15cm (6in)

SHRUB SMALL

Juniperus horizontalis

CREEPING JUNIPER This prostrate, evergreen conifer bears scale-like, blue-green foliage that will spread to provide ground cover. Use it as a foil for colourful bulbs and perennials. Varieties include 'Bar Harbor' with purple-tinted winter foliage.

↕50cm (20in) ↔ indefinite

PERENNIAL MEDIUM

Knautia macedonica

MACEDONICAN SCABIOUS This upright perennial has lobed, green, basal leaves. In summer, it produces a succession of button-shaped, crimson flowers on wiry stems. Easily grown from seed. 'Melton Pastels' (above) has pink and crimson flowers.

↕75cm (30in) ↔ 60cm (24in)

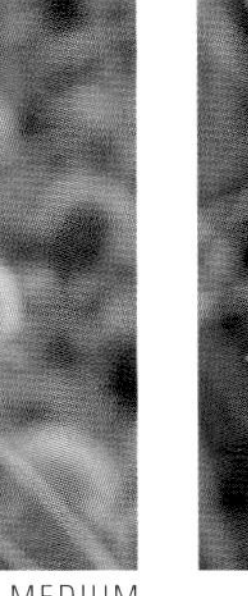

PERENNIAL MEDIUM

Kniphofia triangularis

RED HOT POKER An upright perennial, with grassy, green foliage and clusters of contrasting, bright orange-red, tubular flowers in late summer or early autumn. Plant it in groups with blue- or purple-flowered plants.

↕90cm (36in) ↔ 45cm (18in)

SHRUB SMALL

Lavandula pinnata

FERNLEAF LAVENDER This compact shrub, sometimes sold as annual bedding, has fern-like, silver-green foliage. From summer to early autumn, it produces triangular clusters of scented, violet-blue flowers. Grow it in a sheltered spot.

↕↔ 1m (3ft)

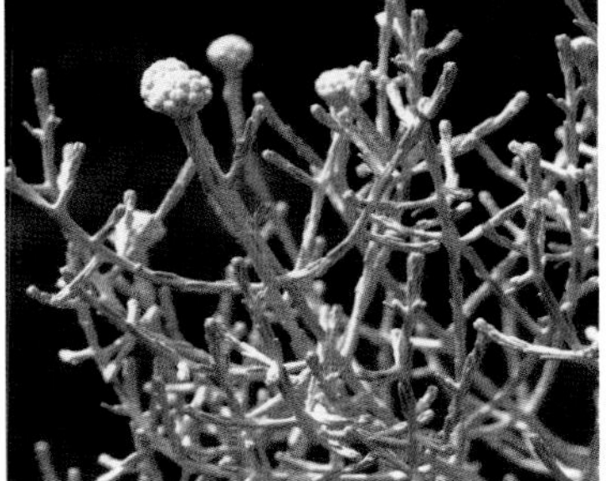

SHRUB SMALL

Leucophyta brownii

CUSHION BUSH A rounded, evergreen shrub, often grown as annual bedding, with silver stems bearing scale-like, aromatic, silver-grey leaves and creamy yellow, button-like flowers in summer. Grow it alongside colourful annuals.

↕75cm (30in) ↔ 90cm (36in)

PERENNIAL SMALL

Lewisia cotyledon

This clump-forming, evergreen perennial has rosettes of lobed, dark green foliage and funnel-shaped, pink, yellow, or orange flowers, which appear from late spring to summer. Grow it at the front of a bed.

↕30cm (12in) ↔ 15cm (6in) or more

PERENNIAL LARGE

Lychnis chalcedonica

MALTESE CROSS An erect perennial, with green basal leaves and tall stems topped with domed clusters of bright scarlet flowers in summer. Allow it to self-seed to create naturalistic groups. The stems require staking.

↕1.2m (4ft) ↔ 45cm (18in)

BULB MEDIUM

Narcissus 'Cheerfulness'

This perennial bulb produces blue-green, strap-like leaves. In mid-spring, it bears small clusters of long-lasting, small, sweetly scented, fully double, white and yellow blooms. This variety is excellent as a cut flower.

↕40cm (16in)

BULB MEDIUM

Nerine bowdenii

BOWDEN CORNISH LILY An upright, perennial bulb, with strap-like, green leaves and, in autumn, clusters of spidery, pale pink flowers on long, slim stems. Use it for late season colour, and plant the bulbs in groups in spring.

↕60cm (24in)

PERENNIAL SMALL

Oenothera speciosa
PINK EVENING PRIMROSE This compact perennial has small, divided, green leaves and a spreading habit. From late spring to early summer, it bears cup-shaped, fragrant, white flowers that age to pink. Plant it at the front of a gravel bed.
↕↔30cm (12in)

PERENNIAL SMALL

Ophiopogon planiscapus 'Nigrescens'
BLACK MONDO GRASS A clump-forming perennial, with grass-like, shiny, black leaves. In summer, clusters of purple-pink flowers appear, followed by black berries. Plant it in groups at the front of a bed.
↕23cm (9in) ↔30cm (12in)

PERENNIAL MEDIUM

Osteospermum 'Buttermilk'
AFRICAN DAISY An evergreen perennial, with slim, mid-green leaves and showy, daisy-like, pale yellow and white summer blooms, with bluish purple centres. Adds interest to the front of a gravel bed.
↕60cm (24in) ↔30cm (12in)

BULB MEDIUM

Puschkinia scilloides
STRIPED SQUILL This perennial bulb has strap-shaped, green leaves and clusters of dark blue-striped bluish white spring flowers. Grow it with other spring bulbs, such as *Muscari* and species tulips, at the front of a bed.
↕15cm (6in)

PERENNIAL LARGE

Phormium tenax
NEW ZEALAND FLAX This evergreen, clump-forming perennial produces arching, sword-shaped, grey-green leaves and, during hot summers, dark red flowers on top of tall stems. Grow it as a focal plant.
↕3m (10ft) ↔2m (6ft)

PERENNIAL MEDIUM

Potentilla thurberi 'Monarch's Velvet'
CINQUEFOIL A spreading perennial, with hairy, strawberry-like, green leaves and abundant, small, saucer-shaped, dark velvety red summer flowers. Combine with spiky plants, such as *Agastache*.
↕75cm (30in) ↔60cm (24in)

BULB MEDIUM

Roscoea cautleyoides
A compact perennial, which produces narrow, dark green leaves and short spikes of orchid-like, yellow, purple, or white flowers in summer. Plant it in groups towards the front of a gravel bed in a sheltered site.
↕25cm (10in)

PERENNIAL SMALL

Scabiosa lucida
GLOSSY SCABIOUS A clump-forming, low-growing perennial, with lance-shaped, silver-grey leaves and a profusion of button-shaped, lilac-pink flowers in summer. Deadhead regularly to prolong the display. This plant can be short-lived.
↕20cm (8in) ↔15cm (6in)

PERENNIAL MEDIUM

Schizostylis coccinea
CRIMSON FLAG LILY An upright perennial, with sword-shaped, light green leaves and red, pink, or white starry flowers from late summer to late autumn. Protect against winter frost; plant in large groups. 'Sunrise' AGM (above) has salmon-pink flowers.
↕60cm (24in) ↔30cm (12in)

PERENNIAL SMALL

Sedum erythrostictum 'Frosty Morn'
This clump-forming perennial forms a mound of fleshy, grey-green leaves, marked with bold white edges and pale pink flowerheads in late summer. A useful plant for a long season of interest.
↕30cm (12in) ↔45cm (18in)

PERENNIAL MEDIUM

Sedum telephium Atropurpureum Group

ORPINE A clump-forming perennial, with dark purple stems and fleshy leaves, and slightly domed, pinkish white flower clusters in late summer and autumn. Combine with green-leaved perennials at the front of beds.
↕60cm (24in) ↔30cm (12in)

PERENNIAL SMALL

Sempervivum arachnoideum

COBWEB HOUSELEEK An evergreen, low-growing perennial, with rosettes of fleshy, green leaves covered with fine, cobweb-like, white hairs, and reddish pink summer flowers. Grow at the front of a gravel bed.
↕12cm (5in) ↔10cm (4in) or more

PERENNIAL SMALL

Silene schafta

AUTUMN CATCHFLY A spreading perennial, it has narrow, dark green leaves and, from late summer to late autumn, it produces simple, star-like, rose-magenta flowers. Plant it at the front of a gravel bed where it can spread.
↕25cm (10in) ↔30cm (12in)

PERENNIAL MEDIUM

Stipa tenuissima

MEXICAN FEATHER GRASS A deciduous, perennial grass, with fine, green leaves and silvery green flowerheads in early summer that turn beige as seeds form. Weave this wispy grass through colourful perennials to create a naturalistic effect.
↕60cm (24in) ↔40cm (16in)

PERENNIAL MEDIUM

Sisyrinchium striatum

This evergreen, upright perennial produces sword-shaped, grey-green leaves and spikes of small, pale yellow flowers throughout summer. Combine it with bearded iris and plant it in drifts in a bed, or use it to line a walkway.
↕60cm (24in) ↔30cm (12in)

PERENNIAL SMALL

Stachys byzantina

LAMBS' EARS An evergreen perennial, with downy, silvery grey foliage and spikes of small, mauve-pink flowers in summer. Grow it at the front of a sunny bed, or use it to line a gravel pathway.
↕38cm (15in) ↔60cm (24in)

BULB MEDIUM

Tulipa clusiana var. *chrysantha*

This perennial bulb has linear, grey-green leaves. In spring, it produces vase-shaped, yellow flowers with a red flash on the outer petals. Plant the bulbs in groups in autumn.
↕30cm (12in)

PERENNIAL LARGE

Verbena bonariensis

An upright perennial, with branching stems of lance-shaped, dark green leaves and domed clusters of scented, purple flowers from midsummer to autumn. Weave it through gravel beds; the slim stems create a see-through veil. Self-seeds freely.
↕1.5m (5ft) ↔60cm (24in)

OTHER SUGGESTIONS

Annuals

Crepis rubra • *Gomphrena* 'Strawberry Fields'

Perennials

Aquilegia caerulea • *Echinacea angustifolia* • *Erigeron* 'Charity' • *Kniphofia* 'Percy's Pride' • *Phlomis tuberosa* • *Sedum* 'Ruby Glow' • *Sisyrinchium* 'E.K. Balls' • *Vernonia lettermannii* 'Iron Butterfly'

Grasses

Cortaderia selloana 'Silver Comet' • *Imperata cylindrica* 'Rubra'

Bulbs

Allium acuminatum • *Allium sphaerocephalon* • *Gladiolus nana* • *Nerine undulata* • *Ornithogalum magnum*

Shrubs

Ceanothus americanus • *Hebe* 'Pink Elephant' • *Hypericum kalmianum*

Plant focus: sedums

Fleshy-stemmed sedums make beautiful rock, roof, and container garden plants, while taller species are perfect for borders.

SEDUMS ARE TOUGH, DROUGHT-RESISTANT SUCCULENTS that have a range of uses in the garden. Taller forms are perfect for gravel gardens or sunny borders, while hardy, low-growing species suit life up on the roof, forming a dense, insulating carpet of foliage that can tolerate sun, wind, and sub-zero temperatures. Mat-forming sedums can also be used in rock gardens, cracks in walls and paving, or in troughs filled with gritty compost. The fleshy foliage, variegated or purple in some cultivars, is highly decorative, as are the heads of star-shaped flowers, which appear from midsummer to autumn. Forms of *S. spectabile* and other large named cultivars are also valued for their ornamental seedheads, which persist through winter. Sedums thrive in full sun, although some will tolerate light shade, and are happiest in free-draining sandy soil – the stems will rot in wet conditions. Divide the larger forms every few years to prevent the stems flopping, leaving an unsightly gap in the centre of the clump.

USING SEDUMS

Try the tallest forms of sedum in beds and borders. The dark, fleshy foliage of *S. telephium* cultivars creates an exciting contrast with variegated and green linear-leaved plants in a border, while the flat-topped flowers of *S.* 'Herbstfreude', a form of *S. spectabile*, add colour to late-season displays, and are followed by graphic seedheads.

The leafy *S. reflexum* and the tender, white-flowered *S. sedoides* make perfect partners for other succulents in a container.

Sedums are practical plants for roofs, requiring the thinnest layer of soil, little fertilizer, and no irrigation once established.

POPULAR SEDUM SPECIES

Sedum acre Creeping stonecrop is an evergreen perennial that forms a dense mat of fleshy, green or variegated foliage and star-shaped, yellow flowers in summer or autumn.

Sedum spathulifolium A low-growing evergreen perennial; most of its cultivars have silver or purple foliage and contrasting yellow summer flowers. It is effective in pots.

Sedum reflexum (syn. *S. rupestre*) This low-growing plant has spreading stems of blue-grey, needle-like foliage that resembles conifer leaves, and yellow summer flowers.

Sedum kamtschaticum A reliable roof plant, producing mounds of semi-evergreen foliage that turns pinkish-red in autumn. It bears yellow flowers in late summer.

Sedum spectabile This perennial features stout stems of fleshy leaves and flat heads of small pink, red, or white flowers from late summer, followed by decorative seedheads.

Sedum telephium This species has knee-high stems of deciduous green leaves, dark purple in some forms, teamed with flat heads of pinkish-purple or white flowers from late summer.

Plants for rock gardens

Rock gardens have undergone a renaissance in recent years as gardeners rediscover the beauty of the diminutive sun-lovers they accommodate.

Rock and alpine plants are easy to care for, requiring very little annual maintenance once established. Create your rock garden in a sunny site, ideally on a naturally occurring slope, and add plenty of grit to wet clay soils, since excellent drainage is a must for all of these plants. Also include flowers for every season, starting in spring with bulbs, such as narcissi and anemones, together with perennial aubretias and arabis. Continue the show with summer-flowering dianthus and helianthemums, and end the year with colchicums and winter heather.

PERENNIAL SMALL

Anemone sylvestris

SNOWDROP WINDFLOWER A dwarf perennial, with divided, mid-green leaves. Fragrant, white flowers with yellow centres appear from spring to early summer. Plant it at the bottom of a sloping rock garden where the soil is moist.

↕↔30cm (12in)

💧 ❄❄❄ (!)

PERENNIAL SMALL

Arabis alpina subsp. *caucasica*

ROCK CRESS A mat-forming, evergreen perennial, with small, hairy, grey-green leaves. It forms a carpet of small, white flowers from early spring to early summer. 'Variegata' (above) has white-edged leaves.

↕↔15cm (6in)

💧 ❄❄❄

PERENNIAL SMALL

Arenaria tetraquetra

SPANISH SOAPWORT An evergreen, cushion-forming perennial, with tiny, grey-green leaves. Stemless, star-shaped, white flowers appear in late spring. It makes a great partner for dwarf spring bulbs, such as *Muscari* and *Narcissus*.

↕2.5cm (1in) ↔15cm (6in) or more

💧 ❄❄❄

PERENNIAL SMALL

Armeria maritima

THRIFT An evergreen, clump-forming perennial that forms rounded tufts of grassy, green leaves. Round heads of small, white to pink flowers appear on slim stems during summer. Suitable as a "dot" plant in a rock garden and in alpine planters.

↕10cm (4in) ↔15cm (6in)

💧 ❄❄❄

PERENNIAL SMALL

Asperula arcadiensis 🏆

ARCADIAN WOODRUFF A cushion-forming perennial, with tiny, hairy, grey leaves. Many tubular, pale pink flowers appear in early summer. Add well-rotted compost to the soil in the planting area to improve moisture retention in summer.

↕8cm (3in) ↔30cm (12in)

💧 ❄❄❄

PERENNIAL SMALL

Aubrieta deltoidea

AUBRETIA An evergreen perennial, with spreading stems of rounded, grey-green leaves. In spring, it becomes covered with tiny blooms in shades of white, pink, red, and purple. Plant it at the edge of a bed where it will slowly spread.

↕5cm (2in) ↔60cm (24in)

💧 ❄❄❄

PERENNIAL SMALL

Calamintha nepeta

LESSER CALAMINT This dwarf, upright perennial has soft, aromatic, dark green basal leaves. From late summer to early autumn, it produces spikes of tiny, lilac-pink flowers that are attractive to bees.

↕45cm (18in) ↔75cm (30in)

💧 ❄❄❄

SHRUB SMALL

Calluna vulgaris

SCOTCH HEATHER An evergreen, bushy, low-growing shrub, with linear, grey-, yellow-, or bright green leaves and bell-shaped flowers in shades of pink, white, and purple from midsummer to late autumn. Grow in acid soil.

↕60cm (24in) ↔45cm (18in)

💧 ❄❄❄ pH

PERENNIAL SMALL

Campanula carpatica 🏆

TUSSOCK BELLFLOWER This clump-forming perennial forms a dense mat of green leaves and open bell-shaped, violet-blue or white flowers throughout summer. Plant it in rock gardens o- trail over the edge of alpine planters.
↕10cm (4in) ↔30cm (12in)

PERENNIAL SMALL

Ceratostigma plumbaginoides 🏆

BLUE-FLOWERED LEADWORT A low-growing, bushy perennial bearing small, oval, green leaves that turn red in autumn and single, blue flower clusters on branched, reddish green stems from late summer to autumn.
↕45cm (18in) ↔20cm (8in)

BULB MEDIUM

Colchicum autumnale

MEADOW SAFFRON This autumn-flowering perennial bulb has strap-shaped leaves that appear after, or at the same time as, the small, vase-shaped, lavender-pink flowers. Plant it where it is easily visible and not lost amongst other plants.
↕15cm (6in)

BULB SMALL

Crocus 'Snow Bunting' 🏆

This perennial corm has grass-like, dark green leaves marked with white lines. In early spring, fragrant, white flowers with mustard-yellow centres and a faint purple blush on the outer petal surfaces appear. Plant the corms in groups in autumn.
↕7cm (3in)

BULB MEDIUM

Crocus speciosus 🏆

AUTUMN CROCUS This perennial corm produces grassy, silver-striped green leaves, which appear in late autumn, just after the vase-shaped, lilac-blue flowers with darker veins form. Plant the corms in late summer in groups between rocks.
↕15cm (6in)

PERENNIAL SMALL

Dianthus 'Doris' 🏆

GARDEN PINK An evergreen, mound-forming perennial, with linear, g-ey-green leaves and an abundance of fragrant, semi-double, pale pink flowers, 3ach with a salmon-red ring in the centre, which appear throughout summer. Deadhead -egularly.
↕↔25cm (10in)

PERENNIAL SMALL

Diascia barberae

A mat-forming, semi-evergreen perennial, sometimes sold as annual bedding, with narrow, oval, green leaves and bell-shaped flowers in a wide range of colours from summer to autumn. 'Blackthorn Apricot' AGM (above) has apricot-pink flowers.
↕25cm (10in) ↔50cm (20in)

PERENNIAL SMALL

Dodecatheon meadia 🏆

AMERICAN COWSLIP This summer-dormant perennial has rosettes of green basal leaves and magenta-pink flowers with reflexed petals in spring. Plant it where it will not overshadow mat-forming plants. Add organic matter to soil before planting.
↕30cm (12in) ↔15cm (6in)

PERENNIAL SMALL

Dryas octopetala 🏆

MOUNTAIN AVENS An evergreen, prostrate perennial that forms mats of oval, lobed, leathery, dark green leaves. In late spring and early summer, it produces cup-shaped, creamy white flowers, followed by attractive, feathery seeds.
↕6cm (2½in) ↔indefinite

SHRUB SMALL

Erica x *darleyensis*

DARLEY DALE HEATH A dwarf, evergreen mound-forming shrub, with linear, cream-tipped dark green foliage and urn-shaped, white or pink flowers from late winter to early spring. *E.* × *darleyensis* f. *albiflora* 'White Glow' (above) is a popular variety.
↕25cm (10in) ↔50cm (20in)

PERENNIAL SMALL

Erinus alpinus

FAIRY FOXGLOVE A semi-evergreen, rosette-forming perennial, with soft, mid-green foliage. From late spring to summer, it produces short, upright stems of purple, pink, or white flowers. It self-seeds and spreads, so deadhead after flowering.
↕↔8cm (3in)

PERENNIAL SMALL

Erodium x *kolbianum* 'Natasha'

HERON'S BILL This mat-forming, evergreen perennial has deeply cut, silver-green leaves and white flowers, with dark purple blotches and veining, from spring to summer. Plant at the front of a rock garden.
↕10cm (4in) ↔30cm (12in)

PERENNIAL SMALL

Euphorbia myrsinites

SPURGE This evergreen, prostrate perennial produces woody stems clothed with small, pointed, fleshy, grey leaves and clusters of bright yellow-green flowers in spring. Allow it to sprawl across rocky surfaces and remove spent flowers.
↕8cm (3in) ↔20cm (8in)

SHRUB SMALL

Euryops acraeus

MOUNTAIN EURYOPS An evergreen, dome-shaped shrub that produces clusters of linear, silvery blue leaves on short stems and daisy-like, bright yellow flowers from late spring to early summer. It makes a good partner for purple-flowered aubretia.
↕↔30cm (12in)

PERENNIAL SMALL

Gentiana sino-ornata

CHINESE GENTIAN This is a mound-forming perennial, with needle-like, green leaves. Funnel-shaped, deep blue flowers, with white stripes within, appear in autumn. Plant at the front of a rock garden in acid soil. Best in moist but well-drained soil.
↕5cm (2in) ↔30cm (12in)

PERENNIAL SMALL

Geranium cinereum

CRANESBILL A spreading perennial, with round, deeply lobed, grey-green leaves. A succession of cup-shaped, pink or mauve flowers, with deep purple centres and veins, appear from late spring to summer. 'Ballerina' AGM (above) is a popular variety.
↕10cm (4in) ↔30cm (12in)

PERENNIAL SMALL

Gypsophila repens

BABY'S BREATH This is a semi-evergreen, spreading perennial that produces mats of narrow, bluish green leaves, above which tiny, round, pale pink flowers, which age to deep pink, appear over a long period from midsummer. Trim back after flowering.
↕20cm (8in) ↔30cm (12in) or more

SHRUB SMALL

Hebe cupressoides 'Boughton Dome'

A slow-growing, evergreen, dome-shaped shrub, with scale-like, grey-green leaves. Paler, larger young leaves give it a two-toned look in summer. Flowers rarely, but is a good backdrop to colourful alpines and bulbs.
↕30cm (12in) ↔60cm (24in)

PERENNIAL SMALL

Helianthemum 'Fire Dragon'

ROCK ROSE An evergreen perennial that bears trailing stems of soft, grey-green leaves and bright red summer flowers with contrasting yellow eyes. 'Henfield Brilliant' AGM has vivid orange flowers.
↕30cm (12in) ↔45cm (18in)

ANNUAL/BIENNIAL SMALL

Iberis umbellata

CANDYTUFT This upright, bushy annual has lance-shaped, mid-green leaves and domed heads of tiny flowers in shades of pink, red, purple, or white from spring to summer. Use it to fill gaps between permanent plantings.

↕↔ 20cm (8in)

BULB MEDIUM

Ipheion uniflorum

SPRING STARFLOWER This bulb has narrow, grass-like, pale green leaves and fragrant, star-shaped, white, violet, or pale blue flowers from late winter to spring. 'Froyle Mill' AGM (above) has pale mauve flowers. Protect with mulch in cold areas.

↕ 15cm (6in)

BULB MEDIUM

Iris reticulata

DWARF IRIS This perennial bulb has upright, grass-like, green foliage and bears small, blue flowers, with yellow and white markings, during early spring. Plant the bulbs in autumn in small clumps where the plants will be seen easily.

↕ 15cm (6in)

BULB SMALL

Iris unguicularis

ALGERIAN IRIS A dwarf, perennial bulb, with grass-like, green leaves. In late winter, violet, lavender-blue, or purple flowers, with white and yellow markings on the lower petals, appear. *I. unguicularis* subsp. *cretensis* (above) has rich blue flowers.

↕ 10cm (4in) ↔ indefinite

PERENNIAL SMALL

Leontopodium alpinum

EDELWEISS A compact perennial, with linear, woolly, grey-green leaves and small, silvery white, starry flowers in late spring or early summer. Plant it in groups with low-growing, spreading perennials, such as *Oxalis*.

↕↔ 20cm (8in)

ANNUAL/BIENNIAL SMALL

Lobularia maritima

SWEET ALYSSUM This ground-hugging annual produces narrow, mid-green leaves and clusters of fragrant, white flowers from summer to early autumn. 'Snow Crystals' (above) is a compact form, ideal for gaps between rocks.

↕ 25cm (10in) ↔ 35cm (14in)

BULB MEDIUM

Narcissus bulbocodium

HOOP PETTICOAT DAFFODIL A perennial bulb, with grass-like, dark green leaves and unusual funnel-shaped, golden-yellow flowers, surrounded by slim petals. Plant the bulbs in autumn; ensure the soil is moist in winter and spring.

↕ 15cm (6in)

BULB MEDIUM

Narcissus 'Pipit'

JONQUIL PIPIT This perennial bulb bears narrow, cylindrical, green leaves. It produces fragrant, lemon-yellow flowers with creamy white cups in mid-spring. Plant the bulbs in groups in autumn.

↕ 25cm (10in)

PERENNIAL SMALL

Oenothera fruticosa 'Fyrverkeri'

EVENING PRIMROSE An upright perennial, with lance-shaped, bronze-green leaves on purple-tinted stems and, from summer to early autumn, evening-fragrant, cup-shaped flowers. Plant at the back of a rock garden.

↕↔ 38cm (15in)

PERENNIAL SMALL

Oxalis adenophylla

SAUER KLEE This low-growing perennial forms a mat of small, distinctive, umbrella-like, grey-green leaves. In spring, it bears rounded, purplish pink flowers with darker purple eyes. Don't allow it to be swamped by neighbouring plants.

↕5cm (2in) ↔10cm (4in)

PERENNIAL MEDIUM

Parahebe perfoliata

DIGGER'S SPEEDWELL An evergreen, woody perennial, with willowy stems, oval, blue-green leaves, and branching sprays of small, blue flowers in late summer. Plant at the back of a rock garden or in a gravel bed. Cut back each spring.

↕60cm (24in) ↔45cm (18in)

PERENNIAL SMALL

Phlox subulata

An evergreen, spreading perennial, with trailing stems of needle-like, green foliage. In summer, it is covered with pink, white, or mauve flowers, forming a mound of colour. 'Emerald Cushion' (above) is a pink-flowered variety.

↕15cm (6in) ↔50cm (20in)

SHRUB SMALL

Picea pungens 'Montgomery'

DWARF BLUE SPRUCE This compact, slow-growing, evergreen conifer makes a mound of needle-like, grey-blue leaves. Ideal for year-round colour, use it as a foil for flowering plants and bulbs.

↕↔60cm (24in)

SHRUB SMALL

Pinus mugo 'Ophir'

MUGO PINE A dwarf, slow-growing, rounded, evergreen conifer, with needle-like, golden-yellow foliage and small cones on mature plants. Use it for year-round colour, and as a foil for flowering plants and dwarf spring bulbs.

↕↔60cm (24in)

PERENNIAL SMALL

Pulsatilla vulgaris

PASQUE FLOWER This clump-forming perennial produces feathery, light green leaves, silky when young. In spring, it bears nodding, cup-shaped flowers in shades of purple, red, pink, or white, followed by silky seedheads.

↕↔23cm (9in)

PERENNIAL SMALL

Rhodanthemum hosmariense

MORROCAN DAISY An evergreen perennial, it forms a dense carpet of finely cut, silvery green leaves. White, daisy-like flowers appear on short stems from late spring to early autumn. Deadhead after flowering.

↕15cm (6in) or more ↔30cm (12in)

PERENNIAL SMALL

Saxifraga Southside Seedling Group

SAXIFRAGE An evergreen, rosette-forming perennial, with pale green leaves that die back after flowering. From late spring to early summer, it bears clusters of open cup-shaped, red-centred white flowers.

↕30cm (12in) ↔20cm (8in)

BULB MEDIUM

Scilla siberica ♡

SIBERIAN SQUILL This perennial bulb has strap-shaped, mid-green leaves and nodding, bell-shaped, blue flowers in spring. Plant the bulbs in groups in autumn, together with other dwarf, spring-flowering bulbs, such as miniature narcissus.

↕20cm (8in)

PERENNIAL SMALL

Scutellaria baicalensis

SKULLCAP An upright perennial, with small, hairy, mid-green leaves and spikes of hooded, purple and white flowers in late summer, followed by bronze-purple seed pods. The blooms are attractive to bees.

↕↔30cm (12in)

PERENNIAL SMALL

Sedum acre

BITING STONECROP This mat-forming, evergreen perennial has tiny, triangular, yellow-green leaves with white tips and, in summer, clusters of star-shaped, yellow-green flowers. Good companion for spring bulbs. It is invasive but easily controlled.

↕5cm (2in) ↔indefinite

PERENNIAL SMALL

Sedum 'Vera Jameson' ♡

STONECROP A spreading perennial, with purple stems and fleshy, oval, purple-pink and grey-green leaves. From late summer to early autumn, it produces rounded clusters of star-shaped, rose-pink flowers.

↕30cm (12in) ↔45cm (18in)

PERENNIAL SMALL

Sempervivum giuseppii ♡

HOUSELEEK This evergreen, ground-hugging perennial produces rosettes of spiky, green leaves with purple-pointed tips. In summer, clusters of star-shaped, deep pink or red flowers form on upright stems.

↕in flower 10cm (4in) ↔10cm (4in)

PERENNIAL SMALL

Sempervivum tectorum ♡

COMMON HOUSELEEK A mat-forming, evergreen perennial, with fleshy rosettes of red-flushed green leaves, and short stems of star-shaped, reddish purple flowers in summer. Plant at the front of a bed, away from spreading plants.

↕15cm (6in) ↔20cm (8in)

PERENNIAL SMALL

Stokesia laevis

STOKES ASTER An evergreen perennial, with slender, green leaves and large, cornflower-like, purple-blue flowers from summer to mid-autumn. Plant to give height to a rock garden. 'Purple Parasols' (above) has violet-purple blooms.

↕↔45cm (18in)

BULB SMALL

Tulipa biflora

TULIP This dwarf, perennial bulb has grey-green leaves. In mid-spring, it bears yellow-centred, white flowers that are flushed greenish pink on the outside. Plant bulbs in late autumn in groups or drifts for the best effect.

↕10cm (4in)

BULB MEDIUM

Tulipa 'Madame Lefèber'

TULIP This perennial bulb has tapered, oval, grey-green leaves and large, bright red, bowl-shaped flowers, which appear from early- to mid-spring. Plant the bulbs in groups in late autumn together with other spring-flowering plants.

↕40cm (16in)

OTHER SUGGESTIONS

Annuals

Omphalodes linifolia ♡

Perennials

Anthyllis vulneraria • *Arabis alpina* • *Arenaria purpurascens* • *Asperula suberosa* • *Cerastium tomentosum* • *Phlox* x *procumbens* • *Ruellia humilis* • *Saxifraga sempervivum* • *Scutellaria incana* • *Silene acaulis* • *Sisyrinchium angustifolium* 'Lucerne'

Grasses

Pennisetum alopecuroides 'Little Bunny'

Bulbs

Narcissus 'Hawera' ♡ • *Narcissus* 'Ice Wings' ♡ • *Tulipa* 'Little Beauty' ♡ • *Tulipa bakeri* 'Little Wonder' • *Tulipa sylvestris*

Shrubs

Juniperus squamata • *Pinus heldreichii* 'Smidtii' ♡

Plants for city gardens

Sheltered by a blockade of tall buildings, city gardens are warmer year-round than more rural plots, broadening your choice of plants.

Plants, such as penstemons and alstroemerias, that often suffer in winter sail through unscathed in snug city enclaves. Pollution, rubble-filled soils, and lack of privacy are the main problems for urban gardeners, but these can all be overcome by digging out the debris and strategic planting. Line the boundaries with well-behaved shrubs, such as *Physocarpus* and *Ceanothus*, to shield your space from prying eyes, and fill flower beds with tough, colourful plants, or create an elegant monotone scheme. In tiny plots, restrict your palette to just a few species, which will produce a more cohesive design.

SHRUB LARGE

Abutilon x *suntense*

INDIAN MALLOW This upright, deciduous shrub has narrow, lobed, felted, grey-green foliage. From late spring to early summer, it produces large, bowl-shaped, pale to deep purple flowers. Grow it close to a south-facing wall.

↕4m (12ft) ↔2.5m (8ft)

PERENNIAL LARGE

Acanthus mollis

BEAR'S BREECHES An upright perennial, with large, deeply lobed, dark green foliage and tall, sturdy spikes of hooded white and pale purple flowers from mid- to late summer. The decorative leaves extend the season of interest.

↕1.2m (4ft) ↔60cm (24in)

TREE LARGE

Acer rubrum 'October Glory'

A deciduous, spreading tree, with glossy, dark green leaves that turn bright red in autumn, particularly on neutral to acid soil. In spring, the bare branches are covered with clusters of tiny, red flowers.

↕20m (70ft) ↔10m (30ft)

BULB MEDIUM

Alstroemeria ligtu hybrids

PERUVIAN LILY An upright, rhizome-forming, perennial, with lance-shaped, grey-green leaves and funnel-shaped flowers in a range of showy colours that appear in summer and are ideal for cutting. Plant in groups in a sheltered spot; mulch in winter.

↕60cm (24in)

TREE SMALL

Amelanchier laevis

ALLEGHENY SERVICEBERRY A small, upright, deciduous tree, with purple-tinted young leaves that mature to dark green, then turn orange-red in autumn. The white flowers in spring are followed by round, edible berries in summer.

↕↔8m (25ft)

SHRUB LARGE

Amelanchier lamarckii

SNOWY MESPILUS This small, upright, deciduous tree or large shrub has bronze-tinted young leaves, dark green when mature, and orange and red in autumn. White flowers in spring are followed by red to purple berries that are loved by birds.

↕5m (15ft) ↔6m (20ft)

PERENNIAL MEDIUM

Artemisia ludoviciana 'Silver Queen'

WESTERN MUGWORT An upright perennial, with slim, toothed-edged, silver-grey leaves. Plumes of small, yellow flowers may form in summer, but remove to maintain the foliage effect. Spreads quickly, but easy to control.

↕↔75cm (30in)

SHRUB LARGE

Buddleja globosa

ORANGE BALL TREE This large, semi-evergreen shrub produces lance-shaped, dark green leaves and spherical, scented, bright orange flowerheads in early summer. A reliable background plant for a border, it will also thrive beside a fence.

↕↔5m (15ft)

SHRUB LARGE

Callistemon citrinus 'Splendens'

CRIMSON BOTTLEBRUSH An evergreen, upright shrub, with arching stems, lance-shaped, glossy, green leaves and, in summer, spikes of bottlebrush, crimson flowerheads. Grow against a wall; protect from frost.

↕↔ 3m (10ft)

TREE LARGE

Catalpa bignonioides

INDIAN BEAN TREE This deciduous, spreading tree has broad, oval, pale green leaves and white flowers in spring that form bean-like seed pods. Cut back the branches hard in late winter to keep compact. Pollard trees to encourage larger leaves.

↕↔ up to 15m (50ft)

SHRUB MEDIUM

Ceanothus x *delileanus* 'Gloire de Versailles'

CALIFORNIAN LILAC This vigorous, bushy, deciduous shrub has oval, mid-green leaves and powdery blue, branching flowerheads from midsummer to early autumn. Grow in a sheltered position; deadhead regularly.

↕↔ 1.5m (5ft)

SHRUB MEDIUM

Ceanothus 'Pershore Zanzibar'

CALIFORNIAN LILAC An evergreen shrub, with oval, dark green and greenish yellow variegated leaves. From late spring to early summer, it bears fluffy, oval, blue flowerheads. Grow in a sheltered site.

↕↔ 2.5m (8ft)

SHRUB LARGE

Ceanothus thyrsiflorus var. *repens*

CREEPING BLUE BLOSSOM A spreading, low, evergreen shrub, with oval, dark green leaves and rounded clusters of light blue flowers from late spring to early summer. Let the stems cascade over a raised bed or wall.

↕↔ 6m (20ft)

SHRUB LARGE

Chionanthus virginicus

FRINGE TREE A vase-shaped, deciduous shrub or small tree, with peeling bark and oval, green leaves, white beneath. Bears white, scented, spidery flowers in summer. Good for year-round interest in a small garden; needs long summers to flower well.

↕↔ 3m (10ft)

PERENNIAL LARGE

Crambe cordifolia

GREATER SEA KALE This robust perennial forms a mound of large, crinkled, lobed leaves. In summer, it produces a cloud of small, fragrant, white flowers in branching sprays. It needs space to be seen at its best.

↕ 2m (6ft) ↔ 1.2m (4ft)

SHRUB SMALL

Cryptomeria japonica 'Globosa Nana'

JAPANESE CEDAR An evergreen, dome-shaped, dwarf conifer, with scaly, green leaves. Makes a great specimen for winter interest in a border or large container, or as a green foil for colourful blooms.

↕↔ 1m (3ft)

TREE LARGE

Davidia involucrata

DOVE TREE This deciduous tree produces oval, green leaves and small, reddish brown spring flowers, which are surrounded by petal-like, creamy white bracts. Best in larger gardens; use it as a specimen in a lawn or border.

↕ 15m (50ft) ↔ 10m (30ft)

PERENNIAL SMALL

Delosperma nubigenum

YELLOW ICE PLANT An evergreen perennial, with trailing stems of small, succulent, pale green leaves. In summer, it produces daisy-like, bright lemon-yellow flowers that attract butterflies. An excellent subject for trailing over raised beds.

↕ 5cm (2in) ↔ 50cm (20in)

SHRUB LARGE

Elaeagnus x *ebbingei* 🏆

This vigorous, evergreen shrub has oval, tapered, dark green leaves, dusted with grey flecks. It makes a good screen or windbreak, and should be clipped in late summer. Variegated varieties include 'Gilt Edge', which has yellow-splashed leaves.

↕↔ 5m (15ft)

💧 ❄❄❄

SHRUB LARGE

Fargesia murieliae 🏆

MURIEL BAMBOO This large, clump-forming bamboo produces arching, yellow-green canes and lance-shaped, bright green leaves. Use it as a screen to divide a garden, or as an accent plant in a border or gravel bed.

↕ 4m (12ft) ↔ indefinite

💧 ❄❄❄

ANNUAL/BIENNIAL SMALL

Gazania Kiss Series

TREASURE FLOWER A dwarf annual, with oval, dark green leaves. From summer to autumn, it produces daisy-like flowers in shades of gold, bronze, pink, and white, with contrasting eyes. Grow it in shallow pots or in raised beds.

↕ 30cm (12in) ↔ 25cm (10in)

💧 ❄

PERENNIAL SMALL

Geranium himalayense 'Plenum'

CRANESBILL A clump-forming, compact perennial, with lobed, mid-green foliage, red in autumn, and double, blue-purple flowers, with darker veins, in summer. Use it to line a path or as ground cover.

↕ 25cm (10in) ↔ 60cm (24in)

💧 ❄❄❄

PERENNIAL MEDIUM

Geranium pratense

MEADOW CRANESBILL This clump-forming perennial has hairy stems and divided, lobed, mid-green leaves, bronze in autumn, and saucer-shaped, violet-blue, veined summer flowers. 'Mrs Kendall Clark' AGM (above) has pearl-grey, pale-pink flushed flowers.

↕↔ 60cm (24in)

💧 ❄❄❄

PERENNIAL SMALL

Geranium sanguineum 'Album'

BLOODY CRANESBILL This clump-forming, spreading perennial has deeply cut, dark green leaves, which turn red in autumn. White flowers appear in early summer. Use it to line a pathway or as ground cover.

↕ 30cm (12in) ↔ 40cm (16in)

💧 ❄❄❄

PERENNIAL SMALL

Gerbera 'Mount Rushmore'

FLORIST GERBERA This upright perennial has oval, lobed, green basal leaves and daisy-like, pink, yellow, or orange flowers from summer to autumn. Plant it in a large container, or use it to add colour to a border.

↕↔ 40cm (16in)

💧 ❄

PERENNIAL SMALL

Geum 'Borisii'

This clump-forming perennial produces round, divided, mid-green leaves and saucer-shaped, bright orange-red flowers on top of slender stems from late spring to early summer. Plant it in groups at the front of a bed or border.

↕ 50cm (20in) ↔ 30cm (12in)

💧 ❄❄❄

TREE LARGE

Ginkgo biloba

MAIDENHAIR TREE A deciduous, conical-shaped tree, with fan-shaped, pale green leaves that turn yellow in autumn. Female plants produce yellow fruits in autumn. Best in larger gardens; use it as a specimen in a lawn or border.

↕ 30m (100ft) ↔ 8m (25ft)

💧 ❄❄❄

TREE MEDIUM

Gleditsia triacanthos 'Sunburst'

HONEY LOCUST A deciduous tree, with large, golden-yellow leaves, divided into leaflets, which turn green in summer and yellow in autumn. With its long season of interest, it looks pretty in a small garden.
↕12m (40ft) ↔10m (30ft)

PERENNIAL LARGE

Lupinus 'The Chatelaine'

LUPIN An upright perennial, with round leaves, divided into lance-shaped leaflets. Tall spikes of dark pink and white flowers appear in early summer, adding colour to the middle of a border. It may need staking. The seeds are toxic.
↕1.2m (4ft) ↔45cm (18in)

SHRUB SMALL

Hypericum calycinum

ROSE OF SHARON A dwarf, semi- or fully evergreen shrub, with dark green leaves. From midsummer to mid-autumn, it bears large, open, bright yellow flowers with a tuft of fluffy stamens in the centre. It can be used for ground cover; can be invasive.
↕60cm (24in) ↔indefinite

PERENNIAL MEDIUM

Leucanthemum x *superbum*

SHASTA DAISY This upright perennial has lance-shaped, dark green leaves and daisy-like, white flowers, with golden-yellow eyes in summer. Grow in groups in the middle of borders The spidery-flowered 'Beauté Nivelloise' (above) is popular.
↕up to 90cm (36in) ↔60cm (24in)

TREE SMALL

Magnolia x *soulangeana*

SAUCER MAGNOLIA A small, spreading, deciduous tree, with oval, dark green leaves and goblet-shaped white, pink, or purple flowers in mid- and late spring. Use it as a feature in a lawn or mixed border; protect flower buds from late frosts.
↕↔6m (20ft)

TREE SMALL

Malus domestica

EATING APPLE A compact tree, when grafted onto a dwarfing rootstock, such as M26. It bears oval, green leaves, pale pink or white flowers in spring, and edible fruits in autumn. Choose from the many varieties available; provide a sheltered site.
↕↔8m (25ft)

TREE MEDIUM

Malus floribunda

JAPANESE CRAB APPLE A deciduous tree, with a round head of oval, green leaves. In spring, crimson buds open to reveal pale pink flowers, followed, in autumn, by edible red and yellow fruits, ideal for making jelly.
↕↔10m (30ft)

PERENNIAL MEDIUM

Malva moschata

MUSK MALLOW An upright perennial, with heart-shaped basal leaves and ferny leaves on the upper stems. Saucer-shaped, pale pink flowers appear from summer to early autumn. Its long flowering period is useful in a small city garden.
↕1m (3ft) ↔60cm (24in)

TREE MEDIUM

Morus nigra

BLACK MULBERRY This tree has a rounded canopy of heart-shaped, green leaves and edible, blackberry-like fruits in autumn. Grow it in a lawn, away from patios, as the fruits will stain paving. Prune from late autumn to midwinter to avoid weeping wounds.
↕12m (40ft) ↔15m (50ft)

PERENNIAL MEDIUM

Papaver orientale

ORIENTAL POPPY An upright perennial, with divided, mid-green leaves. In early summer, it bears large, saucer-shaped blooms with dark eyes. Plant it in groups towards the front of a border. 'Karine' AGM (above) has salmon-pink flowers.
↕↔ up to 90cm (36in)

SHRUB SMALL

Philadelphus 'Belle Etoile'

MOCK ORANGE A deciduous, arching shrub, with small, mid-green leaves. From late spring to early summer, it bears highly fragrant, white flowers, each with a pale purple mark at the base. Plant it near seating to enjoy the fragrance.
↕1.2m (4ft) ↔2.5m (8ft)

SHRUB SMALL

Pleioblastus variegatus

DWARF WHITE-STRIPE BAMBOO
A compact, evergreen bamboo, with pale green canes and broad, creamy white striped green leaves. Plant it in a large container or a border, or use it to line a pathway. It performs best in fertile soil.
↕80cm (32in) ↔ indefinite

SHRUB LARGE

Phyllostachys aureosulcata f. *aureocaulis*

GOLDEN GROOVE BAMBOO A tall, upright, clump-forming bamboo, with yellow canes that are striped at the base, and mid-green, lance-shaped leaves. Good as a boundary screen in a city garden; best in fertile soil.
↕6m (20ft) ↔ indefinite

SHRUB LARGE

Physocarpus opulifolius

This is a fast-growing, deciduous shrub, with lobed, green leaves. In early summer, it produces clusters of small, white blooms. 'Lady in Red' (above) has vivid, dark red leaves, providing a long season of colour.
↕3m (10ft) ↔1.5m (5ft)

TREE MEDIUM

Prunus 'Kanzan'

JAPANESE FLOWERING CHERRY A vase-shaped, deciduous tree producing double, pink spring flowers before the oval, bronze leaves appear. The foliage turns green in summer and orange in autumn. Plant it as a specimen in a lawn or border.
↕↔ 10m (30ft)

PERENNIAL SMALL

Potentilla 'Gibson's Scarlet'

CINQUEFOIL A clump-forming perennial, with lobed, soft green leaves and sprays of small, bright, black-eyed scarlet flowers in summer. It will brighten up the front of a border. Avoid hot, dry sites.
↕45cm (18in) ↔60cm (24in)

TREE SMALL

Prunus dulcis

ALMOND This small, deciduous tree bears bowl-shaped, single or double, pink flowers in spring before the lance-shaped, green leaves appear. Plant it in a border, or as a specimen in a lawn. Edible nuts form in autumn following a warm summer.
↕↔ 8m (25ft)

TREE LARGE

Quercus palustris

PIN OAK A pyramid-shaped, deciduous tree, with oval, deeply lobed, green leaves that turn bright red in autumn. A tall tree, it is best planted at the end of a large town garden and used as a focal point or to provide privacy.
↕20m (70ft) ↔12m (40ft)

SHRUB LARGE

Rhus typhina 'Dissecta'

STAG'S HORN A spreading shrub or small, deciduous tree, with velvety stems and large, dissected, bright green leaves that turn yellow and red in autumn. Female plants have large, maroon, bud-shaped fruit clusters. The suckers can be invasive.

↕↔3m (10ft)

SHRUB MEDIUM

Rosa CRAZY FOR YOU

This floribunda-type rose has dark green, glossy leaves. From summer to autumn, it produces cupped, semi-double, red-splashed cream flowers, with a light, fruity perfume. It is a disease-resistant variety.

↕1.5m (5ft) ↔1.1m (3½ft)

SHRUB SMALL

Rosa RHAPSODY IN BLUE

This modern shrub rose produces disease-resistant, green leaves and clusters of scented, semi-double, purple flowers, which fade to slate-blue, from summer to early autumn. Plant it in a mixed border or large container.

↕1.2m (4ft) ↔90cm (36in)

SHRUB SMALL

Santolina pinnata

COTTON LAVENDER This bushy, evergreen shrub produces slender, finely toothed, grey-green foliage and a profusion of pompon, pale lemon-yellow flowers in summer. *S. pinnata* subsp. *neapolitana* 'Sulphurea' (above) is a popular variety.

↕75cm (30in) ↔1m (3ft)

PERENNIAL LARGE

Solidago 'Goldkind'

An upright, clump-forming perennial, it has slender, pointed green leaves. From mid- to late summer, it produces branching heads of golden-yellow blooms. Deadhead to prolong the display; plants may need staking.

↕1.2m (4ft) ↔1m (3ft)

SHRUB LARGE

Syringa vulgaris

COMMON LILAC A vigorous, deciduous shrub, with mid-green, heart-shaped foliage. In late spring, it produces large, rounded clusters of highly fragrant flowers in shades of purple, pink, and white. Choose compact varieties for smaller plots.

↕↔7m (22ft)

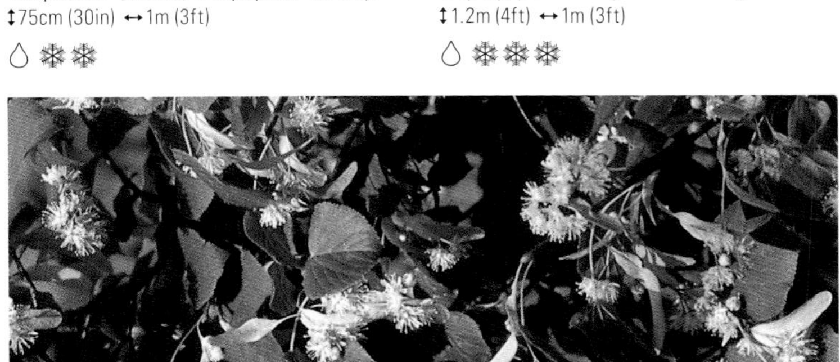

TREE LARGE

Tilia cordata

SMALL-LEAVED LIME This deciduous tree has heart-shaped, glossy, dark green leaves that turn yellow in autumn. Small, yellowish white flowers appear in summer. Keep compact by pleaching or grow it as a formal hedge.

↕30m (100ft) ↔12m (40ft)

PERENNIAL SMALL

Veronica spicata

SPEEDWELL An upright, clump-forming perennial, with lance-shaped, toothed-edged, green leaves, and spikes of dark pink flowers in summer. Grow it in groups at the front of a mixed border with other summer-flowering perennials.

↕↔50cm (20in)

OTHER SUGGESTIONS

Perennials

Delosperma floribunda 'Stardust' • *Macleaya cordata* • *Penstemon* 'Andenken an Friedrich Hahn' • *Salvia microphylla* • *Saponaria officinalis* 'Rosea Plena'

Shrubs

Brachyglottis 'Sunshine' • *Callistemon rigidus* • *Ceanothus arboreus* 'Trewithen Blue' • *Euonymus fortunei* • *Genista lydia* • *Halimium lasianthum* • Hebe 'Blue Chip' • *Nandina domestica* • *Skimmia* x *confusa* • *Yucca flaccida* 'Golden Sword'

Grasses

Muhlenbergia capillaris

Trees

Chionanthus retusus • *Prunus avium* 'Plena' • *Robinia pseudoacacia* 'Bessoniana' • *Styphnolobium japonicum* • *Tilia platyphyllos* 'Rubra'

Plant focus: grasses and sedges

Elegant and versatile, the slender foliage and airy flowers of grasses and sedges contrast beautifully with broad-leaved plants.

ORNAMENTAL GRASSES AND SEDGES comprise a range of shapes and sizes, from tall and slender to short and squat, and form the key ingredient of many garden styles. Those with slim leaves and delicate flowerheads, such as *Calamagrostis*, blend perfectly into naturalistic schemes, while the blue spikes of *Festuca glauca* are ideal for container displays. Sedges, including those in the *Acorus* family, make elegant features in the boggy ground beside natural water features. Although sedges and grasses look similar, they are adapted to different environments. With a few exceptions, grasses are deciduous and prefer sun and free-draining soils, while the majority of sedges are evergreen and happy in sun or shade, and moist soils. Grasses also tend to have more dramatic flower- and seedheads, adding to their decorative value. Both grasses and sedges help to heighten the beauty of winter landscapes: the leaves of deciduous species tend to remain intact throughout the season, providing useful structure in herbaceous borders.

POPULAR GRASS AND SEDGE SPECIES

Acorus Sweet rushes are moisture-loving, evergreen sedges and thrive in shade. They include striped and golden-leaved forms.

Calamagrostis Deciduous feather-reed grasses produce low clumps of leaves teamed with tall stems of feathery summer flowers.

Carex This group of evergreen and deciduous sedges includes moisture-lovers and bronze forms that prefer drier soils. Happy in some shade.

Panicum The most popular form is *P. virgatum*, a deciduous perennial grass with blue-grey or variegated leaves and airy flowerheads.

Pennisetum Long, bristle-like flowerheads and mounds of slender leaves are features of this deciduous grass; some forms are tender.

Stipa The grass *S. gigantea*, with its tall, arching flower stems, and *S. tennuissima*, featuring hair-like flowers, belong to this group.

USING GRASSES AND SEDGES

Grasses are used in prairie-style schemes for their summer foliage and flowers, and winter structure. Use sedges for year-round colour and interest at the front of a bed.

Short, compact grasses, including the vivid *Hakonechloa macra*, create graphic displays in containers. Evergreen sedges are ideal for winter displays.

Tall, slim-leaved grasses, such as *Calamagrostis* x *acutiflora*, make bold accents in mixed herbaceous borders, while their transparency allows welcome glimpses of the plants behind.

Bronze forms of *Carex* and hair-like *Stipa tennuissima* provide contrasting textures and colours to flower displays in both modern and traditional garden designs.

Plants for roof gardens

Bright and breezy, roof gardens demand wind-proof plants that are happy basking in full sun, and thrive in pots or well-drained soil.

Most roof terraces are container gardens, but weight restrictions often apply so opt for lighter pots made from galvanized metal and man-made materials, or try wooden, raised beds and troughs for rustic designs. Also consider investing in an automatic watering system, as the sun and wind will quickly dry out the plants, especially in the height of summer. Despite these limitations, roof terraces offer plenty of scope. Shrubs and small trees provide height, structure, and shade, while roses, perennials, and annuals lend themselves to romantic schemes, and phormiums and grasses complement modern designs.

PERENNIAL SMALL

Agastache aurantiaca

GIANT HYSSOP This upright perennial has grey-green, scented leaves. From mid- to late summer, it bears long-lasting spikes of small, tubular flowers. 'Apricot Sprite' (above) has pale orange blooms. Plant it in groups in a border or container.
↕50cm (20in) ↔1m (3ft)

PERENNIAL MEDIUM

Anthemis tinctoria

DYER'S CHAMOMILE This clump-forming perennial forms a mat of finely cut, green leaves. In summer, it is covered with a mass of daisy-like, white or yellow flowers, held on slender stems. 'E.C. Buxton' (above) has lemon-yellow flowers.
↕↔1m (3ft)

PERENNIAL SMALL

Armeria maritima

THRIFT This mound-forming perennial has grassy, green leaves. In summer, it is dotted with small, round, white or pink flowerheads. Suitable for small containers, raised beds, or on exposed green roofs.
↕10cm (4in) ↔15cm (6in)

SHRUB SMALL

Artemisia 'Powis Castle'

WORMWOOD A semi-evergreen, clump-forming subshrub, with fern-like, silvery grey leaves that make an excellent foil for other plants. Yellow, pompon flowers may appear in summer, but are best removed to retain its silvery foliage effect.
↕↔1m (3ft)

SHRUB SMALL

Buxus sempervirens 'Suffruticosa'

BOX This slow-growing, compact, evergreen shrub has woody stems of small, oval, green leaves, ideal for clipping into architectural shapes and topiary. Grow in a large pot, or use as a low hedge or border edging.
↕1m (3ft) ↔1.5m (5ft)

PERENNIAL LARGE

Calamagrostis acutiflora

FEATHER REED GRASS A deciduous, clump-forming grass, with long, arching, green leaves. Tall, bronze flowerheads borne in summer, fade to buff. The seedheads give winter interest. Use as a screen or backdrop in a raised bed; cut back in early spring.
↕1.8m (6ft) ↔1.2m (4ft)

TREE SMALL

Carpinus betulus

COMMON HORNBEAM This deciduous tree produces oval, prominently veined, dark green leaves that turn yellow and orange in autumn; young stems retain the dried foliage in winter. Cut it into a hedge on a roof terrace.
↕5m (15ft) as a hedge ↔20m (70ft)

SHRUB SMALL

Chamaecyparis obtusa 'Rigid Dwarf'

HINOKI CYPRESS A compact, dwarf, evergreen conifer, it makes a cone of dark green leaves. Use it as a foil for drought-tolerant plants on a roof terrace, or plant in a large pot of soil-based compost.
↕1.2m (4ft) ↔60cm (24in)

PERENNIAL LARGE

Crocosmia masoniorum ♀

MONTBRETIA A clump-forming perennial, with arching, sword-shaped, dark green leaves, pleated lengthways. Orange-red flowers appear in late summer and last for several weeks. Plant in large containers in prime positions in a roof garden.
↕1.5m (5ft) ↔45cm (18in)

SHRUB SMALL

Chamaecyparis pisifera 'Plumosa Compressa'

SAWARA CYPRESS A slow-growing, dwarf, evergreen conifer, with an irregular shape. The light sulphur-yellow foliage turns yellow-green in winter. Plant with small shrubs and bulbs in a large pot of soil-based compost.
↕↔1.2m (4ft)

PERENNIAL SMALL

Erigeron karvinskianus ♀

MEXICAN FLEABANE A spreading, low-growing perennial, with lance-shaped, hairy, green leaves and daisy-like summer flowers that open white, turn pink, and fade to purple. Use it to edge a border or in paving cracks.
↕15cm (6in) ↔indefinite

CLIMBER MEDIUM

Clematis macropetala

DOWNY CLEMATIS A compact climber, with mid-green leaves, divided into leaflets. From late spring to early summer, it is covered with semi-double, mauve-blue flowers, followed by fluffy seedheads. Grow in a pot or through a shrub; keep the roots shaded.
↕3m (10ft)

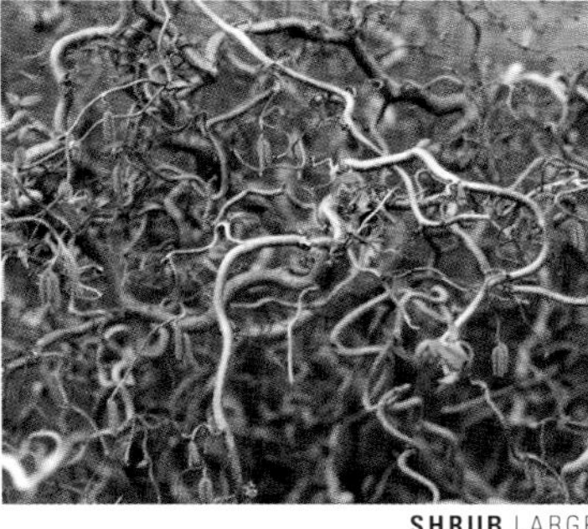

SHRUB LARGE

Corylus avellana 'Contorta' ♀

CORKSCREW HAZEL Deciduous and bushy, with oval, mid-green leaves, this shrub or small tree is grown for its distinctive, contorted stems. In late winter, pale yellow catkins form on bare stems. Grow in a large pot, and use it as a winter focal point.
↕↔5m (15ft)

SHRUB MEDIUM

Fuchsia magellanica var. *molinae*

An upright, deciduous shrub, this hardy fuchsia has small, lance-shaped, green leaves and arching stems of pendent, pale pink flowers from summer to early autumn. Plant in a large pot of soil-based compost.
↕↔2m (6ft)

PERENNIAL SMALL

Hakonechloa macra 'Aureola' ♀

GOLDEN HAKONECHLOA A deciduous, slow-growing grass, with purple stems, green-striped yellow leaves and reddish brown flower spikes from early autumn to winter. Grow in pots or raised beds; water well.
↕40cm (16in) ↔60cm (24in)

SHRUB SMALL

Juniperus procumbens 'Nana' ♀

DWARF JUNIPER This low-growing, mat-forming, evergreen conifer produces dense, prickly leaves that provide good ground cover. Use it to create year-round colour at the edge of a border or raised bed.
↕20cm (8in) ↔75cm (30in)

SHRUB SMALL

Lavandula pedunculata subsp. *pedunculata*

An evergreen shrub, with aromatic, grey-green foliage and small, scented, violet flowers with purple-pink "ears" from late spring to summer. Plant in a pot of soil-based compost, or in a border or raised bed.
↕↔80cm (32in)

BULB MEDIUM

Narcissus 'Ice Follies' ♀

DAFFODIL This is a perennial bulb, with sword-shaped, green leaves. In mid-spring, single, creamy white flowers with yellow cups that fade to near white appear. Plant groups of bulbs in autumn in pots or beds.
↕40cm (16in)

PERENNIAL MEDIUM

Osteospermum 'Buttermilk'
AFRICAN DAISY An evergreen perennial, with slim, mid-green leaves and daisy-like, pale yellow and white summer blooms with bluish purple centres. Plant it in a pot of multi-purpose compost, or along the edge of a path.
↕60cm (24in) ↔30cm (12in)

PERENNIAL SMALL

Pelargonium 'Lady Plymouth' 🏆
SCENTED-LEAVED PELARGONIUM A tender, spreading perennial, grown as an annual, with eucalyptus-scented, lobed, silver-margined green leaves and lavender-pink summer flowers. Grow in pots or raised beds.
↕40cm (16in) ↔20cm (8in)

SHRUB SMALL

Perovskia 'Blue Spire' 🏆
This upright, deciduous subshrub produces white stems of small, aromatic, grey-green leaves and branched spikes of blue-purple flowers from late summer to autumn. Plant it in a mixed bed or in containers of soil-based compost.
↕1.2m (4ft) ↔1m (3ft)

PERENNIAL MEDIUM

Phormium 'Bronze Baby'
NEW ZEALAND FLAX An evergreen, upright perennial, with a fountain-like mound of arching, sword-shaped, purple-bronze leaves. Use it to make a statement in a large container of soil-based compost or in a gravel bed.
↕↔60cm (24in)

SHRUB SMALL

Pinus mugo 'Mops' 🏆
DWARF MOUNTAIN PINE This compact, evergreen conifer produces a mound of needle-like, dark green leaves. Combine it with spring bulbs or other small shrubs in a large container of soil-based compost.
↕up to 1m (3ft) ↔2m (6ft)

SHRUB SMALL

Rosa Flower Carpet Series
These spreading ground-cover roses form a dense mound of disease-resistant, glossy, green leaves. Cupped, semi-double flowers in shades of pink, red, yellow, peach, and white are freely produced from summer to autumn. Plant them in large containers.
↕↔60cm (24in)

SHRUB SMALL

Rosa ICEBERG 🏆
This floribunda bush rose has glossy, green leaves and sprays of pink-tinted buds that open to reveal cupped, fully double, white flowers from summer to early autumn. Grow it along the boundary of a roof garden.
↕75cm (30in) ↔65cm (26in) or more

SHRUB SMALL

Rosa KENT 🏆
A spreading ground-cover rose, with disease-resistant, glossy, mid-green leaves and clusters of flat, semi-double, white flowers from summer to autumn. Allow it to trail over the edge of a large container filled with soil-based compost.
↕80cm (32in) ↔90cm (36in)

SHRUB SMALL

Rosa WARM WISHES 🏆
This bushy, hybrid tea rose has disease-resistant, mid-green leaves. In summer and autumn, it produces pointed, fully double, scented, coral-pink flowers that mature to rose-pink. Plant it in a large pot of soil-based compost, or a border.
↕1m (3ft) ↔80cm (32in)

SHRUB MEDIUM

Rosmarinus officinalis
ROSEMARY An evergreen shrub, with aromatic, needle-like, dark green foliage and small, blue flowers in spring. Grow it in a raised bed with other culinary herbs, or in a container filled with soil-based compost.
↕↔1.5m (5ft)

PERENNIAL SMALL

Salvia greggii Navajo Series

AUTUMN SAGE This woody perennial has oval, mid-green leaves, with clusters of small red, purple, pink, or white flowers from late summer to autumn. Plant it in groups in containers, raised beds, or borders. Protect from hard frosts.

↕↔ 50cm (20in)

PERENNIAL SMALL

Sempervivum arachnoideum

COBWEB HOUSELEEK This low-growing, evergreen perennial forms small rosettes of fleshy, green leaves covered with fine, white webbing and branching heads of reddish pink blooms in summer; good for green roofs.

↕ 12cm (5in) ↔ 10cm (4in) or more

PERENNIAL MEDIUM

Santolina chamaecyparissus

COTTON LAVENDER An evergreen, mound-forming perennial, with finely cut, aromatic, silvery grey leaves and pompon, yellow summer flowers. Use it in a raised bed, or to edge a border or pathway. Trim it back after flowering to keep it neat.

↕ 75cm (30in) ↔ 1m (3ft)

PERENNIAL MEDIUM

Tanacetum coccineum

PYRETHRUM An upright perennial, with feathery, dark green leaves and daisy-like, yellow-centred pink or red flowers in early summer. Plant it in groups in pots, raised beds, or borders. Deadhead to encourage more blooms. This plant can be short-lived.

↕ 60cm (24in) ↔ 45cm (18in) or more

SHRUB LARGE

Sambucus racemosa 'Plumosa Aurea'

RED-BERRIED ELDER An upright, deciduous shrub, with bronze leaflets that mature to golden-yellow in early summer. Scarlet fruits follow star-shaped, yellow spring flowers. Grow in large pots in some shade.

↕↔ 3m (10ft)

PERENNIAL MEDIUM

Sedum 'Herbstfreude'

STONECROP This clump-forming perennial has oval, fleshy, grey-green leaves. It bears small, star-shaped, brick-red, flattened flowerheads in late summer, followed by brown seedheads that persist through winter. Plant it in a border or raised bed.

↕ 60cm (24in) ↔ 50cm (20in)

PERENNIAL SMALL

Sedum rupestre

REFLEXED STONECROP An evergreen, low-growing perennial that forms mats of trailing stems bearing narrow, fleshy leaves. Tiny, bright yellow flowers appear in summer. Use it to trail over raised beds, in shallow containers, or on green roofs.

↕ 20cm (8in) ↔ indefinite

OTHER SUGGESTIONS

Perennials

Coreopsis verticillata 'Moonbeam' • *Diascia* 'Little Dancer' • *Erigeron* 'Dunkelste Aller' • *Eryngium alpinum* • *Eryngium* x *oliverianum* • *Sempervivum ciliosum*

Grasses

Andropogon gerardii • *Festuca glauca* • *Miscanthus sinensis* 'Little Kitten' • *Pennisetum alopecuroides* 'Hameln'

Bulbs

Allium christophii • *Narcissus* 'Cheerfulness'

Shrubs and Climbers

Cordyline australis • *Escallonia* 'Apple Blossom' • *Potentilla fruticosa* 'Abbotswood' • *Pyracantha* 'Teton' • *Rhus typhina* 'Dissecta'

Trees

Olea europaea

Plants for exposed sites

Select resilient plants for gardens where temperatures dip well below freezing in winter, or are buffeted by strong winds or coastal spray.

Trees and shrubs from mountainous regions are ideal for frosty and exposed sites, their names often giving clues to their antecedents – *Alchemilla alpina* and *Geranium himalayense*, for example. Many hardy perennials avoid the worst of the winter weather by disappearing beneath the surface and laying dormant till spring, while hardy evergreens fight the cold with an anti-freeze solution of sugars and amino acids within their cells. Choose a selection of plants, such as snowdrops, that flower in winter to brighten up this bleak time of the year, and use evergreens for permanent structure and colour.

TREE LARGE

Abies concolor

WHITE FIR This evergreen conifer tree is grown for its silver foliage and Christmas tree shape. Plant it as a lawn specimen or part of a mixed, windbreak screen in large gardens. Pruning may spoil the shape.
↕30m (100ft) ↔8m (25ft)

TREE LARGE

Acer negundo

This tree has a rounded head and reddish brown, lobed leaves, which turn mid-green in summer and have striking autumn colour. 'Flamingo' (above) has white and pink variegated leaves. Coppice or pollard to keep it compact.
↕15m (50ft) ↔10m (30ft)

TREE LARGE

Acer saccharinum

SILVER MAPLE A large, lobed, round-headed tree, with dark green foliage that develops a fiery colour in autumn, particularly in colder regions. Use it as a specimen; plant in a sheltered site to avoid wind damage.
↕20m (70ft) ↔12m (40ft)

PERENNIAL SMALL

Alchemilla alpina

ALPINE LADY'S MANTLE This mat-forming perennial has rounded, dark green, divided leaves, with a silver, hairy reverse, and frothy, greenish yellow summer flowers. Hardy and wind-resistant, plant it in a rock garden or wild corner.
↕15cm (6in) ↔60cm (24in) or more

PERENNIAL MEDIUM

Aquilegia vulgaris

GRANNY'S BONNETS A ferny, green-leaved perennial, with wiry, upright stems of bell-shaped, blue, pink, purple, or white flowers from late spring to early summer. Grow in informal and cottage borders in cold sites; weed out unwanted seedlings.
↕60cm (24in) ↔50cm (20in)

SHRUB SMALL

Arctostaphylos uva-ursi

BEARBERRY A creeping, evergreen shrub, with glossy, oval, green leaves, silver beneath. The small, pink or white, urn-shaped flowers appear from spring to early summer, followed by red berries. Cascade it over walls or rock gardens.
↕10cm (4in) ↔50cm (20in)

PERENNIAL SMALL

Armeria maritima

THRIFT A hummock-forming evergreen, with grass-like leaves and stiff stems bearing pompon-shaped, pink flowers from late spring to summer. This hardy plant is salt- and wind-tolerant, ideal for alpine troughs, rock gardens, and raised beds.
↕10cm (4in) ↔15cm (6in)

PERENNIAL SMALL

Aster alpinus

A clump-forming perennial, with mid-green, narrow, lance-shaped leaves. In late summer, it bears large, daisy-like, violet-blue flowers with a yellow central disk. Plant this tough, wind-proof alpine on banks, and in gravel gardens and raised beds.
↕15cm (6in) ↔45cm (18in)

SHRUB LARGE

Berberis darwinii 🏆

DARWIN'S BARBERRY This evergreen shrub has small, glossy, prickly, dark green leaves and, in spring, hanging clusters of dark orange flowers, followed by blue-black berries. It makes intruder-proof windbreaks in informal or coastal gardens.

↕↔ 3m (10ft)

PERENNIAL SMALL

Bergenia cordifolia

ELEPHANT'S EARS This clump-forming evergreen perennial has leathery, mid-green, rounded leaves that tint purple in cold weather. In early spring, red stems bearing clusters of deep pink bells appear. Plant it as weed-suppressing ground cover.

↕↔ 50cm (20in)

TREE LARGE

Betula papyrifera

PAPER BIRCH A tall, sparsely-branched, narrowly conical tree, with white, peeling bark, pale orange beneath. The oval, dark green leaves turn yellow in autumn, and yellow catkins appear in spring. Use multi-stemmed forms as focal points.

↕ 20m (70ft) ↔ 10m (30ft)

TREE LARGE

Betula pendula

SILVER BIRCH A white-barked, deciduous tree, with sparse, upright branches and slender, cascading stems. Catkins appear in spring followed by diamond-shaped leaves, which turn yellow in autumn. It suits exposed, wild gardens.

↕ 25m (80ft) or more ↔ 10m (30ft)

PERENNIAL MEDIUM

Briza media

COMMON QUAKING GRASS This perennial makes a tuft of grassy, blue-green leaves. From late spring to midsummer, upright stems bear green, heart-shaped flowerheads that turn straw-coloured. Plant at the front of a cottage or wildlife border.

↕ 60cm (24in) ↔ 10cm (4in)

SHRUB SMALL

Calluna vulgaris 'Silver Knight'

HEATHER This low-growing evergreen shrub has upright branches covered with grey, scale-like leaves that tinge purple in winter, and lavender flower spikes in late summer. Good for ground cover or border edging.

↕ 30cm (12in) ↔ 40cm (16in)

TREE LARGE

Carpinus betulus 🏆

HORNBEAM This deciduous tree has dark green, oval, toothed leaves, yellow in autumn, with a pleated appearance and hop-like fruits from late summer. 'Fastigiata' AGM (above) is compact and narrowly upright when young, turning flame-shaped.

↕ up to 25m (80ft) ↔ 20m (70ft)

PERENNIAL MEDIUM

Centaurea dealbata

A clump-forming perennial, with grey-green divided leaves. From early- to midsummer, it bears branched stems of white-eyed pink flowers with fringed petals. Most robust on poor soil, this drought-resistant plant suits windy coastal sites; may need staking.

↕ 1m (3ft) ↔ 60cm (24in)

TREE MEDIUM

Chamaecyparis pisifera 'Filifera Aurea' 🏆

SAWARA CYPRESS This slow-growing, broadly conical, evergreen conifer makes a mound of weeping branches that carry golden-yellow, scale-like leaves. Use as a specimen in cold or coastal areas.

↕ 12m (40ft) ↔ 5m (15ft)

SHRUB LARGE

Chionanthus virginicus

FRINGE TREE This large, rounded shrub or small, spreading tree has oval, dark green leaves and, in early summer, pendent, long-petalled, fragrant, white flowers, followed by blue-black fruits. It is cold-tolerant, but needs shelter from strong winds.

↕↔3m (10ft)

PERENNIAL SMALL

Coreopsis 'Rum Punch'

TICKSEED This bushy perennial has narrow, divided leaves and slender stems topped with daisy-like flowers in early summer. It is drought-tolerant and ideal for windy flower borders. Deadhead it regularly.

↕45cm (18in) ↔60cm (24in)

TREE SMALL

Crataegus x *persimilis* 'Prunifolia' ♡

This deciduous tree has thorny branches and glossy, oval, dark green leaves with vibrant autumn colours. Red berries follow the short-lived, white, late spring blossom. Good for wildlife in windy or salt-laden sites.

↕8m (25ft) ↔10m (30ft)

SHRUB LARGE

Cornus alba

DOGWOOD A deciduous, upright shrub, with red young stems and dark green leaves that develop bold autumn tints. Flat heads of white flowers in late spring are followed by white berries. 'Sibirica Variegata' AGM (above) has bold white-edged leaves.

↕↔3m (10ft)

SHRUB LARGE

Corylus avellana

HAZEL A spreading shrub or small tree, this wind-tolerant plant has rounded, dark green leaves that turn yellow in autumn. In early spring, yellow catkins form, and may be followed by hazelnuts. Grow it in a wild or informal garden, or as a hedge.

↕↔5m (15ft)

SHRUB LARGE

Cotinus 'Grace'

SMOKE TREE This large, bushy deciduous shrub has rounded, rich purple-tinged red leaves, becoming flame-red in autumn. Misty, purplish flower plumes appear in summer. Grow it for foliage contrast in a mixed border in a windy, coastal plot.

↕6m (20ft) ↔5m (15ft)

PERENNIAL MEDIUM

Crocosmia 'Lucifer' ♡

MONTBRETIA A clump-forming perennial, with slender, sword-shaped, green foliage. In late summer, it produces branched stems with sprays of bold red, tubular blooms that last several weeks. Wind-tolerant, it is ideal for mild, coastal gardens.

↕1m (3ft) ↔25cm (10in)

SHRUB SMALL

Cytisus x *praecox*

WARMINSTER BROOM A bushy, deciduous shrub, with narrow, sparsely-branched, green stems bearing small leaves. From mid- to late spring, it produces a profusion of pale, creamy yellow, pea-like blooms. 'Warminster' (above) is a popular variety.

↕1.2m (4ft) ↔1.5m (5ft)

PERENNIAL LARGE

Delphinium elatum

This bushy perennials has deeply-cut, dark green leaves and tall spires of blue, pink, cream, or white flowers in summer, depending on the variety. Plant it at the back of borders and support the stems. Most varieties are ideal for cold sites.

↕2m (6ft) ↔60cm (24in)

PERENNIAL SMALL

Dianthus deltoides 'Leuchtfunk'

A cold-tolerant, mat-forming, evergreen perennial, with small, narrow, green leaves and bright cerise-red summer blooms. A good front-of-border plant for seaside, cottage and rock gardens, or gravel areas.
↕20cm (8in) ↔30cm (12in)

PERENNIAL MEDIUM

Dictamnus albus var. *purpureus*

PURPLE-FLOWERED DITTANY A bushy perennial, with glossy, lemon-scented, divided leaves, and butterfly-like, pinky purple, early summer blooms on purple stems. Good for cold sites; prefers rich soil.
↕90cm (36in) ↔60cm (24in)

PERENNIAL LARGE

Echium pininana

TREE ECHIUM This biennial or short-lived perennial has lance-shaped, green leaf rosettes, very tall spires of small, blue blooms and pointed, leafy bracts from mid- to late summer. Ideal for windy, coastal areas; protect in winter in frosty sites.
↕4m (12ft) ↔90cm (36in)

TREE LARGE

Fagus sylvatica

COMMON BEECH This large tree has a broad, spreading crown, and produces oval leaves that are light green in spring, darkening in summer, and turning coppery in autumn. Use it as a hedge in cold areas.
↕25m (80ft) ↔15m (50ft)

SHRUB LARGE

Fargesia nitidia

CHINESE FOUNTAIN BAMBOO A clump-forming, evergreen bamboo, with slender, arching, purple-tinged stems, clothed in narrow, green leaves. It is cold-tolerant and makes a dense hedge or windbreak in fertile, moisture-retentive soil.
↕5m (15ft) ↔1.5m (5ft) or more

PERENNIAL SMALL

Eryngium maritimum

SEA HOLLY This upright perennial has prickly, blue-green basal leaves, with white veins. From early summer to early autumn, it produces metallic blue, thimble-like blooms, each with a collar of spiny, leaf-like bracts. Add it to a seaside shingle garden.
↕50cm (20in) ↔45cm (18in)

SHRUB LARGE

Euonymus hamiltonianus

CHINESE SPINDLE TREE This large, deciduous shrub has lance-shaped, green leaves, with fiery autumn tints. Small summer flowers lead to pendulous, pink fruits that split to reveal orange seeds. Cold-tolerant; prefers shelter from wind.
↕↔8m (25ft)

PERENNIAL MEDIUM

Gaillardia x *grandiflora*

BLANKET FLOWER A bushy perennial, with lance-shaped, or lobed, grey-green leaves and a succession of daisy-like flowers from midsummer to autumn, if deadheaded regularly. 'Kobold' (above) has yellow-tipped red petals. This plant is very cold-tolerant.
↕90cm (36in) ↔45cm (18in)

BULB MEDIUM

Galanthus nivalis 'Flore Pleno'

DOUBLE COMMON SNOWDROP A cold-tolerant, perennial bulb, with grassy, grey-green leaves and fragrant, double, white late winter blooms. Grow under deciduous shrubs, on grassy banks and rock gardens.
↕15cm (6in)

SHRUB SMALL

Genista pilosa

HAIRY GREENWEED A deciduous, wind-tolerant, spreading shrub, with narrow, oval leaves, silky-haired beneath. Bright yellow flowers appear in profusion from late spring to early summer. 'Vancouver Gold' is a prostrate form.
↕↔30cm (12in)

PERENNIAL SMALL

Geranium himalayense 'Plenum'

CRANESBILL A mat-forming perennial, with rounded, lobed, green leaves and bowl-shaped, double, violet-blue summer flowers, with a white eye and dark veining. Very hardy, use it as ground cover.

↕25cm (10in) ↔60cm (24in)

SHRUB LARGE

Griselinia littoralis ♀

NEW ZEALAND BROADLEAF An upright, evergreen shrub, with oval, waxy-textured, apple-green leaves. Inconspicuous flowers on female plants, if fertilized, produce purple autumn fruits. It makes a dense hedge or screen in windy seaside plots.

↕8m (25ft) ↔5m (15ft)

SHRUB SMALL

Juniperus rigida

This prostrate evergreen conifer has needle-like, grey-green leaves, and black fruits with a blue bloom. Tolerant of salt-laden air and drought, use it in low maintenance gardens. *J. rigida* subsp. *conferta* 'Blue Pacific' is a popular choice.

↕30cm (12in) ↔indefinite

SHRUB LARGE

Hippophae rhamnoides ♀

SEA BUCKTHORN This is a deciduous, bushy, arching shrub for windy, coastal sites, with narrow, silvery leaves. In mid-spring, it bears tiny, yellow flowers, which are followed in autumn by bright orange berries on female plants.

↕↔6m (20ft)

TREE SMALL

Juniperus communis

COMMON JUNIPER This cold-tolerant, evergreen conifer's habit varies from prostrate to upright and bushy. It has needle-like, grey-green leaves and blue-tinged black berries. The columnar 'Hibernica' AGM is a popular choice.

↕8m (25ft) ↔4m (12ft)

SHRUB SMALL

Juniperus squamata

FLAKY JUNIPER This evergreen conifer for cold sites varies in habit from prostrate to bushy or upright. It has sharp, scale-like, dark grey-green to blue-green leaves, with black fruits. 'Blue Carpet' AGM (above) has attractive pale, blue-grey foliage.

↕30cm (12in) ↔3m (10ft)

PERENNIAL LARGE

Lavatera maritima ♀

This shrub-like, semi-evergreen perennial has rounded, lobed, grey-green, felted leaves. From late summer, it produces saucer-shaped, pale pinky lilac blooms, with darker veins. Grow it in windy seaside gravel gardens or borders.

↕1.5m (5ft) ↔1m (3ft)

PERENNIAL LARGE

Liatris spicata

GAYFEATHER This upright, cold- and salt-tolerant perennial produces clumps of linear leaves and, from late summer to early autumn, poker-like spikes of fluffy, pinky purple or white flowers. 'Kobold' (above) has vivid purple-pink blooms.

↕1.5m (5ft) ↔45cm (18in)

PERENNIAL MEDIUM

Limonium platyphyllum

This perennial forms a basal rosette of oval, wavy-edged, green foliage. In late summer, airy, branched heads bear tiny, lavender-blue to violet flowers. Grow this salt- and cold-tolerant plant in gravel gardens by the sea or in rock gardens.

↕60cm (24in) ↔45cm (18in)

PERENNIAL SMALL

Linum perenne

PERENNIAL FLAX This clump-forming, perennial has wiry stems, narrow, lance-shaped, blue-grey tinged leaves and, from early- to midsummer, sky-blue flowers that close in late afternoon. Cold-, heat-, and wind-tolerant; it may self-seed.
↕30cm (12in) ↔15cm (6in)

SHRUB MEDIUM

Lupinus arboreus

TREE LUPIN A bushy, semi-evergreen shrub, with divided, grey-green, silky-haired leaves and spires of fragrant, pea-like, pale yellow blooms in late spring and summer. Wind-tolerant; it will naturalize in grass or gravel in mild, coastal gardens.
↕↔2m (6ft)

PERENNIAL MEDIUM

Lupinus 'The Page'

LUPIN This bushy, short-lived perennial forms clumps of divided, dark green leaves with slender "fingers". Tall stems of pink-red, pea-like blooms appear in early summer. It is very cold-tolerant, but needs plenty of moisture in spring.
↕1m (3ft) ↔75cm (30in)

PERENNIAL MEDIUM

Lysimachia clethroides

GOOSENECK LOOSESTRIFE This upright, clump-forming perennial has lance-shaped, mid-green leaves and arching, tapered, grey-white flower clusters from mid- to late summer. Very cold-hardy, grow it in semi-wild areas or towards the back of borders.
↕↔1m (3ft)

PERENNIAL LARGE

Molinia caerulea

PURPLE MOOR-GRASS This grass forms a clump of narrow, mid-green, arching leaves, with rich autumn colour. From summer to autumn, slender stems bear airy, purple flowerheads. Ideal for coastal gardens; plant it *en masse*. Cut back in spring.
↕1.5m (5ft) ↔40cm (16in)

ANNUAL/BIENNIAL SMALL

Malcolmia maritima

VIRGINIAN STOCK A low-growing annual, with slender, branched stems of small, narrow, green leaves. From spring to autumn, it bears masses of pink, white, red, or purple, fragrant blooms. Salt- and wind-tolerant, use it at the front of flowerbeds.
↕20cm (8in) ↔8cm (3in)

PERENNIAL MEDIUM

Malva moschata

MUSK MALLOW A bushy perennial, with woody stems, deeply lobed, dark green leaves, and saucer-shaped, pale pink blooms from summer to autumn. Ideal for exposed sites. White-flowered *M. moschata* f. *alba* AGM is a popular form.
↕1m (3ft) ↔60cm (24in)

SHRUB LARGE

Olearia macrodonta

DAISY BUSH An upright evergreen shrub, with glossy, dark green leaves, silver felted beneath, similar to holly. Masses of small, white, fragrant, daisy-like flowers appear in summer. Use it as a windbreak or hedge in coastal gardens.
↕6m (20ft) ↔5m (15ft)

PERENNIAL MEDIUM

Paeonia officinalis

This bushy perennial has divided, dark green leaves and bowl-shaped flowers, with pink, red, or white petals, from early- to midsummer. 'Rubra Plena' AGM (above) is a double, red form. Cold-tolerant, but it needs shelter from strong winds.
↕↔75cm (30in)

SHRUB LARGE

Philadelphus 'Virginal'
MOCK ORANGE This large, upright shrub has oval, dark green leaves and highly fragrant, double white flowers from early- to midsummer. Very hardy and wind-resistant, this shrub creates shelter in exposed gardens at the back of a border.
↕3m (10ft) ↔2.5m (8ft)

TREE LARGE

Picea breweriana
BREWER'S SPRUCE This large, cone-shaped, evergreen conifer has drooping stems covered with dark, grey-green needles. Female cones are red-brown. Thrives in some shade and sheltered from wind; makes a striking lawn specimen in cold gardens.
↕15m (50ft) ↔4m (12ft)

TREE LARGE

Picea pungens 'Koster'
COLORADO SPRUCE A medium-sized, evergreen conifer with stiff, silvery blue, needle-like foliage and grey, scaly bark. On mature plants, upright cones form towards the ends of the branches. Perfect for a cold garden; plant in neutral to acid soil.
↕15m (50ft) ↔5m (15ft)

TREE MEDIUM

Pseudotsuga menziesii
BLUE DOUGLAS FIR This evergreen conifer has dark green needles, and makes a columnar tree, topped by a spreading crown. Younger plants are compact and conical, and all forms grow well in exposed sites.
↕↔10m (30ft)

TREE LARGE

Quercus robur f. fastigiata
This is a large, deciduous tree, with a flame-shaped profile, lobed, dark green leaves and autumn acorns. Slow-growing and cold-tolerant, use it as lawn specimen in larger gardens. 'Koster' AGM (above) has a columnar habit.
↕↔15m (50ft)

TREE LARGE

Pinus nigra
BLACK PINE When mature, this is a large, domed-headed evergreen tree, with long dark green needles and yellow-brown cones. It can be planted as a windbreak, and will tolerate coastal sites. It is best grown on larger plots.
↕30m (100ft) ↔8m (25ft)

SHRUB SMALL

Potentilla fruticosa
CINQUEFOIL This cold- and wind-tolerant shrub has divided, dark to grey-green leaves. Small, yellow, white, pink, orange, and red flowers open from summer to autumn. Varieties include the yellow-flowered 'Goldfinger' (above).
↕1m (3ft) ↔1.5m (5ft)

SHRUB SMALL

Rosmarinus officinalis
ROSEMARY An aromatic, evergreen, wind-tolerant shrub, with dark green, needle-like leaves and small, blue, pink, or white blooms from spring to early summer. Prostratus Group (above) will cascade over walls.
↕15cm (6in) ↔1.5m (5ft)

PERENNIAL LARGE

Rudbeckia laciniata

CONEFLOWER This upright, very hardy perennial has dark green leaves and tall stems, topped with yellow, daisy-like, greenish yellow, cone-centred flowers from midsummer to autumn. Varieties include 'Goldquelle' (above) with double flowers.
↕2m (6ft) ↔75cm (30in)

TREE LARGE

PERENNIAL LARGE

Sanguisorba canadensis

CANADIAN BURNET This is an upright, clump-forming perennial, with divided green leaves and bottlebrush, white flowerheads from summer to early autumn. Plant in swathes for best effect in moist soil in cold, exposed gardens.
↕2m (6ft) ↔60cm (24in)

PERENNIAL LARGE

PERENNIAL SMALL

Senecio cineraria

CINERARIA This evergreen, bushy subshrub, usually grown as an annual, is good for mild, windy coasts. It has deeply lobed, silver leaves and daisy-like, yellow flowerheads, which are best removed in summer.
↕↔30cm (12in)

Taxus baccata

YEW This bushy, evergreen conifer, variable in habit, has dark green needles, and small, yellow spring flowers, followed by red berries. Use it as formal hedging in cold regions; it can be pruned hard if necessary. All parts are toxic.
↕15m (50ft) ↔10m (30ft)

TREE SMALL

Taxus x *media* 'Hicksii'

HICKS YEW This is an evergreen conifer with flattened, needle-like, dark green leaves. In summer, it produces tiny, white blooms, followed, on female plants, by red berries. Faster growing than *T. baccata*; all parts are poisonous.
↕6m (20ft) ↔4m (12ft)

Thalictrum aquilegiifolium

MEADOW RUE An upright, clump-forming perennial, with divided, blue-green leaves and slender, branched stems, topped with fluffy, mauve-pink or white flowers from early- to midsummer. Grow in coastal or cold gardens in soil that does not dry out.
↕1.2m (4ft) ↔45cm (18in)

TREE LARGE

Thuja occidentalis

WHITE CEDAR A slow-growing, cold-tolerant conifer with orange-brown bark, and flat sprays of scale-like, green leaves, greyish green beneath, smelling of apples when crushed. Ovoid cones are yellow-green, ripening to brown.
↕15m (50ft) ↔5m (15ft)

PERENNIAL MEDIUM

Veronica austriaca subsp. *teucrium*

SAW-LEAVED SPEEDWELL This mound-forming perennial has oval, grey-green leaves and blue summer flower spikes. The deep blue 'Crater Lake Blue' AGM (above) is both hardy and salt-tolerant.
↕↔60cm (24in)

PERENNIAL SMALL

Viola tricolor

HEARTSEASE An annual or short-lived perennial, with oval, serrated, green leaves and, from spring to autumn, a succession of flowers with purple, yellow, and white petals, streaked with dark purple. A hardy self-seeder, it suits coastal gardens.
↕15cm (6in) ↔15cm (6in) or more

OTHER SUGGESTIONS

Annuals
Myosotis sylvatica

Perennials
Achillea millefolium • *Cardamine pratensis* 'Flore Pleno' • *Maianthemum racemosum* syn. *Smilacina racemosa* • *Potentilla nitida* • *Tanacetum vulgare*

Grasses
Miscanthus sinensis 'Zebrinus'

Shrubs
Comptonia peregrina • *Escallonia rubra* • *Euonymus europaeus* 'Red Cascade' • *Genista tinctoria* • *Kerria japonica* • *Viburnum sargentii* 'Onondaga'

Trees
Fraxinus excelsior • *Juniperus conferta* • *Pinus sylvestris* Aurea Group • *Prunus maritima* • *Sorbus aucuparia* • *Tamarix ramosissima* 'Pink Cascade'

GARDENS in SHADE

Woodlanders and large-leaved plants that thrive in shade can transform a gloomy garden into a lush oasis. Few plants tolerate permanent deep shade, but many flourish in partial or dappled shade, where brighter spells boost growth and encourage blossom. Foliage plants, such as ferns and hostas, are the stars of shady schemes, but if you are looking for brightly coloured flowers, opt for rhododendrons, camellias, daffodils, and geraniums, which will sing out against the darkness, acting as beacons of colour.

Plants for clay soil

Cool, shady borders on heavy clay soils may seem unpromising, but they can provide the fertile conditions that many plants enjoy.

Clay-rich soils offer a rich supply of nutrients and moisture for most of the year. There is plenty of choice for those with lively colour schemes in mind: dazzling maples (*Acer*), pink asters, orange crocosmias, and rhododendrons of all hues will be happy in partial shade. Light up the darker areas with the bright blooms of camellias and bergenias, or choose a more subdued palette of hellebores, leafy hostas, and *Glaucidium*, which will also tolerate deep shade. To add an element of surprise to a design, spice up your beds with sweetly scented *Daphne* or sweet box.

TREE LARGE

Acer davidii

SNAKEBARK MAPLE This large, multi-stemmed, deciduous tree has white-streaked green bark and dark purple shoots. The green, oval, tapered leaves turn red in autumn. Hanging clusters of yellow flowers appear in spring. Grow in a sheltered site.
↕↔ 15m (50ft)

SHRUB MEDIUM

Acer palmatum var. *dissectum*

JAPANESE MAPLE This slow-growing shrub has a domed to cascading habit, and produces deep purple, lobed leaves, with finely divided, tapering "fingers". The foliage also sports rich autumn colour. Plant in a sheltered site.
↕ 1.5m (5ft) ↔ 1m (3ft)

PERENNIAL MEDIUM

Aconitum 'Bressingham Spire'

MONKSHOOD A clump-forming perennial, with glossy, green, rounded, and deeply cut leaves. Erect spires of violet-purple flowers appear from midsummer to early autumn. May need staking. Toxic if ingested.
↕ 1m (3ft) ↔ 50cm (20in)

PERENNIAL LARGE

Aconitum napellus

MONKSHOOD An upright perennial, with deeply divided, dark green foliage and tall spires of indigo-blue blooms that appear from mid- to late summer and need staking. Plants tolerate poor drainage. Toxic if ingested.
↕ 1.5m (5ft) ↔ 30cm (12in)

PERENNIAL LARGE

Actaea racemosa

BLACK COHOSH This upright perennial has deeply divided leaves, above which appear bottlebrush-like spikes of white flowers in midsummer. The dried brown seedheads are also attractive. Water well during dry spells.
↕ 1.5m (5ft) ↔ 60cm (24in)

PERENNIAL SMALL

Actaea rubra

RED BANEBERRY A clump-forming perennial, with leaves divided into lobed and toothed-edged leaflets, and oval, fluffy, white flowerheads from mid-spring to early summer, followed by shiny, red, poisonous berries. Water well in dry spells.
↕ 50cm (20in) ↔ 30cm (12in)

PERENNIAL LARGE

Actaea simplex 'Brunette'

BUGBANE An upright perennial, with dark purple divided foliage. From early- to mid-autumn, narrow, arching, bottlebrush flowers appear on slender stems. The tiny, off-white blooms are tinged purple. Plant in drifts; water well during dry spells.
↕ 1.2m (4ft) ↔ 60cm (24in)

PERENNIAL SMALL

Ajuga reptans

BUGLE This is an evergreen, ground-cover perennial, spreading freely by runners. It has small rosettes of glossy, deep bronze-purple leaves. Short spikes of blue flowers appear in spring. Ideal for moist areas, it attracts bumblebees.
↕ 15cm (6in) ↔ 90cm (36in)

PERENNIAL SMALL

Anemanthele lessoniana

NEW ZEALAND WIND GRASS This semi-evergreen grass forms a mound of narrow, arching leaves and purple-tinged flowers from midsummer to early autumn. Summer foliage is green-tinged orange, but the whole plant turns orange-brown in winter.

↕↔ 1m (3ft)

PERENNIAL LARGE

Anemone hupehensis

An upright perennial, with dark green, divided leaves. Large, simple, pink or white flowers with yellow eyes open over a long period from mid- to late summer. 'Hadspen Abundance' AGM (above) has dark pink blooms.

↕ 1.2m (4ft) ↔ 45cm (18in)

PERENNIAL MEDIUM

Anemone rivularis

RIVERSIDE WINDFLOWER A clump-forming perennial, with hairy, dark green, three-lobed leaves. It flowers in late spring and early summer, and occasionally in autumn. The slender branched stems carry 10–20 white blooms, often with a blue reverse.

↕ 60cm (24in) ↔ 30cm (12in)

PERENNIAL SMALL

Aruncus aethusifolius

GOAT'S BEARD A compact, clump-forming, ground-cover perennial, with deeply cut, mid-green, fern-like leaves, which turn yellow in autumn. Sprays of tiny, creamy white blooms appear from early- to midsummer. Water well during dry spells.

↕↔ 40cm (16in)

PERENNIAL LARGE

Aruncus dioicus

GOAT'S BEARD An upright, bushy perennial, with long, mid-green, divided leaves. It flowers from early- to midsummer. In the showier male plants, feathery, creamy white plumes appear above the foliage. Water well during dry spells.

↕ 2m (6ft) ↔ 1.2m (4ft)

PERENNIAL MEDIUM

Aquilegia canadensis

ROCK BELLS A mound-forming perennial, with ferny foliage. Slender, upright flower stems bear nodding, bell-shaped blooms, which are coloured red and pale yellow, between mid-spring and midsummer. Plants spread via self-seeding.

↕ 60cm (24in) ↔ 30cm (12in)

PERENNIAL SMALL

Asarum caudatum

WILD GINGER Evergreen, ground-covering perennial, with rich green, heart-shaped leaves, ginger-scented when crushed, and purple, mid- to late spring blooms with three tapering petals. Plants need rich acid soil and tolerate full shade.

↕ 8cm (3in) ↔ 25cm (10in) or more

PERENNIAL LARGE

Aster cordifolius

BLUE WOOD ASTER This upright, bushy perennial has dark green, oval, toothed-edged leaves. From late summer to autumn, it produces sprays of small, pink-tinged white, daisy-like blooms on arching stems.

↕ 1.2m (4ft) ↔ 1m (3ft)

SHRUB LARGE

Berberis x *stenophylla*

BARBERRY This evergreen shrub has a rounded habit, spiny stems, and small, leathery, dark green, lance-shaped leaves. The light orange-yellow flowers open in late spring from red-tinged buds and are followed by small, blue-black berries.

↕3m (10ft) ↔5m (15ft)

PERENNIAL SMALL

Astilbe x *arendsii*

FALSE GOAT'S BEARD This clump-forming, ferny-leaved perennial produces frothy plumes of white, pink, lilac, or red flowers from early to late summer. Brown seedheads feature in winter. 'Fanal' AGM has red flowers and dark foliage.

↕45cm (18in) ↔30cm (12in)

PERENNIAL MEDIUM

Astrantia major

MASTERWORT A clump-forming perennial, with mid-green, divided leaves and sprays of greenish white, pink, or red flowers from early- to midsummer. 'Sunningdale Variegated' AGM (above) has cream-splashed leaves. Deadhead regularly.

↕90cm (36in) ↔45cm (18in)

PERENNIAL MEDIUM

Bergenia 'Ballawley'

A clump-forming, evergreen perennial, with glossy, rounded, mid-green leaves that turn bronze-purple in winter. During mid- and late spring, clusters of bell-shaped, crimson flowers appear on reddish stems. Best sheltered from wind; remove untidy leaves.

↕↔60cm (24in)

PERENNIAL SMALL

Bergenia cordifolia

ELEPHANT'S EARS A clump-forming to spreading, evergreen perennial, with large, rounded, leathery, dark green leaves that develop purple tints in winter. Stems bear pink or white flowers in early spring. 'Purpurea' (above) is a popular variety.

↕↔50cm (20in)

PERENNIAL SMALL

Bergenia 'Silberlicht'

ELEPHANT'S EARS A clump-forming, evergreen perennial, with leathery, rounded leaves. Red-tinged stems carry clusters of white. bell-shaped spring blooms that sometimes develop pink tinges. Flowers have contrasting deep red bud cases.

↕45cm (18in) ↔50cm (20in)

PERENNIAL MEDIUM

Blechnum spicant

HARD FERN This evergreen fern has a tufted habit and divided, stiff, leathery leaves with narrow leaflets, and forms low, arching hummocks. The upright, spore-bearing fronds resemble fish bones. Trim old foliage in spring. Tolerant of full shade.

↕75cm (30in) ↔45cm (18in)

PERENNIAL SMALL

Brunnera macrophylla

With clumps of heart-shaped, green leaves, this perennial makes low, spreading ground cover. In spring, airy sprays of small, light blue flowers, similar to forget-me-nots, appear above the foliage. 'Dawson's White' (above) has variegated leaves.

↕45cm (18in) ↔60cm (24in)

SHRUB LARGE

Camellia japonica

COMMON CAMELLIA An upright, evergreen shrub, with glossy, rounded, dark green leaves. It bears single or double flowers in shades of white, pink, and red in spring. 'Blood of China' (above) has red, semi-double flowers. Grow in a sheltered site.

↕3m (10ft) ↔2m (6ft)

pH

SHRUB LARGE

Camellia x *williamsii*

CAMELLIA The williamsii cultivars are evergreen shrubs, with leathery, glossy, oval leaves and, single or double, white to deep pink spring flowers. 'Donation' AGM (above) is upright, with semi-double, rose-pink blooms. All forms tolerate full shade.
↕5m (15ft) ↔2.5m (8ft)

BULB LARGE

Cardiocrinum giganteum

GIANT LILY An upright, bulbous perennial, with heart-shaped, green leaves. Trumpet-shaped, white flowers with purple throats top sturdy stems in summer. Large seed pods follow the blooms. Plant in a sheltered site and rich soil.
↕3m (10ft)

PERENNIAL MEDIUM

Chelone obliqua

TURTLEHEAD A bushy, erect perennial, with broadly lance-shaped, dark green, heavily-veined, and toothed-edged leaves. From late summer to early autumn, short but showy spikes of purple-pink blooms appear. Plants can tolerate boggy ground.
↕1m (3ft) ↔50cm (20in)

TREE SMALL

Cornus 'Norman Hadden'

A deciduous, spreading tree, with oval, pointed leaves that turn red-green in autumn. From late spring to early summer, it produces flowers consisting of showy petal-like white bracts that age to pink. Strawberry-like, pink fruits follow.
↕↔8m (25ft)

PERENNIAL SMALL

Corydalis flexuosa

A clump-forming perennial, with finely divided, blue-green leaves, often tinged purple. From spring to early summer, blue, tubular flowers appear on wiry stems; the foliage dies back later in summer. Grow in rich, moisture-retentive soil.
↕30cm (12in) ↔20cm (8in)

PERENNIAL MEDIUM

Crocosmia 'Lucifer'

MONTBRETIA A clump-forming perennial, with narrow, sword-shaped, mid-green foliage. Branched stems holding sprays of vivid red, flared, tubular blooms appear over several weeks in late summer. Best in light shade; deadhead to prevent self-seeding.
↕1m (3ft) ↔25cm (10in)

SHRUB SMALL

Cryptomeria japonica 'Globosa Nana'

JAPANESE CEDAR A dwarf, evergreen conifer with a dense, bushy, rounded habit and green foliage. The scale-like leaves closely overlap on the short, arching, cord-like branchlets. Shelter from wind.
↕↔1m (3ft)

SHRUB SMALL

Deutzia x *elegantissima*

A deciduous rounded shrub, with green leaves and, in most cases, flaking bark. Clusters of star- or cup-shaped fragrant flowers appear from late spring to early summer. 'Rosealind' AGM (above) produces pink-flushed white blooms.
↕1.2m (4ft) ↔1.5m (5ft)

PERENNIAL SMALL

Dicentra cucullaria

DUTCHMAN'S BREECHES A clump-forming perennial, with ferny, blue-green leaves. Curious, twin-spurred, yellow-tipped white flowers appear on arching stems in early spring. Shelter young growth from frost damage. Leaves die back in summer.
↕15cm (6in) ↔30cm (12in)

SHRUB MEDIUM

Enkianthus cernuus f. *rubens* ♀

DROOPING RED ENKIANTHUS A deciduous, bushy shrub, with dense, oval, bright green purple-tinged foliage, turning purple-red in autumn. In late spring, hanging bunches of dark red, fringed, bell-shaped flowers form.

↕↔ 2.5m (8ft)

PERENNIAL MEDIUM

Euphorbia griffithii 'Fireglow'

This spreading perennial makes dense strands of red, upright stems bearing narrow, oblong, copper-tinged dark green leaves that intensify in colour in autumn. In early summer, domed clusters of orange-red flowers appear. Best in fertile soil.

↕ 75cm (30in) ↔ 90cm (36in)

SHRUB SMALL

Gaultheria mucronata

This low, suckering, evergreen shrub carries small, dark green, prickle-tipped leaves. 'Wintertime' AGM (above) is a female form with tiny, white, bell-shaped late-spring flowers. It bears white berries in autumn, if a male variety is grown nearby.

↕↔ 1.2m (4ft)

PERENNIAL MEDIUM

Geranium sylvaticum

WOOD CRANESBILL A perennial that produces clumps of lobed, mid-green leaves and round, blue-purple flowers, with white centres, from late spring to early summer. Grow it in groups in moist soil at the front of a flower border.

↕ 75cm (30in) ↔ 60cm (24in)

PERENNIAL LARGE

Gillenia trifoliata ♀

BOWMAN'S ROOT A spreading perennial, with erect, sparsely-branched, reddish stems bearing three-lobed, dark green, prominently veined leaves. From late spring to late summer, airy sprays of starry, white flowers open from contrasting red buds.

↕ 1.2m (4ft) ↔ 60cm (24in)

PERENNIAL SMALL

Glaucidium palmatum ♀

PALMATE GLAUCIDIUM This clump-forming perennial has maple-like, lobed and cut, light green leaves. Pale pinky lilac, saucer-shaped blooms appear from late spring to early summer. Grow in a sheltered spot in moist, fertile soil. It tolerates full shade.

↕↔ 50cm (20in)

PERENNIAL SMALL

Hacquetia epipactis ♀

This clump-forming perennial has glossy, round, deeply lobed leaves. Tiny, yellow flowers surrounded by greenish yellow, petal-like bracts appear from late winter to early spring. Ensure that this small species isn't overwhelmed by neighbouring plants.

↕ 6cm (2½in) ↔ 23cm (9in)

SHRUB LARGE

Hamamelis mollis

CHINESE WITCH HAZEL A spreading shrub, with broad, dark green leaves, yellow in autumn. Fragrant, yellow, spidery flowers sprout from bare branches in late winter. Plant where its scent can be enjoyed. Some varieties have orange blooms.

↕↔ 4m (12ft) or more

PERENNIAL SMALL

Helleborus x *ericsmithii*

HELLEBORE This clump-forming, semi-evergreen perennial has toothed-edged, silver-veined green leaves and pink-tinged or white, saucer-shaped blooms, with green-striped petals on reddish green stems in midwinter. Trim old leaves in autumn.

↕38cm (15in) ↔45cm (18in)

PERENNIAL MEDIUM

Helleborus x *hybridus*

LENTEN ROSE This clump-forming, semi-evergreen perennial has dark green, lobed leaves and bears single or double, plain or speckled, white, green, yellow, pink, purple, or red flowers from midwinter to mid-spring. Remove old leaves in autumn.

↕↔60cm (24in)

PERENNIAL SMALL

Hosta 'Fire and Ice'

PLANTAIN LILY A clump-forming perennial, bearing mounds of heart-shaped, ribbed, dark green foliage, with a bold white splash in the centre of each leaf. Tall spires of lilac blooms appear in summer; cut them back after flowering. Protect from slugs.

↕40cm (16in) ↔indefinite

PERENNIAL SMALL

Helleborus niger

CHRISTMAS ROSE This semi-evergreen perennial forms clumps of lobed, dark green leaves. White, ageing to pinkish, saucer-shaped blooms with yellow stamens appear from midwinter to early spring. Tolerates heavy clay mixed with organic matter.

↕↔30cm (12in)

PERENNIAL SMALL

Hepatica nobilis

LIVERLEAF This slow-growing, semi-evergreen perennial forms a neat dome of rounded, lobed, fleshy, green leaves. In early spring, it bears cup-shaped blooms in shades of violet or purple. Plant it at the front of a shady border.

↕8cm (3in) ↔12cm (5in)

SHRUB MEDIUM

Hydrangea arborescens

A large, rounded shrub, with green, tapered leaves. In midsummer it bears flattened clusters of white flowers, which persist to provide interest in winter. Varieties include 'Annabelle' AGM (above), with extra large flowerheads.

↕↔1.5m (5ft)

SHRUB MEDIUM

Hydrangea macrophylla

LACECAP HYDRANGEA A rounded shrub, with large, oval, light green leaves. Tiny, blue flowers and larger, pale blue to pink, petal-like florets form lace-cap flowerheads from midsummer to early autumn. The dried flowers provide winter interest.

↕2m (6ft) ↔2.5m (8ft)

SHRUB LARGE

Hydrangea paniculata

PANICULATE HYDRANGEA A broadly upright shrub, with oval, pointed, glossy, mid- to dark green leaves, often deeply veined. In late summer, cone-shaped white or pink flowerheads develop. 'Phantom' AGM has extra large, white blooms.

↕↔3m (10ft)

SHRUB MEDIUM

Hydrangea quercifolia

OAK-LEAVED HYDRANGEA This mound-forming shrub has lobed, mid-green leaves, with striking red and purple autumn colour. From midsummer to autumn, it bears cone-shaped, cream flowerheads that turn pink-tinged white. Prefers neutral to acid soil.
↕2m (6ft) ↔2.5m (8ft)

PERENNIAL MEDIUM

Inula hookeri

This clump-forming perennial has upright stems, with soft hairs that bear abundant lance-shaped, mid-green, deeply veined leaves and large, yellow, daisy-like blooms with long narrow petals from late summer to autumn. Water during dry spells.
↕75cm (30in) ↔45cm (18in)

PERENNIAL SMALL

Jeffersonia dubia

A woodland perennial, with lobed, kidney-shaped or rounded, blue-green leaves, often initially tinged purple. From late spring to early summer, wiry stems hold lavender-blue, cup-shaped blooms above the emerging leaves. Tolerant of full shade.
↕15cm (6in) ↔23cm (9in)

BULB MEDIUM

Leucojum aestivum

SUMMER SNOWFLAKE This perennial bulb makes an upright clump of strap-shaped, narrow, glossy leaves. In spring, leafless stems bear small clusters of bell-shaped, drooping, white blooms with green tips. 'Gravetye Giant' AGM (above) is popular.
↕60cm (24in)

SHRUB MEDIUM

Leucothöe fontanesiana

DOG HOBBLE This evergreen shrub forms a mound of arching, red-tinged branches, clothed in glossy, lance-shaped leaves that are mottled and streaked with cream, pink, and maroon. In spring, clusters of cream flowers hang below the branches.
↕up to 1.5m (5ft) ↔2m (6ft)

PERENNIAL MEDIUM

Lysimachia punctata

GARDEN LOOSESTRIFE This vigorous, upright, tall perennial with mid-green leaves spreads rapidly to form large clumps. In summer, it produces spikes of bright yellow flowers. Best suited to larger plots, it can be invasive.
↕1m (3ft) ↔60cm (24in)

SHRUB LARGE

Magnolia x *soulangeana*

SAUCER MAGNOLIA A large, upright shrub, with a bushy habit and oval, dark green leaves. From mid- to late spring, as the leaves unfurl, goblet-shaped, pink to violet-purple or white blooms appear. 'Rustica Rubra' (above) has purple-red blooms.
↕↔6m (20ft)

PERENNIAL MEDIUM

Maianthemum racemosum

FALSE SPIKENARD This is an arching perennial, with oval, light green leaves that turn yellow in autumn. Feathery sprays of white flowers appear from late spring to early summer, followed by fleshy, red fruits.
↕90cm (36in) ↔60cm (24in)

PERENNIAL MEDIUM

Onoclea sensibilis

SENSITIVE FERN This deciduous, creeping, ground cover fern has deeply lobed, toothed-edged, light green fronds. Stiff, brown fronds stand erect in late summer, lasting into winter. Requires neutral to acid soil. Water if dry in summer; can be invasive.
↕60cm (24in) ↔1m (3ft)

PERENNIAL SMALL

Phlox divaricata subsp. *laphamii*

A semi-evergreen, creeping perennial, with narrow, lance-shaped, mid-green leaves on purplish, upright stems. From mid- to late spring, small, lavender-blue blooms appear. 'Chattahoochee' AGM is popular.
↕30cm (12in) ↔20cm (8in)

SHRUB LARGE

Pieris japonica

LILY OF THE VALLEY BUSH This evergreen, compact shrub has glossy, oval, dark green leaves, often brightly coloured at first. Small, creamy white, flask-shaped blooms hang in tassels in spring. The popular 'Mountain Fire' AGM has red young leaves.
↕4m (12ft) ↔3m (10ft)

SHRUB MEDIUM

Skimmia japonica

A compact evergreen shrub, with glossy, oval to lance-shaped, dark green leaves. Male plants have attractive, cone-shaped, green or red bud clusters through winter, opening to fragrant white flowers in spring. Females (above) produce red berries.
↕↔1.5m (5ft)

PERENNIAL MEDIUM

Tradescantia Andersoniana Group

SPIDERWORT This clump-forming perennial has upright stems of long, narrow, arching leaves, and pink, blue, purple, or white flower clusters. 'Concord Grape' (above) has violet-purple blooms and bluish green leaves.
↕↔60cm (24in)

TREE LARGE

Picea pungens 'Koster'

COLORADO SPRUCE This evergreen, cone-shaped conifer produces stiff, needle-like, bright silvery blue foliage. Upright cones form towards the ends of the branches. It requires neutral to acid soil, shelter, and prefers only light shade.
↕15m (50ft) ↔5m (15ft)

PERENNIAL MEDIUM

Polemonium caeruleum

JACOB'S LADDER A perennial that forms clumps of divided leaves comprising narrow, lance-shaped, green leaflets. Upright stems bears clusters of open, bell-shaped, white or lavender-blue early summer blooms. May self-seed, becoming invasive.
↕↔60cm (24in)

PERENNIAL MEDIUM

Tricyrtis formosana

TOAD LILY This slow-spreading, upright, clump-forming perennial has lance-shaped, mid-green leaves. It bears upturned clusters of intricately-shaped, purple-spotted white blooms in autumn. Mulch annually with compost. It tolerates full shade.
↕1m (3ft) ↔45cm (18in)

SHRUB SMALL

Sarcococca confusa ♀

SWEET BOX This bushy, evergreen shrub has glossy, oval to lance-shaped, dark green leaves. Tiny, white, sweetly fragrant, winter blooms are followed by shiny black fruits. Plant where the scent can be enjoyed. Tolerates deep shade.
↕↔1m (3ft)

OTHER SUGGESTIONS

Perennials

Anemone leveillei • *Aquilegia flabellata* f. *alba* var. *pumila* ♀ • *Asarum europaeum* ♀ • *Aster pilosus* var. *pringlei* 'Monte Cassino' ♀ • *Bergenia* 'Bressingham White' ♀ • *Cardamine pentaphylla* ♀ • *Disporum sessile* • *Euphorbia* 'Excalibur' • *Euphorbia schillingii* ♀ • *Hemerocallis fulva* 'Flore Pleno' • *Silene dioica* 'Firefly'

Shrubs

Calycanthus floridus • *Corylopsis glabrescens* • *Daphne pontica* • *Rhododendron williamsianum* ♀ • *Viburnum davidii* ♀ • *Viburnum tinus* 'Eve Price' ♀

Grasses

Carex elata 'Aurea' ♀ • *Miscanthus sinensis* 'Kleine Fontäne' ♀

Trees

Betula albosinensis 'Burkii' • *Carpinus caroliniana*

Plant focus: hostas

Invaluable in cool, shady sites, the dramatic, sculptural leaves of hostas add colour and texture to many areas of the garden.

THEIR DRAMATIC FOLIAGE and love of cool, damp conditions make hostas popular plants for shady gardens. Plant them as ground cover at the edge of tree canopies, include them in a bog garden beside a pond, or use colourful forms to create architectural container displays. Most hostas are knee-high, although taller and dwarf forms are also available, and they are grown for their dramatic blue, golden, or green leaves, which can be a solid colour or variegated. The foliage is joined in summer by tall spires of lily-like, often fragrant, flowers. Although fairly undemanding and tolerant of periods of drought, all hostas suffer to some degree from slug and snail damage, with thin-leaved forms being the worst affected. To prevent shredded foliage, protect plants by covering tender new growth with horticultural grit and sand, or surround mature plants with copper rings. Also look out for slug-resistant cultivars, such as 'Krossa Regal', 'Great Expectations', 'Halcyon', and 'Big Daddy'.

POPULAR HOSTA FORMS

Large-leaved hostas With leaves growing up to 40cm (16in) long, popular choices include *H. crispula*, *H.* 'Sagae', and 'Sum and Substance'.

Dwarf forms Less than 45cm (18in) in height, dwarf hostas include *H. venusta*, *H.* 'Dorset Blue', 'Hadspen Heron', and 'Pilgrim'.

Variegated forms Large or small, these colourful and variable plants feature white or golden margins, or pale or darker central markings.

Golden-leaved forms Requiring some sunlight, the golden-leaved forms include a range of plant sizes and foliage shapes. Popular forms include 'Yellow River', 'Yellow Splash', 'Piedmont Gold', 'Golden Prayers', 'Gold Drop', *H. fortunei* var. *aureomarginata*, and *H. albopicta*.

Blue-leaved forms Grown for its blue, textured foliage, this group ranges from the large-leaved 'Blue Angel' to the dainty 'Blue Moon'.

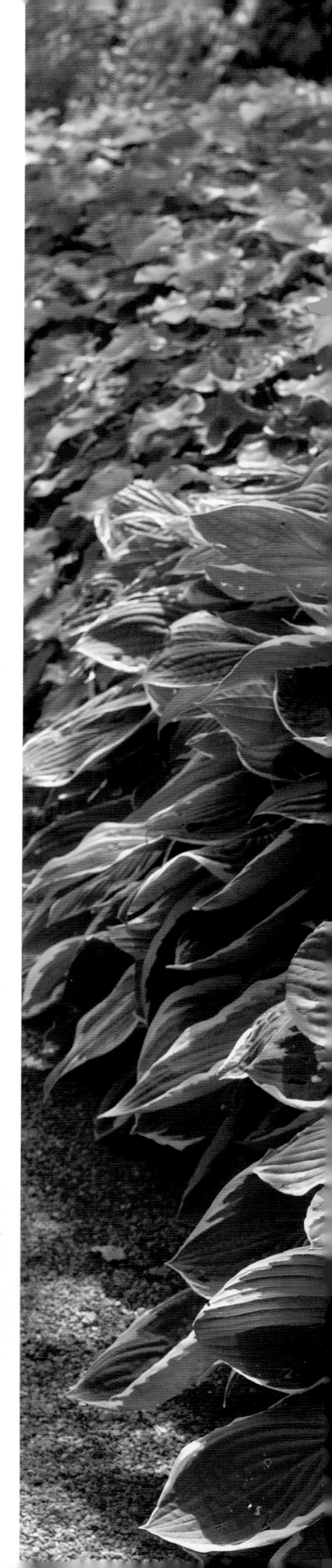

USING HOSTAS

Dense, overlapping hosta leaves make useful ground cover, and these plants thrive in shade, where bright variegated forms add a splash of colour. Ideal for planting at the edge of woodland or beneath larger shrubs.

Pots of hostas of contrasting sizes, leaf colours, and shapes add a cool, elegant note to a shady patio display.

Bright, variegated hostas, such as *H.* 'Patriot', combine well with astilbes, ferns, and alchemilla in shady beds and borders in a cottage garden.

Hostas thrive beside ponds and streams and although shade-lovers, they will tolerate more sun where the soil is reliably damp in summer.

Plants for sandy soil

Free-draining sandy sites are generally more suited to sun-lovers, but there are a good range of plants that will cope with these conditions in shade.

Most plants prefer dappled shade or sun for part of the day, but you can try experimenting with hardy geraniums, epimediums and *Alchemilla mollis* in darker areas. Also consider painting walls and fences white or cream to reflect light into the garden to broaden your scope, and trim off the lower branches of shrubs and trees to allow more sun to filter through. Choose from a selection of shrubs, such as viburnums, hazels (*Corylus*), and *Philadelphus* for background structure, and seasonal bulbs and perennials to inject some colour into your garden.

PERENNIAL SMALL

Alchemilla mollis

LADY'S MANTLE This perennial forms spreading clumps of rounded, velvety, pale green leaves, with crinkled edges. It produces small sprays of tiny, acid greenish yellow flowers in midsummer. Use it as ground cover or border edging.
↕↔ 50cm (20in)

BULB LARGE

Allium stipitatum

ORNAMENTAL ONION This upright, perennial bulb produces strap-shaped, grey-green leaves that wither as tall, sturdy stems topped with spherical, white flowerheads appear in late spring. 'Violet Beauty' has violet-pink flowerheads.
↕ 1.4m (4½ft)

TREE SMALL

Amelanchier arborea

SNOWY MESPILUS This is a small, deciduous tree, with a dense, oval habit, and copper-red young leaves that mature to green, and then turn vivid red in autumn. It also produces masses of white flowers in spring that lead to attractive red berries.
↕ 8m (25ft) ↔ 12m (40ft)

SHRUB LARGE

Amelanchier canadensis

SNOWY MESPILUS This deciduous, upright, dense shrub has oval, white-haired leaves that age to dark green and turn orange-red in autumn. It produces starry, white flowers from mid- to late spring, and edible, maroon, sweet and juicy fruits in summer.
↕ 6m (20ft) ↔ 3m (10ft)

TREE SMALL

Amelanchier laevis

ALLEGHENY SERVICEBERRY A deciduous, spreading tree or large shrub, with oval, bronze leaves that turn dark green in summer, then red and orange in autumn. It produces sprays of white flowers in spring, followed by round, juicy, red fruits.
↕↔ 8m (25ft)

SHRUB LARGE

Amelanchier lamarckii

SNOWY MESPILUS This deciduous, spreading shrub's bronze new leaves and airy sprays of starry, white flowers open together from mid- to late spring. The mature foliage is dark green, but becomes brilliant red and orange in autumn.
↕ 5m (15ft) ↔ 6m (20ft)

PERENNIAL MEDIUM

Anaphalis margaritacea

PEARL EVERLASTING A bushy perennial, with lance-shaped, white-edged grey-green or silvery grey leaves, felted on the undersides. In late summer, papery, pearl-like, white flower clusters form in profusion on erect stems. Plant in light shade.
↕ 75cm (30in) ↔ 60cm (24in)

PERENNIAL SMALL

Anemone nemorosa

WOOD ANEMONE A vigorous, carpeting perennial, with deeply divided, mid-green leaves. From spring to early summer, masses of starry, white flowers appear. Varieties include 'Vestal' AGM, which has double flowers with button-like centres.
↕ 15cm (6in) ↔ 30cm (12in)

PERENNIAL MEDIUM

Aquilegia vulgaris

GRANNY'S BONNETS A clump-forming, upright perennial, with grey-green leaves, divided into rounded leaflets. Pink, crimson, purple, blue, and white, bell-shaped flowers appear in early summer. 'William Guiness' (above) has maroon and white flowers.
↕ 1m (3ft) ↔ 50cm (20in)

SHRUB LARGE

Aronia x *prunifolia*

PURPLE CHOKEBERRY A deciduous, upright shrub, with oval, dark green leaves that turn bright red in autumn. Edible red or purple-black berries, packed with antioxidants, follow the white, spring flowers. 'Brilliant' (above) is a popular variety.
↕ 3m (10ft) ↔ 2.5m (8ft)

PERENNIAL SMALL

Bergenia cordifolia

ELEPHANT'S EARS An evergreen, clump-forming perennial, with large, rounded, leathery, green leaves, tinted purple in winter, and spikes of cup-shaped, light pink flowers in spring. 'Purpurea' (above) has bright magenta-pink flowers on red stems.
↕↔ 50cm (20in)

SHRUB LARGE

Buddleja alternifolia

BUTTERFLY BUSH This deciduous, arching shrub has slender, drooping stems and narrow, grey-green leaves. Long clusters of fragrant, lilac-purple flowers clothe the pendent branches in early summer. It can be trained as a weeping tree.
↕↔ 4m (12ft)

❄❄❄

PERENNIAL MEDIUM

Campanula persicifolia

An upright perennial, with rosettes of narrow, lance-shaped, bright green leaves. Nodding, papery, bell-shaped flowers in white or blue appear in summer. 'Chettle Charm' AGM (above) has large, white flowers with violet-blue tinted edges.
↕ 60cm (24in) ↔ 30cm (12in)

❄❄❄

SHRUB LARGE

Corylus avellana

HAZEL This large shrub or small tree has a spreading habit, and bears rounded, green leaves that turn yellow in autumn. In spring, it produces dangling, yellow catkins, which may be followed by edible nuts in autumn. Plant it to encourage wildlife.
↕↔ 5m (15ft)

❄❄❄

PERENNIAL MEDIUM

Dictamnus albus

BURNING BUSH An upright, clump-forming perennial, with glossy, divided, light green leaves. From late spring to early summer, it produces fragrant, white or pink flowers, followed by ornamental, star-shaped seedheads in autumn.
↕ 90cm (36in) ↔ 60cm (24in)

PERENNIAL LARGE

Crocosmia x *crocosmiiflora*

MONTBRETIA This perennial corm forms clumps of upright, sword-shaped, pleated, mid-green leaves. Branched stems bear large, trumpet-shaped, yellow, orange, or red flowers from mid- to late summer. 'George Davidson' (above) is a popular form.
↕ 1.2m (4ft)

❄❄

PERENNIAL LARGE

Deschampsia cespitosa

TUFTED HAIR GRASS A deciduous, tuft-forming perennial that produces clouds of tiny, golden-yellow flowers on long stems in summer. Both the seedheads and foliage turn golden-yellow in autumn. Varieties include 'Goldtau' (above).
↕ up to 2m (6ft) ↔ 50cm (20in)

❄❄❄

PERENNIAL SMALL

Dicentra formosa

BLEEDING HEART A spreading perennial, with fern-like, grey-green leaves. From late spring to early summer, slender, arching stems bearing nodding, heart-shaped, pink or dusky red flowers appear. Plant it in groups in borders or cottage gardens.
↕ 45cm (18in) ↔ 30cm (12in)

❄❄❄

PERENNIAL SMALL

Epimedium x *rubrum*

RED BARRENWORT A carpeting perennial, with abundant, heart-shaped, mid-green leaves that flush brownish red in spring and autumn, and last into winter. Clusters of cup-shaped, crimson flowers with yellow spurs appear in spring on wiry stems.
↕ 30cm (12in) ↔ 20cm (8in)

❄❄❄

PERENNIAL SMALL

Epimedium x *warleyense*

BARRENWORT This carpeting, evergreen perennial has heart-shaped, light green leaves that are tinged purple-red in spring and autumn. Cup-shaped, coppery orange flower clusters appear on wiry stems in spring. It makes a good ground-cover plant.
↕↔ 30cm (12in)

❄❄❄

PERENNIAL MEDIUM

Erysimum 'Bowles's Mauve'

PERENNIAL WALLFLOWER This evergreen, border perennial has narrowly oval, grey-green leaves. From late winter to summer, spikes of small, mauve flowers open from dark purple buds. Deadhead regularly.
↕ 60cm (24in) ↔ 40cm (16in)

❄❄❄

SHRUB MEDIUM

Escallonia 'Apple Blossom' ♡

A compact, evergreen shrub, with small, glossy, leathery, dark green leaves. From early- to midsummer, it produces a profusion of pink flowers on leafy stems. It makes a beautiful flowering hedge or back-of-border shrub.

↕↔2.5m (8ft)

PERENNIAL MEDIUM

Euphorbia amygdaloides var. *robbiae*

MRS ROBB'S BONNET An evergreen, spreading perennial, with rosettes of shiny, leathery, oblong, dark green leaves and, in spring, rounded heads of small, lime-green flowers and bracts. May be invasive.

↕↔60cm (24in)

SHRUB MEDIUM

Exochorda x *macrantha*

PEARL BUSH This deciduous shrub forms a dense mound of arching stems covered with dark green foliage, which turns yellow and orange in autumn. A profusion of white flowers appears in late spring and early summer.

↕2m (6ft) ↔3m (10ft)

SHRUB LARGE

Forsythia x *intermedia*

A vigorous, upright, deciduous shrub that produces long stems covered with small, star-shaped, yellow flowers from late winter to mid-spring before the green leaves appear. 'Lynwood Variety' AGM (above) has larger, bright yellow blooms.

↕↔3m (10ft)

SHRUB MEDIUM

Fuchsia magellanica ♡

A deciduous, upright shrub, with oval to tapering, mid-green leaves. Throughout summer, it produces an abundance of small, pendent, tubular, red and purple flowers, followed by black fruits. Use it in a wildlife border.

↕↔2m (6ft)

SHRUB SMALL

Gaultheria mucronata

An evergreen, bushy shrub, with prickly, glossy, dark green leaves and tiny, white flowers from late spring to early summer. Only female plants bear sprays of showy, long-lasting berries. 'Wintertime' AGM (above) has large, white berries.

↕↔1.2m (4ft)

SHRUB SMALL

Gaultheria procumbens

CHECKERBERRY An evergreen shrub, with oval, leathery, aromatic leaves that flush red in winter. Small, bell-shaped, pink-flushed white flowers appear in summer, followed by red berries. 'Very Berry' (above) has bright scarlet berries.

↕15cm (6in) ↔indefinite

PERENNIAL SMALL

Geranium 'Ann Folkard' ♡

CRANESBILL A spreading perennial, with swathes of deeply dissected, lobed, yellowish green leaves that mature to green. In midsummer and sometimes in autumn, it bears saucer-shaped, magenta flowers with black centres and veins.

↕50cm (20in) ↔1m (3ft)

PERENNIAL SMALL

Geranium macrorrhizum

CRANESBILL A semi-evergreen, carpeting perennial, with soft, rounded, divided, aromatic leaves, which take on bright autumn hues. Open, magenta flowers appear in early summer. Makes decorative ground cover next to hedges and walls.

↕38cm (15in) ↔60cm (24in)

PERENNIAL SMALL

Geranium x *oxonianum*

CRANESBILL A clump-forming, evergreen perennial, with divided, toothed, green leaves. Rounded, dark-veined, pink flowers with notched petals appear from late spring to midsummer. 'Wargrave Pink' AGM (above) has pale salmon-pink blooms.

↕45cm (18in) ↔60cm (24in)

PERENNIAL MEDIUM

Geranium pratense

MEADOW CRANESBILL A clump-forming perennial, with deeply lobed and divided green leaves that develop autumnal tints. In summer, saucer-shaped, violet-blue flowers with attractive veins appear. 'Mrs Kendall Clark' AGM (above) is a popular variety.

↕↔60cm (24in)

PERENNIAL SMALL

Geranium wallichianum

CRANESBILL A perennial with divided, toothed, white-marbled mid-green leaves. Saucer-shaped, lilac or pink-purple, veined flowers appear from midsummer to autumn. 'Buxton's Variety' AGM (above) has white-centred, light violet-blue flowers.

↕45cm (18in) ↔90cm (36in)

SHRUB LARGE

Hamamelis x *intermedia*

A vase-shaped shrub, with broadly oval leaves that turn yellow in autumn. Lightly scented, spidery flowers appear on bare stems from early- to midwinter. Varieties include 'Jelena' AGM (above), with large, coppery orange flowers.

↕↔4m (12ft)

PERENNIAL MEDIUM

Helleborus argutifolius

HOLLY-LEAVED HELLEBORE This clump-forming, evergreen perennial has leathery, divided, spiny-edged, dark green leaves. Clusters of nodding, bowl-shaped, pale green flowers appear from late winter to spring. Grow it in a lightly shaded border.

↕60cm (24in) ↔45cm (18in)

SHRUB SMALL

Hypericum 'Hidcote'

ST JOHN'S WORT This tough, evergreen or semi-evergreen shrub forms a dense bush of narrowly oval, dark green leaves. Masses of saucer-shaped, golden-yellow flowers appear from midsummer to early autumn. Site it mid-border in part shade.

↕1.2m (4ft) ↔1.5m (5ft)

PERENNIAL MEDIUM

Iris foetidissima

STINKING IRIS This evergreen, rhizome-forming perennial has glossy, strap-shaped, green leaves and yellow-tinged dull purple or yellow flowers from early- to midsummer. It is grown for its scarlet, berry-like fruits that appear in winter from long seed pods.

↕90cm (36in) ↔indefinite

PERENNIAL MEDIUM

Lamprocapnos spectabilis

syn. ***Dicentra spectabilis*** This perennial has fern-like, mid-green foliage. From late spring to early summer, heart-shaped, rose-red and white flowers hang from wiry, arching stems. Plant in small groups in a cottage garden or informal border.

↕↔1m (3ft)

PERENNIAL SMALL

Lathyrus vernus

SPRING VETCHLING This perennial forms a dense, sprawling clump, and has soft, pointed, dark green leaves. In spring, it bears small, sweet pea-like, red-veined purple and blue flowers. Easy to grow from seed, it is best at the front of a border.

↕↔30cm (12in)

PERENNIAL MEDIUM

Leucanthemum x *superbum*

SHASTA DAISY This clump-forming perennial has slightly toothed, dark green leaves and white, daisy-like flowers, with yellow centres, borne on tall stems from early summer to autumn. Varieties include 'Wirral Pride' (above), with double flowers.

↕1m (3ft) ↔60cm (24in)

BULB LARGE

Lilium henryi

This is an upright, perennial bulb, with scattered, narrow, lance-shaped leaves. In late summer, it produces nodding, turks-cap, black-spotted orange flowers in dramatic arching sprays. It grows best on acid soil.

↕1m (3ft)

BULB LARGE

Lilium lancifolium

LANCE-LEAVED LILY This upright perennial bulb has long, narrow, lance-shaped leaves and, from summer to early autumn, nodding, pink- to red-orange, purple-spotted, turks-cap flowers. 'Splendens' AGM (above) has larger, black-spotted red-orange blooms.

↕1.5m (5ft)

BULB LARGE

Lilium longiflorum

EASTER LILY An upright perennial bulb, with long, shiny, lance-shaped leaves. In summer, this classic lily produces sprays of fragrant, large, outward-facing, funnel-shaped, pure white flowers, with petals that curve back slightly. Excellent for cutting.

↕1m (3ft)

BULB LARGE

Lilium nepalense

HIMALAYAN LILY An upright or arching perennial bulb, with scattered, lance-shaped leaves. Its summer blooms are large, nodding, greenish white or greenish yellow trumpets, with dark reddish purple throats and reflexed petals.

↕1m (3ft)

BULB LARGE

Lilium pardalinum

LEOPARD LILY This is an upright perennial bulb, with long, narrow leaves. In summer, it produces nodding, turks-cap, crimson flowers, with orange markings and maroon spots, on tall stems. The blooms are often scented.

↕1.5m (5ft)

BULB LARGE

Lilium speciosum

This upright, perennial bulb has broad, lance-shaped leaves. Sprays of large, scented, turks-cap, pale pink or white flowers, splashed deeper pink, appear from late summer to early autumn. *L. speciosum* var. *rubrum* is a popular lily.

↕1.2m (4ft)

BULB LARGE

Lilium superbum

AMERICAN TURKSCAP LILY An upright perennial bulb, with mottled stems bearing lance-shaped, green leaves. From late summer to early autumn, it bears nodding, turks-cap, orange blooms, with maroon spots and green stars in the throats.

↕2m (6ft)

PERENNIAL SMALL

Liriope muscari

LILYTURF An evergreen, spreading perennial that forms dense clumps of glossy, grass-like, dark green leaves. In autumn, upright spikes of thickly clustered, small, lavender- or purple-blue flowers appear, followed by black berries.

↕30cm (12in) ↔45cm (18in)

PERENNIAL SMALL

Luzula sylvatica 'Aurea'

GREATER WOODRUSH A clump-forming, evergreen perennial, with glossy, grass-like leaves that are yellow-green in winter and spring, and green in summer. Tiny, brown flowers appear from late spring to early summer. It tolerates deep shade.

↕40cm (16in) ↔45cm (18in)

PERENNIAL MEDIUM

Lysimachia clethroides

GOOSENECK LOOSESTRIFE This vigorous, clump-forming, spreading perennial has narrow, pointed leaves. Nodding spikes of small, white flowers become upright as they open in late summer. The tall stems may need support. It is potentially invasive.

↕↔1m (3ft)

PERENNIAL MEDIUM

Lysimachia punctata

GARDEN LOOSESTRIFE Large and vigorous, this upright, clump-forming perennial has mid-green, oval leaves, and produces spikes of small, bright yellow flowers over a long period in summer. It can be invasive. 'Alexander' (above) has variegated foliage.

↕1m (3ft) ↔60cm (24in)

SHRUB MEDIUM

Mahonia x *media*

An evergreen, upright shrub, with large leaves formed of holly-like, dark green leaflets. Long, slender spikes of small, scented, lemon-yellow flowers appear from late autumn through winter. 'Charity' (above) has upright, then spreading spikes.

↕1.8m (6ft) ↔4m (12ft)

SHRUB MEDIUM

Paeonia delavayi var. *delavayi* f. *lutea*

TREE PEONY An upright, deciduous shrub, with deeply lobed, dark green leaves, blue-green beneath, and nodding, cup-shaped, yellow flowers in late spring. Grow it at the back of a border.

↕2m (6ft) ↔1.2m (4ft)

CLIMBER LARGE

Parthenocissus quinquefolia

VIRGINIA CREEPER This deciduous, woody climber has divided, toothed, dull green leaves that blaze crimson in autumn, when blue-black berries also appear. Grow this plant up a large fence or wall at the back of a border.

↕ 15m (50ft) or more

PERENNIAL MEDIUM

Persicaria bistorta 'Superba' 🏆

COMMON BISTORT Forming a dense clump, this vigorous perennial produces oval, tapering, green leaves and spikes of soft pink blooms from early- to late summer. Deadhead regularly for repeat-flowering.

↕ 75cm (30in) ↔ 60cm (24in)

SHRUB MEDIUM

Philadelphus 'Beauclerk' 🏆

MOCK ORANGE This deciduous shrub produces slightly arching stems of oval leaves and very fragrant, single, white, early summer flowers, with pale purple centres. Grow it towards the back of a border in part shade.

↕↔ 2.5m (8ft)

SHRUB LARGE

Pittosporum tobira 🏆

JAPANESE PITTOSPORUM This evergreen, dense shrub has a neat, bushy-headed shape and leathery, dark green leaves. In late spring, it produces clusters of very sweetly scented, small, starry white flowers that age to creamy yellow.

↕ 10m (30ft) ↔ 3m (10ft)

SHRUB MEDIUM

Ribes sanguineum

FLOWERING CURRANT This deciduous, spreading shrub has lobed, aromatic leaves and small, pink, red, or white, bell-shaped flower clusters in spring, followed by white-coated, black fruits. 'Pulborough Scarlet' AGM (above) has deep crimson blooms.

↕ 2m (6ft) ↔ 2.5m (8ft)

SHRUB LARGE

Sambucus nigra f. *porphyrophylla* 'Eva' 🏆

A rounded, deciduous shrub, grown for its dissected, fern-like, almost black foliage. It produces flat heads of scented, pale pink flowers from late spring to early summer, followed by blackish red berries in autumn.

↕↔ 6m (20ft)

PERENNIAL MEDIUM

Tradescantia Andersoniana Group

SPIDERWORT This clump-forming perennial has arching, lance-shaped leaves and, from early summer to autumn, flowers in shades of white, red, blue, purple, or pink. 'Concord Grape' (above) has bright purple blooms.

↕↔ 60cm (24in)

SHRUB MEDIUM

Vaccinium corymbosum

HIGHBUSH BLUEBERRY A deciduous, soft fruit bush, with an upright, slightly arching habit. The foliage reddens in autumn. Small, white or pinkish flowers from late spring to early summer are followed by sweet, edible, blue-black berries.

↕↔ up to 1.5m (5ft)

PERENNIAL LARGE

Veronicastrum virginicum

CULVER'S ROOT An upright perennial, with narrow, lance-shaped, dark green leaves, punctuated in late summer by slender spikes of small, starry, purple-blue, pink, or white flowers, held on long stems. It is suitable for light shade only.

↕ 2m (6ft) ↔ 45cm (18in)

SHRUB MEDIUM

Viburnum davidii 🏆

This evergreen, domed shrub has leathery, deeply veined, dark green leaves. Clusters of small, white flowers appear above the foliage in late spring. For decorative, metallic blue berries on females, grow plants of both sexes.

↕↔ 1.5m (5ft)

SHRUB LARGE

Viburnum rhytidophyllum

This is a large, evergreen shrub, with long, broad, corrugated, veined, dark green leaves. Large, dense, domed heads of small, creamy white flowers appear in spring, and are followed by oval, red fruits, which ripen to black.

↕ 5m (15ft) ↔ 4m (12ft)

OTHER SUGGESTIONS

Perennials

Acanthus mollis • *Aster divaricatus* • *Bergenia* 'Overture' • *Centaurea montana* • *Dicentra eximia* • *Galega officinalis* 'Alba' • *Geranium wallichianum* 'Wisley Jewel' • *Paeonia lutea* • *Persicaria affinis*

Bulbs

Allium 'Purple Sensation' 🏆

Shrubs

Buddleja davidii • *Chaenomeles* x *superba* • *Corylus avellana* 'Contorta' 🏆 • *Cotoneaster conspicuous* 'Decorus' • *Escallonia rubra* 'Crimson Spire' 🏆 • *Euonymus fortunei* • *Hebe* 'Neil's Choice' 🏆 • *Hebe* 'Sapphire' 🏆 • *Jasminum nudiflorum* 🏆 • *Pittosporum tenuifolium* 'Tom Thumb' 🏆 • *Viburnum* x *bodnantense* 'Charles Lamont' 🏆

Plants for pond perimeters

Shady areas close to ponds are perfect for some of the most dramatic plants, the cool moist environment providing ideal growing conditions.

Combine plants such as *Gunnera*, *Filipendula*, *Rheum*, and hostas to produce lush, tropical-style displays – all except the *Gunnera* are fully hardy – and add splashes of colour with bright yellow or orange *Trollius* and *Mimulus*, together with crimson and pink astilbes. Alternatively, opt for a cool fernery around your water feature, with bold shuttlecock (*Matteuccia struthiopteris*) and royal ferns (*Osmunda regalis*) beefing up the design. Spring-flowering primulas complement ferns beautifully; choose a range of candelabra and drumstick forms to grace your pondsides with their elegant shapes and compelling colours.

PERENNIAL SMALL

Ajuga reptans

BUGLE A tough, evergreen, fast-spreading perennial, with small rosettes of glossy, dark green leaves. Short spikes of dark blue flowers appear from late spring to early summer. 'Multicolor' has bronze-purple leaves, splashed cream and pink.
↕15cm (6in) ↔90cm (36in)

PERENNIAL SMALL

Arisaema triphyllum

JACK-IN-THE-PULPIT An upright perennial, with a few large leaves, divided into broad, lance-shaped leaflets. Grown for its striking, green or purple, hooded, goblet-shaped, summer blooms, formed by fleshy bracts. These are followed by bright red berries.
↕50cm (20in) ↔45cm (18in)

PERENNIAL MEDIUM

Astilbe 'Venus'

This leafy perennial produces divided leaves and feathery, tapering plumes of tiny, pale pink flowers on tall stems in midsummer. The flowerheads dry and remain on the plant, providing interest well into winter. Best in fertile soil.
↕↔1m (3ft)

PERENNIAL SMALL

Aruncus aethusifolius

GOAT'S BEARD A clump-forming, compact perennial, with fern-like, dissected, green foliage, which turns yellow in autumn before falling. In early summer, it produces sprays of tiny, creamy white flowers above the leaves.
↕↔40cm (16in)

PERENNIAL SMALL

Astilbe x *arendsii*

FALSE GOAT'S BEARD A leafy perennial, with lush, fern-like, green foliage. In summer, it produces long, feathery plumes of tiny, starry, pinkish white flowers, which dry to brown and last into winter. Many varieties are available.
↕45cm (18in) ↔30cm (12in)

PERENNIAL MEDIUM

Astilbe chinensis

This perennial forms loose clumps of deeply cut, toothed, dark green leaves. In late summer, soft plumes of tiny, pinkish white flowers appear on tall stems. *A. chinensis* var. *pumila* AGM is shorter, with fluffy, raspberry-red spikes.
↕60cm (24in) ↔20cm (8in)

PERENNIAL SMALL

Carex elata

SEDGE An evergreen perennial that forms tufts of long, gently arching, narrow, green leaves. Thin, grass-like flower spikes form in summer. 'Aurea' AGM (above) is popular, and has golden-yellow leaves and blackish brown flower spikes.
↕40cm (16in) ↔15cm (6in)

PERENNIAL MEDIUM

Chelone obliqua

TURTLE HEAD This upright perennial produces toothed, lance-shaped, green leaves. From late summer to autumn, spikes of two-lipped, dark pink or purple flowers form. Pinch out stem tips in spring for bushier growth.
↕1m (3ft) ↔50cm (20in)

SHRUB LARGE

Cornus amomum

SILKY DOGWOOD This large, deciduous shrub has purplish winter shoots and lance-shaped, dark green leaves that turn red and orange in autumn. In late spring, it produces creamy white flowerheads, followed by attractive purple-blue fruits.

↕3m (10ft) ↔4m (12ft)

PERENNIAL LARGE

Darmera peltata

UMBRELLA PLANT This spreading perennial forms clumps of dinner plate-sized leaves on long stalks that burnish red in autumn. Flat clusters of white or pale pink flowers appear in spring before the foliage.

↕1.2m (4ft) ↔60cm (24in)

PERENNIAL MEDIUM

Filipendula ulmaria

MEADOWSWEET This leafy perennial has divided, blue-green leaves, downy white beneath, that turn brilliant red in autumn. Upright stems bear lacy clusters of tiny, fragrant, white flowers in summer. 'Aurea' (above) has golden-yellow spring foliage.

↕90cm (36in) ↔60cm (24in)

PERENNIAL LARGE

Eupatorium purpureum

JOE PYE WEED A stately, upright perennial, with oval, green leaves, held on purplish green stems. From late summer to autumn, it produces tall stems of fluffy, pinkish purple flowers. It is a perfect back-of-border plant for a large bog garden.

↕2.2m (7ft) ↔1m (3ft)

PERENNIAL LARGE

Filipendula purpurea

PURPLE MEADOWSWEET This upright, clump-forming perennial has divided, toothed, green leaves and large, airy sprays of tiny, reddish purple flowers atop tall, branching stems in late summer. Plant it at the back of a pond perimeter.

↕1.2m (4ft) ↔60cm (24in)

PERENNIAL LARGE

Filipendula rubra

QUEEN OF THE PRAIRIE This upright perennial has large, aromatic leaves. In midsummer, it produces branching stems of feathery, pink plumes that fade as they age. Use it to spread through a boggy site.

↕up to 2.5m (8ft) ↔1.2m (4ft)

PERENNIAL MEDIUM

Geum rivale

WATER AVENS This perennial has rosettes of scalloped, green foliage, above which rise slim stems of nodding, bell-shaped, pink to dark orange flowers from late spring to summer. 'Leonard's Variety' has coppery pink, double blooms.

↕↔60cm (24in)

PERENNIAL MEDIUM

Hosta sieboldiana

PLANTAIN LILY This clump-forming perennial has large, broad, blue- and grey-green leaves, with a puckered texture, and bell-shaped, lilac-grey early summer flower spikes. *H. sieboldiana.* var. *elegans* AGM has larger, bluer leaves, and pale lilac flowers.

↕1m (3ft) ↔1.2m (4ft)

PERENNIAL MEDIUM

Iris ensata

JAPANESE FLAG An upright, clump-forming perennial, with sword-shaped, grey-green leaves. From early- to midsummer, it bears beardless flowers in shades of purple, pink, and white. 'Rose Queen' AGM has large, soft pink flowers, with deeper pink veining.

↕90cm (36in) ↔60cm (24in)

PERENNIAL MEDIUM

Kirengeshoma palmata

An upright perennial, with reddish stems and large, rounded, lobed, bright green, serrated-edged leaves, above which clusters of narrow, bell-shaped, creamy yellow flowers appear from late summer to autumn on strong stems.

↕1m (3ft) ↔60cm (24in)

PERENNIAL LARGE

Ligularia dentata

LEOPARD PLANT An upright, clump-forming perennial, with large, leathery, rich green, heart-shaped leaves. From midsummer to early autumn, it bears large, daisy-like, orange-yellow flowers. 'Desdemona' AGM (above) has brownish green leaves.

↕1.2m (4ft) ↔60cm (2ft)

PERENNIAL MEDIUM

Lysimachia ephemerum 🏆

WILLOW-LEAVED LOOSESTRIFE A clump-forming perennial, with elegant, willow-like, rough-textured, grey-green leaves. Upright, tapering spikes of starry, greyish white summer flowers are followed by light green seedheads. Add a mulch in winter.

↕1m (3ft) ↔30cm (12in)

PERENNIAL MEDIUM

Lysimachia punctata

GARDEN LOOSESTRIFE A vigorous, clump-forming perennial, with soft haired, mid-green leaves and, in summer, tall spikes of bright yellow flowers nestling between the leaves. It is good for cutting, and attracts butterflies, but can be invasive.

↕1m (3ft) ↔60cm (24in)

PERENNIAL LARGE

Ligularia stenocephala

LEOPARD PLANT This clump-forming perennial has large, toothed, green leaves and, in summer, slender, daisy-like, yellow flower spikes on dark brown, tall stems. 'The Rocket' AGM (above) has ragged-edged foliage and black stems.

↕1.8m (6ft) ↔1m (3ft)

SHRUB LARGE

Lindera benzoin

BENJAMIN This deciduous shrub has dark green leaves, yellow in autumn, and aromatic when crushed. Showy clusters of fragrant, tiny, greenish yellow flowers form in spring. For small, scarlet berries on females, grow plants of both sexes.

↕↔3m (10ft)

pH

TREE MEDIUM

Magnolia virginiana

SWEET BAY This conical, semi-evergreen or deciduous shrub or tree has glossy, dark green leaves, bluish white beneath. It bears vanilla-scented, cup-shaped, creamy white flowers from early summer to early autumn. Grow it at the back of a boggy border.

↕10m (30ft) ↔6m (20ft)

PERENNIAL MEDIUM

Matteuccia struthiopteris 🏆

OSTRICH-FEATHER FERN This deciduous fern has lance-shaped, deeply divided, green fronds that sprout from a central crown, producing a "shuttlecock" effect. Some fronds dry and last into winter; cut back all growth in spring.

↕1m (3ft) ↔45cm (18in)

PERENNIAL SMALL

Mimulus guttatus

MONKEY FLOWER A spreading, mat-forming perennial, with toothed or ragged-edged leaves. Clusters of snapdragon-like, vivid yellow flowers, spotted reddish brown on the lower petals, appear in succession over summer and early autumn. It attracts bees.

↕30cm (12in) ↔60cm (24in)

PERENNIAL SMALL

Onoclea sensibilis 🏆

SENSITIVE FERN This deciduous, creeping fern covers the ground with deeply divided, toothed, light green fronds. Stiff, brown fronds stand erect in late summer, and last into winter. Plants prefer moist soil in dappled shade.

↕60cm (24in) ↔1m (3ft)

pH

PERENNIAL LARGE

Osmunda regalis 🏆

ROYAL FERN This large, deciduous fern forms upright clumps of elegant, deeply divided, bright green fronds, pinkish when young and red-brown in autumn. Mature plants bear tassel-like spikes of rust-brown spores at the ends of taller fronds.

↕2m (6ft) ↔1m (3ft)

PERENNIAL SMALL

Primula japonica

JAPANESE PRIMROSE This is a deciduous perennial, with rosettes of toothed-edged, pale green basal leaves. Tubular, deep red flowers appear in early summer. 'Postford White' AGM (above) has white flowers. Plant in groups around a part-shaded pond.

↕↔45cm (18in)

PERENNIAL LARGE

Rodgersia pinnata

A clump-forming perennial, with large, corrugated, divided, dark green leaves. Conical spires of small, pink, red, or yellow-white flowers appear in summer. 'Superba' AGM (above) has bronze-tinged emerald leaves and bright pink flowers.

↕1.2m (4ft) ↔75cm (30in)

PERENNIAL MEDIUM

Primula pulverulenta 🏆

This upright perennial produces rosettes of large, oblong, green leaves and upright, white-coated stems, studded with clusters of reddish purple flowers in early summer. Grow it in groups around a part-shaded pool.

↕1m (3ft) ↔60cm (24in)

PERENNIAL SMALL

Primula veris 🏆

COWSLIP This evergreen or semi-evergreen perennial forms rosettes of oval to lance-shaped, toothed leaves. Tight clusters of nodding, fragrant, tubular, butter-yellow flowers appear on stout stems in spring. Try it in boggy meadows by pools.

↕↔25cm (10in)

PERENNIAL LARGE

Rheum palmatum

CHINESE RHUBARB This upright perennial has huge, jaggedly lobed, dark green leaves, purple-red beneath. In early summer, sturdy stems bear large plumes of fluffy, cream to red flowers. 'Atrosanguineum' (above) has crimson blooms.

↕↔2m (6ft)

PERENNIAL MEDIUM

Saururus cernuus

SWAMP LILY This perennial marginal water or bog plant has clumps of heart-shaped, green leaves and slender, nodding spires of tiny, cream flowers in summer. It can become aggressive, so divide it regularly if growing in a bog garden.

↕1m (3ft) ↔indefinite

TREE LARGE

Taxodium distichum

SWAMP CYPRESS This conical, deciduous conifer has slender, fresh green leaves that turn rich orange-brown in autumn. It has green female and red male cones. Plant it in boggy soil in a large garden, some distance from open water.

↕40m (130ft) ↔9m (28ft)

PERENNIAL LARGE

Thalictrum delavayi 🏆

CHINESE MEADOW RUE A clump-forming perennial, with fern-like, mid-green leaves. From late summer to autumn, it produces large, billowing panicles of tiny, lavender blooms. 'Hewitt's Double' AGM (above) has dainty, double flowers.

↕1.5m (5ft) ↔60cm (24in) or more

PERENNIAL MEDIUM

Tradescantia Andersoniana Group

SPIDERWORT This tufted, clump-forming perennial has erect, branching stems and arching, narrow, lance-shaped, mid-green leaves. 'Concord Grape' (above) has purple flowers. Grow in part shade around a pond.

↕↔60cm (24in)

PERENNIAL MEDIUM

Trollius chinensis

GLOBEFLOWER A clump-forming perennial, with lobed, toothed, green leaves and bowl-shaped, golden-orange flowers, with upright central stamens, in summer. It tolerates light shade and looks best planted in groups around a pond edge.

↕90cm (36in) ↔45cm (18in)

PERENNIAL MEDIUM

Trollius x *cultorum*

GLOBEFLOWER This clump-forming perennial has deeply divided, fern-like, green foliage and cup-shaped, double, orange-yellow flowers from mid-spring to early summer. Use it to brighten up the banks of a pond or stream.

↕75cm (30in) ↔45cm (18in)

OTHER SUGGESTIONS

Annuals

Mimulus Magic Series

Perennials

Actaea matsumurae 'White Pearl' • *Aruncus dioicus* 🏆 • *Astilbe chinensis* var. *taquetii* 'Superba' 🏆 • *Cardamine bulbifera* • *Eupatorium cannabinum* • *Eupatorium maculatum* • *Eupatorium rugosum* 'Chocolate' • *Gunnera manicata* 🏆 • *Hosta* 'Frances Williams' • *Gunnera tinctoria* 🏆 • *Ligularia* 'Britt Marie Crawford' 🏆 • *Ligularia przewalskii* • *Miscanthus sinensis* 'Kleine Silberspinne' 🏆 • *Miscanthus sinensis* 'Zebrinus' 🏆 • *Phalaris arundinacea* var. *picta* • *Podophyllum hexandrum* var. *chinense* • *Primula alpicola* 🏆 • *Primula beesiana* 🏆 • *Primula denticulata* 🏆 • *Rheum* 'Ace of Hearts' 🏆 • *Rodgersia sambucifolia* • *Salvia uliginosa* 🏆 • *Trollius europaeus*

Plants for ponds

The selection of plants for shady pools is much smaller than those for ponds in sun, and most need some light to put on their best performance.

Unless you have a natural pond in deep shade, choose the brightest site available for an artificial feature, or your options will be limited to the spring-flowering skunk cabbage (*Lysichiton*), which, although beautiful, did not acquire its common name for no reason. However, in partly shaded ponds, your choice of irises, marsh marigolds (*Caltha*), rushes, and other marginals will create a dynamic display of colourful reflections from spring until autumn. The only evergreens are water figwort (*Scrophularia*) and horsetail (*Equisetum*); keep the latter's roots confined to a pond basket.

PERENNIAL MEDIUM

Acorus calamus

SWEET FLAG This upright perennial can be either deciduous or semi-evergreen. It is grown for its clumps of grass-like, green leaves. In summer, it produces insignificant, brown flowers. Tolerates light shade.

↕↔ 60cm (24in)

❄❄❄

PERENNIAL SMALL

Calla palustris

BOG ARUM A deciduous or semi-evergreen, spreading marginal perennial, with heart-shaped, glossy leaves. In spring, it produces large, elegant, vase-shaped, white bracts, each surrounding a fleshy spike of minute flowers, followed by red or orange berries.

↕ 25cm (10in) ↔ 30cm (12in)

❄❄❄ (!)

PERENNIAL SMALL

Caltha palustris

MARSH MARIGOLD This perennial marginal produces rounded, heart-shaped, green leaves and single or double, golden-yellow flowers on lax stems in early spring. It tolerates light shade, with some sun during the day. It will self-seed and spread.

↕ 40cm (16in) ↔ 45cm (18in)

❄❄❄

PERENNIAL SMALL

Caltha palustris 'Flore Pleno' 🏆

MARSH MARIGOLD This small perennial has the same dark green leaves as the species and clusters of double, bright golden spring flowers. It is also suitable for pond perimeters.

↕↔ 25cm (10in)

❄❄❄

PERENNIAL SMALL

Cardamine pratensis

LADY'S SMOCK This perennial marginal or bog plant has divided leaves, and bears clusters of pinkish mauve flowers in late spring. 'Flore Pleno' (above) has double, pale pink flowers. Grow it in very shallow water or a bog garden in shade.

↕ 45cm (18in) ↔ 30cm (12in)

PERENNIAL MEDIUM

Cyperus involucratus 🏆

UMBRELLA PLANT An evergreen, perennial marginal sedge, it forms tall, upright, slender stems, each ending in a whorl of horizontal, grassy bracts. Airy sprays of tiny, yellowish green flowers arise from the "umbrellas" in summer. Protect from frost.

↕ 1m (3ft) ↔ 30cm (12in)

❄

PERENNIAL SMALL

Houttuynia cordata

A vigorous, spreading perennial, with heart-shaped, blue-green, orange-scented leaves. Yellow-centred white flowers appear in late spring. Plant in the shallows; restrict its roots in a pond basket. 'Chameleon' (above) has yellow- and red-splashed foliage.

↕ 30cm (12in) ↔ indefinite

❄❄❄

PERENNIAL MEDIUM

Iris ensata

JAPANESE FLAG This upright perennial has sword-shaped, grey-green leaves. From early- to midsummer, it produces flowers in shades of purple, pink, and white, depending on the variety. Plant it in shallow water.

↕ 90cm (36in) ↔ 60cm (24in)

❄❄❄ (!)

PERENNIAL MEDIUM

Lysichiton americanus 🏆

YELLOW SKUNK CABBAGE This vigorous, perennial marginal produces large, green leaves, followed by bright yellow bracts in spring. Avoid planting it near streams, as it spreads rapidly, or too close to seating, as it has a rancid odour.

↕75cm (30in) ↔1.2m (4ft)

PERENNIAL MEDIUM

Lysichiton camtschatcensis 🏆

ASIAN SKUNK CABBAGE A clump-forming perennial marginal, with large, bright green leaves and white, spring flowers, which comprise petal-like bracts around spikes of tiny flowers. Its scent may be unpleasant.

↕75cm (30in) ↔60cm (24in)

❄❄❄ (!)

PERENNIAL MEDIUM

Peltandra virginica

This perennial marginal is grown for its bold, arrow-shaped, glossy, dark green leaves, formed on long stalks. In late summer, it produces minute flowers inside upright bracts that look like green chilli peppers.

↕90cm (36in) ↔60cm (24in)

❄❄❄

PERENNIAL MEDIUM

Pontederia cordata 🏆

PICKEREL WEED This upright perennial bears large, spear-shaped, glossy, green leaves. In summer, it produces sturdy spikes of blue flowers. Remove spent growth in autumn. The variety 'Alba' has white flowers.

↕75cm (30in) ↔45cm (18in)

❄❄❄

WATER PLANT SMALL

Persicaria amphibia

AMPHIBIOUS BISTORT A vigorous perennial aquatic, with lance-shaped, dark green, floating leaves. From midsummer to autumn, it produces clusters of pink flowers on upright stems above the foliage, followed by glossy, brown fruits.

↕30cm (12in) ↔2m (6ft)

❄❄❄

PERENNIAL MEDIUM

Scrophularia auriculata

VARIEGATED WATER FIGWORT This is a tall, evergreen, clump-forming perennial marginal, with oval, rough-textured, cream-edged dark green leaves. In summer, it produces small, brown flowers, which are attractive to bees.

↕1m (3ft) ↔30cm (12in) or more

❄❄❄

PERENNIAL MEDIUM

Petasites japonicus var. *giganteus*

GIANT JAPANESE BUTTERBUR A spreading perennial, with large, round leaves, edible stems, and lime-green spring flower spikes. Can be invasive. 'Nishiki-buki' (above) has cream-splashed, green leaves.

↕1.1m (3½ft) ↔1.5m (5ft)

❄❄❄

PERENNIAL LARGE

Saururus cernuus

SWAMP LILY A perennial marginal with spreading clumps of heart-shaped leaves and soft, tapering spires of tiny, creamy white flowers in summer. Confine it in a basket in small ponds and water features, as it can be aggressive.

↕1m (3ft) ↔indefinite

❄❄❄

PERENNIAL MEDIUM

Zantedeschia aethiopica

ARUM LILY A deciduous or semi-evergreen perennial, with upright, arrow-shaped, dark green leaves. From early- to midsummer, it produces vase-shaped, hooded, white, petal-like bracts around short, yellow flower spikes. Tolerates light shade only.

↕↔90cm (36in)

❄❄

OTHER SUGGESTIONS

Water plants

Aponogeton distachyos • *Nuphar lutea* • *Nymphoides peltata*

Perennial marginals

Acorus gramineus 'Hakuro-nishiki' • *Acorus gramineus* 'Ōgon' • *Caltha palustris* 'Marilyn' • *Carex riparia* 'Variegata' • *Cyperus alternifolius* • *Equisetum arvense* • *Houttuynia cordata* 'Variegata' • *Juncus ensifolius* • *Mentha aquatic* • *Myosotis scorpioides* • *Oenanthe javanica* 'Flamingo' • *Sagittaria graminea* • *Sparganium erectum* • *Zantedeschia* 'Picasso'

Plants for boundaries, hedges, and windbreaks

Hedging and wall plants are both decorative and useful, helping to create shelter, mark boundaries, and divide up large gardens.

Conifers make excellent hedges, but most prefer a sunny site – hemlock (*Tsuga canadensis*) is an exception and grows happily in partial shade. Other evergreens to consider are *Pyracantha*, box (*Buxus*), and laurel (*Prunus*), while *Elaeagnus* and privet can provide colourful leaves, too. However, all these plants need sun for part of the day to put on a good show. For areas in dense shade, opt for a holly (*Ilex*) or *Aucuba* to create your hedge. If you are using large shrubs, clip them to size and shape, while smaller shrubs can be used as edging or low hedges.

SHRUB MEDIUM

Aucuba japonica

SPOTTED LAUREL A rounded, evergreen shrub, with glossy, oval, green leaves. Small, purple spring flowers are followed by red berries on female plants, if a male is grown close by. 'Crotonifolia' AGM (above) has spotted, yellow leaves.

↕↔ 2m (6ft)

SHRUB LARGE

Buxus sempervirens

COMMON BOX A compact, evergreen shrub, with small, oval, glossy, dark green foliage and tiny, yellowish spring flowers. A classic hedging and topiary plant, clip in late spring or early summer to allow time for the new growth to mature to avoid frost damage.

↕↔ 5m (15ft)

TREE LARGE

Carpinus betulus

HORNBEAM This deciduous tree has oval, corrugated, dark green leaves that turn yellow and orange in autumn. Green catkins in spring precede clusters of winged nuts. A classic formal or informal hedging plant, trim it in late summer.

↕ 25m (80ft) ↔ 20m (70ft)

SHRUB LARGE

Elaeagnus x ebbingei

This vigorous, evergreen shrub has oval, dark green leaves that are dusted with grey flecks. It produces small, white, scented flowers in autumn. Plant it as a screen or windbreak. It tolerates light shade only; trim in late summer.

↕↔ 5m (15ft)

SHRUB LARGE

Fatsia japonica

JAPANESE ARALIA An evergreen shrub, with large, palm-shaped, glossy, dark green leaves. Round, white flowerheads in autumn are followed by black fruits. Grow it as an informal screen; prune in spring, if necessary.

↕↔ 4m (12ft)

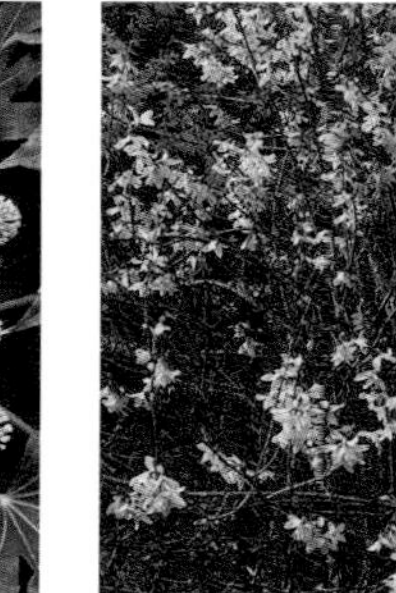

SHRUB LARGE

Forsythia x intermedia

This deciduous, vigorous shrub has long stems of golden-yellow flowers in early spring before the small, green foliage appears. Grow it as an informal hedge; prune after flowering. 'Lynwood Variety' AGM (above) has larger, golden blooms.

↕↔ 3m (10ft)

SHRUB MEDIUM

Hydrangea quercifolia

SNOW QUEEN

OAK-LEAVED HYDRANGEA This deciduous shrub has large, lobed, green leaves, purple and red in autumn. In late summer, white, conical flowerheads form. Use as a screen; clip it in spring to keep within bounds.

↕ 2m (6ft) ↔ 2.5m (8ft)

TREE LARGE

Ilex aquifolium

ENGLISH HOLLY This is an evergreen shrub, or tree, with wavy-edged, spiny, dark green leaves and scarlet berries on pollinated female plants. Thornless and variegated forms are also available. Trim with secateurs in late summer.

↕ up to 20m (70ft) ↔ 6m (20ft)

SHRUB LARGE

Kalmia latifolia

MOUNTAIN LAUREL An evergreen, dense shrub, with glossy, dark green foliage. In early summer, large clusters of pale pink flowers open from darker pink, prettily crimped buds. Grow it as a flowering screen, and prune lightly after flowering.

↕↔ 3m (10ft)

SHRUB MEDIUM

Lonicera nitida

SHRUBBY HONEYSUCKLE A popular formal or informal hedging plant, this evergreen, dense shrub has arching shoots and small, dark green leaves. 'Baggesen's Gold' AGM (above) has greenish yellow leaves. Cut it thrice a year between spring and autumn.

↕2m (6ft) ↔3m (10ft)

SHRUB SMALL

Lonicera pileata

BOX-LEAVED HONEYSUCKLE A spreading, evergreen, dense shrub, grown mainly for its narrow, dark green, glossy leaves. Purple berries follow the small, tubular, cream flowers in late spring. It is easy to train into a low hedge; trim in late spring.

↕60cm (24in) ↔2.5m (8ft)

SHRUB LARGE

Osmanthus heterophyllus

FALSE HOLLY An evergreen, dense, rounded shrub, with holly-like, prickly, dark green foliage and tiny, fragrant, white flowers in autumn. 'Aureomarginatus' (above) has gold-edged leaves. Use as a low screen in a sheltered, warm garden; trim in late spring.

↕↔5m (15ft)

SHRUB LARGE

Phyllostachys nigra

BLACK BAMBOO This evergreen, clump-forming shrub has grooved, greenish brown stems, which turn black when mature, and long, narrow leaves. Grow it as a screen in part shade; cut out dead and old canes in spring.

↕8m (25ft) ↔indefinite

SHRUB LARGE

Prunus laurocerasus

CHERRY LAUREL A dense, evergreen shrub, with large, glossy, oval, green leaves. From mid- to late spring, upright spikes of small, single, white flowers appear, followed by cherry-red, later black, fruits. Prune hedges hard in spring.

↕8m (25ft) ↔10m (30ft)

SHRUB LARGE

Prunus lusitanica

PORTUGAL LAUREL This dense, evergreen shrub has reddish purple shoots and glossy, dark green foliage. Slender, frothy, fragrant, white flower spikes appear in early summer, followed by deep purple fruits. Prune hedges in late spring.

↕↔10m (30ft)

SHRUB LARGE

Pyracantha 'Mohave'

FIRETHORN This dense, evergreen shrub has small, green leaves and very spiny stems. In summer, it produces clusters of small, white blooms that lead to bold, orange-red berries. Use it to deter intruders; prune in spring.

↕↔4m (12ft)

SHRUB MEDIUM

Ribes odoratum

BUFFALO CURRANT This deciduous shrub has lobed, bright green leaves that flush red and purple in autumn. Clusters of starry, spicily scented, golden flowers appear in spring; purple berries follow. Grow it as an informal screen; trim after flowering.

↕↔2m (6ft)

SHRUB MEDIUM

Symphoricarpos x *doorenbosii*

SNOWBERRY This vigorous, deciduous shrub forms a dense barrier of twiggy stems and small, green leaves. It produces small, white blooms in summer, followed by round, white fruits in autumn. Use it for informal and wildlife hedges; trim in spring.

↕2m (6ft) ↔indefinite

TREE LARGE

Taxus baccata

YEW This bushy, evergreen tree produces needle-like, dark green leaves and red berries in autumn. It makes a superb hedge for boundaries and is useful for dividing up a garden. Clip it in spring and summer. All parts are highly toxic.

↕15m (50ft) ↔10m (30ft)

SHRUB LARGE

Viburnum prunifolium

BLACKHAW VIBURNUM A deciduous, round-headed shrub, with elliptical, toothed, shiny, dark green leaves that turn reddish purple in autumn. White flowers in late spring are followed by edible pink fruits that ripen to bluish black. Trim after flowering.

↕5m (15ft) ↔4m (12ft)

OTHER SUGGESTIONS

Shrubs and trees

Acer campestre • *Aucuba japonica* 'Salicifolia' • *Berberis julianae* • *Berberis* x *ottawensis* 'Auricoma' • *Berberis stenophylla* • *Buxus microphylla* • *Corylus avellana* • *Elaeagnus pungens* 'Maculata' • *Escallonia* 'Peach Blossom' • *Euonymus fortunei* 'Silver Queen' • *Ilex aquifolium* 'Ferox Argentea' • *Ligustrum delavayanum* • *Ligustrum lucidum* 'Excelsum Superbum' • *Ligustrum ovalifolium* 'Aureum' • *Phyllostachys aureosulcata* f. *aureocaulis* • *Phyllostachys bissetii* • *Poncirus trifoliata* • *Rhododendron fulvum* • *Rosa rugosa* • *Rosa scabrosa* • *Sambucus nigra* • *Symphoricarpos orbiculatus* • *Taxus baccata* 'Semperaurea' • *Tsuga canadensis* • *Viburnum opulus* 'Roseum' • *Viburnum tinus* 'Eve Price'

Plants for beside hedges, walls, and fences

Areas shaded by vertical structures or hedges do not make natural homes for many plants, but a few tolerate the dark and drought.

Apart from plants that can naturally survive in such areas, you can increase your choices by offering additional moisture. This can be done either by improving the structure of your soil with annual applications of well-rotted manure, or by installing a leaky hose irrigation system to top-up soil water levels. Deep shade can also be a limiting factor, although *Lamium*, *Tolmiea*, and some ferns will cope. To cover a wall or fence, consider a wall shrub, such as Japanese quince (*Chaenomeles*), and those that will climb when confronted with a vertical surface, such as *Euonymus fortunei*.

PERENNIAL LARGE

Acanthus mollis

BEAR'S BREECHES This upright, semi-evergreen perennial has long, deeply cut, bright green basal leaves. Tall spikes of funnel-shaped, white flowers nestle between dusky purple bracts in summer. Tolerates dry soil and partial shade.

↕1.2m (4ft) ↔45cm (18in)

◊ ❄❄❄

PERENNIAL LARGE

Aconitum carmichaelii

MONK'S HOOD An upright perennial, with deeply divided, mid-green leaves and tall spikes of hooded, lavender-blue autumn flowers. Use it to inject late-season colour at the back of a border or along a boundary. 'Arendsii' (above) has rich blue flowers.

↕1.5m (5ft) ↔30cm (12in)

◊ ❄❄❄ (!)

PERENNIAL SMALL

Arum italicum subsp. *italicum*

This low-growing perennial is grown for its large, arrow-shaped, dark green, glossy leaves from winter to spring. Orange-red berries follow spikes of yellow flowers in early summer. 'Marmoratum' (above) has attractively veined leaves.

↕25cm (10in) ↔30cm (12in)

PERENNIAL SMALL

Centaurea montana

PERENNIAL CORNFLOWER This perennial has oval, tapering, green leaves, hairy beneath. In early summer, it bears thistle-like flowers in shades of purple, blue, white, or pink. A cottage garden favourite, leave it to spread beside a wall or hedge.

↕50cm (20in) ↔60cm (24in)

◐ ❄❄❄

SHRUB MEDIUM

Chaenomeles x *superba*

JAPANESE QUINCE A deciduous shrub, with spiny branches and narrow, oval, glossy, green leaves. In late spring, it bears clusters of cup-shaped flowers in shades of white, pink, orange, or red, depending on the variety. Train it on a wall or fence.

↕1.5m (5ft) ↔2m (6ft)

◊ ❄❄❄

SHRUB LARGE

Crinodendron hookerianum 🏆

CHILE LANTERN TREE An evergreen, architectural shrub, with stiff branches and narrow, lance-shaped, dark green leaves. From late spring to early summer, lantern-like, red flowers hang from the stems. Train it against a sheltered wall.

↕6m (20ft)

◐ ❄❄ pH

PERENNIAL MEDIUM

Dryopteris dilatata 🏆

BROAD BUCKLER FERN This deciduous or semi-evergreen fern forms a rosette of arching fronds, with triangular to oval, serrated segments and dark brown stems. It tolerates the dry soil next to hedges, if watered well in the first season.

↕1m (3ft) ↔45cm (18in)

◐ ❄❄❄

PERENNIAL SMALL

Epimedium perralderianum

BARRENWORT A semi-evergreen perennial, with large, toothed, glossy, heart-shaped, dark green leaves. Airy spikes of small, pendent, bright yellow flowers form on wiry stalks in spring before new foliage appears. An effective ground cover in deep shade.

↕30cm (12in) ↔45cm (18in)

◐ ❄❄❄

PERENNIAL MEDIUM

Lamium galeobdolon

DEAD NETTLE This vigorous, mat-forming evergreen perennial has creeping stems of oval, silver-marked dark green leaves and, in summer, spikes of brown-spotted yellow flowers. 'Hermann's Pride' (above) is less vigorous; suited to most gardens.

↕60cm (24in) ↔indefinite

◊ ❄❄❄

PERENNIAL SMALL

Lamium maculatum

This semi-evergreen, mat-forming perennial has white-variegated, mid-green foliage and spikes of hooded, white flowers from late spring to summer. Varieties include 'White Nancy' (above), which produces pure white flowers above silver leaves.

↕15cm (6in) ↔1m (3ft)

PERENNIAL LARGE

Macleaya microcarpa

PLUME POPPY An upright perennial, with large, deeply lobed, grey-green leaves, white beneath. In summer, sprays of tiny, buff-pink flowers form on tall stems above the foliage. Grow it near the back of a border in light shade only.

↕2.5m (8ft) ↔1.2m (4ft)

BULB MEDIUM

Narcissus 'Actaea'

DAFFODIL This late-spring-flowering perennial bulb bears narrow, strap-shaped, grey-green leaves and white flowers with shallow, red-rimmed yellow cups. Plant the bulbs in groups close to a partly shaded wall, fence, or hedge in autumn.

↕40cm (16in)

SHRUB MEDIUM

Paeonia delavayi

TREE PEONY This is an upright, deciduous shrub, with deeply lobed, dark green leaves that are blue-green beneath. It produces nodding, cup-shaped, dark red flowers in late spring. It prefers the shelter found next to a wall or fence.

↕↔2m (6ft)

PERENNIAL SMALL

Pulmonaria officinalis

LUNGWORT This semi-evergreen perennial bears oval, pointed, green leaves, with white spots, and clusters of funnel-shaped, pink, blue, or white flowers in spring. It is shade- and drought-tolerant. 'Sissinghurst White' AGM (above) has white flowers.

↕30cm (12in) ↔60cm (24in)

PERENNIAL SMALL

Saxifraga 'Aureopunctata'

A low-growing perennial that forms rosettes of mid- to dark green- and yellow-variegated leaves. From late spring to early summer, masses of tiny, star-shaped, pink flowers appear. Grow it at the front of a bed or border.

↕30cm (12in) ↔indefinite

PERENNIAL MEDIUM

Tolmiea menziesii

PICK-A-BACK PLANT This small, semi-evergreen perennial has textured, ivy-shaped, green leaves and spikes of nodding, tubular, greenish yellow flowers in spring. 'Taff's Gold' (above) has cream and green variegated foliage.

↕60cm (24in) ↔2m (6ft)

PERENNIAL MEDIUM

Tricyrtis formosana

TOAD LILY This upright perennial produces dark green leaves with purplish green spots. Star-shaped, purple-spotted cream flowers appear from late summer to early autumn on upright stems. Grow it in the shelter of a wall in full or part shade.

↕80cm (32in) ↔45cm (18in)

PERENNIAL MEDIUM

Uvularia grandiflora

BELLWORT This clump-forming perennial has lance-shaped leaves and clusters of long, bell-shaped, yellow mid- to late spring flowers, with slightly twisted petals, that hang from slender stems. It is useful for deep shade beside walls and hedges.

↕60cm (24in) ↔30cm (12in)

SHRUB MEDIUM

Viburnum acerifolium

MAPLE LEAF VIBURNUM This deciduous, upright shrub has bright green, corrugated, jagged-edged leaves, turning orange, red, and purple in autumn. Small, creamy white early summer flower clusters are followed by berries that ripen to purple-black.

↕2m (6ft) ↔1.2m (4ft)

SHRUB LARGE

Viburnum x *bodnantense*

A deciduous, upright shrub with bronze leaves that later turn dark green. From late autumn to early spring, clusters of sweetly scented, pink to white-pink flowers appear along bare branches. 'Dawn' AGM (above) has dark pink buds and pink flowers.

↕3m (10ft) ↔2m (6ft)

OTHER SUGGESTIONS

Perennials

Aconitum hemsleyanum • *Dryopteris affinis* • *Epimedium grandiflorum* • *Geranium* 'Johnson's Blue' • *Geranium nodosum* • *Geranium phaeum* 'Variegatum' • *Geranium* ROZANNE • *Vinca minor* 'Argenteovariegata'

Bulbs

Allium 'Purple Sensation' • *Narcissus* 'February Gold' • *Narcissus* 'Jenny' • *Nectaroscordum siculum*

Shrubs

Chaenomeles x *superba* 'Nicoline' • *Crinodendron patagua* • *Euonymus fortunei* 'Emerald Gaiety' • *Euonymus fortunei* 'Silver Queen' • *Euphorbia amygdaloides* var. *robbiae* • *Forsythia* x *intermedia* • *Garrya* x *issaquahensis* • *Kerria japonica* • *Lindera obtusiloba* • *Viburnum plicatum* 'Grandiflorum' • *Vinca major*

Plants for walls, fences, and vertical surfaces

Plants that scramble up vertical surfaces are chiefly useful in small gardens, where they bolster colour and interest, while taking up little ground space.

Many climbers and wall shrubs like a shady root run, while their stems clamber or are trained towards the sun. *Clematis* are a perfect example, and most are happy if their top growth can bathe in sunlight for a few hours each day in summer. Some, including *C.* 'Niobe', bloom on north-facing walls, while others like 'Nelly Moser' prefer shade as the sun can bleach out its subtle flower colours. Deep shade is not a problem for natural forest dwellers, such as *Parthenocissus* and ivy, while those used to life on the edge of a woodland, like hydrangeas, honeysuckle, and *Schizophragma*, need sunshine for part of the day.

CLIMBER MEDIUM

Clematis alpina

ALPINE CLEMATIS A deciduous climber, with divided, mid-green leaves. From early- to late spring, it bears lantern-shaped, blue, pink, or white flowers, depending on the variety, followed by fluffy, silvery seedheads. Keep the roots shaded.

↕3m (10ft)

CLIMBER LARGE

Clematis 'Bill MacKenzie'

This deciduous climber, with dark green leaves, is grown for its bell-shaped, waxy, yellow flowers that appear from midsummer to late autumn. The flowers are followed by fluffy seedheads. Requires a large support. Cut back in late winter or early spring.

↕7m (22ft)

CLIMBER MEDIUM

Clematis cirrhosa

This evergreen climber has jagged-edged, green leaves. From late winter to early spring, it produces small, bell-shaped, cream flowers that are spotted red inside. Provide a sturdy support and a sheltered site; shade the roots.

↕3m (10ft)

CLIMBER LARGE

Clematis montana

This is a vigorous, deciduous climber, with divided, mid-green foliage. From late spring to early summer, it produces an abundance of scented, white flowers, with yellow centres. Provide a large support, such as a tree, large wall, or pergola.

↕12m (40ft)

CLIMBER MEDIUM

Clematis 'Nelly Moser'

This medium-sized, deciduous climber has dark green leaves. In early summer, it produces large, rose-mauve blooms, with a carmine stripe on each petal, followed by globular seedheads. Trim lightly in late winter or early spring.

↕3m (10ft)

CLIMBER MEDIUM

Clematis 'Niobe'

This is a deciduous climber, with dark green leaves. Throughout summer, it produces masses of single, velvety, deep red flowers that have contrasting yellow anthers. Trim lightly in late winter or early spring.

↕3m (10ft)

SHRUB LARGE

Forsythia suspensa

WEEPING FORSYTHIA A deciduous shrub, with slender, arching stems and small, toothed, green leaves. Nodding, narrow, slim-petalled, bright yellow flowers form on bare branches in early spring. Train it as a wall shrub; prune after flowering.

↕↔3m (10ft)

CLIMBER LARGE

Holboellia coriacea

SAUSAGE VINE An evergreen climber, with dark green leaves, divided into three leaflets. In spring, clusters of fragrant, pale purple male and greenish white female, flowers appear, sometimes followed by sausage-shaped, purple seedpods.

↕5m (15ft)

CLIMBER MEDIUM

Clematis 'White Swan'

This large deciduous clematis produces divided, green foliage and nodding, open bell-shaped, white flowers from late spring to early summer. Grow it on wires or through a large shrub. Prune after flowering.

↕3m (10ft)

CLIMBER LARGE

Hydrangea anomala subsp. *petiolaris*

CLIMBING HYDRANGEA A self-clinging, deciduous climber, with woody stems and rounded, dark green leaves. In summer, it bears flattened, white flowerheads. Grow against a support, such as a fence or wall.

↕15m (50ft)

SHRUB LARGE

Itea virginica

This spreading, evergreen shrub produces arching stems that bear oval, spiny, dark green leaves. From midsummer to early autumn, it produces decorative, greenish white catkins. Train the stems on a sheltered wall or fence.

↕3m (10ft) ↔1.5m (5ft)

CLIMBER MEDIUM

Jasminum humile

ITALIAN JASMINE An evergreen, climbing shrub, with glossy, bright green leaves, divided into leaflets, and fragrant, tubular, bright yellow flower clusters from early spring to early summer. Grow in a sheltered site. 'Revolutum' AGM (above) is popular.

↕2.5m (8ft)

CLIMBER MEDIUM

Lathyrus odoratus

SWEET PEA This annual, self-clinging climber has oval, green leaves. In summer, it bears white, pink, purple, and red, sweetly scented blooms. Deadhead regularly to encourage a prolonged display. Best in light shade.

↕3m (10ft)

CLIMBER LARGE

Lonicera x *brownii*

SCARLET TRUMPET HONEYSUCKLE This semi-evergreen or deciduous, twining climber has rounded, blue-green leaves. 'Dropmore Scarlet' (above) has unscented, tubular, scarlet blooms from summer to early autumn. Provide a large support.

↕4m (12ft)

CLIMBER LARGE

Lonicera periclymenum

HONEYSUCKLE A large, twining climber, with dark green leaves. Red berries follow fragrant, white to yellow summer flowers. Prune it in early spring. 'Serotina' (above) has creamy white flowers, streaked dark red-purple.

↕7m (22ft)

CLIMBER LARGE

Lonicera x *tellmanniana*

This deciduous, woody-stemmed, twining climber has oval leaves; the upper leaves are joined and resemble saucers. Bright yellowish orange flowers are carried in clusters at the ends of shoots in late spring and summer.

↕5m (15ft)

CLIMBER LARGE

Parthenocissus henryana

A vigorous, deciduous, self-clinging climber, with palm-shaped, cream-veined dark green leaves. In autumn, the foliage turns fiery crimson when blue-black berries also form. Ideal for a house wall; it performs best in partial shade.

↕10m (30ft) or more

CLIMBER LARGE

Parthenocissus quinquefolia

VIRGINIA CREEPER This is a vigorous, deciduous, self-clinging climber, grown for its rounded, divided, green leaves that turn vibrant red and orange in autumn. Provide it with a large support, such as a house wall or boundary fence.

↕15m (50ft) or more

CLIMBER LARGE

Parthenocissus tricuspidata

BOSTON IVY This vigorous, deciduous, self-clinging climber has large, lobed, green leaves that turn spectacular shades of crimson in autumn. Use it to cover large expanses of a wall or boundary fence. It requires a sturdy support.

↕20m (70ft)

CLIMBER LARGE

Rosa 'Albéric Barbier'

This vigorous, semi-evergreen, rambler rose has disease-resistant, bright green leaves. The clusters of slightly fragrant, double, creamy white blooms are produced in a single flush and last several weeks during summer.

↕5m (15ft) ↔3m (10ft)

CLIMBER MEDIUM

Rosa CONSTANCE SPRY

This climbing rose produces greyish green leaves and an abundance of large, bowl-shaped, pink, double flowers, with a rich myrrh-like scent, which appear for about a month during summer. Add a mulch of organic matter each spring.

↕2m (6ft) ↔1.5m (5ft)

CLIMBER LARGE

Schizophragma integrifolium ♡

CHINESE HYDRANGEA VINE A deciduous, self-clinging climber, with heart-shaped or oval, green leaves and clusters of tiny blooms, with petal-like bracts in summer. It requires a large support.

↕12m (40ft)

CLIMBER LARGE

Solanum crispum

CHILEAN POTATO TREE A semi-evergreen, scrambling climber, with oval, green leaves and fragrant, yellow-eyed violet-blue flower clusters from summer to autumn. Grow in a warm, lightly shaded spot. 'Glasnevin' AGM (above) has purple-blue flowers.

↕6m (20ft)

CLIMBER LARGE

Rosa 'Mermaid' ♡

This is a slow-growing climbing rose, with glossy, green, disease-resistant foliage. Repeat-flowering from summer to autumn, it bears flat, single, primrose-yellow blooms. The stems have large, hooked thorns; plant it to deter intruders.

↕↔6m (20ft)

CLIMBER LARGE

Rosa 'New Dawn' ♡

This climbing rose has glossy, green, disease-resistant foliage, and blooms repeatedly from summer to autumn, producing clusters of fragrant, double, pale pink flowers. Deadhead regularly to prolong the display.

↕↔5m (15ft)

SHRUB LARGE

Stachyurus praecox ♡

A deciduous, spreading shrub, with purplish red shoots and pale greenish yellow, bell-shaped flowers that hang from bare stems from late winter to early spring. The slim, dark green leaves appear soon after. Train it against a wall or fence as a wall shrub.

↕4m (12ft) ↔3m (10ft)

CLIMBER LARGE

Trachelospermum jasminoides ♡

STAR JASMINE This evergreen, twining climber has sweetly fragrant, white flower clusters in summer. Its glossy, dark green leaves turn bronze in autumn and winter, and it is ideal for partly shaded walls.

↕9m (28ft)

CLIMBER MEDIUM

Tropaeolum speciosum ♡

FLAME CREEPER A perennial climber, with twining stems and rounded, blue-green leaves, divided into oval leaflets. In summer, it bears scarlet flowers, followed by spherical, bright blue fruits. Best in fertile soil; it tolerates light shade only.

↕3m (10ft)

OTHER SUGGESTIONS

Climbers

Asteranthera ovata • *Berchemia racemosa* • *Billardiera longiflora* ♡ • *Celastrus orbiculatus* • *Celastrus scandens* • *Clematis* 'Blue Bird' • *Clematis macropetala* • *Clematis montana* 'Freda' • *Clematis tangutica* • *Clematis* 'White Columbine' ♡ • *Fallopia baldschuanica* • *Hedera colchica* • *Hedera helix* 'Glacier' ♡ • *Hedera hibernica* • *Hedera nepalensis* • *Holboellia latifolia* • *Lathyrus latifolius* ♡ • *Lonicera* x *italica* • *Lonicera japonica* 'Halliana' • *Lonicera sempervirens* • *Rosa* WHITE CLOUD • *Schizophragma hydrangeoides* • *Vitis vinifera* 'Purpurea' ♡ • *Wisteria sinensis*

Plants for cracks in walls and paving

Dull walls and paved patios in shady gardens can be rebuilt into decorative features with some leafy ferns or small flowering plants between the cracks.

While some plants, such as small ferns, may seed themselves in dry stone walls, creeping campanulas will grow in the most inhospitable of places, their stems of small, blue flowers enlivening structures in deep shade. Shady paving offers a good home for a range of pretty saxifrages, and the bright yellow flowers of *Lysimachia nummularia* are perfect for lightening paving or walls in gloomy gardens. For year-round colour in shaded walls and patios, opt for evergreen bugles (*Ajuga*), the hart's tongue fern (*Asplenium*), *Leptinella potentillina*, and small-leaved ivies.

PERENNIAL SMALL

Alchemilla alpina

ALPINE LADY'S MANTLE A mound-forming perennial, with lobed, white-edged green leaves, silky-haired beneath. Spikes of tiny, greenish yellow flowers appear in summer. Ideal for both cracks in walls and paving. Cut back untidy foliage and flowers.

↕15cm (6in) ↔60cm (24in) or more

PERENNIAL MEDIUM

Asplenium scolopendrium 🏆

HART'S TONGUE FERN This evergreen fern has an upright rosette of tongue-shaped, long, leathery, bright green fronds that have spores across the undersides. Plant in shady cracks in walls and paving. Crispum Group AGM (above) has crinkled-edged leaves.

↕70cm (28in) ↔60cm (24in)

PERENNIAL SMALL

Athyrium niponicum var. *pictum* 🏆

JAPANESE PAINTED FERN This deciduous fern has deeply cut, grey-green, purple- and silver-tinted leaves. Plant it in damp shaded paving or walls. Thrives in deep shade; protect from hard frost.

↕30cm (12in) ↔indefinite

PERENNIAL SMALL

Ajuga reptans

BUGLE A tough, evergreen perennial that spreads by creeping stems to form mats of dark green leaves. Short spikes of small, deep blue flowers are produced in late spring and early summer. Plant it in a shady paving or cracks in a wall.

↕15cm (6in) ↔90cm (36in)

PERENNIAL SMALL

Campanula poscharskyana

TRAILING BELLFLOWER This spreading perennial forms low mounds of round, serrated-edged, mid-green leaves. Starry, violet flowers appear on leafy stems from summer to early autumn. Use it to cascade from walls or spread between paving.

↕15cm (6in) ↔indefinite

PERENNIAL SMALL

Corydalis lutea

This evergreen perennial forms a mound of delicate, ferny, grey-green leaves. Slender, yellow flowers, with spurred petals, appear from late spring to summer. Sprinkle the seeds in paving and on top of stone walls.

↕↔30cm (12in)

PERENNIAL SMALL

Erigeron aureus

FLEABANE A clump-forming perennial, with small, spoon-shaped, hairy, grey-green leaves and large, daisy-like, golden-yellow blooms in summer. Plant it in cracks in walls and paving; tolerates light shade only. 'Canary Bird' AGM (above) has pale blooms.

↕10cm (4in) ↔15cm (6in)

PERENNIAL SMALL

Geranium asphodeloides

ASPHODEL CRANE'S BILL A spreading, deciduous perennial, with small, rounded, deeply cut leaves and masses of starry, white or light pink, magenta-veined flowers in early summer. Grow it between paving stones or in cracks in walls in part shade.

↕↔30cm (12in)

PERENNIAL SMALL

Phlox subulata

This evergreen perennial has narrow, pointed, green leaves. In early summer, it becomes carpeted with starry flowers in shades of purple, red, pink, lilac, or white, depending on the variety. It is ideal for cracks in walls in dappled shade.

↕15cm (6in) ↔50cm (20in)

PERENNIAL SMALL

Geranium dalmaticum

DALMATIAN GERANIUM An evergreen, except in severe winters, mat-forming perennial, with glossy, aromatic, dissected, dark green leaves. Round, shell-pink flowers appear in summer. It grows taller in shade, and is good for paving cracks.

↕10cm (4in) or more ↔20cm (8in)

PERENNIAL SMALL

Geranium sanguineum

BLOODY CRANESBILL This hummock-forming, spreading perennial has deeply dissected, dark green foliage and abundant round, magenta-pink summer flowers. Use it to fill gaps in paving or walls; deadhead regularly to prolong flowering.

↕25cm (10in) ↔30cm (12in) or more

PERENNIAL SMALL

Mentha requienii

CORSICAN MINT This semi-evergreen, mat-forming perennial has diminutive, rounded, apple-green leaves that emit a peppermint scent when crushed. Tiny, lavender-purple flowers appear in summer. Grow it in full or part shade in paving.

↕1cm (1/2in) ↔indefinite

PERENNIAL SMALL

Saxifraga fortunei

SAXIFRAGE A semi-evergreen perennial that forms rosettes of rounded, frilly, green leaves, red beneath. In autumn, it produces airy sprays of tiny, starry, white flowers. Grow it in cracks in paving in part or deep shade.

↕↔30cm (12in)

PERENNIAL SMALL

Saxifraga Southside Seedling Group

SAXIFRAGE An evergreen perennial, with pale green, leafy rosettes, which die back after flowering, and open, cup-shaped, red-banded white blooms from late spring to early summer. Plant in paving or on wall tops.

↕30cm (12in) ↔20cm (8in)

PERENNIAL SMALL

Silene schafta

AUTUMN CATCHFLY A spreading perennial, with narrow, lance-shaped, green leaves. From late summer to late autumn, it bears small, upright sprays of starry, rose-magenta flowers with notched petals. Grow it in wall or paving cracks.

↕25cm (10in) ↔30cm (12in)

PERENNIAL SMALL

Viola odorata

SWEET VIOLET This semi-evergreen, spreading perennial forms loose mats of heart-shaped, toothed-edged leaves and fragrant, flat-faced, violet or white flowers from late winter to early spring. Leave it to spread through paving cracks in part shade.

↕20cm (8in) ↔30cm (12in)

PERENNIAL SMALL

Viola sororia

BLUE VIOLET This small perennial has rounded, dark green leaves and dainty, blue flowers in summer. 'Freckles' (above) has white flowers, marked with violet-purple speckles. Like most violas, it self-seeds in paving and wall cracks.

↕12cm (5in) ↔15cm (6in)

OTHER SUGGESTIONS

Perennials

Ajuga reptans 'Chocolate Chip' • *Anemone sylvestris* • *Arenaria purpurascens* • *Asplenium trichomanes* • *Campanula carpatica* • *Campanula gaganica* • *Dodecatheon meadia* f. *album* • *Galium odoratum* • *Geranium cinereum* 'Ballerina' • *Leptinella potentillina* • *Lysimachia nummularia* 'Aurea' • *Oxalis enneaphylla* • *Phlox* 'Chattahoochee' • *Sanguinaria canadensis* f. *multiplex* 'Plena' • *Saxifraga hypnoides* • *Saxifraga* 'Major'

Bulbs

Narcissus bulbocodium • *Narcissus* 'Pipit' • *Scilla siberica*

Shrubs and climbers

Hedera helix 'Little Diamond' • *Juniperus squamata* 'Blue Star'

Plants for patios, balconies, and windowsills

While most colourful bedding plants require sun to bloom, shady patio and balcony displays need not be restricted to leaves alone.

Many annuals will flower in light shade, but if your area is darker, choose begonias, fuchsias, and impatiens for summer-long displays. For autumn and winter troughs on windowsills, include some young shrubs, such as skimmias and *Gaultheria*, which produce pretty buds and berries and can be planted in pots of their own when they outgrow their first home; shrubs can be kept small by clipping. In autumn, squeeze daffodil and muscari bulbs between the shrubs for a burst of colour in early spring, and if you have space, also add a few large shrubs in bold containers to lend permanent structure to your design.

PERENNIAL SMALL

Begonia Olympia Series

This is an evergreen perennial, often sold as annual summer bedding. It produces glossy, green leaves, and small, white, pink, or red flowers with golden eyes. Plant it in pots and baskets of multi-purpose compost.
↕↔25cm (10in)

PERENNIAL SMALL

Bellis perennis

DAISY This evergreen perennial produces rosettes of spoon-shaped, dark green leaves and white daisies in late spring. Tasso Series has pink, white, or red, pompon flowers. Plant in pots, baskets, and windowboxes in multi-purpose compost.
↕↔20cm (8in)

ANNUAL/BIENNIAL MEDIUM

Calendula officinalis

POT MARIGOLD A bushy annual, with lance-shaped, aromatic, pale green leaves and single or double, daisy-like flowers in shades of yellow and orange from spring to autumn. Sow seeds directly in containers of multi-purpose compost.
↕↔60cm (24in)

SHRUB LARGE

Camellia x *williamsii*

This evergreen shrub has oval, pointed, glossy, bright green leaves and, in spring, large clusters of pink or white blooms. Varieties include 'Donation' AGM (above), with pink, semi-double flowers. Plant it in large pots of ericaceous compost.
↕5m (15ft) ↔2.5m (8ft)

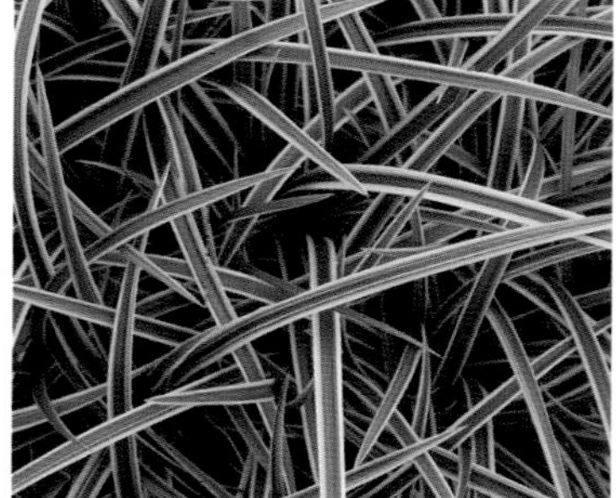

PERENNIAL MEDIUM

Carex 'Ice Dance'

JAPANESE SEDGE An evergreen perennial sedge, featuring mounds of grassy, green leaves, with creamy white margins, and small, white spring flowers. Plant it in pots and windowboxes in soil-based compost; water well. Potentially invasive.
↕60cm (24in) ↔75cm (30in)

PERENNIAL SMALL

Carex oshimensis

ORNAMENTAL SEDGE This evergreen perennial has glossy, arching, cream-striped dark green leaves and insignificant flower spikes in summer. Yellow-striped 'Evergold' (above) gives bright winter interest in pots and windowboxes of soil-based compost.
↕↔20cm (8in)

PERENNIAL SMALL

Carex siderosticha

BROAD-LEAVED SEDGE A deciduous or semi-evergreen perennial, with arching, grassy, green leaves and small, brown flower spikes in spring. Plant it in pots or baskets of soil-based compost. 'Variegata' (above) has cream-edged foliage.
↕30cm (12in) ↔40cm (16in)

SHRUB MEDIUM

Clethra alnifolia

SWEET PEPPER BUSH This deciduous shrub has oval, serrated, mid-green leaves. From late summer to early autumn, it produces spikes of fragrant, bell-shaped, white flowers. Plant it in a large pot of ericaceous compost in full or part shade.
↕↔2.5m (8ft)

BULB SMALL

Cyclamen coum

EASTERN CYLAMEN A dwarf perennial bulb, with rounded, dark green, often silver-marbled leaves. In late winter, deep pink flowers, with a purple blotch at the base of each twisted, swept back petal, appear. Plant in soil-based compost mixed with grit.
↕10cm (4in)

BULB SMALL

Cyclamen hederifolium

IVY-LEAVED CYCLAMEN A dwarf perennial with silvery green patterned leaves that appear just after the autumn flowers. Its pink, swept-back flower petals are darker at the mouth. Ideal for containers filled with gritty, soil-based compost.

↕10cm (4in)

PERENNIAL SMALL

Dichondra argentea

SILVER PONYFOOT An evergreen perennial, 'Silver Falls' (above) is sold as summer bedding. With its cascading, silver stems of round, shiny, silvery green foliage, it is ideal for a windowbox or basket. Tolerates light shade only; plant in multi-purpose compost.

↕50cm (20in) ↔ indefinite

SHRUB SMALL

Fuchsia 'Genii'

A deciduous, bushy shrub, with oval, pointed, gold-green foliage. It produces single or double, pendent, cerise and purple flowers from summer to autumn. Plant it in a pot of soil-based compost. Shelter from cold, drying winds.

↕↔75cm (30in)

SHRUB SMALL

Fuchsia 'Tom Thumb'

This dwarf, deciduous shrub produces oval, dark green leaves. From summer to autumn, a profusion of pendent, bell-shaped, red and mauve-blue flowers appears. Grow it in windowboxes, pots, and baskets in soil-based compost.

↕↔50cm (20in)

BULB MEDIUM

Galanthus nivalis

SNOWDROP A perennial bulb, with grass-like, green foliage and nodding, white flowers, which appear in late winter. Plant it with other early-flowering bulbs and violas in pots or windowboxes filled with soil-based compost.

↕15cm (6in)

SHRUB SMALL

Gaultheria procumbens

CHECKERBERRY A dwarf, evergreen shrub with oval, leathery leaves that are tinged red in winter. Small, pink-flushed white summer flowers are followed by scarlet berries. Plant it in a winter basket or windowbox in ericaceous compost.

↕15cm (6in) ↔ indefinite

PERENNIAL LARGE

Gaura lindheimeri

WAND FLOWER This upright perennial has small, lance-shaped, green leaves and tall, slim stems, dotted with starry, butterfly-shaped, pink-budded white flowers, which appear all summer. Grow it in a sheltered spot in large pots of soil-based compost.

↕1.5m (5ft) ↔ 90cm (36in)

SHRUB SMALL

Heliotropium arborescens

CHERRY PIE A bushy, evergreen shrub, grown as an annual, with oval, dark green, crinkled leaves and clusters of fragrant, purple or white flowers throughout summer. Plant it in pots of multi-purpose compost near seating or paths.

↕↔45cm (18in)

PERENNIAL SMALL

Heuchera 'Plum Pudding'

This evergreen perennial is grown for its rounded, lobed and veined, maroon-purple leaves. It produces slim stems of small, white flowers in summer. Grow it as a foil for colourful flowers in windowboxes and pots filled with soil-based compost.

↕50cm (20in) ↔ 30cm (12in)

PERENNIAL SMALL

x *Heucherella* 'Tapestry'

CORAL BELLS This evergreen or semi-evergreen perennial has deeply lobed, green foliage, with purple centres and veins. In early summer, sprays of small, pink flowers appear. Plant it in windowboxes or pots in a sheltered spot. Remove spent blooms.

↕↔30cm (12in)

CLIMBER MEDIUM

Lathyrus odoratus

SWEET PEA This annual climbs by tendrils. It has divided, mid-green leaves and scented pink, blue, purple, or white flowers from summer to early autumn. Plant it in large pots of multi-purpose compost, with a tripod support. Deadhead regularly.

↕ up to 3m (10ft)

BULB MEDIUM

Narcissus 'Jack Snipe'

DAFFODIL This early- to mid-spring-flowering, dwarf perennial bulb has narrow, dark green leaves and creamy white flowers, with short, yellow cups. Plant it in groups in containers and windowboxes filled with multi-purpose compost.

↕ 23cm (9in)

PERENNIAL SMALL

Impatiens New Guinea Group

This shrubby perennial, grown as an annual, has glossy, lance-shaped, green leaves and vibrant, pink, red, lavender, and white flowers, which appear in summer. It is suitable for containers filled with multi-purpose compost.

↕ 35cm (14in) ↔ 30cm (12in)

ANNUAL/BIENNIAL SMALL

Lobelia erinus

TRAILING LOBELIA This annual produces branching stems of dark green leaves and masses of tiny, blue, pink, or white flowers from summer to autumn. Trailing and bushy varieties are available. Grow it in multi-purpose compost; water daily in summer.

↕ 20cm (8in) ↔ 15cm (6in)

BULB MEDIUM

Narcissus 'Tête-à-tête'

DAFFODIL This spring perennial bulb has strap-shaped leaves. It is a dwarf, early-flowering form, bearing bright golden-yellow blooms. Plant it in windowboxes and baskets filled with multi-purpose compost.

↕ 30cm (12in)

BULB MEDIUM

Muscari armeniacum

GRAPE HYACINTH A spring-flowering perennial bulb, with narrow, grassy leaves, and short spikes of small, fragrant, bell-shaped, blue flowers held in cone-shaped clusters. Combine with other spring bulbs in pots of soil-based compost and grit.

↕ 20cm (8in)

ANNUAL/BIENNIAL SMALL

Nemophila maculata

FIVE-SPOT BABY A fast-growing, spreading annual, with lobed leaves. It bears small, bowl-shaped, white flowers, with purple petal tips, throughout summer, provided it is watered well. Grow it in multi-purpose compost as edging in mixed containers.

↕↔ up to 30cm (12in)

PERENNIAL SMALL

Impatiens walleriana

BUSY LIZZIE This evergreen perennial, grown as annual, has oval, green leaves, and a proliferation of rounded blooms in shades of red, pink, purple, violet, orange, or white from spring to autumn. Plant it in pots of multi-purpose compost.

↕↔ up to 30cm (12in)

BULB MEDIUM

Narcissus 'Hawera'

DAFFODIL This perennial bulb has upright, green, strap-like foliage, and is grown for its small, nodding, yellow flowers, held on upright stems in mid-spring. Best in light shade only; plant it in containers and windowboxes.

↕ 15cm (6in)

ANNUAL/BIENNIAL SMALL

Nicotiana Domino Series

TOBACCO PLANT This tall perennial, grown as an annual, has long, sticky, mid-green leaves. Heads of long, tubular, white flowers appear in summer. Plant in large pots and support the stems; place where its evening fragrance can be enjoyed.

↕ 45cm (18in) ↔ 40cm (16in)

PERENNIAL LARGE

Nicotiana sylvestris ♡

FLOWERING TOBACCO This tall perennial, grown as an annual, has long, sticky, mid-green leaves. Heads of long, tubular, white flowers appear in summer. Plant it in large pots of multi-purpose compost, where you can enjoy its evening fragrance.

↕1.5m (5ft) ↔75cm (30in)

PERENNIAL SMALL

Ophiopogon planiscapus 'Nigrescens'

BLACK MONDO GRASS An evergreen, clump-forming perennial, with leathery, grass-like, black leaves. Black berries follow small, bell-shaped, white or mauve summer flowers. Plant it in pots of soil-based compost.

↕23cm (9in) ↔30cm (12in)

PERENNIAL SMALL

Petunia Surfinia Series

This group of trailing petunias produces dark green leaves and a profusion of trumpet-shaped flowers in shades of pink, purple, red, white, and yellow. Use them to trail from windowboxes, baskets, and pots of multi-purpose compost.

↕40cm (16in) ↔90cm (36in)

SHRUB MEDIUM

Rhododendron yakushimanum ♡

YAKUSHIMA RHODODENDRON An evergreen, compact shrub, with leathery, green leaves, felted beneath, and bell-shaped, pink-budded white flowers in spring. Grow it in ericaceous compost; water in dry periods.

↕↔up to 2m (6ft)

ANNUAL/BIENNIAL SMALL

Phlox drummondii

ANNUAL PHLOX A bushy annual, with neat, green leaves and clusters of small, pink, red, purple, blue, and white flowers, which appear throughout summer. Use it to edge a container display, and grow it in multi-purpose compost.

↕45cm (18in) ↔25cm (10in)

SHRUB LARGE

Physocarpus opulifolius

NINEBARK A deciduous shrub, with lobed, green, yellow, or purple-red leaves. Brown fruits follow domed clusters of pale pink summer blooms. Grow it in a large pot of soil-based compost; prune after flowering. 'Lady in Red' (above) has bold red foliage.

↕up to 3m (10ft) ↔1.5m (5ft)

PERENNIAL SMALL

Primula Crescendo Series

POLYANTHUS This perennial, grown as an annual, has rosettes of corrugated, dark green leaves. From winter to spring, it bears clusters of yellow-eyed flowers in many colours. Plant it in pots of multi-purpose compost.

↕↔20cm (8in)

SHRUB SMALL

Skimmia japonica

A bushy, evergreen shrub, with oval, mid- to dark green leaves and clusters of fragrant, white spring flowers, followed by red berries on females (above). The male 'Rubella' has crimson clusters of winter buds only. Plant it in pots of soil-based compost.

↕↔up to 1m (3ft) in a pot

ANNUAL/BIENNIAL SMALL

Torenia fournieri

WISHBONE FLOWER This bushy annual has serrated, light green leaves and, from summer to early autumn, flared blooms, with pale lilac and blue-purple petals and white and yellow throats. Plant in a shaded windowbox in multi-purpose compost.

↕30cm (12in) ↔20cm (8in)

OTHER SUGGESTIONS

Perennials

Ipomoea batatas 'Blackie' • *Lamium maculatum* 'White Nancy' • *Primula vulgaris* 'Miss Indigo' • *Solenostemon scutellarioides* • *Sutera* 'Snowflake' • *Teucrium chamaedrys*

Bulbs

Cyclamen repandum subsp. *repandum* ♡ • *Galanthus plicatus* ♡

Shrubs, trees, and climbers

Acer palmatum 'Bloodgood' ♡ • *Buxus sempervirens* 'Suffruticosa' • *Choisya* SUNDANCE • *Fatsia japonica* ♡ • *Fuchsia* 'Doctor Foster' ♡ • *Fuchsia* 'Mrs Popple' • *Hebe* 'Red Edge' • *Hedera helix* cvs • *Heliotropium* 'Princess Marina' ♡ • *Hydrangea* Endless Summer Series • *Rhododendron impeditum* • *Skimmia* x *confusa* 'Kew Green' ♡

Plant focus: begonias

Sporting colourful flowers and patterned leaves, begonias are ideal plants for shady, summer container and bedding displays.

MOST BEGONIAS ARE TENDER PERENNIALS, grown in shady summer gardens. The rex and rhizomatous forms, with their colourful, intricately patterned foliage, create dramatic displays in pots. The trailing pendula group offers both foliage and flower interest, and makes excellent hanging basket plants, while the taller, upright, cane-forming hybrids, and large-flowered tuberous types are perfect for larger summer containers. The compact Semperflorens begonias, often sold as bedding, produce a frill of colourful flowers around beds and borders. Most begonias thrive in warm, sheltered, shady gardens; where temperatures dip below 10°C (50°F) for long periods in summer, grow plants in a conservatory or indoors. Avoid overwatering, as many dislike wet soil, and apply a balanced fertilizer weekly when plants are in full growth. Water evergreens sparingly in winter and store the dormant tubers of deciduous plants in a cool, dry, frost-free place – repot in late spring and plant out when the risk of frost has passed.

USING BEGONIAS

In cool areas, begonias are best grown on an outdoor windowsill, where the heat of the building provides additional warmth and encourages the plants to bloom. Trailing begonias are suitable for hanging baskets and tall pots, but give them a sheltered position to prevent damage to the brittle leaves and stems. Keep plants well watered during summer.

Semperflorens begonias tolerate some sun and can be used as decorative edging plants or as the ingredients for a traditional bedding scheme.

Rex begonias, with their large ornamental leaves, offer a textural contrast to colourful foliage plants, such as flame nettles, in a contemporary patio container display.

POPULAR BEGONIA SPECIES

Rex begonias These perennials, mostly evergreens, have colourful foliage, which is often arranged on plants in a spiral pattern. Small, inconspicuous flowers appear in spring.

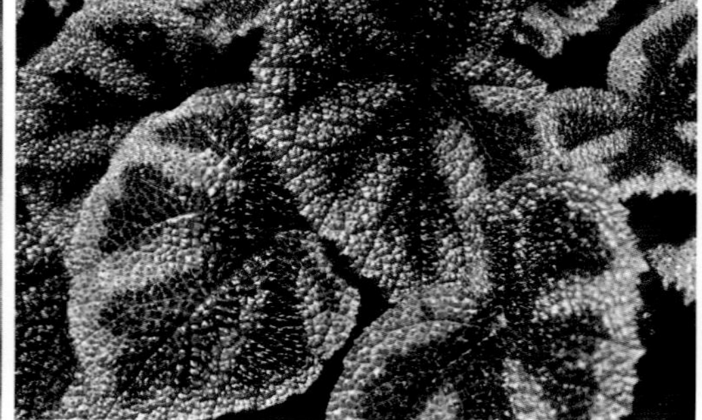

Rhizomatous begonias These are grown for their decorative, mostly evergreen, leaves, which are often arranged in spirals, have unusual textures, and feature colourful markings.

Semperflorens begonias These compact, bushy evergreen perennials, often grown as annual bedding, have rounded, bronze or green leaves and small, single or double summer flowers.

Tuberous begonias Either upright and bushy, or trailing and pendulous, these perennials offer tear-shaped foliage and bright summer flowers from winter-dormant tubers.

Shrub-like begonias Often grown as house plants, these evergreens have ornamental leaves and small flowers from spring to summer. They can be grown outside in summer in warm areas.

Cane-stemmed begonias Tough, bamboo-like stems bear silver-spotted or splashed leaves, some finely cut or with wavy margins, and showy, pink or white summer flowers.

Plants for productive patios

You'll struggle to create a productive patio in deep shade, but partial shade affords the cool growing conditions preferred by a number of leafy crops.

Lettuces, chard, rocket, herbs, such as mint and parsley, and other leafy crops thrive in lightly shaded sites. Fruit crops that will ripen in light or dappled shade include blackberries, alpine strawberries, currants, and some apple varieties, so although the choice is not extensive, there is still plenty of scope to grow your own. Plant fruit trees and bushes in large tubs of soil-based compost, and replace the top layer with fresh compost mixed with an all-purpose granular fertilizer each spring. Apart from peas and beans, which need big containers, vegetables and herbs will be happy in medium-sized pots of multi-purpose compost.

BULB MEDIUM

Chives

Allium schoenoprasum This herb forms a clump of slender, upright, green leaves that have a mild, onion-like flavour. Its pale pink summer flowers can also be cropped. Plant it in windowboxes and pots in multi-purpose compost.
↕30cm (12in)

ANNUAL/BIENNIAL MEDIUM

Celeriac

Apium graveolens This crop is grown for its swollen roots that have a mild nutty, celery-like flavour. Best in large containers, sow seeds in spring and harvest baby roots in summer or mature crops in autumn. Water plants regularly.
↕90cm (36in) ↔45cm (18in)

ANNUAL/BIENNIAL SMALL

Land cress

Barbarea verna A low-growing annual, this alternative to watercress has pepper-flavoured leaves. Sow seed in windowboxes and pots in spring and summer; harvest the leaves and stems when large enough. Re-sow regularly; water and feed well.
↕↔30cm (12in)

ANNUAL/BIENNIAL SMALL

Swiss chard

Beta vulgaris This is grown for its leaves, which can be eaten raw or steamed, and its fleshy stems that are best cooked. Sow seeds in spring and harvest during summer. Ideal for containers and windowboxes; water the plant regularly.
↕↔45cm (18in)

ANNUAL/BIENNIAL SMALL

Mustard greens

Brassica juncea Grown for its peppery leaves, which can be repeat-harvested when young or left to mature. Sow seeds in spring and summer. Grow it in containers and windowboxes in light shade only. Water regularly.
↕↔30cm (12in)

ANNUAL/BIENNIAL SMALL

Kale

***Brassica oleracea* Acephala Group**
This upright, leafy crop is planted in summer and can be harvested from autumn to winter. When grown in containers, it is best picked as individual leaves when required. Water well and stake if required.
↕↔45cm (18in)

ANNUAL/BIENNIAL SMALL

Mizuna

Brassica rapa* subsp. *nipposinica* var. *laciniata Sown from seed from spring to summer, this leaf crop has a peppery flavour. Harvest after a few weeks as a "cut-and-come-again" salad, or cut whole plants at the base. Ideal for containers.
↕↔15cm (6in)

PERENNIAL MEDIUM

Chicory

Cichorium intybus This green-leaved chicory has rounded, lettuce-like heads of crisp, bitter-tasting leaves. Planted in early summer, it can be harvested until spring, if protected. Grow in containers of multi-purpose compost; keep well watered.

↕↔30cm (12in) when harvested

PERENNIAL SMALL

Radicchio

Cichorium intybus This red-leaved chicory forms rounded heads of crisp, bitter-tasting leaves. Planted in early summer, it can be harvested until spring, if protected. Grow it in containers of multi-purpose compost; water well.

↕↔30cm (12in) when harvested

ANNUAL/BIENNIAL SMALL

Coriander

Coriandrum sativum This quick-growing, annual herb has rounded, green, aromatic leaves. The leaves, stems, and seeds can be used in cooking. Grown from seed, it is ideal for containers and windowboxes in light shade; water well to prevent bolting.

↕50cm (20in) ↔20cm (8in)

ANNUAL/BIENNIAL SMALL

Rocket salad

Eruca sativa This annual forms a rosette of long, lobed leaves, which have a peppery flavour. Sow it directly in pots and windowboxes in spring and summer, and cut the leaves as required, leaving the base to regrow. Water well.

↕↔20cm (8in) when harvested

PERENNIAL LARGE

Sweet fennel

Foeniculum vulgare This tall, upright perennial produces fern-like, aromatic leaves that taste of aniseed and clusters of tiny, yellow flowers, which are followed by edible seeds. Grow it in large containers of soil-based compost in light shade.

↕1.8m (6ft) ↔45cm (18in)

ANNUAL/BIENNIAL SMALL

Florence fennel

Foeniculum vulgare* var. *azoricum Grown for its swollen leaf stems, this annual forms an aniseed-flavoured "bulb" at the base of the plant. Sow in summer and harvest young if large enough, or leave to reach full-size. Grow in multi-purpose compost.

↕↔30cm (12in)

PERENNIAL SMALL

Alpine strawberries

Fragaria vesca Forming a low mound of toothed-edged leaves, this type of strawberry is ideal for containers. It bears small, richly flavoured fruits from early- to late summer. Water and feed plants regularly; best in light shade only.

↕30cm (12in) ↔indefinite

ANNUAL/BIENNIAL SMALL

Lettuce

Lactuca sativa Suitable for cropping young as "cut-and-come-again" leaves or left to mature into full heads, lettuce comes in many colours and textures. Sow or plant in spring and keep plants well watered. Best in light shade only.

↕↔30cm (12in)

TREE SMALL

Cooking apples

Malus domestica This tree bears sour fruits from late summer to autumn. Buy plants on dwarfing rootstocks to restrict their size. Plant in a large container of soil-based compost; feed and water regularly from spring to summer. Best in light shade.

↕↔up to 8m (25ft)

TREE SMALL

Dessert apples

Malus domestica Many varieties are available, but choose one grafted onto a dwarfing rootstock to restrict its size. Ideal for large containers of soil-based compost; feed and water regularly. You may also need a pollinator variety.

↕up to 6m (20ft) ↔up to 4m (12ft)

PERENNIAL MEDIUM

Lemon balm

Melissa officinalis This herb has toothed-edged, textured, green leaves that release a rich lemon scent when brushed. Remove the flowers in summer to encourage leaf growth. Grow in pots and windowboxes on its own; water regularly.

↕90cm (36in) ↔45cm (18in)

PERENNIAL SMALL

Peppermint

Mentha x piperita A spreading perennial, with spikes of pink-purple summer flowers and textured, dark green, aromatic leaves, which can be used to flavour savoury dishes or as an infusion to make tea. Plant in pots of soil-based compost; water regularly.

↕50cm (20in) ↔1m (3ft)

PERENNIAL SMALL

Spearmint

Mentha spicata A spreading perennial, with textured, bright green, aromatic foliage and short spikes of pink summer flowers. Infuse the leaves to make tea, or use to flavour savoury dishes. Plant in a container to prevent the roots spreading.

↕50cm (20in) ↔indefinite

PERENNIAL SMALL

Marjoram

Origanum vulgare This clump-forming perennial forms a dense mat of aromatic, rounded, green leaves, which can be used to flavour savoury dishes. It may also bear tiny, mauve flowers in summer. Plant it in pots of gritty, soil-based compost.

↕↔45cm (18in)

ANNUAL/BIENNIAL MEDIUM

Parsley

Petroselinum crispum This clump-forming biennial, best grown as an annual, has deeply divided, aromatic green leaves, which are used to flavour a wide range of savoury dishes, and for decorative, edible garnishes.

↕80cm (32in) ↔60cm (24in)

CLIMBER SMALL

Runner beans

Phaseolus coccineus This climbing bean bears long, flattened green pods all summer. Harvest regularly, before the seeds swell, for the best crop. Provide tall supports, and water and feed well. Ideal for large containers in light shade.

↕1.8m (6ft)

ANNUAL/BIENNIAL MEDIUM

Garden peas

Pisum sativum These traditional, podded peas crop from late spring to autumn when sown in spring or early summer. Plant in large containers in light shade and water well. Dwarf and climbing varieties are available; harvest regularly.

↕70cm (28in) ↔50cm (20in)

ANNUAL/BIENNIAL MEDIUM

Mangetout peas

Pisum sativum These peas are grown for their crisp, flattened pods, which are produced from late spring to autumn, and eaten whole. Sow or plant in large containers in spring or early summer. Water well and pick regularly.

↕90cm (36in) ↔10cm (4in)

ANNUAL/BIENNIAL MEDIUM

Sugar snap peas

Pisum sativum This short, climbing plant bears crisp pods from late spring to autumn, if sown in spring or early summer. The pea pods are eaten whole. Best planted in large containers in light shade; keep well watered and pick regularly.

↕60cm (24in) ↔50cm (20in)

TREE SMALL

Acid cherries

Prunus cerasus This small tree bears white or pink flowers in spring, followed by sharp-tasting red berries in summer. Select self-fertile varieties on dwarfing rootstocks. Best in large pots of soil-based compost in light shade; water and feed regularly.

↕↔4m (12ft)

SHRUB MEDIUM

Blackcurrants

Ribes nigrum This upright, branching shrub flowers in spring and bears strings of glossy, black, sharp-tasting fruits in summer. Grow it in a large container of soil-based compost in light shade. Mulch in spring; water well in summer.

↕↔1.5m (5ft)

ANNUAL/BIENNIAL SMALL

Summer radishes

Raphanus sativus This type of radish is grown for its crisp, rounded, peppery roots. Sow seeds in pots and windowboxes from spring to early autumn. Pull the roots once large enough, after five weeks, then re-sow. Ideal for light shade; water well.

↕↔20cm (8in)

PERENNIAL LARGE

Rhubarb

Rheum x hybridum A vigorous perennial, with large, glossy, green leaves. It is grown for its sharp-tasting stems, which are eaten cooked. Plant in a large container of soil-based compost; water and feed well. Harvest stems from spring to midsummer.

↕↔up to 2m (6ft)

SHRUB MEDIUM

Redcurrants

Ribes rubrum This upright, branching shrub flowers in spring and produces strings of glossy, sharp-tasting, red fruits. Plant it in a large container of soil-based compost. Mulch with compost in spring, and water regularly during summer.

↕↔1.5m (5ft)

CLIMBER MEDIUM

Blackberries

Rubus fruticosus This vigorous, climbing shrub flowers in spring and bears clusters of sweet, black berries in summer. It is best trained against a support; plant in a large container of soil-based compost. Be careful of its sharp thorns.

↕2.5m (8ft)

CLIMBER MEDIUM

Thornless blackberries

Rubus fruticosus This climbing shrub has dissected leaves and thornless stems and, from late summer to autumn, clusters of large, black berries. Plant in a large container of soil-based compost in light shade, and provide support for its stems.

↕2.5m (8ft)

PERENNIAL SMALL

Sweet violets

Viola odorata A semi-evergreen perennial, with heart-shaped leaves and, from late winter to early spring, fragrant, flat-faced, violet flowers, which can be added to salads or crystallized to make cakes and desserts. 'Alba' (above) has white flowers.

↕20cm (8in) ↔30cm (12in)

OTHER SUGGESTIONS

Fruit trees

Apple 'Ashmead's Kernel' • Apple 'Royal Gala'

Soft fruits

Blackberry 'Loch Ness' • Redcurrant 'Rovada'

Herbs

Apple mint • Garlic chives • Moroccan mint

Vegetables

Chop suey greens • Kale 'Nero di Toscana' • Kale 'Redbor' • Lambs lettuce • Lettuce 'Little Gem' • Lettuce 'Red Salad Bowl' • Lettuce 'Winter Density' • Mustard Oriental 'Giant Red' • Pea 'Ambassador' • Pea 'Waverex' • Radish 'Cherry Belle' • Radish 'Mantanghong' • Runner bean 'Hestia' • Spinach 'Mikado' • Spring onion 'White Lisbon' • Swiss chard 'Bright Lights' • Wild rocket

Plant focus: leafy salads

Easy to grow and delicious when fresh, lettuces and other salad leaves are the perfect crop for small gardens and patios.

Despite their high cost in the shops, lettuces and salad leaves are inexpensive and easy to grow from seed, and taste twice as nice when picked fresh. They are also pretty plants and can be used in combination with edible flowers, such as nasturtiums and violas, as part of a container display. To grow salad leaves, sow seeds in modules under cover in early spring, and plant outside in moist but free-draining soil in sun or part shade when a good root system has developed. Sow a few seeds every couple of weeks to extend the harvest period, sowing directly outside from late spring, and keep the plants well watered and protected from slugs. Where space allows, grow heart-forming lettuce and harvest the heads when mature. On smaller plots, grow rocket, mustard, and loose-leaf lettuce varieties as a "cut-and-come-again crop", cutting the leaves about 2cm (¾in) from the soil when plants are 10cm (4in) in height. Keep them well watered, and the plants will regrow to give you repeated summer crops.

USING LEAFY SALADS

Grow salad crops in raised beds, which are close to hand and make picking easy, with edible flowers and herbs. Incorporate varieties with coloured or frilly leaves into bedding schemes as foliage plants, or use to edge beds. Crops can also be grown in individual pots on a windowsill.

Loose-leaved salads can be harvested as "microgreens", grown in pots and trays. Where space is limited, grow them in a green-wall growing system.

Swiss chard grows best in light shade, where it is less prone to bolting. Cut individual leaves when young and tender, leaving the base to regrow.

TYPES OF LEAFY SALAD

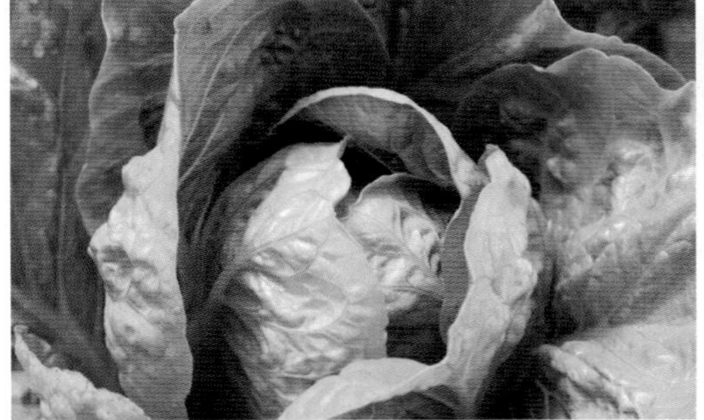

Cos lettuces Also known as romaine lettuces, they produce a tall head of crisp leaves with a rib down the centre of each leaf. Leave plants to form a head before harvesting.

Butterhead lettuces Plants produce heads of tender leaves, with the sweetest at the centre forming a loose heart. Provide them with enough space to mature fully.

Loose-leaf lettuces This group produces loose green or red leaves without a central heart, and includes oak- and frilly-leaved forms, such as 'Lollo Rosso' (*above*).

Crisphead lettuces Known as iceberg lettuces, these plants develop a solid heart of firm, rounded, densely packed pale leaves. These are the traditional summer lettuces.

Rocket Short plants produce dark green, oval or divided leaves with a peppery flavour and edible flowers. Grow it as a "cut-and-come-again" crop or as "microgreens". Sow seeds regularly.

Mustard leaves These hardy plants develop loose leaves, often red-tinged, with a peppery flavour, which intensifies as the plants mature. Harvest the leaves individually when required.

Plants for woodland gardens

Cool, shady glades are home to a range of elegant shrubs, climbers, bulbs, and perennials – visit native woodlands for inspiration.

Use seasonal variations to the best effect in your woodland garden. Spring is the highlight of the year, when bulbs and flowers come into bloom, making the most of the light before the tree canopies darken the woodland floor. As summer approaches, use foliage plants such as ferns to provide interest, and grow foxgloves and other flowers in the partial shade at the edge of the tree line. For an autumn display, choose maples, whose red and orange foliage burns brightly, and berried plants for added colour. Winter stems make an impact when foliage has fallen, creating a graphic picture as temperatures tumble.

PERENNIAL MEDIUM

Actaea pachypoda

WHITE BANEBERRY This clump-forming perennial has divided, bright green leaves. Spikes of fluffy, white flowers in midsummer are followed by oval, black-tipped white berries on stiff, bright red stalks. Plant it in moist soil just beyond tree canopies.

↕1m (3ft) ↔50cm (20in)

PERENNIAL SMALL

Adonis amurensis

A clump-forming perennial, with ferny, mid-green leaves. In early spring, just as the foliage unfolds, single, golden-yellow, cup-shaped blooms, with many small petals, appear. Grow it in soil that does not dry out, in part or full shade.

↕↔30cm (12in)

SHRUB LARGE

Aesculus parviflora

BOTTLEBRUSH BUCKEYE This deciduous, suckering shrub has palm-shaped, bronze leaves, which age to dark green and turn yellow in autumn, and spires of fluffy, white flowers from mid- to late summer. Plant it in gaps between trees.

↕3m (10ft) ↔5m (15ft)

BULB MEDIUM

Allium moly

GOLDEN GARLIC This is a clump-forming perennial bulb, with narrow, lance-shaped, grey-green leaves. In summer, domed heads of star-shaped, golden-yellow blooms appear. It is an ideal plant for naturalizing in deciduous woodland.

↕35cm (14in)

BULB MEDIUM

Anemone blanda 'Violet Star'

WINTER WINDFLOWER A tuber-forming, spreading perennial bulb, with dark green, divided leaves and daisy-like, amethyst-violet flowers, with white centres, in early spring. Plant the tubers in autumn. Keep the plant well watered.

↕15cm (6in)

BULB SMALL

Anemone blanda 'White Splendour'

WINTER WINDFLOWER This tuber-forming perennial bulb makes carpets of divided leaves. It has daisy-like, white flowers that appear in early spring. Plant the tubers in groups in autumn.

↕10cm (4in)

PERENNIAL SMALL

Anemone nemorosa

WOOD ANEMONE This vigorous, carpeting perennial has deeply cut, mid-green leaves and, from spring to early summer, daisy-like flowers with a central tuft of yellow stamens. Suitable for naturalizing, it is available in many colours and flower forms.

↕15cm (6in) ↔30cm (12in)

PERENNIAL SMALL

Anemone sylvestris

SNOWDROP ANEMONE This carpeting perennial has divided, mid-green leaves and, from spring to early summer, white, bowl-shaped flowers, with yellow centres. With its spreading roots, it may become invasive, so is better suited to wild gardens.

↕↔30cm (12in)

PERENNIAL SMALL

Anemonella thalictroides

RUE ANEMONE This tuberous-rooted perennial has delicate, dark blue-green leaves, divided into rounded leaflets and, from spring to early summer, cup-shaped, white or pink blooms. It thrives in dappled shade; plant *en masse* under trees.

↕10cm (4in) ↔4cm (1½in) or more

BULB MEDIUM

Arisaema triphyllum

JACK-IN-THE-PULPIT This perennial bulb has triangular-shaped leaves, divided into three leaflets. Green or purple- and white-striped, hooded summer spathes surround a central column of tiny blooms. Bright red berries follow. Plant the tubers in late winter.

↕50cm (20in)

PERENNIAL SMALL

Arisarum proboscideum

MOUSE PLANT A clump-forming perennial, with arrow-shaped leaves. In spring, it bears strange, maroon blooms with a white base that consists of a "hood", drawn out into a mouse-like tail. The hood conceals the tiny flowers. Dies back after flowering.

↕15cm (6in) ↔30cm (12in)

PERENNIAL SMALL

Asarum caudatum

WILD GINGER This evergreen perennial forms clumps of glossy, heart-shaped, dark green leaves. Curious, pitcher-shaped, reddish to purple-brown flowers, with tail-like lobes, appear in early summer. It thrives in deep shade beneath trees.

↕8cm (3in) ↔25cm (10in) or more

PERENNIAL MEDIUM

Asplenium scolopendrium

HART'S TONGUE FERN This evergreen fern has rosettes of long, slender, crimp-edged, green leaves. It makes a beautiful addition to understorey plantings beneath deciduous trees. Tolerates deep shade. Crispum Group AGM (above) has crinkle-edged leaves.

↕70cm (28in) ↔60cm (24in)

PERENNIAL SMALL

Asplenium trichomanes

MAIDENHAIR SPLEENWORT This compact, evergreen fern produces rosettes of dark-stemmed fronds, divided into rounded segments. It provides a textural carpet beneath deciduous trees, and tolerates deep shade.

↕15cm (6in) ↔30cm (12in)

PERENNIAL SMALL

Blechnum penna-marina

ALPINE WATER FERN A fast-growing, evergreen, carpeting fern, with narrow, ladder-like, dark green fronds that are attractively tinged red when young. Grow it in consistently moist soil, between tree canopies in part or deep shade.

↕30cm (12in) ↔45cm (18in)

PERENNIAL SMALL

Brunnera macrophylla

A ground-cover perennial, with oval leaves, often speckled or edged with silvery white, and sprays of bright blue spring flowers. 'Dawson's White' (above) bears heart-shaped, creamy white-edged leaves. Grow it beneath trees in a sheltered garden.

↕45cm (18in) ↔60cm (24in)

SHRUB LARGE

Callicarpa bodinieri var. *giraldii*

BEAUTY BERRY An upright shrub, with lance-shaped leaves and tiny, pink summer flowers. 'Profusion' AGM (above) has violet berries and purple-tinged leaves, rosy pink in autumn. For more berries, plant in groups.

↕3m (10ft) ↔2.5m (8ft)

SHRUB LARGE

Camellia x *williamsii*

This is a bushy, evergreen shrub, with glossy, oval, dark green leaves. It bears showy, single or double blooms in shades of pink or white from early- to mid-spring. The upright 'Donation' AGM (above) has pink flowers with golden stamens.
↕5m (15ft) ↔2.5m (8ft)

PERENNIAL MEDIUM

Carex grayi

MORNING STAR SEDGE An evergreen perennial, with tufts of narrow, grassy, bright green leaves and, in summer, upright stems topped with clusters of green flowers, followed by star-shaped, greenish brown fruits. Plant on the edge of a woodland.
↕60cm (24in) ↔20cm (8in)

PERENNIAL MEDIUM

Carex buchananii

SEDGE This evergreen perennial has narrow, grass-like, upward then arching, copper-coloured leaves that are red towards the base. Triangular stems bear brown flower clusters in summer. Grow in groups on the edge of a woodland garden.
↕60cm (24in) ↔20cm (8in)

PERENNIAL MEDIUM

Chelone glabra

TURTLE HEAD An upright perennial, with oval, mid-green foliage and unusual, two-lipped, white flowers that appear from late summer to early autumn. It is best planted just beyond the tree canopies.
↕up to 1m (3ft) ↔45cm (18in)

BULB MEDIUM

Chionodoxa forbesii

GLORY OF THE SNOW This perennial bulb has grass-like foliage and small, purple-blue flowers, with white centres, in early spring. Plant the bulbs in large groups in autumn in soil that will not dry out in summer.
↕15cm (6in)

SHRUB MEDIUM

Clethra alnifolia

SWEET PEPPER BUSH This bushy shrub has oval, serrated, mid-green leaves that turn yellow in autumn. In late summer, slender spires of small, fragrant, bell-shaped, white blooms form. It is a good choice for damp soil, just beyond the tree canopies.
↕↔2.5m (8ft)

SHRUB LARGE

Cornus mas

CORNELIAN CHERRY This spreading, open shrub has oval, dark green leaves that turn reddish purple in autumn. Rounded clusters of yellow, star-shaped blooms form on bare stems in late winter, followed by edible, red fruits. Suitable for light shade only.
↕↔5m (15ft)

PERENNIAL SMALL

Corydalis flexuosa

An upright perennial, with light grey-blue-tinged, ferny leaves. From late spring to summer, sprays of tubular, blue flowers, with white throats, form. The plants die back completely in summer. Plant it in dappled shade beneath deciduous trees.
↕30cm (12in) ↔20cm (8in)

SHRUB MEDIUM

Corylopsis pauciflora

BUTTERCUP WITCH HAZEL This bushy, spreading shrub has oval, serrated, bright green leaves, bronze- or red-tinged when young. In spring, short clusters of bell-shaped, fragrant, pale yellow blooms hang from bare stems. Plant it in dappled shade.
↕1.5m (5ft) ↔2.5m (8ft)

BULB SMALL

Cyclamen coum

EASTERN CYLAMEN This small perennial bulb has rounded leaves, often marked with silver marbling, and pink flowers, which bloom from late winter to early spring. Plant the bulbs in bold groups beneath trees in autumn.
↕10cm (4in)

BULB SMALL

Cyclamen hederifolium

IVY-LEAVED CYCLAMEN This perennial bulb has variable, ivy-shaped, dark green leaves, usually overlaid with silvery green marbling. The pink or white flowers appear in autumn, before or with the foliage. Tolerates dry shade beneath trees.
↕10cm (4in)

PERENNIAL LARGE

Dicksonia antarctica

TREE FERN An evergreen or, in colder areas, deciduous, tree-like fern, with a stout, brown, fibrous "trunk" and palm-like crown of arching, much divided fronds. Grow it in a sheltered spot, between trees with light canopies.
↕up to 6m (20ft) ↔4m (12ft)

PERENNIAL LARGE

Digitalis ferruginea

RUSTY FOXGLOVE This perennial has a basal rosette of oval, dark green leaves. In midsummer, long, slender, upright spikes of funnel-shaped, orange-brown and white blooms appear. Plant in groups in partial shade beneath trees.
↕1.2m (4ft) ↔30cm (12in)

PERENNIAL MEDIUM

Digitalis grandiflora

LARGE YELLOW FOXGLOVE This evergreen, clump-forming perennial has a basal rosette of oval, dark green leaves and spires of tubular, downward-pointing, creamy yellow blooms in summer. Plant it beneath trees in part shade; it can be short-lived.
↕1m (3ft) ↔45cm (18in)

PERENNIAL SMALL

Dodecatheon dentatum

TOOTHED AMERICAN COWSLIP This clump-forming perennial makes rosettes of oval, mid-green leaves and, in late spring, slim stems of nodding white flowers, with swept-back petals. Grow it between tree canopies in a woodland in rich soil.
↕↔20cm (8in)

PERENNIAL MEDIUM

Dryopteris affinis

GOLDEN MALE FERN This virtually evergreen perennial makes a "shuttlecock" of lance-shaped, divided fronds, which are pale green as they unfurl in spring, contrasting with the scaly, golden-brown midribs. Grow beneath tree canopies; tolerates drought.
↕↔90cm (36in)

PERENNIAL SMALL

Dryopteris erythrosora

COPPER SHIELD FERN A semi-evergreen, clump-forming fern, with broadly triangular, arching, divided fronds, copper-red when young, ageing to pinkish green or bronze tinged, and then bright green. Tolerates dry shade, but water well in first year.
↕40cm (16in) ↔30cm (12in)

PERENNIAL LARGE

Dryopteris filix-mas ♡

MALE FERN This deciduous or semi-evergreen fern forms "shuttlecocks" of arching, upright, mid-green fronds that arise from crowns of large, upright, brown-scaled rhizomes. It thrives in deep shade beneath deciduous trees.

↕1.2m (4ft) ↔1m (3ft)

PERENNIAL SMALL

Epimedium grandiflorum ♡

BARRENWORT A clump-forming perennial, with heart-shaped, green leaves that are bronze-tinged when young. In spring, nodding flowers in shades of white, yellow, pink, or purple appear. Varieties include 'Lilafee' (above), with purple blooms.

↕25cm (10in) ↔30cm (12in)

PERENNIAL SMALL

Erythronium dens-canis ♡

DOG'S-TOOTH VIOLET This tuberous-rooted perennial has paired, oval, maroon-mottled basal leaves. In spring, single stems bear nodding, mauve-pink flowers, with swept-back petals, and red-brown and yellow shading at the base. Plant it between trees.

↕25cm (10in) ↔10cm (4in)

BULB LARGE

Fritillaria pallidiflora ♡

SIBERIAN FRITILLARY An upright perennial bulb, with lance-shaped, blue-green leaves and nodding, bell-shaped, greenish yellow flowers in early summer. Plant the bulbs at four times their own depth in autumn just beyond the tree canopies.

↕70cm (28in)

BULB MEDIUM

Galanthus nivalis ♡

SNOWDROP This perennial bulb produces clumps of grass-like, grey-green leaves. In late winter, it bears single nodding, white flowers, with green-tipped inner petals. Plant it in groups in deciduous woodland and leave to naturalize.

↕15cm (6in)

PERENNIAL SMALL

Galium odoratum

SWEET WOODRUFF A spreading perennial, with emerald-green foliage formed of rounded clusters of lance-shaped leaves. From late spring to midsummer, it produces a froth of tiny, starry, white blooms. Use it as ground cover beneath trees.

↕15cm (6in) ↔30cm (12in) or more

PERENNIAL MEDIUM

Geranium phaeum

DUSKY CRANESBILL This erect, clump-forming perennial has rounded, deeply cut, mid-green leaves. From late spring to early summer, dusky, purple, lilac, or white flowers appear. It thrives in dappled shade under trees.

↕75cm (30in) ↔45cm (18in)

PERENNIAL MEDIUM

Geranium sylvaticum

WOOD CRANESBILL This bushy perennial has lobed, deeply-cut, mid-green leaves and, from late spring to early summer, pink, blue, white, or purple, saucer-shaped blooms. 'Album' AGM (above) has white flowers. It thrives in dappled shade.

↕75cm (30in) ↔60cm (24in)

PERENNIAL SMALL

Hacquetia epipactis ♡

This clump-forming perennial has lobed, green leaves that unfurl after the late winter to early spring flowers, which are yellow-green and surrounded by notched, apple-green, petal-like bracts. Plant it in dappled shade.

↕6cm ($2^1/_2$in) ↔23cm (9in)

pH

PERENNIAL MEDIUM

Helleborus foetidus ♡

STINKING HELLEBORE A bushy, evergreen perennial, with upright stems and glossy, palm-shaped leaves formed of narrow leaflets. From late winter to early spring, it has nodding, cup-shaped, pale green, red-margined flowers. Grow in dappled shade.

↕↔ 60cm (24in)

PERENNIAL MEDIUM

Helleborus orientalis

LENTEN ROSE This evergreen woodland perennial has leathery, dark green leaves, divided into slim leaflets. From midwinter to spring, it bears saucer-shaped, white or greenish cream flowers, which age to dark pink. Deadhead to prevent self-seeding.

↕60cm (24in) ↔45cm (18in)

PERENNIAL SMALL

Hosta 'Revolution' ♡

This clump-forming perennial has dark green leaves, with irregular creamy white central markings and green speckles. Spikes of lavender flowers appear briefly in summer. Grow it in dappled shade to add bright highlights to beds and borders.

↕50cm (20in) ↔1.1m ($3^1/_2$ft)

SHRUB LARGE

Kalmia latifolia

MOUNTAIN LAUREL A bushy, evergreen shrub, with glossy, oval, green leaves and, from late spring to early summer, clusters of pink, cup-shaped flowers that open from attractively crimped buds. Grow it in dappled shade between tree canopies.

↕↔ 3m (10ft)

BULB MEDIUM

Hyacinthoides non-scripta

ENGLISH BLUEBELL A vigorous perennial bulb, with narrow, strap-shaped, glossy, dark green leaves. In spring, mid-blue or, occasionally, white, bell-shaped, fragrant flowers hang from arching stems. Grow it in large swathes beneath deciduous trees.

↕40cm (16in)

BULB MEDIUM

Leucojum aestivum

SUMMER SNOWFLAKE This perennial bulb makes an upright clump of narrow, strap-shaped, glossy, dark green leaves. Small clusters of green-tipped white, drooping, bell-shaped blooms form in spring. 'Gravetye Giant' AGM (above) is a popular variety.

↕60cm (24in)

PERENNIAL SMALL

Hylomecon japonica

This vigorous, spreading perennial has soft, dark green leaves, divided into toothed leaflets. From late spring to summer, slender stems bear single, cup-shaped, bright yellow, poppy-like flowers. Plant it in a large woodland garden.

↕30cm (12in) ↔20cm (8in)

PERENNIAL MEDIUM

Kirengeshoma palmata

This upright perennial has rounded, lobed, bright green, slightly hairy leaves on dark purple-red stems. From late summer to autumn, small clusters of narrow, funnel-shaped, creamy yellow, pendulous flowers appear. Grow it in moist soil.

↕1m (3ft) ↔60cm (24in)

BULB LARGE

Lilium martagon ♡

TURKSCAP LILY A clump-forming perennial bulb, with lance-shaped leaves on erect stems and nodding, dark-spotted, pink or purple summer flowers, with swept-back petals and protruding stamens. Grow on the edge of a woodland garden.

↕up to 2m (6ft)

BULB LARGE

Lilium nepalense

HIMALAYAN LILY This upright perennial bulb has lance-shaped, green leaves and trumpet-shaped, scented, greenish cream flowers, with chocolate-red throats. Plant it in part shade in deep, fertile soil, just beyond the tree canopies.

↕1m (3ft)

SHRUB LARGE

Mahonia x *media*

An architectural, evergreen shrub, with long, dark green leaves, divided into spiny leaflets. In winter, it produces long clusters of scented, yellow flower spikes. 'Charity' (above) is a popular variety. Ideal for dry shade.
↕5m (15ft) ↔ 4m (12ft)

PERENNIAL MEDIUM

Maianthemum racemosum

FALSE SPIKENARD An upright perennial, with oval, prominently veined, green leaves. It bears scented, fluffy, cream plumes from mid- to late spring, followed by red berries. Though hardy, it needs shelter from wind. Best in larger gardens.
↕90cm (36in) ↔ 60cm (24in)

PERENNIAL MEDIUM

Matteuccia struthiopteris

OSTRICH-FEATHER FERN This large, deciduous, shuttlecock-shaped fern produces long, pale green fronds, divided into toothed segments. It is ideal for growing in the damp soil just beyond tree canopies in a woodland setting.
↕1m (3ft) ↔ 45cm (18in)

PERENNIAL LARGE

Meconopsis betonicifolia

HIMALAYAN BLUE POPPY This short-lived perennial forms a rosette of oval, serrated, bluish green leaves. In early summer, it bears saucer-shaped, clear blue blooms on upright stems. Grow it in dappled shade, just beyond the tree canopies.
↕1.2m (4ft) ↔ 45cm (18in)

PERENNIAL MEDIUM

Milium effusum 'Aureum'

BOWLES' GOLDEN GRASS This semi-evergreen perennial has narrow, arching, golden-yellow leaves. Slender stems bear airy sprays of yellow summer flowerheads. Allow it to self-seed and form carpets in dappled shade just beyond tree canopies.
↕60cm (24in) ↔ 30cm (12in)

BULB MEDIUM

Narcissus 'February Gold'

DAFFODIL This perennial bulb has narrow, mid-green leaves and, from late winter to early spring, golden-yellow blooms, with swept-back petals and long trumpets. Plant the bulbs in groups in autumn, and allow to naturalize beneath deciduous trees.
↕30cm (12in)

BULB MEDIUM

Narcissus pseudonarcissus

WILD DAFFODIL This perennial bulb has erect, strap-shaped, mid-green leaves, tinged blue-grey, and nodding early spring flowers, with overlapping, straw-yellow petals and darker yellow trumpets. Plant the bulbs in autumn.
↕30cm (12in)

PERENNIAL SMALL

Omphalodes cappadocica

NAVELWORT This clump-forming perennial has deeply veined, oval to heart-shaped, mid-green leaves. In early spring, it forms airy heads of white-eyed sky-blue blooms. 'Cherry Ingram' AGM (above) is a more compact, deep blue form.
↕30cm (12in) ↔ 60cm (24in)

SHRUB LARGE

Osmanthus x *burkwoodii*

This large, rounded, evergreen shrub has glossy, oval, dark green leaves. Clusters of small, highly-scented, tubular, white flowers in spring are followed by black berries. Grow it in dappled shade beneath trees.
↕↔3m (10ft)

PERENNIAL SMALL

Pachysandra procumbens

This clump-forming, evergreen perennial has rounded, lobed, dark green leaves. Spikes of fragrant, white, bottlebrush-like flowers are produced in spring as new leaves develop. Use it as ground cover in deep shade beneath trees.
↕25cm (10in) ↔ 60cm (24in)

PERENNIAL MEDIUM

Polemonium caeruleum

JACOB'S LADDER This clump-forming perennial has finely divided, fern-like leaves. Loose clusters of small, cup-shaped, lavender-blue blooms, with orange-yellow stamens, open in early summer. Plant it in dappled shade beneath deciduous trees.

↕↔ 60cm (24in)

PERENNIAL MEDIUM

Persicaria bistorta

This vigorous, clump-forming perennial has large, prominently veined, green leaves and dense spikes of small, pale pink flowers from summer to early autumn. Use it sparingly as it spreads rapidly. 'Superba' AGM (above) is a popular variety.

↕ 75cm (30in) ↔ 60cm (24in)

SHRUB MEDIUM

Pieris 'Flaming Silver'

This small, upright, evergreen shrub produces leathery, lance-shaped, bright red young leaves that age to green with white margins. In spring, branching clusters of bell-shaped, creamy white flowers appear.

↕ 1.5m (5ft) ↔ 1m (3ft)

PERENNIAL SMALL

Podophyllum versipelle

This unusual perennial produces large, umbrella-shaped, bright green leaves, under which crimson flowers form in summer. It makes an eye-catching addition to a shady border. 'Spotty Dotty' (above) has red-spotted foliage.

↕ 45cm (18in) ↔ 60cm (24in)

PERENNIAL LARGE

Polygonatum x *hybridum*

SOLOMON'S SEAL An arching, leafy perennial, with oval, green leaves. In early summer, it produces clusters of small, tubular, green-tipped white flowers that hang from the stems. It thrives in full shade beneath trees.

↕ up to 1.2m (4ft) ↔ 1m (3ft)

PERENNIAL MEDIUM

Polygonatum odoratum

ANGLED SOLOMON'S SEAL This arching perennial has oval to lance-shaped, mid-green leaves. From late spring to early summer, it bears fragrant, hanging, tubular to bell-shaped, green-tipped white flowers. It thrives in deep shade beneath trees.

↕ 60cm (24in) ↔ 30cm (12in)

PERENNIAL MEDIUM

Polystichum polyblepharum

This evergreen fern forms "shuttlecocks" of spreading, lance-shaped, divided, green fronds, covered with golden hairs when they unfurl. The oval frond segments have spiny-toothed margins. Prefers full shade.

↕ 80cm (32in) ↔ 90cm (36in)

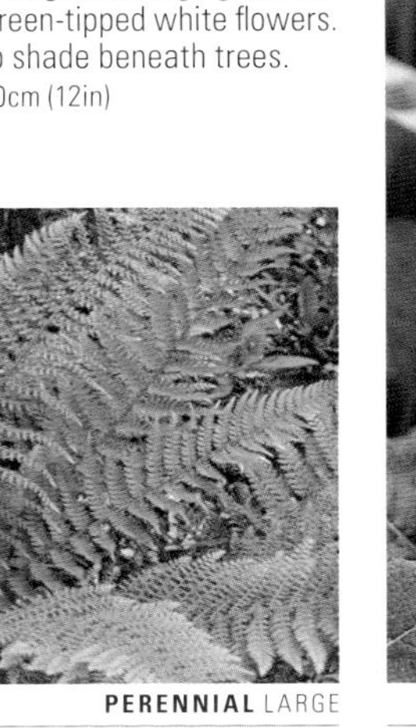

PERENNIAL LARGE

Polystichum setiferum

SOFT SHIELD FERN This clump-forming fern produces large, lance-shaped, mid-green fronds, divided into segments. The leafstalks feature orange-brown scales. Grow it beneath trees in partial or deep shade to produce a naturalistic effect.

↕ 1.2m (4ft) ↔ 90cm (36in)

PERENNIAL MEDIUM

Primula beesiana

BEE'S PRIMROSE This is a deciduous or semi-evergreen, rosette-forming perennial, with toothed, mid-green leaves. In summer, it produces whorls of tubular, yellow-eyed reddish pink flowers, carried on upright stems.

↕ 60cm (24in) ↔ 30cm (12in)

PERENNIAL SMALL

Primula japonica

JAPANESE PRIMROSE This perennial has basal rosettes of wrinkled, oval, pale green leaves. Clusters of white, pink, or red flowers are held on upright stems in late spring. Varieties include the white-flowered 'Postford White' AGM (above).
↕↔ 45cm (18in)

PERENNIAL SMALL

Primula vulgaris

PRIMROSE This perennial forms basal rosettes of oval, pale green leaves. It bears scented, primrose-yellow flowers in early spring. Plant it with woodland spring bulbs in dappled shade beneath deciduous trees.
↕ 20cm (8in) ↔ 35cm (14in)

SHRUB SMALL

Rhododendron 'Yaku Prince'

This rounded, evergreen shrub produces olive-green leaves, with orange-brown hairs on the undersides. Large clusters of pink flowers with orange-red spots appear from late spring to early summer. Plant it in acid soil on a woodland edge.
↕↔ 1m (3ft)

PERENNIAL SMALL

Pulmonaria officinalis

LUNGWORT This ground-cover perennial has pointed, oval, green leaves, with silvery spots. In spring, it bears funnel-shaped, blue and purple, or pink and white blooms. Varieties include the white 'Sissinghurst White' (above). It tolerates deep shade.
↕ 30cm (12in) ↔ 60cm (24in)

PERENNIAL SMALL

Sanguinaria canadensis f. *multiplex*

A clump-forming perennial, with scalloped, kidney-shaped, blue-grey leaves and single, cup-shaped, white spring flowers. 'Plena' AGM (above) has double blooms. Plant in dappled shade.
↕ 15cm (6in) ↔ 30cm (12in)

SHRUB SMALL

Sarcococca confusa

SWEET BOX A bushy, evergreen shrub, with glossy, lance-shaped, dark green leaves. In winter, it bears clusters of tiny, sweetly fragrant, white blooms, followed by shiny, black fruits. It tolerates full shade and the dry soil beneath trees.
↕↔ 1m (3ft)

PERENNIAL SMALL

Saxifraga fortunei

SAXIFRAGE A semi-evergreen, clump-forming perennial, with rounded, jaggedly lobed, fleshy, brownish green leaves, red beneath. Airy heads of small, white-tinged pink, starry flowers open in late summer. Use it as ground cover beneath trees.
↕↔ 30cm (12in)

PERENNIAL MEDIUM

Tellima grandiflora

FRINGE CUPS A semi-evergreen, clump-forming perennial, with serrated, heart-shaped, hairy, green-tinted purple leaves. From late spring to midsummer, long spikes of small, bell-shaped, fringed, cream flowers form. Allow it to naturalize in dappled shade.
↕↔ 60cm (24in)

PERENNIAL MEDIUM

Tricyrtis formosana

TOAD LILY An upright, clump-forming, slow-spreading perennial, with lance-shaped, mid-green leaves and upturned clusters of intricately-shaped, purple-spotted white autumn blooms. Autumn frost may kill the flowers. Mulch annually with compost.
↕ 1m (3ft) ↔ 45cm (18in)

PERENNIAL MEDIUM

Silene dioica

RED CAMPION This semi-evergreen, clump-forming perennial has oval, dark green basal leaves. From late spring to midsummer, it bears clusters of small, pink flowers. Use it to provide colour in the dappled shade beneath tree canopies.
↕ 80cm (32in) ↔ 45cm (18in)

PERENNIAL SMALL

Thalictrum kiusianum

MEADOW RUE This mat-forming perennial has small, fern-like, dark blue-green leaves. In early summer, it produces loose clusters of fluffy, pale purple-pink flowers with prominent stamens. Plant in part shade in soil that does not dry out.
↕ 8cm (3in) ↔ 15cm (6in)

PERENNIAL SMALL

Trillium chloropetalum

GIANT WAKEROBIN This clump-forming perennial has grey- or maroon-marbled, dark green leaves that form in groups of three. The large, purplish pink to white flowers stand erect just above the leaves in spring. Thrives in partial shade.
↕↔ 45cm (18in)

TREE MEDIUM

Styrax obassia

FRAGRANT SNOWBELL A spreading tree, with rounded, dark green leaves, blue-grey beneath, that turn yellow in autumn. In early summer, it bears long, pendent clusters of fragrant, bell-shaped, white flowers. Use it to create a woodland setting.
↕ 12m (40ft) ↔ 7m (22ft)

PERENNIAL MEDIUM

Uvularia grandiflora 🏆

BELLWORT A clump-forming perennial, with arching stems, clothed in lance-shaped, bright green leaves. From mid- to late spring, clusters of long, bell-shaped, yellow flowers with twisted petals hang from the stems. Thrives in deep shade beneath trees.
↕ 60cm (24in) ↔ 30cm (12in)

PERENNIAL SMALL

Tiarella cordifolia var. *collina*

FOAM FLOWER This evergreen, spreading perennial has lobed, pale green leaves, with veins that turn bronze-red in winter. Spikes of fluffy, white flowers appear in late spring and early summer. It makes excellent ground cover in deep shade.
↕ 20cm (8in) ↔ 30cm (12in) or more

OTHER SUGGESTIONS

Perennials

Anemone 'Wild Swan' • *Brunnera macrophylla* 'Jack Frost' 🏆 • *Cardamine pratensis* 'Flore Pleno' • *Dactylorhiza fuchsii* • *Deinanthe caerulea* • *Digitalis purpurea* • *Prosartes smithii* • *Veratrum formosanum*

Bulbs

Allium ursinum • *Chionodoxa siehei* 🏆 • *Fritillaria camschatcensis* • *Hyacinthoides hispanica*

Shrubs

Camellia x *williamsii* 'Mary Christian' • *Euonymus alatus* • *Mahonia* x *media* 'Winter Sun' 🏆 • *Mahonia* x *wagneri* 'Pinnacle' 🏆 • *Osmanthus heterophyllus* • *Pieris* 'Forest Flame' 🏆 • *Rhododendron* 'Persil' 🏆

Plant focus: azaleas and rhododendrons

Big and bold or small and dainty, azaleas and rhododendrons offer a great choice of spring to early summer blooms.

Azaleas and rhododendrons are part of the rhododendron family but, confusingly, while all azaleas are rhododendrons, not all rhododendrons are azaleas. Azaleas can be deciduous or evergreen, while rhododendrons are strictly evergreen, and they also have slightly different flowers. However, both groups produce abundant, often scented, spring and early summer blooms, and make excellent specimens for beds, borders, and containers. The larger rhododendrons are also useful as hedging. Some species can be invasive, so choose those in this book or ask a specialist nursery for advice on the best plants for your garden. All require acid soil, so test your soil before planting, and if you have alkaline conditions, grow compact or dwarf forms in pots of ericaceous compost. Happiest in moist soil and dappled or partial shade, trim plants lightly to keep them in shape in early summer, after flowering.

USING AZALEAS AND RHODODENDRONS

Tree-like and large, shrubby rhododendrons make beautiful flowering evergreen screens. Trim them to shape in midsummer after flowering. Compact or dwarf rhododendrons are perfect for large containers or barrels, but they must be planted in acid soil or ericaceous compost. Water container-grown plants regularly, especially during summer.

The compact *R. yakushimanum* hybrids are invaluable for shady borders, producing masses of medium-sized flowers in late spring. There is a large range of colourful varieties to choose.

Ideal for woodlands, azaleas and rhododendrons thrive in the cool, dappled shade cast by neighbouring trees. Plant them in groups for the best display.

TYPES OF AZALEA AND RHODODENDRON

Compact rhododendrons Forming mounds less than 1m (3ft) tall, this useful group consists of *R. yakushimanum* and *R. williamsianum* hybrids, which flower in late spring.

Shrub rhododendrons Most rhododendrons are rounded shrubs reaching hip- to head-height. Their often fragrant, colourful flowers appear from mid-spring to early summer.

Evergreen hybrid azaleas Growing 1m (3ft) tall, this group includes dwarf and cascading varieties. The flowers, mostly unscented, appear from mid-spring to early summer.

Deciduous hybrid azaleas These reach up to 1.5m (5ft) in height, and include *Ghent* and *Knaphill* forms. The flowers, often scented, appear from mid-spring to early summer.

Dwarf rhododendrons These tough, alpine evergreen shrubs flower mainly in mid-spring, and most grow to just 50cm (20in) in height. They are ideal for rockeries and containers.

Plants for rock gardens

Rock features are normally associated with sunny sites, but a surprising number of alpines and low-growing plants will thrive in some shade.

Most alpine plants require a few hours of sun each, and may not flower if left to languish in deep shade, so site your rock garden carefully. You can maintain year-round colour and interest with a succession of seasonal displays, starting in spring with a range of bulbs and progressing into early summer with early-flowering perennials, such as sedums, aquilegias, and *Dicentra*. Midsummer highlights include geraniums, *Silene alpestris*, and *Roscoea*, followed by autumn-flowering gentians. Evergreen junipers and hebes will provide colour and interest in the winter months.

PERENNIAL SMALL

Adiantum aleuticum

MAIDENHAIR FERN A compact, deciduous or semi-evergreen fern, with pale green fronds, divided into oblong segments. Ideal for rock gardens with rich soil. The dwarf form 'Subpumilum' AGM has black stalks and fronds with overlapping segments.
↕↔ 45cm (18in)

PERENNIAL SMALL

Alchemilla alpina

ALPINE LADY'S MANTLE This mound-forming perennial has lobed, white-edged green leaves that are silky-haired beneath. Spikes of tiny, greenish yellow flowers with larger, green outer petals appear in summer. Trim untidy foliage.
↕ 15cm (6in) ↔ 60cm (24in) or more

BULB MEDIUM

Allium schoenoprasum

CHIVES A clump-forming, upright, perennial bulb, with narrow, hollow, dark green, edible leaves. In summer, it produces fluffy, pompon, pale purple flowerheads. It is ideal for rock gardens; plant in dappled shade.
↕ 30cm (12in)

PERENNIAL SMALL

Anemone blanda

WINTER WINDFLOWER This perennial corm has divided leaves and daisy-like blue, pink, or white flowers in early spring. 'Violet Star' (above) has amethyst-violet flowers. Plant the tubers in autumn where they will receive some sun.
↕↔ 15cm (6in)

PERENNIAL SMALL

Anemone nemorosa

WOOD ANEMONE This dwarf perennial has deeply cut, mid-green leaves. From spring to early summer, a profusion of daisy-like, yellow-eyed white flowers appears. Plant it in groups at the front of a partly shaded rock garden.
↕ 15cm (6in) ↔ 30cm (12in)

PERENNIAL SMALL

Anemone sylvestris

SNOWDROP ANEMONE This dwarf perennial has divided, mid-green leaves. From spring to early summer, it produces fragrant, bowl-shaped, white flowers, with yellow centres. Prefers neutral to alkaline soil.
↕↔ 30cm (12in)

PERENNIAL SMALL

Anemonella thalictroides

RUE ANEMONE A dwarf perennial, with dark blue-green leaves, divided into rounded leaflets. From spring to early summer, it produces cup-shaped, white or pink blooms. Grow it at the front of a rock garden in a sheltered position.
↕ 10cm (4in) ↔ 4cm (1½in) or more

PERENNIAL SMALL

Aquilegia fragrans

This upright perennial has finely divided, bluish green leaves and nodding, bell-shaped, fragrant, creamy white summer flowers, with bluish- or pinkish white outer sepals. Deadhead spent flowers to prevent excessive self-seeding.
↕ 40cm (16in) ↔ 20cm (8in)

PERENNIAL SMALL

Asperula arcadiensis

ARCADIAN WOODRUFF This clump-forming perennial makes a mound of loose stems, with tiny, hairy, grey leaves. It bears tiny, tubular, pale pink flowers in early summer. Dislikes winter wet and is best planted in gritty, free-draining soil.
↕ 8cm (3in) ↔ 30cm (12in)

PERENNIAL SMALL

Campanula raineri

RAINER'S HAREBELL A ground-cover perennial, with oval, toothed-edged, grey-green leaves. Bell-shaped, pale lavender flowers appear in summer. Allow it to spread over the rocks in areas where it will receive some sun during the day.
↕ 8cm (3in) ↔ 20cm (8in)

BULB MEDIUM

Chionodoxa forbesii

GLORY OF THE SNOW An early spring-flowering perennial bulb, with linear, mid-green foliage, and star-shaped, blue flowers with white eyes. Ideal for rock gardens with sharp drainage. Grow in light shade. Plant the bulbs in groups in autumn.
↕15cm (6in)

BULB SMALL

Crocus banaticus

BYZANTINE CROCUS This autumn-flowering perennial corm produces cup-shaped, pale violet flowers before the narrow, basal leaves appear in spring. Plant it in groups where it will receive some sun during the day.
↕10cm (4in)

PERENNIAL SMALL

Dicentra formosa

BLEEDING HEART This spreading perennial forms finely divided, grey-green leaves and clusters of pendent, heart-shaped, pink flowers from late spring to early summer. Varieties include 'Bacchanal' AGM (above), with deep crimson blooms.
↕45cm (18in) ↔30cm (12in)

PERENNIAL SMALL

Dodecatheon hendersonii

SAILOR CAPS A clump-forming perennial, with a rosette of oval, green leaves, above which deep pink flowers with reflexed petals appear in late spring. It prefers rich soil; all growth dies down in summer. Keep soil moist during the growing season.
↕30cm (12in) ↔25cm (10in)

PERENNIAL SMALL

Epimedium x *versicolor*

BISHOP'S HAT A clump-forming, evergreen perennial, with heart-shaped, green leaves, copper-red when young. Sprays of small, pink and yellow flowers appear in spring. Tolerates full shade. 'Sulphureum' AGM (above) has primrose-yellow blooms.
↕↔30cm (12in)

BULB SMALL

Eranthis hyemalis

WINTER ACONITE This clump-forming tuber produces stalkless, cup-shaped, yellow flowers, with leaf-like ruffs, from late winter to early spring. The leaves are lobed and green. Grow it in pockets between rocks, and combine with spring bulbs.
↕10cm (4in)

BULB MEDIUM

Erythronium californicum

FAWN LILY This clump-forming perennial bulb has mottled, dark green leaves and creamy white flowers with reflexed petals in spring. 'White Beauty' AGM (above) has creamy white flowers with red-brown throats. Plant bulbs in autumn.
↕35cm (14in)

BULB MEDIUM

Galanthus nivalis

SNOWDROP This dwarf perennial bulb has grassy, grey-green leaves that disappear in summer. From late winter to early spring, 'Flore Pleno' AGM (above) has double, fragrant, green-tinged white blooms. Plant it in soil that does not dry out in summer.
↕15cm (6in)

BULB MEDIUM

Galanthus 'S. Arnott'

SNOWDROP This dwarf perennial bulb produces grass-like, grey-green leaves and, in late winter, small, nodding, white flowers, with green marks on the inner segments. Plant it in groups in soil that does not dry out in summer.
↕15cm (6in)

PERENNIAL SMALL

Galium odoratum

SWEET WOODRUFF This spreading perennial has small, palm-shaped, divided, dark green leaves and small, starry, white flowers in summer. Can be used in large, partly shaded rock gardens where it will spread freely between the rocks.
↕15cm (6in) ↔30cm (12in) or more

SHRUB SMALL

Gaultheria procumbens

CHECKERBERRY A dwarf, evergreen shrub, with oval, leathery, dark green leaves, tinged red in winter. Small, pink-flushed white flowers in summer are followed by scarlet berries. Grow it for winter colour in part or deep shade.
↕15cm (6in) ↔ indefinite

PERENNIAL SMALL

Gentiana verna

SPRING GENTIAN An evergreen perennial, with rosettes of oval, dark green leaves and tubular, bright blue flowers, with white throats, in early spring. Grow in some sun; provide shade in summer. *G. verna* subsp. *balcanica* (above) has star-shaped flowers.

↕↔5cm (2in)

PERENNIAL SMALL

Geranium cinereum

CRANESBILL A spreading perennial, with round, deeply lobed, grey-green leaves. Cup-shaped, pink or mauve flowers, with deep purple centres and veins, appear from late spring to summer. 'Ballerina' AGM (above) has pale pink blooms.

↕10cm (4in) ↔30cm (12in)

SHRUB SMALL

Hebe ochracea 'James Stirling'

A dense, dome-shaped, dwarf shrub, with bright yellow-green, scale-like foliage. In late spring, small, white flowers appear. Use it to provide structure in a rock garden and combine with low-growing perennials.

↕45cm (18in) ↔60cm (24in)

PERENNIAL SMALL

Helleborus purpurascens

LENTEN ROSE A clump-forming perennial, with dark green leaves, divided into lance-shaped leaflets. Small, cup-shaped, nodding, deep purple flowers with cream stamens appear in early spring. Trim untidy foliage in autumn or late winter.

↕↔30cm (12in)

PERENNIAL SMALL

Hepatica nobilis

LIVERLEAF A clump-forming, semi-evergreen perennial, with silky-haired leaves featuring three rounded lobes. In early spring, it produces rounded, violet or purple blooms with white stamens. Grow it in soil that does not dry out in summer.

↕8cm (3in) ↔12cm (5in)

SHRUB SMALL

Juniperus procumbens

CREEPING JUNIPER This compact, mat-forming, evergreen conifer produces dense, prickly leaves that provide good ground cover. It bears small, round, brown or black fruits. Plant it in a rock garden in dappled or light shade.

↕20cm (8in) ↔75cm (30in)

PERENNIAL SMALL

Lewisia cotyledon

SISKIYOU LEWISIA This clump-forming, evergreen perennial has rosettes of dark green foliage. From late spring to summer, funnel-shaped flowers in shades of pink, yellow, or orange appear. Grow it in light shade in free-draining soil.

↕30cm (12in) ↔15cm (6in)

BULB MEDIUM

Muscari armeniacum

GRAPE HYACINTH This spring-flowering bulb produces grass-like, green leaves and short spikes of small, fragrant, bell-shaped, deep blue flowers, held in cone-shaped clusters. Plant it in cracks between rocks.

↕20cm (8in)

BULB MEDIUM

Narcissus 'Jumblie'

DAFFODIL This bulb has strap-shaped leaves and single, golden flowers, with orange-yellow cups, in early spring. Other dwarf daffodils come in shades of orange or white, sometimes bicoloured. Plant the bulbs in groups in autumn.

↕20cm (8in)

BULB MEDIUM

Narcissus 'Tête-à-tête'

DAFFODIL A spring bulb with strap-shaped leaves and up to three yellow blooms per stem. Other dwarf varieties come in shades of orange or white, sometimes bicoloured. Plant the bulbs in groups in autumn throughout a rock garden.

↕30cm (12in)

PERENNIAL SMALL

Origanum vulgare

WILD MARJORAM This clump-forming perennial forms a dense mat of aromatic, rounded, green leaves. It may also bear tiny, mauve summer flowers. Use it to fill gaps in a partly shaded rock garden.

↕↔up to 45cm (18in)

PERENNIAL SMALL

Oxalis adenophylla

A mat-forming perennial, with rounded, grey-green leaves, divided into narrow, wavy leaflets. In spring, rounded, purplish pink flowers, with darker purple eyes, form. Use it to spread over the rocks in a shady rock garden.

↕5cm (2in) ↔10cm (4in)

PERENNIAL SMALL

Phlox stolonifera

CREEPING PHLOX An evergreen, low-growing perennial, with oval, pale green leaves and saucer-shaped, purple blooms in early summer. 'Ariane' (above) has white blooms. It is ideal for a shaded rock garden; cut back after flowering.

↕15cm (6in) ↔30cm (12in)

PERENNIAL SMALL

Primula marginata

ALPINE PRIMROSE This evergreen or semi-evergreen perennial has rosettes of oblong, toothed, mid-green leaves. Grow with spring bulbs at the front of a rock garden. Lavender-blue spring flowers top upright stems in 'Kesselring's Variety' (above).

↕15cm (6in) ↔30cm (12in)

BULB MEDIUM

Puschkinia scilloides var. *libanotica*

STRIPED SQUILL This perennial bulb has strap-shaped, green leaves and, in spring, clusters of bell-shaped, white flowers. Grow it with other spring bulbs, such as *Muscari* and species tulips, in light, dappled shade.

↕15cm (6in)

PERENNIAL SMALL

Sedum kamtschaticum var. *kamtschaticum*

STONECROP A semi-evergreen, spreading perennial, with fleshy, oval, cream-edged leaves and orange-flushed yellow early autumn flower clusters. 'Variegatum' AGM (above) has pink-tinted, mid-green leaves.

↕8cm (3in) ↔20cm (8in)

BULB MEDIUM

Roscoea cautleyoides

A compact, summer-flowering, tuberous perennial bulb, which produces linear, dark green leaves and short spikes of orchid-like, yellow, purple, or white flowers. Plant it in groups in a sheltered site in dappled shade.

↕25cm (10in)

PERENNIAL SMALL

Silene alpestris

ALPINE CAMPION This evergreen perennial has narrow, lance-shaped, mid-green leaves and small, rounded, fringed, white, occasionally pink-flushed, flowers from late spring to early summer. Grow it in light, dappled shade in gritty soil.

↕15cm (6in) ↔20cm (8in)

PERENNIAL SMALL

Viola cornuta

HORNED VIOLET This small, spreading, evergreen perennial has oval, toothed leaves and flat-faced, purplish blue, occasionally white, flowers from spring to late summer. Grow it in groups in a rock garden.

↕20cm (8in) ↔20cm (8in) or more

BULB SMALL

Scilla mischtschenkoana

SQUILL An early spring-flowering perennial bulb, with strap-shaped basal leaves. Cup-shaped, pale blue flowers, with darker blue veins, appear from late winter to early spring. Plant the bulbs in groups in autumn throughout a rock garden.

↕10cm (4in)

OTHER SUGGESTIONS

Perennials

Alchemilla erythropoda • *Aquilegia flabellate* var. *pumila* f. alba • *Campanula carpatica* • *Campanula garganica* 'W.H. Paine' • *Campanula poscharskyana* • *Convallaria majalis* • *Dodecatheon pulchellum* • *Gentiana acaulis* *Gentiana* 'Inverleith' • *Gentiana* 'Strathmore' • *Geranium pyrenaicum* • *Helleborus odorus* • *Hosta* 'Blue Mouse Ears' • *Oxalis enneaphylla* • *Roscoea humeana* • *Saxifraga fortunei* 'Black Ruby'

Bulbs

Chionodoxa sardensis • *Puschkinia scilloides* • *Scilla bifolia*

Shrubs

• *Juniperus procumbens* 'Nana'

Plants for city gardens

Urban gardens are generally warmer than more exposed areas and provide suitable growing conditions for a wide range of plants.

Even though most city gardens are small, they can be made to look larger by disguising the boundaries with leafy trees and shrubs, such as *Physocarpus*, magnolias, and viburnums. Always remember that the soil next to walls and fences will be sheltered from the rain and therefore dry, so plant at least 30cm (12in) away from vertical surfaces where the roots will find more moisture. You can also create year-round colour with evergreens, and add a selection of deciduous plants for their seasonal flowers, fruit, and autumn tints.

SHRUB LARGE

Amelanchier canadensis

SNOWY MESPILUS A deciduous, upright, dense shrub, with oval, white-haired leaves that age to dark green and turn orange-red in autumn. It produces starry, white flowers from mid- to late spring, and edible, maroon, sweet and juicy fruits in summer.

↕6m (20ft) ↔3m (10ft)

PERENNIAL LARGE

Anemone x *hybrida*

JAPANESE ANEMONE This is an upright perennial, with divided, dark green basal leaves. From late summer to early autumn, it produces large, single or double, pink or white flowers. 'Königin Charlotte' AGM (above) is a popular, double, pink variety.

↕1.2m (4ft) ↔indefinite

PERENNIAL MEDIUM

Aquilegia vulgaris Vervaeneana Group

GRANNY'S BONNET This upright perennial has rounded, divided, orange- and gold-splashed olive-green leaves. In spring, two-tone, flowers in white, red, pink, or blue appear. Deadhead to prevent self-seeding.

↕90cm (36in) ↔45cm (18in)

PERENNIAL MEDIUM

Aspidistra elatior

This evergreen perennial has upright, narrow, glossy, dark green leaves. Cream to purple summer flowers occasionally appear near soil level. 'Variegata' AGM (above) has cream-striped leaves. Ideal for summer borders or pots in deep shade.

↕60cm (24in) ↔45cm (18in)

PERENNIAL MEDIUM

Astrantia major

MASTERWORT A clump-forming perennial, with divided, mid-green leaves. From midsummer to early autumn, it produces sprays of small, green and white, pink or red flowers. 'Hadspen Blood' (above) has dark red flowers.

↕90cm (36in) ↔45cm (18in)

SHRUB MEDIUM

Aucuba japonica

SPOTTED LAUREL This evergreen, bushy shrub has glossy, dark green leaves. Small, purplish flowers in mid-spring are followed, on female plants, by red berries. 'Crotonifolia' AGM (above) has yellow-spotted foliage. Trim it in spring.

↕↔2m (6ft)

PERENNIAL SMALL

Brunnera macrophylla

This ground-cover perennial has large, heart-shaped leaves that form a low carpet. In spring, it bears delicate sprays of small, bright blue flowers. Shelter it from drying winds. 'Dawson's White' (above) has cream-variegated foliage.

↕45cm (18in) ↔60cm (24in)

SHRUB LARGE

Buxus sempervirens

COMMON BOX This evergreen shrub has oval, glossy, dark green leaves. It tolerates close clipping and makes a dense low hedge or screen in a city garden. Disinfect paving tools regularly to prevent the spread of fungal diseases.

↕↔5m (15ft)

SHRUB LARGE

Camellia japonica

COMMON CAMELLIA This evergreen, rounded shrub has oval, glossy, dark green leaves and red flowers in early spring. 'Blood of China' (above) has semi-double, salmon-red blooms. Tolerant of full shade, it makes a beautiful specimen in a border.

↕3m (10ft) ↔2m (6ft)

SHRUB LARGE

Camellia x williamsii

CAMELLIA The williamsii camellias are evergreen shrubs, with glossy, dark green leaves and single or double, white to deep pink spring flowers. Varieties include 'Anticipation' AGM with double, crimson blooms. It tolerates full shade.

↕5m (15ft) ↔2.5m (8ft)

PERENNIAL MEDIUM

Carex comans

NEW ZEALAND SEDGE This clump-forming perennial has arching, grass-like, blue-green leaves that curl at the ends and resemble hair. Grow it in a gravel bed or container. The bronze-leaved form (above) needs some sun for the best colour.

↕60cm (24in) ↔45cm (18in)

PERENNIAL SMALL

Chiastophyllum oppositifolium

LAMB'S TAIL An evergreen perennial, with large, oval, scalloped, mid-green leaves and arching stems of small, bell-shaped, bright yellow flowers from spring to summer. Plant it in part shade.

↕20cm (8in) ↔15cm (6in)

SHRUB LARGE

Daphne bholua

NEPALESE PAPER PLANT An evergreen shrub with leathery, oval, dark green leaves. In late winter, clusters of sweetly fragrant, pink and white flowers appear, followed by black berries. Grow it in a border to provide winter scent.

↕3m (10ft) ↔1.5m (5ft)

PERENNIAL LARGE

Deschampsia cespitosa

TUFTED HAIR GRASS This deciduous, tuft-forming perennial bears clouds of tiny, golden-yellow summer flowers on long stems. Both the seedheads and the linear, sharp-edged, green leaves turn golden in autumn. 'Goldtau' (above) is a popular form.

↕up to 2m (6ft) ↔50cm (20in)

PERENNIAL SMALL

Deschampsia flexuosa

WAVY HAIR GRASS This evergreen grass produces tufts of bluish green leaves and silvery bronze or purple flowerheads held on long slim stems in summer. 'Tatra Gold' (above) has bright yellow-green leaves and bronze flowerheads.

↕up to 50cm (20in) ↔30cm (12in)

SHRUB SMALL

Deutzia gracilis

A compact, deciduous shrub, with lance-shaped, green leaves. In early summer, it produces star-shaped, white or pink, sometimes fragrant, flowers. 'Nikko' (above) has white blooms and purple-tinted, autumn leaves. Grow it in a mixed border.

↕↔1m (3ft)

PERENNIAL SMALL

Dicentra 'Stuart Boothman'

BLEEDING HEART A compact perennial, with finely cut, fern-like, grey-green leaves. From late spring to summer, it produces arching stems of pendent, heart-shaped, carmine-pink flowers. Plant it with later flowering plants in front of a border.

↕30cm (12in) ↔40cm (16in)

PERENNIAL MEDIUM

Digitalis parviflora

FOXGLOVE This deciduous perennial has glossy, dark green basal leaves and, in early summer, tall spikes of orange-brown flowers, each with a purple lip. Grow it in groups in dappled shade. All parts are toxic.

↕60cm (24in) ↔30cm (12in)

PERENNIAL MEDIUM

Dryopteris affinis

GOLDEN MALE FERN An evergreen or semi-evergreen fern that produces a "shuttlecock" of tall, lance-shaped, divided, pale green fronds, which mature to dark green, with scaly, golden-brown midribs. Grow it beneath trees and shrubs.

↕↔90cm (36in)

PERENNIAL LARGE

Ensete ventricosum

ABYSSINIAN BANANA With its large, paddle-shaped leaves, texturing cream midribs, and red undersides, this evergreen perennial has a palm-like appearance. Plant it in a large pot of soil-based compost and grit, and protect from frost.

↕2m (6ft) ↔1m (3ft)

PERENNIAL SMALL

Epimedium x *youngianum* 'Niveum' ♀

SNOWY BARRENWORT A ground-cover perennial, with heart-shaped, serrated, bronze-tinted leaves that turn green in late spring, when small, cup-shaped, white blooms appear. Grow under trees.

↕↔30cm (12in)

SHRUB LARGE

Fargesia murielae ♀

UMBRELLA BAMBOO This clump-forming shrub has arching, yellow-green canes and lance-shaped, bright green leaves. Varieties include 'Simba', a compact form ideal to grow as a focal plant in a small city garden. Best in fertile soil.

↕4m (12ft) ↔indefinite

SHRUB MEDIUM

x *Fatshedera lizei* ♀

TREE IVY An upright, evergreen shrub, often trained as a climber, with deeply lobed, glossy, dark green leaves. Sprays of small, white flowers appear in autumn. Train it up a pillar, or against a wall or fence.

↕2m (6ft) ↔3m (10ft)

SHRUB LARGE

Fatsia japonica ♀

JAPANESE ARALIA An evergreen, rounded, dense shrub, with stout shoots and very large, rounded, deeply lobed, glossy, dark green leaves. Spherical clusters of tiny, white flowers, produced in mid-autumn, are followed by black fruits.

↕↔4m (12ft)

PERENNIAL MEDIUM

Geranium phaeum

DUSKY CRANESBILL A clump-forming, upright perennial, with lobed, green leaves. From late spring to early summer, rounded, maroon or white flowers, with curved-back petals, appear on lax stems. Drought- and shade-tolerant; grow beside a wall or fence.

↕75cm (30in) ↔45cm (18in)

PERENNIAL SMALL

Hakonechloa macra ♀

GOLDEN HAKONECHLOA A slow-growing, deciduous grass, with purple stems and green-striped yellow leaves, ageing to reddish brown. From early autumn to winter, it bears reddish brown flower spikes. Use it to brighten a partly shaded border or path.

↕40cm (16in) ↔60cm (24in)

SHRUB SMALL

Hebe topiaria ♀

A compact, rounded, evergreen shrub, with dense stems bearing small, grey-green leaves, ideal for clipping into shapes. Clusters of small, white flowers appear in summer. Use it to edge a border, or for a small topiary.

↕60cm (24in) ↔90cm (36in)

PERENNIAL MEDIUM

Helleborus foetidus ♀

STINKING HELLEBORE This evergreen, upright perennial has palm-like, dark green leaves. From late winter to early spring, clusters of cup-shaped, red-margined pale green flowers, with an unpleasant odour, appear. Plant it away from seating.

↕↔60cm (24in)

PERENNIAL MEDIUM

Helleborus x *hybridus*

LENTEN ROSE A semi-evergreen perennial, with dark green leaves, divided into lance-shaped, toothed leaflets. Nodding, saucer-shaped flowers appear in shades of white, pink, yellow, green, and purple from midwinter to spring.

↕↔60cm (24in)

PERENNIAL SMALL

Heuchera 'Plum Pudding'

This compact, evergreen or semi-evergreen perennial is grown for its rounded, lobed, maroon leaves. Tiny, white flowers appear on wiry stems in summer. Trim off spent flowers to keep the plant tidy.

↕50cm (20in) ↔30cm (12in)

PERENNIAL MEDIUM

Hosta 'Honeybells'

This is a clump-forming perennial, grown for its large, pale green leaves that are blunt at the tips and have wavy margins. In late summer, it bears short-lived, fragrant, pale lilac flowers, which should be removed as soon as they fade.

↕60cm (24in) ↔1.2m (4ft)

BULB MEDIUM

Hyacinthus orientalis 🏆

HYACINTH A late spring-flowering bulb, with linear, bright green leaves and spikes of highly scented, bell-shaped flowers in a wide range of colours. Plant the bulbs in autumn at the front of a border or in pots.

↕30cm (12in)

SHRUB MEDIUM

Hydrangea macrophylla

MOPHEAD HYDRANGEA A deciduous shrub, with broadly oval, dark green leaves that turn yellow in autumn, and flat clusters of mauve-pink or blue flowers in summer. Grow as a specimen, or at the back of a border. Blue forms (above) need acid soil.

↕2m (6ft) ↔2.5m (8ft)

SHRUB SMALL

Hypericum calycinum

ROSE OF SHARON This is a dwarf, semi- or fully evergreen shrub, with dark green leaves. From midsummer to mid-autumn, it produces large, open, bright yellow flowers, with a tuft of fluffy stamens in the centre. Ideal for ground cover; it can be invasive.

↕60cm (24in) ↔indefinite

SHRUB MEDIUM

Leucothöe fontanesiana

DOG HOBBLE An evergreen, arching shrub, with toothed-edged, leathery, dark green leaves and small, bell-shaped, white flowers in spring. 'Rainbow' (above) has reddish green stems and mottled leaves. It tolerates deep shade and needs acid soil.

↕up to 1.5m (5ft) ↔2m (6ft)

PERENNIAL SMALL

Liriope muscari 🏆

LILYTURF This evergreen, spreading perennial has grass-like, glossy, dark green leaves. In autumn, spikes of tiny, lavender or purple-blue flowers appear. Use it to edge a shady border, or as ground cover beneath shrubs.

↕30cm (12in) ↔45cm (18in)

SHRUB LARGE

Itea virginica

This spreading evergreen shrub has arching stems that bear oval, spiny, dark green leaves. From midsummer to early autumn, it produces decorative, greenish white catkins. Grow it in a sheltered spot beside a fence or wall in acid soil.

↕3m (10ft) ↔1.5m (5ft)

SHRUB SMALL

Juniperus squamata 🏆

FLAKY JUNIPER This dense, rounded, evergreen conifer produces a low mound of slender, bright blue-grey foliage, and makes a good foil for pastel- and hot-coloured flowers in a city garden. It is suitable for large containers of soil-based compost.

↕40cm (16in) ↔1m (3ft)

SHRUB LARGE

Magnolia x soulangeana

SAUCER MAGNOLIA A spreading, deciduous shrub, with oval, dark green leaves and, from mid- to late spring, pink, white, or purple, goblet-shaped flowers. Use it as a feature in a lawn or mixed border; protect flower buds from late frosts.

↕↔6m (20ft)

BULB MEDIUM

Narcissus 'Jack Snipe' 🏆

DAFFODIL This early- to mid-spring-flowering perennial bulb has narrow, dark green leaves and creamy white flowers, with short, bright yellow cups. Plant groups of bulbs in autumn at the front of a border or in containers.

↕23cm (9in)

BULB MEDIUM

Narcissus jonquilla

JONQUIL A mid-spring-flowering perennial bulb, with narrow, dark green leaves and clusters of richly fragrant, yellow flowers, with shallow, dark golden-yellow cups. Plant the bulbs in groups in autumn at the front of a border or in pots.

↕30cm (12in)

PERENNIAL MEDIUM

Onoclea sensibilis

SENSITIVE FERN A deciduous fern, with arching, almost triangular, divided, pale green fronds that are tinted pinkish brown in spring. In autumn, the fronds turn a decorative yellowish brown. It thrives in shady borders; can be invasive.

↕60cm (24in) ↔1m (3ft)

PERENNIAL SMALL

Ophiopogon planiscapus 'Nigrescens'

BLACK MONDO GRASS This clump-forming, evergreen perennial has narrow, grass-like leaves and small, bell-shaped, white or mauve summer flowers, followed by black berries. Plant it to edge borders.

↕23cm (9in) ↔30cm (12in)

SHRUB LARGE

Osmanthus heterophyllus

FALSE HOLLY This evergreen shrub has holly-like, glossy, bright green leaves and, in autumn, tiny, fragrant, white flowers. Grow it as a screen, or in a border; protect it from hard frosts. 'Aureomarginatus' (above) has gold-edged leaves.

↕↔5m (15ft)

PERENNIAL SMALL

Pachysandra procumbens

A clump-forming, evergreen perennial, with rounded, lobed, dark green leaves and spikes of mint-scented, small, white flowers in spring. Tolerates full shade; grow it as ground cover beneath trees and shrubs.

↕25cm (10in) ↔60cm (24in)

PERENNIAL MEDIUM

Paeonia lactiflora 'Sarah Bernhardt'

This clump-forming perennial has divided leaves and large, fragrant, double flowers, with ruffled, rose-pink petals, fading to silvery blush white at the margins. Grow it in a mixed border in fertile soil; feed well.

↕↔1m (3ft)

SHRUB MEDIUM

Paeonia suffruticosa

TREE PEONY This is an upright, deciduous shrub, with deeply lobed, dark green leaves. In late spring, bowl-shaped, sometimes scented, pink, white, red, or purple, flowers appear. Grow it next to a wall or fence in fertile soil.

↕↔2.2m (7ft)

SHRUB MEDIUM

Physocarpus opulifolius

NINEBARK A deciduous shrub, with peeling bark and lobed, mid-green leaves. It bears domed, white flower clusters in late spring, followed by red-brown fruits. Use at the back of a border, or to mask a fence. 'Dart's Gold' AGM (above) has yellow young foliage.

↕2m (6ft) ↔2.5m (8ft)

PERENNIAL MEDIUM

Physostegia virginiana

OBEDIENT PLANT This erect, compact perennial has toothed, mid-green leaves. In late summer and early autumn, it bears spikes of tubular, white, pink, or purple flowers that can be placed in position. 'Vivid' AGM (above) has purple flowers.

↕↔60cm (24in)

PERENNIAL SMALL

Pulmonaria saccharata

BETHLEHEM SAGE A clump-forming, semi-evergreen perennial, with bristly, white-spotted blue-green leaves. In spring, it bears clusters of blue-purple, funnel-shaped flowers. Use it as ground cover beneath trees or shrubs, or to edge a flowerbed.

↕30cm (12in) ↔60cm (24in)

PERENNIAL MEDIUM

Salvia nemorosa

BALKAN CLARY A compact perennial, with lance-shaped, wrinkled, mid-green leaves, and spikes of pink, white, or purple flowers from summer to early autumn. Plant it in light shade. 'Ostfriesland' AGM (above) has violet-blue flowers with pink bracts.

↕ up to 75cm (30in) ↔ 60cm (24in)

SHRUB SMALL

Salvia officinalis

SAGE An evergreen or semi-evergreen shrub, with aromatic, grey-green or pale green leaves. It sometimes bears tubular, purple-blue flowers in summer. 'Icterina' AGM (above) has yellow-variegated leaves. Use it as ground cover or to edge a border.

↕ 80cm (32in) ↔ 1m (3ft)

SHRUB SMALL

Sarcococca confusa

SWEET BOX A bushy evergreen shrub, with glossy, lance-shaped, dark green leaves. In winter, clusters of tiny, white, sweetly fragrant blooms appear, followed by shiny, black fruits. Plant it next to a walkway in part or deep shade.

↕↔ 1m (3ft)

TREE MEDIUM

Taxus baccata 'Fastigiata'

IRISH YEW This upright, slender, evergreen conifer is better suited to small gardens than the species. It has needle-like, dark green leaves, and female plants bear red berries. It makes a neat focal point, and tolerates deep shade. All parts are toxic.

↕ up to 10m (30ft) ↔ 4m (12ft)

SHRUB MEDIUM

Sarcococca hookeriana var. *digyna*

SWEET BOX A bushy evergreen shrub, with glossy, lance-shaped, dark green leaves. Clusters of tiny, white, fragrant winter blooms precede shiny, black fruits. Plant it next to a walkway to enjoy the scent.

↕ 1.5m (5ft) ↔ 2m (6ft)

SHRUB LARGE

Skimmia japonica

This evergreen shrub has dark green, glossy leaves. In spring, clusters of pink-budded, small, white flowers appear. Female plants (above) bear bright red berries, if both sexes are grown nearby. Grow it in moist but well-drained soil.

↕↔ 6m (20ft)

PERENNIAL MEDIUM

Spigelia marilandica

INDIAN PINK This upright perennial has oval, green leaves. In early summer, it bears tubular, red flowers, with star-shaped yellow tips. 'Wisley Jester' (above) has crimson and pale yellow blooms. Provide a sheltered site in part shade.

↕ 60cm (24in) ↔ 45cm (18in)

TREE SMALL

Trachycarpus fortunei

CHUSAN PALM This evergreen palm has an unbranched stem and head of large, deeply divided, fan-like, mid-green leaves. Sprays of fragrant, creamy yellow flowers appear in early summer. Plant it in a sheltered area, away from drying winds.

↕ up to 2m (6ft) ↔ 2.5m (8ft)

SHRUB MEDIUM

Viburnum davidii

A dome-shaped, evergreen shrub, with dark green, deeply veined foliage. In late spring, it bears clusters of small, white flowers. Female plants produce decorative, metallic blue fruits, if both sexes are grown. Use it as ground cover, or grow under trees.

↕↔ 1.5m (5ft)

SHRUB LARGE

Viburnum farreri

This deciduous, upright shrub has dark green foliage that is bronze when young. In late autumn and during mild periods in winter and early spring, it bears clusters of fragrant, white or pale pink flowers, followed by red, blue, or black berries.

↕ 3m (10ft) ↔ 2.5m (8ft)

OTHER SUGGESTIONS

Perennials

Astelia chathamica • *Begonia* Illumination Series • *Bergenia purpurascens* • *Pachysandra terminalis* 'Green Carpet'

Shrubs, trees, and climbers

Acer palmatum 'Atropurpureum' • *Buddleja davidii* • *Daphne laureola* subsp. *philippi* • *Daphne pontica* • *Euonymus japonicus* 'Microphyllus Albovariegatus' • *Hebe rakaiensis* • *Juniperus chinensis* 'Pyramidalis' • *Juniperus squamata* 'Blue Star' • *Lapageria rosea* • *Lonicera nitida* • *Physocarpus Diabolo* • *Salix* 'Hakuro Nishiki' • *Skimmia* x *confusa*

Plant focus: ferns

Perfect for shady gardens, ferns form carpets of textured foliage that provide a decorative foil for woodland flowers.

RANGING FROM TINY CREEPING PLANTS, such as the Himalayan maidenhair fern *Adiantum venustrum*, to the tall, stately royal fern, *Osmunda regalis*, which reaches up to 2m (6ft) in height, this group of sun-shy woodlanders offers the gardener many exciting design opportunities. Ferns are grown primarily for their textured foliage, which provides a lacy backdrop to spring bulbs and other shade-tolerant blooms. For continuous colour and interest throughout the year, mix evergreens, such as *Polystichum*, with deciduous ferns, which produce coiled shoots as their leaves unfurl in spring and turn rusty brown in autumn. Although most ferns prefer damp soil, a few, such as the *Polypodium* family and male fern, *Dryopteris filix-mas*, do well in drier conditions and are invaluable for providing understorey planting beneath trees and large shrubs.

Ferns create a luscious green blanket in cool, shady areas where few others plants will thrive

USING FERNS

Ferns look delicate, with their fine, lacy foliage, but most are fully hardy and have many uses. They are also highly varied, offering solid or finely divided, glossy or matt, green leaves, some decorated with purple or silver variegation. They look most effective planted together in shady borders, but many also make good specimen plants in containers, such as *Polystichum setiferum*, with its elegant, arching fronds.

Shade-loving astilbes and heucheras provide the perfect planting partners for *Dryopteris* and other ferns in a mixed border where the soil is moist.

Ferns of different sizes and shapes, such as *Asplenium* and *Dryopteris*, make excellent subjects for a textural foliage container on a shady patio.

Ferns often naturalize in cracks in shady walls, softening the edges and adding interest.

TYPES OF HARDY FERN

Adiantum Hardy forms include *A. aleuticum* (above), with deciduous finger-like fronds; *A. pedatum* with black-stalked segmented fronds; and evergreen *A. venustrum.*

Asplenium The harts tongue fern, *A. scolopendrium* (above), bears rosettes of long, glossy, evergreen, fronds. The foliage of *A. trichomanes* is divided into oblong segments.

Athyrium The lady fern, *A. filix-femina* (above), has deciduous fronds, made up of long segments. The deciduous Japanese painted fern, *A. niponicum,* is variegated.

Blechnum Hard fern, *B. spicant* (above), has narrow, evergreen dark green, divided fronds. *B. penna-marina* is shorter, and bears reddish young fronds that mature to green.

Dryopteris Most species of buckler fern are fully hardy, and produce large, deciduous fronds, crested in some forms, arranged like an elegant shuttlecock.

Matteuccia The large-sized ostrich or shuttlecock fern, *M. struthiopteris* (above), bears large, deciduous, pale green fronds, divided into finger-like segments.

Osmunda The royal fern, *O. regalis,* has very tall, deciduous fronds, divided into long segments, with rust-coloured spore-bearing segments at the tips of some segments.

Polypodium This genus includes many hardy, evergreen ferns; the most common is the low-growing *P. vulgare* (above), with triangular-shaped, dark-green fronds.

Polystichum Most holly or shield ferns are hardy, with large, dark green, triangular, evergreen fronds, divided into feathery segments, with shuttlecock-like crowns.

Plants for exposed sites

Grow spring bulbs, summer-flowering perennials, deciduous trees and shrubs for autumn tints, and frost-proof evergreens to dress winter schemes.

Many ferns are unfazed by sub-zero temperatures and add a textural backdrop to colourful bulbs and perennials, such as scillas, irises, and aquilegias, while scented plants lend a sensory note to planting schemes; mahonias, lily of the valley (*Convallaria majalis*), and spring-flowering narcissus will all help to perfume your plot. Large shrubs and trees can help shelter an exposed garden, but plant them on the north side to ensure that they don't cast even more shade. Also use tough hedging plants, such as *Crataegus* and yew (*Taxus*) to create a warmer seating area; a shoulder-high screen will protect a patio.

PERENNIAL SMALL

Adiantum aleuticum

MAIDENHAIR FERN A deciduous or semi-evergreen fern, with black-stalked, mid-green fronds, divided into oblong segments. New fronds may be tinged pink. It is tolerant of low temperatures, but avoid planting in deep shade.

↕↔45cm (18in)

PERENNIAL SMALL

Alchemilla mollis

LADY'S MANTLE This clump-forming perennial has rounded, pale green leaves, with crinkled edges. In summer, small sprays of tiny, bright greenish yellow flowers appear. Ideal as a ground-cover plant; trim back untidy leaves.

↕↔50cm (20in)

PERENNIAL MEDIUM

Aquilegia vulgaris

GRANNY'S BONNETS This upright perennial has green leaves, divided into rounded, lobed leaflets. From late spring to early summer, it bears pendent, single or double flowers in shades of purple and blue to pink and white. Deadhead after flowering.

↕1m (3ft) ↔50cm (20in)

PERENNIAL MEDIUM

Astilbe 'Fanal'

A clump-forming perennial, with attractive, deeply divided, fern-like, dark green foliage. In summer, it produces dense spikes of feathery, dark pink flowers on slim, dark stems. Plant it in moist, fertile soil in partial shade.

↕60cm (24in) ↔45cm (18in)

pH

PERENNIAL LARGE

Athyrium filix-femina

LADY FERN This deciduous fern produces arching, lance-shaped, pale green fronds, divided into narrow segments, with toothed margins. It thrives in deep shade in cold, frosty borders in moist, acid soil, but avoid planting in windy sites.

↕1.2m (4ft) ↔1m (3ft)

pH

TREE LARGE

Betula utilis var. *jacquemontii*

HIMALAYAN BIRCH This vase-shaped, deciduous tree is grown for its stunning white bark. Its diamond-shaped leaves turn yellow in autumn. Use it as a specimen or to make a small copse.

↕18m (60ft) ↔10m (30ft)

PERENNIAL MEDIUM

Blechnum spicant

HARD FERN This evergreen fern has narrow, lance-shaped, dark green fronds, divided into linear segments. Good in deep shade in cold and exposed gardens where the soil is reliably moist year round. Remove old foliage in spring.

↕75cm (30in) ↔45cm (18in)

SHRUB SMALL

Calluna vulgaris

SCOTS HEATHER An evergreen, bushy, low-growing shrub, with tiny, grey, yellow, or bright green leaves. It bears stems of bell-shaped, single or double flowers in shades of pink, white, and purple from midsummer to late autumn.

↕30cm (12in) ↔35cm (14in)

pH

TREE SMALL

Crataegus x *persimilis*

A small, deciduous tree, 'Prunifolia' (above) has thorny branches and glossy, dark green leaves that develop rich autumnal shades. It flowers for a short period in late spring, and bears red berries that are attractive to birds. Suitable for windy or coastal sites.

↕8m (25ft) ↔10m (30ft)

PERENNIAL MEDIUM

Crocosmia x *crocosmiiflora*

MONTBRETIA An upright perennial, with arching, sword-shaped, green leaves. From summer to early autumn, clusters of small, trumpet-like blooms appear on tall stems. 'Star of the East' AGM (above) has apricot-yellow flowers. Suitable for coastal areas.

↕70cm (28in) ↔8cm (3in)

PERENNIAL LARGE

Digitalis ferruginea

RUSTY FOXGLOVE An upright perennial, with basal rosettes of oval, rough, dark green leaves. Long, slender spikes of funnel-shaped, orange-brown and white flowers appear in midsummer. An excellent border plant for cold and exposed gardens.
↕1.2m (4ft) ↔30cm (12in)

PERENNIAL MEDIUM

Dryopteris wallichiana

WALLICH'S WOOD FERN A deciduous, clump-forming fern, with a "shuttlecock" of lance-shaped, divided, bright yellow-green fronds, ageing to dark green, and scaly, brownish black stems. Best in fertile soil; it tolerates deep shade and dry conditions.
↕90cm (36in) ↔75cm (30in)

BULB LARGE

Fritillaria pallidiflora

SIBERIAN FRITILLARY A perennial bulb, with lance-shaped, blue-green leaves that die back after flowering. It bears nodding, bell-shaped, greenish yellow flowers, chequered brownish red within, in early summer. Plant the bulbs in groups in autumn.
↕70cm (28in)

PERENNIAL SMALL

Helleborus niger

CHRISTMAS ROSE This evergreen, clump-forming perennial has divided, leathery, dark green leaves and, from winter to early spring, cup-shaped, nodding, white or pink-flushed flowers with green "eyes". 'Potter's Wheel' (above) has pure white blooms.
↕↔30cm (12in)

TREE SMALL

Juniperus scopulorum

ROCKY MOUNTAIN JUNIPER A rounded or spreading, cold-tolerant conifer, with scale-like, sharply pointed leaves and small cones. 'Skyrocket' (above) has a narrow, columnar habit and blue-grey foliage; ideal for a small garden in light shade.
↕6m (20ft) ↔75cm (30in)

PERENNIAL MEDIUM

Hosta 'Krossa Regal'

This is a clump-forming, vase-shaped perennial, with arching, deeply ribbed, greyish blue leaves. It produces spikes of short-lived, lilac flowers in summer. Suitable for cold sites; shelter it from drying winds, and protect from slugs.
↕70cm (28in) ↔75cm (30in)

PERENNIAL MEDIUM

Iris sibirica

SIBERIAN IRIS A rhizome-forming perennial, with upright, sword-like, blue-green leaves and large, beardless, pink, blue, white, or yellow flowers from late spring to early summer. 'Butter and Sugar' AGM (above) has yellow and white blooms.
↕1m (3ft) ↔indefinite

SHRUB SMALL

Juniperus x *pfitzeriana*

JUNIPER This flat-topped, shrubby, evergreen conifer has scale-like, grey-green leaves and dark, fading to pale purple, cones. 'Pfitzeriana Aurea' (above) has golden foliage, which turns yellow-green in winter. Best in dappled shade.
↕90cm (36in) ↔2m (6ft)

SHRUB LARGE

Kalmia latifolia

MOUNTAIN LAUREL A bushy, evergreen shrub, with glossy, oval, green leaves. From late spring to early summer, large clusters of pink, cup-shaped flowers appear from crimped buds. Prune lightly after flowering to keep it neat.
↕↔3m (10ft)

PERENNIAL MEDIUM

Lamprocapnos spectabilis f. *alba*

syn. *Dicentra spectabilis* var. *alba*
A clump-forming perennial, with fern-like foliage and, from late spring to early summer, pendent, white, heart-shaped blooms; it dies back after flowering.
↕75cm (30in) ↔60cm (24in)

SHRUB SMALL

Mahonia aquifolium

OREGON GRAPE A spreading, evergreen shrub, with glossy, spiny, dark green leaves that turn bronze in cold weather. In spring, it bears clusters of scented, yellow flowers, followed by black berries. It is ideal for year-round colour.
↕1m (3ft) ↔1.5m (5ft)

PERENNIAL MEDIUM

Maianthemum racemosum ♀

FALSE SPIKENARD This upright perennial has prominently veined, oval, green leaves. From mid- to late spring, it bears scented, fluffy, cream plumes, which are sometimes followed by red berries. Ideal for cold sites, but shelter it from strong, drying winds.
↕90cm (36in) ↔60cm (24in)

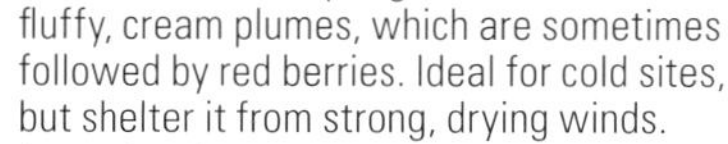

PERENNIAL MEDIUM

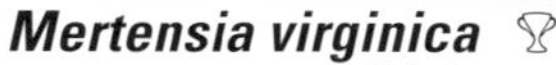

Mertensia virginica ♀

VIRGINIA BLUEBELLS This is a compact perennial, with soft blue-green leaves, and clusters of nodding, funnel-shaped, blue flowers in spring. The plant dies down in summer. It tolerates low temperatures, but only light shade.
↕60cm (24in) ↔45cm (18in)

BULB MEDIUM

Narcissus bulbocodium

HOOP PETTICOAT DAFFODIL A dwarf perennial bulb, with grass-like, dark green leaves and unusual, funnel-shaped, golden-yellow flowers surrounded by slim petals. Plant the bulbs in autumn in soil that is reliably moist in winter and spring.
↕15cm (6in)

PERENNIAL LARGE

Persicaria amplexicaulis

RED BISTORT This clump-forming perennial has oval, green leaves, and slender spikes of bell-shaped, white or pink flowers from summer to early autumn. Varieties include 'Firetail' (above), with bright red blooms. It is ideal for cold or coastal gardens.
↕↔1.2m (4ft)

PERENNIAL MEDIUM

Polygonatum odoratum

ANGLED SOLOMON'S SEAL An arching perennial, with oval to lance-shaped, mid-green leaves and fragrant, tubular, green-tipped white flowers in late spring. Grow it in full or part shade in cold gardens, sheltered from drying winds.
↕60cm (24in) ↔30cm (12in)

PERENNIAL MEDIUM

Polystichum acrostichoides

CHRISTMAS FERN An evergreen perennial, with slender, lance-shaped, dark green fronds, divided into segments. It will produce clumps over time and is tolerant of cold, shady sites, including deep shade. It thrives in most soils.
↕60cm (24in) ↔45cm (18in)

PERENNIAL MEDIUM

Polystichum munitum ♀

SWORD FERN This evergreen, "shuttlecock" perennial produces large, erect, leathery, triangular-shaped, dark green fronds, consisting of small, spiny-margined segments. It is tolerant of deep shade and cold, frosty sites, where the soil is moist.
↕90cm (36in) ↔1.2m (4ft)

SHRUB SMALL

Potentilla fruticosa

CINQUEFOIL A deciduous shrub, with small, dark green leaves, divided into narrow leaflets. From late spring to early autumn, it produces saucer-shaped flowers in shades of yellow, red, pink, orange, or white, depending on the cultivar.
↕1m (3ft) ↔1.5m (5ft)

PERENNIAL SMALL

Primula vulgaris ♀

PRIMROSE This perennial forms basal rosettes of oval, pale green leaves. In early spring, it produces clusters of often scented, primrose-yellow flowers. It lives happily in cold gardens in dappled shade.
↕20cm (8in) ↔35cm (14in)

PERENNIAL SMALL

Sagina subulata var. *glabrata*

HEATH PEARLWORT This mat-forming perennial has small, pointed, green leaves and tiny, white summer flowers. 'Aurea' (above) has yellow-green foliage. Grow it as ground cover or a lawn alternative in moist, acid soil, and a site shaded at midday.
↕10cm (4in) ↔30cm (12in)

PERENNIAL SMALL

Saxifraga fortunei

SAXIFRAGE A deciduous or semi-evergreen perennial, with lobed, mid-green leaves, red-purple beneath. Upright, red stems supporting pendent clusters of white flowers appear in late summer or autumn. Ideal for a shaded cold garden.

↕↔30cm (12in)

BULB MEDIUM

Scilla siberica

SIBERIAN SQUILL A dwarf perennial bulb, with grass-like, mid-green leaves, and nodding, bell-shaped, blue flowers in spring. Plant the bulbs in groups in autumn. If left undisturbed, plants will slowly naturalize.

↕20cm (8in)

TREE LARGE

Taxus baccata

YEW This bushy, evergreen conifer has dark green needles and small, yellow spring flowers, followed by red berries. Use it as formal hedging, or a windbreak in cold regions, or to attract wildlife. All parts of yew are toxic.

↕15m (50ft) ↔10m (30ft)

PERENNIAL SMALL

Tiarella wherryi

FOAM FLOWER A compact, clump-forming perennial, with deeply lobed, maroon-tinted green leaves. From late spring to early summer, clusters of tiny, frothy, white or pink flowerheads appear. It makes a good edging plant in deep shade.

↕20cm (8in) ↔15cm (6in)

SHRUB LARGE

Tsuga canadensis

EASTERN HEMLOCK This evergreen conifer has needle-like, blue-green leaves, white beneath, held on pendent stems, and produces brown cones. 'Pendula' AGM (above) is a slow-growing, spreading, mound-forming shrub.

↕↔5m (15ft)

SHRUB SMALL

Vaccinium vitis-idaea Koralle Group

COWBERRY A spreading, evergreen shrub, with glossy, dark green leaves. From late spring to early summer, bell-shaped, white or pink, nodding flowers appear, followed by edible, red berries.

↕25cm (10in) ↔indefinite

SHRUB LARGE

Viburnum plicatum f. *tomentosum*

JAPANESE SNOWBALL A wide-spreading, deciduous shrub, with tiered branches that carry prominently veined, dark green leaves, that turn purple in autumn. In late spring, it produces white flowerheads.

↕3m (10ft) ↔4m (12ft)

SHRUB LARGE

Viburnum tinus

LAURUSTINUS An evergreen, bushy shrub, with oval, dark green leaves. From late winter to spring, it produces flat heads of small, white blooms, followed by metallic blue berries. Varieties include 'Eve Price' AGM, with pink-budded white flowers.

↕↔3m (10ft)

PERENNIAL SMALL

Viola cornuta

HORNED VIOLET This small, spreading, evergreen perennial has oval, toothed leaves and, from spring to late summer, flat-faced, purplish blue, occasionally white, flowers. Grow it in a wind-blown garden or border; ideal for ground cover.

↕20cm (8in) ↔20cm (8in) or more

SHRUB MEDIUM

Weigela florida

A deciduous, arching shrub, with oval, toothed, mid-green leaves. From late spring to early summer, it produces dark pink flowers, pale pink to white inside. Good for cold sites. 'Variegata' AGM has colourful, variegated foliage.

↕↔2.5m (8ft)

OTHER SUGGESTIONS

Perennials

Amsonia tabernaemontana • *Convallaria majalis* • *Gentiana septemfida* • *Geranium pyrenaicum* 'Bill Wallis' • *Tierella cordifolia*

Bulbs

Chionodoxa luciliae • *Galanthus* 'Atkinsii' • *Galanthus nivalis* 'Flore Pleno' • *Leucojum vernum*

Shrubs

Gaultheria cuneata • *Mahonia* x *wagneri* • *Physocarpus opulifolius* • *Rhododendron yakushimanum* • *Ribes alpinum* • *Sambucus racemosa* • *Viburnum opulus* 'Roseum'

Trees

Betula nigra • *Juniperus communis* • *Sorbus hybrida* 'Gibbsii'

PLANTS for SPECIAL EFFECTS

PLANTS FOR
GARDEN STYLES

PLANTS FOR
SEASONAL INTEREST

PLANTS FOR
COLOUR AND SCENT

PLANTS FOR
SHAPE AND TEXTURE

PLANTS FOR
GARDEN PROBLEMS

Choosing plants for special effects

When planning a design, choose plants for their colour, texture, shape, and scent, and ring in seasonal changes with species that perform at different times. Add evergreens and plants with a bold structure for year-round appeal.

PLANNING YOUR SCHEME

The first section of this book helps you to pinpoint plants that thrive in your particular garden conditions, while this second section takes that idea one step further by providing you with a selection of plants that can be used to create a particular design or achieve a specific effect. For example, you may be looking for ideas to create a Japanese-style garden, or plants that will draw a range of birds and insects to a wildlife border. Also look at pp.14–17, which explain how to group plants to achieve certain effects, such as a formal or contemporary design.

Once you have chosen a style, you should think about introducing seasonal interest to keep the colour and interest going throughout the year. This is particularly important in small gardens, where the whole space is on view at all times. When selecting seasonal plants, try to combine those that flower or fruit consecutively to avoid peaks and troughs – the description of each plant featured tells you whether it is an early-, mid-, or late-season performer. Also fuse your blooming plants with leafy species and evergreens for long-lasting interest.

GARDEN STYLES

Creating a style To produce a chosen garden style, select an appropriate palette of plants. The Asian-style garden above uses architectural plants in muted colours to offset the bold features.

Using plants Combine plants within a style so that they contrast or complement each other. Here yellow foxtail lilies contrast with the adjacent purple dianthus (*see left*).

SEASONAL INTEREST

Seasonal highlights To capture the essence of a season, use plants that celebrate that time of year, such as spring bulbs or trees with fiery autumn leaves, like the Japanese maple above.

Retaining interest The flowers of many plants, such as sea holly (*left*) and hydrangeas, dry *in situ*, and last into winter, providing interest for more than one season.

APPEALING TO THE SENSES

Plants with scented leaves and flowers are prized because they introduce a sensory quality that can lift the spirits at all times of the year. The perfume from a mahonia or witch hazel in the depths of winter is tempting enough to venture outside for, while the scent of honeysuckle is reason enough to spend summer evenings relaxing in the garden. You will find a wealth of scented plants of all types outlined on pp.340–345.

You can also add a designer element to to your garden using plants that have different shapes, foliage colour, and textures. Such plants also provide a permanent decorative framework for the more transitory elements.

The final section of the book focuses on problem areas, such as hot, dry sites, slopes, and waterlogged areas, as well as recommends plants to choose for gardens where pests are becoming a nuisance. This section also offers low-maintenance solutions for large areas by offering a selection of weed-suppressing ground-cover plants.

COLOUR AND SCENT

Sensory pleasure Plants that produce colourful or sweetly scented flowers or leaves, such as lavender (*above*), form a key element of the sensory experience you can enjoy in your garden.

Choice plants Some plants, such as peonies (*left*), have exquisite, but short-lived blooms. Position them with other plants that provide longer lasting colour.

SHAPE AND TEXTURE

Enduring appeal Plants with bold stem and leaf structures and textures, such as the silvery verbascums above, will sustain interest over a longer period than plants with transient flower forms.

Creating shapes Many plants are ideal for training, shaping, or using as topiary (*left*), allowing you to create enduring dramatic and artistic effects in your garden.

PROBLEM SOLVING

Plant allies Ground-cover plants, such as pulmonaria (*above*) and leafy geraniums, are perfect for suppressing unwanted weeds in shady areas.

Recipes for garden styles

TRADITIONAL COTTAGE BORDER

This medley of classic cottage garden favourites in pastel hues provides the perfect setting for a romantic border in sun. The tall delphiniums provide height and structure behind a blend of roses and astrantias. Other plants to consider if you have space include lupins, both tall and spreading forms of *Campanula*, and geraniums. Cottage border plants perform best on moisture-retentive soil. They require feeding annually in spring with an all-purpose granular fertilizer and a mulch of well-rotted organic matter, such as manure.

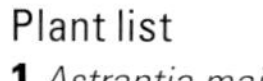

Plant list
1 *Astrantia major*
2 *Rosa* - pink shrub variety
3 *Delphinium elatum*

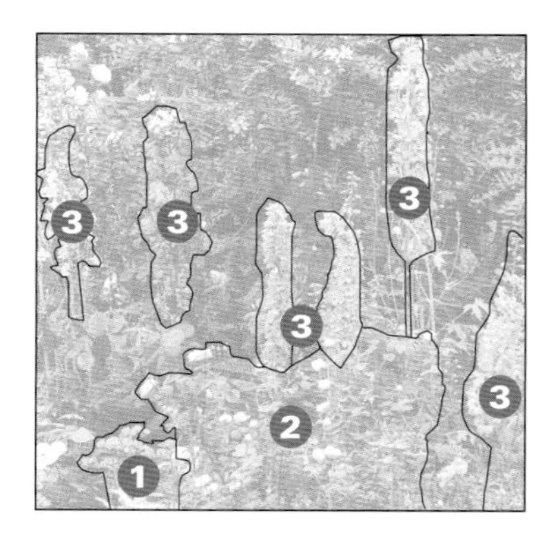

CHIC MODERNIST DESIGN

The trick to producing an elegant modernist scheme is to keep your planting designs simple, selecting varieties for their structural qualities, and minimizing colour choices. In this scheme, white summer-flowering alliums provide accents above crisply clipped box balls, while ferns and *Stipa tenuissima* offer textural contrasts. All the plants like free-draining conditions, except for the ferns, which require more moisture.

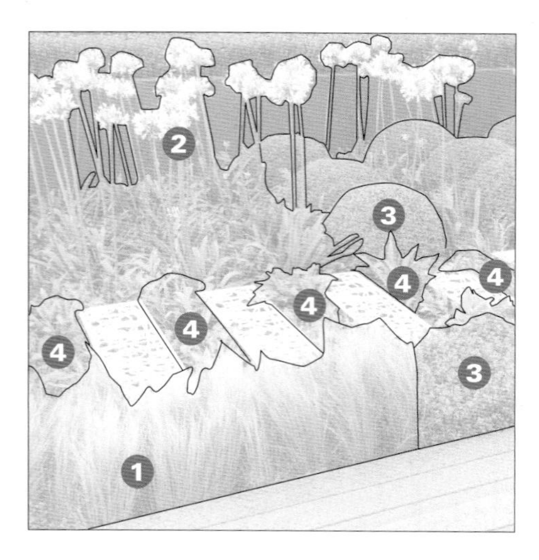

Plant list
1 *Stipa tenuissima*
2 *Allium* – white form
3 *Buxus sempervirens* (box balls)
4 *Asplenium scolopendrium* (ferns)

PERENNIAL BORDER FOR A SMALL SPACE

Fuse naturalistic swathes of just a few plant species to create a small-scale prairie. Here, flat-topped achillea is teamed with spires of purple salvias and allium globes to give an exciting contrast of flower shapes and textures. Leafy grasses provide a cool and long-lasting foil for the flowers. To extend the season, add groups of rudbeckia and asters in the same colours, and leave the spent flowerheads to stand over winter. Plant in moisture-retentive but free-draining soil for the best results.

Plant list
1 *Salvia* 'Ostfriesland'
2 *Allium caeruleum* 'Azureum'
3 *Achillea* 'Credo'

Recipes for colour, texture, and scent

STEM INTEREST IN A SPRING BORDER

Few plants surpass the Himalayan birch (*Betula*) for year-round stem interest. Set the tree's startling white trunks against a dark backdrop to produce the most eye-catching effect, and choose a multi-stemmed form to increase the impact. Underplant the tree with shade-tolerant alliums, geraniums, alchemilla, and ferns to provide colour and texture through spring and summer. These plants also cope with the well-drained soil conditions beneath a tree. In early spring, apply a mulch of well-rotted organic matter, such as manure.

Plant list

1 *Alchemilla mollis*
2 *Geranium himalayense* 'Gravetye'
3 *Galega x hartlandii*
4 *Allium* 'Purple Sensation'
5 *Geranium clarkei* 'Kashmir White'
6 *Betula utilis* var. *jacquemontii*

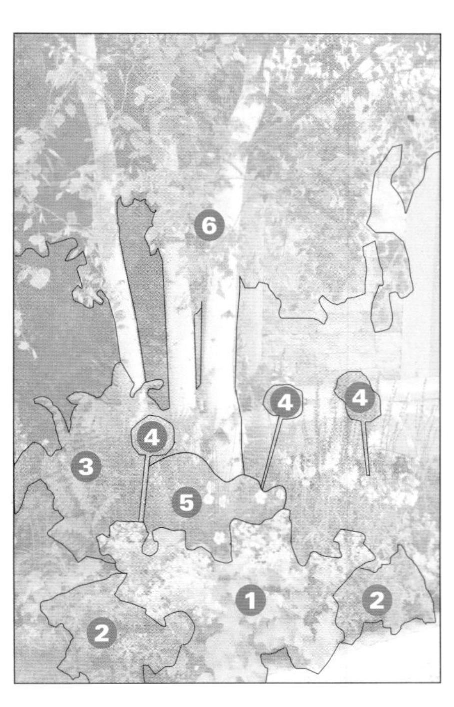

COLOURFUL FOLIAGE EFFECTS

Create a medley of colourful foliage with this collection of trees, shrubs, and perennials. Place the Japanese maple (*Acer*) in pride of place, as this will produce the main focal point in the border, with its canopy of finely cut, red foliage, which fires up further in autumn. The fatsia adds a large-leaved, green contrast, while low-growing heuchera and persicaria lend foliage colour at ground level. These plants enjoy moist but free-draining soil, but ensure the persicaria is mulched and kept well watered in summer.

Plant list
1 *Persicaria affinis* 'Superba'
2 *Fatsia japonica*
3 *Acer palmatum* 'Atropurpureum'
4 *Heuchera* 'Palace Purple'

FRAGRANT BORDER

Choose scented plants, such as lavender, to edge a seating area or pathway, where they will emit their scent when brushed against. There is a wealth of fragrant shrub roses to choose from, so select those that suit your colour scheme, and also consider climbing forms to create a scented wall behind the border. You could also weave in *Lilium regale* at the back of the scheme, and garden pinks (*Dianthus*) to edge the front. Provide plants with a sunny site and free-draining soil.

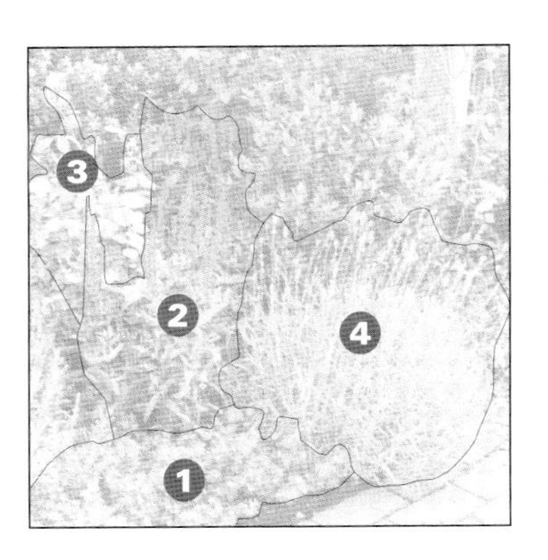

Plant list
1 *Geranium sanguineum*
2 *Salvia nemorosa*
3 Fragrant shrub rose, e.g. *Rosa* 'Claire Austin'
4 *Lavandula angustifolia* 'Munstead'

Recipes for seasonal effects

COLOURFUL SUMMER BORDER

For midsummer colour, combine this group of perennials and grasses for a contemporary border. The hot orange and yellow achillea and euphorbia are tempered by the grass, *Anemanthele lessoniana,* and blue sea hollies, which also introduce different shapes and forms. Repeat these plants throughout your border to produce a cohesive design, and also include some alliums, astrantia, and irises to create interest earlier in the season. Plant this collection in a sunny site in free-draining soil.

Plant list
1 *Eryngium* x *zabelii* (sea holly)
2 *Anemanthele lessoniana*
3 *Achillea* 'Walther Funcke'
4 *Euphorbia* species
5 *Trifolium ochroleucon*
6 *Achillea* 'Terracotta'

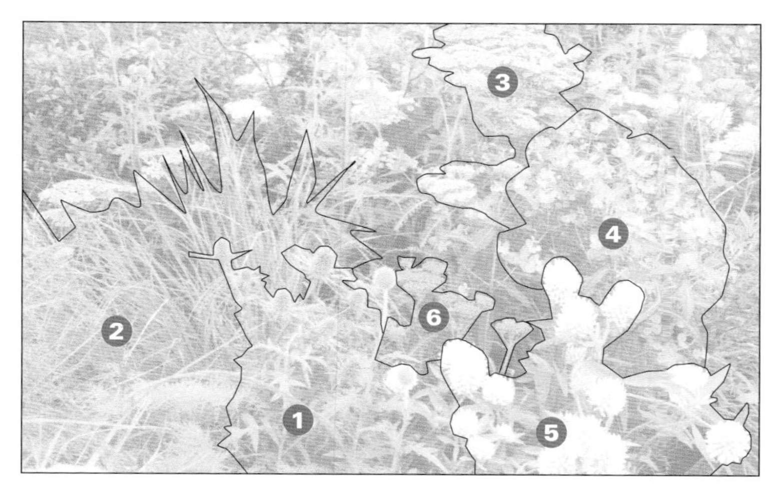

LATE SPRING BORDER

Ideal for a bog garden or moisture-retentive clay, this border of leafy ferns and perennials, interspersed with colourful flowers, will perform well in part-shade. The ostrich fern and *Gunnera manicata* are large plants, so ensure you allow them sufficient space, and do not let them swamp the dainty primulas and irises in front. To extend the colour into summer, also include astilbe and filipendula. Mulch in spring with well-rotted compost.

Plant list
1 *Primula prolifera*
2 *Dicentra formosa*
3 *Iris sibirica*
4 *Gunnera manicata*
5 *Matteuccia struthiopteris* (ostrich fern)
6 *Hosta fortunei*

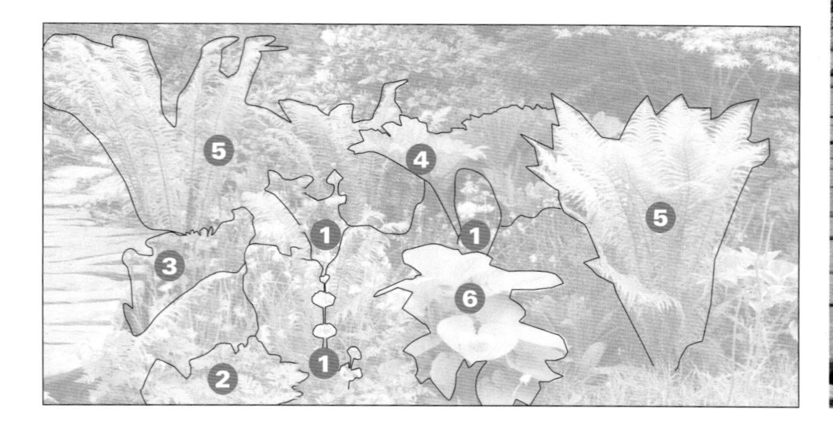

WINTER COLOUR AND INTEREST

Bolster winter interest with a border filled with evergreen conifers. Choose a variety of shapes, colours, and textures for maximum impact, and check plant labels for heights and spreads carefully before buying, as even slow-growing types can eventually form large plants. All these conifers prefer a sunny site and moist but free-draining soil.

Plant list

1 *Picea glauca*
2 *Juniper communis*
3 *Picea pungens*
4 *Thuja plicata* 'Collyer's Gold'
5 *Cryptomeria japonica* 'Elegans Compacta'
6 *Cupressus glabra*
7 *Chamaecyparis obtusa* 'Nana Aurea'

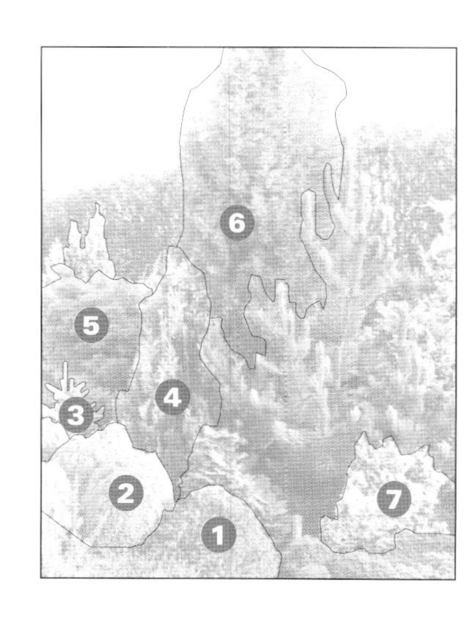

Recipes for problem areas

SLUG-PROOF BORDER

All plants can be attacked by slugs and snails, but in a mixed border, the pests are more likely to take the easy option and target those with soft, succulent leaves. A selection of the most slug- and snail- resistant plants are illustrated on pp.386–389, and others include plants with tough foliage, such as evergreen box, and many aromatic herbs – molluscs find their pungent flavours and downy leaves unpalatable. Oregano, thyme, and santolina here are all resistant to attack. You can also try growing sacrificial plants, such as the pot marigolds in this scheme, which tempt slugs and snails and draw them away from your prized ornaments or crops.

Plant list

1 *Buxus sempervirens* (box)
2 *Thymus vulgaris* (thyme)
3 *Origanum vulgare* 'Aureum' (golden marjoram)
4 *Phlomis russeliana*
5 *Calendula officinale* (pot marigold)
6 *Santolina pinnata* (cotton lavender)

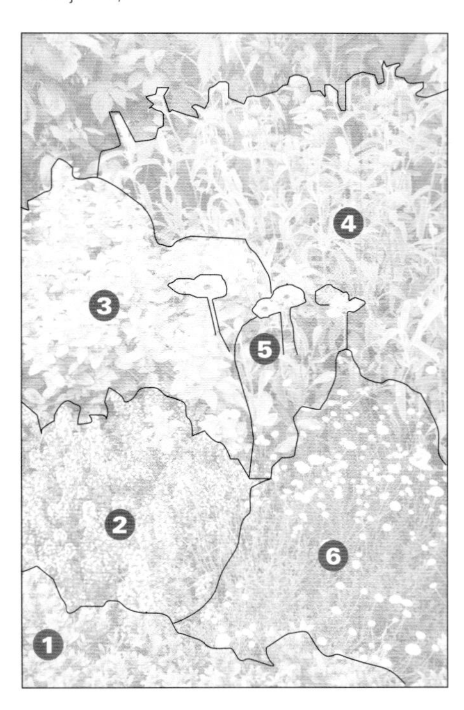

GROUND COVER IN SUN

The range of plants suitable for ground cover in a sunny site is large, and in addition to those illustrated on pp.380–383, consider those shown here. Ideal for sandy soil, sage and thyme are both aromatic herbs, and if cropped regularly by pinching off their tips, they will form dense, bushy clumps. For further colour, plant green- or yellow-leaved thyme varieties and variegated sage. The euphorbia is a tender plant, useful for adding summer colour, while also suppressing weeds.

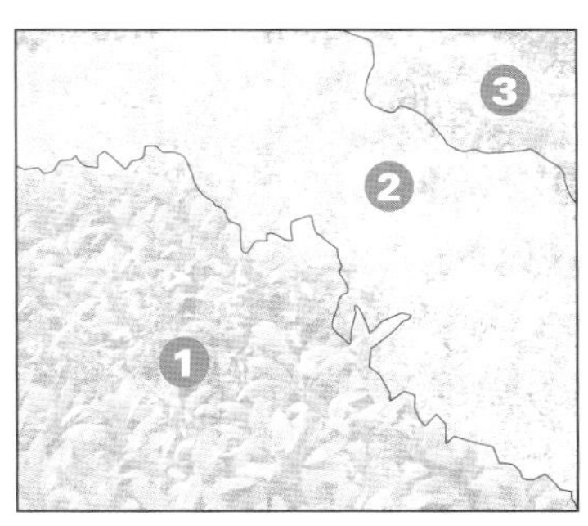

Plant list

1 *Salvia officinalis* 'Purpurascens' (purple sage)

2 *Thymus pulegioides* 'Aureus' (golden thyme)

3 *Euphorbia hypericifolia* DIAMOND FROST

HOT, DRY BORDER

Sunny, dusty sites can produce beautiful flowering displays if you choose your plants carefully. Field poppies (*Papaver rhoeas*) and larkspur (*Consolida*), which are sown annually in spring, together with marguerites (*Argyranthemum*), will produce a long show of colour in summer, while the perennial *Sisyrinchium* adds colour and structure in between. The poppies and larkspur may self-seed if you turn the soil over with a fork each year to encourage germination. Any unwanted seedlings can easily be removed.

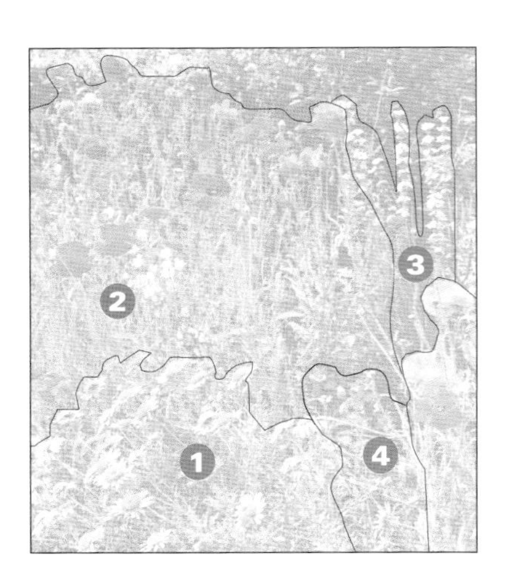

Plant list

1 *Argyranthemum* 'Vancouver'

2 *Papaver rhoeas* (field poppy)

3 *Sisyrinchium striatum*

4 *Consolida ajacis* (larkspur)

PLANTS for GARDEN STYLES

Whatever your taste or garden size, there's a style to suit you, from the clean lines of a formal garden to the relaxed informality of a cottage scheme. You may also find it easier to create a unified planting plan by focusing on a theme. Consider, too, how much work you want to do in the garden, and choose a style that matches your needs. For example, contemporary perennials and wildlife gardens use tough plants that need little attention, while lush tropical plantings and Asian-style gardens demand more time.

Plants for contemporary perennial schemes

Contemporary perennial schemes focus on tough, hardy perennials and grasses that are combined to form waves of colour and texture.

When creating your own scheme, use a limited palette of species and knit together large swathes of contrasting heights and shapes to produce beautiful patterns. Although this planting style tends to suit big gardens, it can work in smaller spaces if you reduce the number of species but still plant in large groups. Try "see-through" plants like *Verbena bonariensis* towards the front of your scheme, and include key plant shapes, such as flat-topped achilleas, spiky salvias, and daisy-like echinaceas, together with a few grasses for foliage interest. Leave the seedheads and stems to stand over winter, then cut them down in early spring.

PERENNIAL LARGE

Achillea filipendulina

This upright perennial has deeply divided, pale green leaves and, in summer, dense, domed heads of tightly packed golden-yellow flowers. The flower stems can be left to dry *in situ* for autumn colour; also good for cutting or drying indoors.

↕1.4m (4½ft) ↔60cm (24in)

PERENNIAL LARGE

Actaea racemosa

BLACK COHOSH This upright perennial has deeply divided leaves, above which branched, bottlebrush-like spikes of white flowers form in midsummer. The dried, brown seedheads are also attractive. Grow it at the back of a border.

↕1.5m (5ft) ↔60cm (24in)

PERENNIAL LARGE

Actaea simplex

BUGBANE This upright perennial has divided foliage and bottlebrush, white-tinged purple flowers in autumn. Plant with contrasting green-leaved perennials in shade or behind sun-lovers. Purple-leaved 'Brunette' AGM (above) is popular.

↕1.2m (4ft) ↔60cm (24in)

PERENNIAL LARGE

Agastache foeniculum

ANISEED HYSSOP This upright perennial has leaves that smell and taste of liquorice. Plant it in swathes towards the front of a border to show off its spikes of fluffy, lavender-blue summer flowers, which attract bees and butterflies.

↕1.2m (4ft) ↔30cm (12in)

BULB LARGE

Allium 'Purple Sensation'

This perennial bulb produces sturdy stems topped with spherical heads of purple flowers in early summer. Plant the bulbs in groups in autumn between perennials that will disguise the leaves, which fade as the flowers appear.

↕80cm (32in)

ANNUAL/BIENNIAL LARGE

Angelica archangelica

ANGELICA A tall biennial, with an upright, branching habit and large, decorative, deeply cut, green leaves. It forms a clump of foliage in the first year and flowers the next, bearing domed heads of tiny, green blooms during summer. It self-seeds freely.

↕2m (6ft) ↔1m (3ft)

PERENNIAL LARGE

Aster cordifolius

BLUE WOOD ASTER This perennial has small leaves and, from late summer to autumn, sprays of daisy-like, blue-tinted white flowers that add colour and texture towards the back of a border. 'Sweet Lavender' AGM has lavender-blue blooms.

↕1.2m (4ft) ↔45cm (18in)

PERENNIAL MEDIUM

Astilbe 'Fanal'

This perennial has broad, green leaves, divided into smaller leaflets. In summer it bears upright, tapering, feathery heads of tiny, crimson flowers, which turn brown and keep their shape in winter. It is ideal for boggy areas and pond margins.

↕60cm (24in) ↔45cm (18in)

PERENNIAL LARGE

Calamagrostis brachytricha

KOREAN FEATHER REED GRASS An upright, deciduous grass, with arching, grey-green leaves that turn straw-coloured in winter, and tall, feather-shaped, pink-tinted silver flowerheads. Use it as a backdrop to colourful blooms from summer to autumn.

↕1.4m (4½ft) ↔50cm (20in)

PERENNIAL MEDIUM

Centranthus ruber

RED VALERIAN This perennial has grey-green foliage and, from late spring to early autumn, rounded clusters of small, reddish pink flowers. It self-seeds freely, creating naturalistic groups in a border, but can become invasive.

↕90cm (36in) ↔60cm (24in) or more

PERENNIAL LARGE

Crambe cordifolia ♡

GREATER SEA KALE This tall perennial produces mounds of dark green foliage below clouds of tiny, fragrant, white blooms that appear on sturdy stems in summer. Its see-through flowers mean it can be used towards the front of a border.

↕2m (6ft) ↔1.2m (4ft)

PERENNIAL LARGE

Cynara cardunculus ♡

CARDOON This upright perennial has spiny, silver-grey foliage and large, thistle-like, purple flowers. Use it to provide interest at the back of a border from summer to autumn. Combine it with other sun-lovers.

↕2m (6ft) ↔1m (3ft)

PERENNIAL MEDIUM

Echinacea purpurea 'Rubinstern'

An erect perennial, with oval, dark green leaves and, from summer to mid-autumn, large, dark pink, daisy-like flowers. The spent flowerheads look attractive, but can be removed to encourage further flowering.

↕80cm (32in) ↔45cm (18in)

PERENNIAL MEDIUM

Echinacea purpurea 'White Swan'

An upright perennial, with dark green leaves and large, white flowers, with prominent spiky centres, from late summer to autumn. Plant it near the front of a border, and use to contrast with darker-flowered plants.

↕60cm (24in) ↔45cm (18in)

PERENNIAL LARGE

Echinops ritro ♡

GLOBE THISTLE This upright perennial has prickly, divided, dark green leaves and globe-shaped, spiky, metallic blue flowerheads in late summer. Use it to plug the gaps left by tall alliums as they begin to fade. Varieties include 'Veitch's Blue'.

↕1.2m (4ft) ↔75cm (30in)

PERENNIAL LARGE

Eryngium giganteum ♡

SEA HOLLY An upright, short-lived perennial, with marbled, heart-shaped, grey-green foliage and tall stems topped with silvery grey, cone-like summer flowers surrounded by spiny bracts. It self-seeds freely in free-draining soil to create natural swathes.

↕90cm (36in) ↔30cm (12in)

PERENNIAL LARGE

Eupatorium purpureum

JOE PYE WEED This bold, upright perennial has coarse, green leaves on tall, purple-flushed stems and, from late summer to early autumn, fluffy, domed, purple-pink flowerheads. Plant it at the back of a border for screening or to create a backdrop.

↕2.2m (7ft) ↔1m (3ft)

PERENNIAL LARGE

Gaura lindheimeri ♡

WAND FLOWER This upright perennial produces spoon-shaped, green leaves and pink-tinged buds that open to reveal small, white flowers throughout summer. Weave it in groups between flat-headed flowers, such as sedums and achilleas.

↕1.5m (5ft) ↔90cm (36in)

PERENNIAL MEDIUM

Geranium sylvaticum

WOOD CRANESBILL This perennial produces clumps of lobed, mid-green leaves and, from late spring to early summer, round, white-centred blue-purple flowers. Grow it in groups in moist soil at the front of a border.

↕75cm (30in) ↔60cm (24in)

PERENNIAL MEDIUM

Geum rivale

WATER AVENS This perennial forms neat rosettes of rounded, green leaves and, from late spring to summer, bell-shaped, pink or dark orange flowers on slender stems. Plant it at the front of a bed in moist soil; it may self-seed.

↕↔60cm (24in)

PERENNIAL LARGE

Gillenia trifoliata 🏆

BOWMAN'S ROOT A spreading perennial, with lobed, prominently veined, dark green leaves and, from late spring to late summer, airy sprays of red-budded, starry, white flowers. Use it mid-border with groups of colourful, upright flowers.

↕1.2m (4ft) ↔60cm (24in)

PERENNIAL MEDIUM

Helenium 'Moerheim Beauty' 🏆

SNEEZEWEED This upright perennial has daisy-like, dark-centred coppery red blooms. Plant it in swathes in the middle of a sunny border, where it will lend an eye-catching focal point to a summer scheme.

↕1m (3ft) ↔60cm (24in)

PERENNIAL MEDIUM

Hemerocallis 'All American Chief' 🏆

DAYLILY This upright, clump-forming perennial has arching, strap-like leaves and trumpet-shaped, yellow-throated red blooms. Flowers last a day, but are borne over many weeks. Resents root disturbance.

↕80cm (32in) ↔23cm (9in)

PERENNIAL MEDIUM

Knautia macedonica

MACEDONICAN SCABIOUS This upright perennial has lobed, green basal leaves, and produces a succession of button-like, crimson flowers on wiry stems in summer. Weave small groups of this plant through the front of a sunny border.

↕75cm (30in) ↔60cm (24in)

PERENNIAL LARGE

Ligularia przewalskii

This upright perennial has large, round, deeply cut, dark green leaves, with an architectural appeal. From mid- to late summer, it bears tall, narrow spires of spidery, daisy-like, yellow flowers. Provide support and deadhead spent flower stems.

↕2m (6ft) ↔1m (3ft)

PERENNIAL MEDIUM

Monarda 'Cambridge Scarlet'

BERGAMOT A clump-forming perennial, with aromatic, dark green leaves and trumpet-shaped, rich red summer blooms, arranged in whorls. Use it as a foil for paler-coloured plants or to attract bees. May need staking.

↕1m (3ft) ↔45cm (18in)

PERENNIAL SMALL

Nepeta racemosa 🏆

DWARF CATMINT A spreading perennial, with aromatic foliage and masses of violet-blue summer flower spikes. A good front-of-border plant, combine it with contrasting red- and yellow-flowered plants. 'Walker's Low' AGM (above) is a popular form.

↕30cm (12in) ↔45cm (18in)

PERENNIAL MEDIUM

Osmunda cinnamomea 🏆

CINNAMON FERN This deciduous fern has tall, mid-green, upright fronds that emerge from a central base, giving it a shuttlecock-like appearance. In summer, reproductive fronds, covered with rusty brown spores, appear. It will eventually form a small colony.

↕1m (3ft) ↔45cm (18in)

PERENNIAL MEDIUM

Penstemon digitalis 'Husker Red'

This bushy, upright perennial has lance-shaped, dark red young leaves and stems. Contrasting, tubular, white flowers form on branching stems from late spring to midsummer. Plant near the front of a border.

↕75cm (30in) ↔30cm (12in)

PERENNIAL LARGE

Persicaria amplexicaulis

RED BISTORT This vigorous perennial has oval, green leaves and slender spikes of white or pink flowers from summer to early autumn. It is ideal for a large garden with moist soil. Varieties include 'Firetail' (above), with bright red blooms.

↕↔1.2m (4ft)

PERENNIAL LARGE

Phlox paniculata

An upright perennial, with lance-shaped, green leaves and large clusters of white or lilac flowers, which create a sea of colour from summer to autumn in a sunny border. Varieties include 'Blue Paradise' (above), with violet-blue flowers.

↕1.2m (4ft) ↔1m (3ft)

PERENNIAL MEDIUM

Rudbeckia hirta

An upright, branching, short-lived perennial, usually grown as an annual, with lance-shaped, mid-green leaves and large, daisy-like, brown-centred golden flowers from summer to autumn. 'Becky Mixed' (above) is a popular variety.

↕up to 90cm (36in) ↔45cm (18in)

PERENNIAL MEDIUM

Salvia nemorosa

BALKAN CLARY A compact, upright perennial, with lance-shaped, wrinkled, green leaves and spikes of violet-blue flowers. Weave long swathes of this plant through a border, where it will inject colour from summer to early autumn.

↕up to 75cm (30in) ↔60cm (24in)

PERENNIAL LARGE

Sanguisorba officinalis

GREAT BURNET This clump-forming perennial has green leaves, divided into oval leaflets, and slim stems that carry oval clusters of tiny, dark red flowers from summer to early autumn. Grow it towards the front of a border.

↕1.2m (4ft) ↔60cm (24in)

PERENNIAL MEDIUM

Typha minima

A deciduous, perennial marginal water plant, with grass-like leaves. Spikes of rust-brown flowers in late summer are followed by decorative brown, rounded seedheads. Confine the plant in a basket to keep it under control.

↕60cm (24in) ↔30cm (12in)

PERENNIAL MEDIUM

Sedum 'Herbstfreude' 🏆

STONECROP This clump-forming perennial has oval, fleshy, grey-green leaves and flat, star-shaped, brick-red flowerheads in late summer, followed by brown seedheads that persist through winter. Combine with contrasting flower spires, such as salvias.

↕60cm (24in) ↔50cm (20in)

PERENNIAL LARGE

Verbena bonariensis 🏆

PURPLETOP VERVAIN A tall perennial, with branching stems of small, dark green leaves. From midsummer to autumn, domed clusters of scented, purple flowers create accents at the front or back of a border when planted in groups.

↕1.5m (5ft) ↔60cm (24in)

PERENNIAL MEDIUM

Sedum telephium Atropurpureum Group

ORPINE This perennial has domed, pinkish white flower clusters from late summer to autumn, and dark purple stems and leaves, which contrast beautifully with green-leaved perennials at the front of a border.

↕60cm (24in) ↔30cm (12in)

PERENNIAL LARGE

Veronicastrum virginicum

CULVER'S ROOT An upright perennial, with tall stems of lance-shaped, dark green leaves, topped with slender spikes of purple-blue, white, or pink flowers in late summer. Use it as a backdrop to rounded flowers in a sunny or part-shaded border.

↕2m (6ft) ↔45cm (18in)

PERENNIAL LARGE

Stipa gigantea 🏆

GOLDEN OATS A clump-forming perennial grass, with arching, green leaves. Tall spikes of oat-like, purple-tinted flowers shoot up in summer and ripen to gold in autumn. The see-through flowering stems work well towards the front of a border.

↕2.5m (8ft) ↔1m (3ft)

OTHER SUGGESTIONS

Perennials

Achillea 'Madder' • *Achillea* 'Terracotta' • *Artemisia absinthium* 'Lambrook Silver' • *Arundo donax* • *Aster amellus* 'Sonora' • *Aster* 'Herfstweelde' • *Aster* 'Octoberlight' • *Astilbe thunbergi* • *Astrantia major* 'Claret' • *Calamagrostis* 'Karl Foerster' • *Cirsium rivulare* 'Atropurpureum' • *Echinacea purpurea* 'August Konigin' • *Echinacea purpurea* 'Green Edge' • *Echinops sphaerocephalus* • *Epilobium angustifolium* var. *album* • *Eupatorium maculatum* 'Riesenschirm' 🏆 • *Helenium* 'Rubinzwerg' 🏆 • *Miscanthus sinsensis* • *Molinia caerulea* subsp. *caerulea* 'Variegata' 🏆 • *Monarda* 'Cherokee' • *Monarda* 'Fishes' • *Nepeta racemosa* 'White Cloud' • *Panicum virgatum* • *Phlomis taurica* • *Salvia nemorosa* 'East Friesland' • *Sedum telephium* 'Matrona' 🏆 • *Stipa pulcherrimum*

Plants for formal and modern schemes

Formal gardens conform to a symmetrical plan, while modern designs follow an asymmetrical format, but both employ similar plants.

Closely clipped hedging is a common feature in formal schemes and is used to carve out screens, parterres, and knot gardens. Box and yew are traditional choices, but you can also try shrubby honeysuckle (*Lonicera*) or *Berberis* to update the look. Modern designs often include blocks of leafy plants, such as bamboo, grasses, or hebes, to produce graphic slabs of colour and texture. Also popular are pleached trees, where the stems are left bare and the branches trained horizontally to produce a slim hedge on stilts – the small-leaved lime (*Tilia cordata*) will create this effect. Use flowers sparingly to add splashes of seasonal colour.

SHRUB SMALL

Buxus sempervirens

BOX This compact, slow-growing, evergreen shrub produces woody stems of small, oval, green leaves that are ideal for clipping into architectural shapes and topiary. Grow it in a large container, or use as a low hedge or border edging.

↕1m (3ft) ↔1.5m (5ft)

TREE MEDIUM

Laurus nobilis 🏆

BAY LAUREL This evergreen tree has stems of aromatic, dark green leaves that can be clipped into shapes to form topiary. Grow it in a sheltered area, and plant in large containers of soil-based compost or as a centrepiece in a parterre.

↕12m (40ft) ↔10m (30ft)

SHRUB SMALL

Hebe topiaria 🏆

A compact, rounded, evergreen shrub, with dense stems of small, grey-green leaves that are ideal for clipping. Clusters of small, white flowers appear in summer. Use it to edge a formal geometric bed or as a small topiary specimen.

↕60cm (24in) ↔90cm (36in)

SHRUB MEDIUM

Lonicera nitida

SHRUBBY HONEYSUCKLE This evergreen, bushy shrub, with arching stems of tiny, dark green leaves is used as a compact knee- or waist-high hedge. It can also be trimmed into topiary shapes. 'Baggesen's Gold' AGM (above) has golden foliage.

↕2m (6ft) ↔3m (10ft)

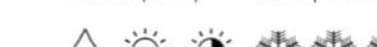

SHRUB MEDIUM

Hydrangea macrophylla

LACECAP HYDRANGEA This deciduous, compact shrub has broad, oval leaves and flattened clusters of tiny, pink or blue flowers surrounded by larger, white blooms. It makes an elegant mid-border plant or edging for a path.

↕↔1.5m (5ft)

PERENNIAL SMALL

Pelargonium 'Lady Plymouth' 🏆

SCENTED-LEAVED PELARGONIUM Grown as an annual, this spreading perennial has eucalyptus-scented, silver-margined green leaves and lavender-pink summer flower clusters. Grow in an urn in a parterre.

↕40cm (16in) ↔20cm (8in)

SHRUB SMALL

Rosa GRAHAM THOMAS 🏆

A shrub rose or short climber, with disease-resistant green leaves and cup-shaped, highly fragrant, fully double, yellow blooms from summer to autumn. Grow it in a rose bed underplanted with geraniums, or on an arch.

↕1.2m (4ft) ↔1.5m (5ft)

SHRUB SMALL

Rosa KENT

A spreading, ground-cover rose, with disease-resistant, glossy, mid-green leaves and clusters of flat, semi-double, white flowers from summer to autumn. Plant it in a large container of soil-based compost, or in a mixed border.

↕80cm (32in) ↔90cm (36in)

SHRUB SMALL

Rosa 'Penelope'

A hybrid musk shrub rose, with long, arching stems, glossy, dark green foliage, and large clusters of fragrant, semi-double, cream and pale pink flowers from summer to autumn. Plant it in a formal rose bed or formal border.

↕↔1m (3ft), more if lightly pruned

SHRUB SMALL

Rosa WARM WISHES

This bushy, hybrid tea rose has disease-resistant, mid-green leaves and, from summer to autumn, pointed, fully double, scented, orange-pink flowers that age to rose-pink. Plant in groups to fill parterres or use as a specimen plant in formal designs.

↕1m (3ft) ↔80cm (32in)

ANNUAL/BIENNIAL SMALL

Salvia splendens

SCARLET SAGE A tender, upright perennial, grown as an annual, with spear-shaped, dark green leaves and, in summer, compact spikes of tubular, pink, red, and purple blooms. Plant at the front of a bed for a bold splash of colour. Deadhead spent spikes.

↕25cm (10in) ↔35cm (14in)

PERENNIAL MEDIUM

Salvia x *sylvestris*

WOOD SAGE This compact perennial has small, dark green, aromatic leaves and, in summer, branched spikes of violet blooms. 'Mainacht' AGM has indigo-blue flowers. It is an excellent front-of-border plant or filler for a parterre.

↕80cm (32in) ↔30cm (12in)

TREE LARGE

Taxus baccata

YEW An evergreen tree, with needle-like, dark green leaves and red berries in autumn. It makes a superb hedge for boundaries or for dividing up a garden. Clip it in spring and summer; it tolerates hard pruning, if necessary. All parts are highly toxic.

↕15m (50ft) ↔10m (30ft)

TREE SMALL

Taxus x *media*

HICKS YEW An evergreen conifer, with needle-like, dark green leaves, lighter green below, and red berries in autumn. Faster growing than *T. baccata* AGM, it makes a beautiful formal hedge. Trim in early spring. All parts are highly toxic.

↕↔6m (20ft)

TREE LARGE

Tilia cordata

SMALL-LEAVED LIME This deciduous tree has heart-shaped, glossy, dark green leaves that turn yellow in autumn. Small yellowish white flowers appear in summer. It is perfect for pleaching or as formal hedges; trim in summer.

↕30m (100ft) ↔12m (40ft)

BULB MEDIUM

Tulipa 'Queen of Night'

TULIP This late spring-flowering tulip has grey-green leaves and cup-shaped, single, dark purple flowers. Plant the bulbs in groups in autumn in beds or parterres, or in pots of soil-based compost mixed with grit.

↕60cm (24in)

OTHER SUGGESTIONS

Perennials and grasses

Miscanthus sinensis 'Gracillimus' • *Miscanthus sinensis* 'Morning Light' • *Pelargonium* Maverick Series • *Salvia splendens* Vista Series

Bulbs

Tulipa 'Angélique' • *Tulipa* 'Greenland'

Shrubs and trees

Berberis thunbergii f. *atropurpurea* 'Rose Glow' • *Buxus sempervirens* 'Suffruticosa' • *Carpinus betulus* 'Purpurea' • *Hebe* 'Pewter Dome' • *Hydrangea paniculata* • *Ilex crenata* • *Laurus nobilis* 'Aurea' • *Ligustrum jonandrum* • *Rosa* 'Margaret Merrill' • *Tilia cordata* 'Greenspire'

Plants for tropical gardens

Brightly-coloured flowers and lush foliage create a tropical theme. While some plants hail from hot countries, many hardy types also create the look.

Choose plants with very large leaves that lend a sculptural quality to your designs. In cool climates, plant a backdrop of hardy foliage plants, such as hostas, bamboos, phormiums, and *Fatsia*, and inject colour in summer with a range of exotic-looking flowers. Dahlias, ginger lilies, *Melianthus*, and cannas will survive mild winters if covered with a deep mulch of chipped bark or rotted manure – or simply bring the plants indoors if your climate is too cold. Likewise, many species of eucomis will survive outside in a pot, if turned on its side to keep the compost dry, and placed in a sheltered area.

SHRUB MEDIUM

Abutilon 'Kentish Belle'

This shrub, semi-evergreen in warm areas, produces small, triangular leaves and bell-shaped, orange and red flowers in summer. Plant it close to a sunny wall, or use it as a screen to lend a tropical flavour to a sheltered garden.

↕↔2.5m (8ft)

PERENNIAL SMALL

Adiantum venustum

A deciduous fern, evergreen in mild areas, with triangular fronds, divided into fan-shaped segments. The fronds turn rusty brown in autumn and winter. Plant it in a shady area and combine with contrasting foliage plants, such as *Solenostemon*.

↕23cm (9in) ↔30cm (12in)

PERENNIAL LARGE

Agapanthus Headbourne hybrids

This perennial has arching, strap-like leaves and clusters of funnel-shaped, blue flowers from late summer to early autumn. Plant near the front of a border. In cold areas, mulch with compost; remove spent blooms.

↕1.2m (4ft) ↔60cm (24in)

PERENNIAL LARGE

Agave americana 'Variegata'

CENTURY PLANT An evergreen perennial, with lance-shaped, sharply pointed, cream-edged grey-green leaves. In hot summers, cream flowers appear atop tall stems. Grow in a pot in cold areas; overwinter it indoors.

↕↔up to 1.5m (5ft)

PERENNIAL MEDIUM

Aloe vera

An evergreen perennial, with fleshy, spine-edged, grey-green leaves. It is often grown as a house plant in frost-prone areas, but makes an exotic contribution to sunny summer gardens when grown in a pot; overwinter it indoors.

↕60cm (24in) ↔indefinite

PERENNIAL SMALL

Begonia 'Escargot'

An evergreen perennial, this variety has spiral-shaped, purple-tinted green leaves, with swirly, silver markings, and small, pink autumn flowers. Grow it as a house plant in winter and bring outdoors in a pot during summer.

↕25cm (10in) ↔50cm (20in)

SHRUB LARGE

Brugmansia x *candida*

ANGELS' TRUMPETS An evergreen shrub or small tree, with oval, green leaves and, from summer to autumn, large, exotic-looking, trumpet-shaped, night-scented blooms. 'Grand Marnier' AGM (above) has apricot blooms. Protect it from frost.

↕5m (15ft) ↔2.5m (8ft)

PERENNIAL LARGE

Canna 'Wyoming'

An upright perennial, with large, paddle-shaped, purple-bronze leaves and, from midsummer to early autumn, pale orange, gladiolus-like flowers. Varieties with yellow, orange, or red blooms and colourful foliage are also available. Keep frost-free in winter.

↕1.8m (6ft) ↔50cm (20in)

SHRUB LARGE

Chamaerops humilis

DWARF FAN PALM A compact, evergreen palm, with fan-shaped, divided, green leaves. Mature plants bear clusters of small, yellow summer flowers. Relatively hardy, it lends an exotic look to sheltered tropical-style gardens.

↕3m (10ft) ↔2m (6ft)

ANNUAL/BIENNIAL LARGE

Cleome hassleriana

SPIDER FLOWER This tall annual produces spiny stems of divided foliage and rounded clusters of small, spidery, white, pink, or purple summer flowers, with a light fragrance. Use it to add a tropical touch to beds and borders.
↕1.2m (4ft) ↔45cm (18in)

CLIMBER LARGE

Cobaea scandens ♀

CUP-AND-SAUCER VINE This evergreen, perennial climber, grown as an annual, has dark green leaves and scented, cup-shaped, creamy green flowers that age to purple. It produces a wall of colour from summer to autumn when grown on trellis or wires.
↕5m (15ft)

PERENNIAL LARGE

Colocasia esculenta ♀

BLACK-STEM ELEPHANT'S EAR A marginal perennial, with large, spear-shaped leaves. Plant it in a pot in boggy soil and overwinter it indoors. In cold areas, grow it outside only during warm summers. 'Fontanesii' has dark red leaf stalks and veins.
↕1.5m (5ft) ↔60cm (24in)

PERENNIAL MEDIUM

Crocosmia 'Lucifer' ♀

MONTBRETIA This clump-forming perennial has narrow, sword-shaped, mid-green foliage. In summer, it produces bright red flowers. Yellow-, red-, and orange-flowered varieties are also available.
↕1m (3ft) ↔25cm (10in)

SHRUB MEDIUM

Cycas revoluta ♀

JAPANESE SAGO PALM An elegant evergreen palm, with a rough-textured trunk and arching, oval, glossy, green leaves, divided into needle-like leaflets. It is hardy to -8°C and will survive winters outside in a sheltered area.
↕↔2m (6ft)

BULB MEDIUM

Dahlia 'Yellow Hammer' ♀

This dwarf bedding dahlia grows from a tuber. It produces dark bronze foliage and contrasting, single, orange-streaked bright yellow flowers. Plant it in containers of gritty, soil-based compost, or in groups at the front of a border.
↕60cm (24in)

PERENNIAL LARGE

Ensete ventricosum ♀

ABYSSINIAN BANANA This evergreen, palm-like perennial has large, paddle-shaped leaves, with cream midribs and red undersides. Grow it in a pot of soil-based compost and grit. 'Maurelii' AGM has dark red-splashed leaves with red midribs.
↕2m (6ft) ↔1m (3ft)

BULB MEDIUM

Eucomis bicolor ♀

PINEAPPLE LILY This summer-flowering, exotic-looking perennial bulb has wavy-edged basal leaves and clusters of greenish white flowers, with purple-edged petals, on spotted stems. The flowers, with their leaf-like bracts, resemble pineapples.
↕50cm (20in) ↔60cm (24in)

PERENNIAL LARGE

Hedychium densiflorum

GINGER LILY This perennial forms large, lance-shaped, green leaves and, in summer, torch-like clusters of orange flowers. Grow it in groups in moist soil to create a spectacular summer border. Apply a thick mulch to protect it from frost.
↕2m (6ft) ↔60cm (24in)

ANNUAL/BIENNIAL SMALL

Helichrysum petiolare ♀

LIQUORICE PLANT This annual bedding plant is grown for its trailing stems of downy, grey-green leaves, and is perfect for edging containers or beds. 'Limelight' AGM (above) has lime-green foliage, and makes a good partner for red-flowered plants.
↕15cm (6in) ↔30cm (12in)

PERENNIAL MEDIUM

Hemerocallis 'Chicago Apache'

DAYLILY This perennial has strap-shaped foliage and trumpet-like, deep scarlet flowers throughout summer, with each bloom lasting just one day. Plant it in groups of three in a tropical-style scheme.
↕65cm (26in) ↔50cm (20in)

SHRUB LARGE

Hibiscus syriacus

ROSE MALLOW A deciduous, upright shrub, with lobed, dark green foliage and single, blue, violet, or white flowers, with dark centres from late summer to early autumn. 'Oiseau Blue' AGM (above) is a popular blue-flowered variety.
↕3m (10ft) ↔2m (6ft)

PERENNIAL SMALL

Hosta 'Lakeside Cha Cha'

This perennial has large, rounded, mid-green leaves, with cream edges, heavily textured with radiating veins. Pendent, tubular, lilac flower spikes appear briefly in summer; remove after fading. Provides a lush, leafy understorey to tall plants. Protect from slugs.
↕45cm (18in) ↔45cm (18in)

CLIMBER MEDIUM

Ipomoea coccinea

RED MORNING GLORY This tender annual has heart-shaped, green leaves, and small, fragrant, tubular, scarlet flowers with yellow throats. Grow this twining climber up a tripod where it will produce colour throughout summer and early autumn.
↕3m (10ft)

PERENNIAL MEDIUM

Lobelia cardinalis

CARDINAL FLOWER This deciduous perennial has narrow, lance-shaped, glossy, green leaves and, in summer, spires of exotic-looking scarlet flowers. Ideal for a pond margin or bog garden. All parts of the plant are toxic.
↕75cm (30in) ↔23in (9in)

PERENNIAL SMALL

Lotus berthelotii

PARROT'S BEAK The grey-green, ferny foliage and beak-shaped, orange-red summer flowers of this trailing perennial, grown as annual bedding, offer a decorative edge to a container display.
↕20cm (18in) ↔indefinite

SHRUB LARGE

Melianthus major

HONEYBUSH This evergreen subshrub, grown for its striking foliage, has blue-grey leaves, divided into toothed-edged leaflets, and small, brownish red flowers in late spring. Shelter from cold winds, and provide a dry mulch in winter.
↕↔up to 3m (10ft)

CLIMBER LARGE

Passiflora caerulea

BLUE PASSION FLOWER This evergreen or semi-evergreen climber has glossy, lobed, dark green leaves. From summer to autumn, exotic white flowers with purple filaments appear, followed by egg-shaped, orange fruits. Clings using tendrils; provide support.
↕10m (30ft) or more

PERENNIAL LARGE

Phormium tenax

NEW ZEALAND FLAX An evergreen, clump-forming perennial that produces arching, sword-shaped, grey-green leaves and, during hot summers, tall stems of dark red flowers. It is hardy, and lends a tropical note to a border or pot.
↕3m (10ft) ↔2m (6ft)

SHRUB LARGE

Phyllostachys vivax f. *aureocaulis*

GOLDEN GROOVE BAMBOO A tall, upright, clump-forming bamboo, with yellow canes, sometimes striped green, and lance-shaped, mid-green leaves. Combine it with green foliage plants and colourful flowers.
↕6m (20ft) ↔3m (10ft) or more

SHRUB LARGE

Pittosporum tobira 🏆

JAPANESE PITTOSPORUM This evergreen shrub has a neat, bushy-headed shape and long, oval, dark green leaves. From late spring to early summer, clusters of sweetly-scented, small, starry, white flowers, ageing to creamy yellow, add to its appeal.

↕10m (30ft) ↔3m (10ft)

SHRUB SMALL

Plectranthus argentatus 🏆

SILVER SPURFLOWER This spreading, evergreen subshrub has silver stems, grey-green leaves, and small spikes of bluish white summer flowers. Use it as decorative edging in a large container or bed of contrasting colourful flowers.

↕↔1m (3ft)

SHRUB MEDIUM

Sasa palmata

BROAD-LEAVED BAMBOO A small, vigorous bamboo, with broad, lance-shaped, dark green leaves, the tips and margins of which turn brown in winter, creating a two-tone effect. It lends a tropical look to a garden; can be invasive.

↕2m (6ft) ↔indefinite

ANNUAL/BIENNIAL SMALL

Tropaeolum majus

NASTURTIUM This spreading annual has round, green leaves and red, yellow, or orange, trumpet-shaped flowers from summer to autumn. Use it to provide colour beneath taller plants. Alaska Series AGM (above) has cream-splashed foliage.

↕30cm (12in) ↔45cm (18in)

PERENNIAL SMALL

Solenostemon scutellarioides

FLAME NETTLE A bushy perennial, grown as an annual, with spear-shaped foliage in a variety of colours including pink, red, green, and yellow. It creates a leafy edge to paths and borders, or in containers.

↕45cm (18in) ↔30cm (12in) or more

SHRUB SMALL

Zamia pumila

This unusual, slow-growing, frost tender shrub resembles a palm and has robust, green, frond-like foliage. It produces brown flower cones when mature. Grow in a large container. Move under cover for winter; shift outside once the risk of frost has passed.

↕1.2m (4ft) ↔2m (6ft)

ANNUAL/BIENNIAL LARGE

Tithonia rotundifolia

A tall, branching annual, with coarse, spear-shaped leaves. Bright orange, daisy-like blooms, with yellow centres appear from late summer to autumn. Plant it near the back of a border, provide support, and deadhead regularly to prolong the display.

↕1.2m (4ft) ↔60cm (24in)

ANNUAL/BIENNIAL SMALL

Zinnia elegans Dreamland Series

This bushy annual has oval, dark green leaves, above which daisy-like, red, yellow, purple, pink, or green flowers appear from summer to early autumn. It offers masses of colour; grow in a container or flower bed.

↕↔30cm (12in)

TREE SMALL

Trachycarpus fortunei 🏆

CHUSAN PALM This evergreen palm has an unbranched stem and head of large, deeply divided, fan-like, mid-green leaves. Sprays of fragrant, creamy yellow flowers appear in early summer. Plant it in a sheltered area away from drying winds.

↕up to 2m (6ft) ↔2.5m (8ft)

OTHER SUGGESTIONS

Annuals

Ipomoea batatas • *Tropaeolum majus* • *Zinnia elegans*

Perennials and bulbs

Adiantum pedatum • *Agapanthus* 'Bressingham White' • *Alstroemeria* 'Butterfly Hybrids' • *Begonia rex* • *Calathea mekoyana* • *Canna* 'Durban' • *Crocosmia masoniorum* 🏆 • *Dahlia* 'Bishop of Llandaff' 🏆 • *Dryopteris affinis* 🏆 • *Eucomis comosa* 🏆 • *Gunnera manicata* 🏆 • *Helichrysum petiolare* 'Limelight' 🏆 • *Hemerocallis* 'Burning Daylight' 🏆 • *Hosta* 'Sum and Substance' 🏆 • *Lampranthus roseus* • *Lysimachia* 'Firecracker' 🏆 • *Matteuccia struthiopteris* 🏆 • *Musa basjoo* 🏆 • *Ricinus communis*

Shrubs and trees

Abutilon megapotamicum 🏆 • *Magnolia grandiflora*

Plants for oriental gardens

Reflecting Buddhist philosophies, Japanese gravel, stroll, and tea gardens are admired by people of all races and religions for their beauty and tranquility.

Japanese gravel gardens comprise large boulders and minimal planting – traditional dry gardens include only moss – with miniature pines and clipped Japanese hollies used to decorate these spare schemes. The more exuberant stroll and tea gardens allow a greater planting range, including cloud-pruned topiary, colourful maples (*Acer*), bamboos, irises, and seasonal flowering shrubs, such as rhododendrons and peonies. Combine these plants to create a miniaturized woodland or water landscape, and include winding paths and open spaces to achieve an authentic look.

TREE SMALL

Acer palmatum 'Bloodgood' ♈

JAPANESE MAPLE A small tree with maple-like, reddish purple leaves that turn bright red in autumn. Winged, red fruits follow the small, purple spring flowers. Grow it as a feature in a Japanese stroll-style garden.
↕↔ 5m (15ft)

SHRUB LARGE

Callicarpa bodinieri var. *giraldii* ♈

BEAUTY BERRY An upright shrub, with lance-shaped leaves, tiny, pink summer flowers, and purple berries. 'Profusion' AGM (above) has purple-tinged young leaves that turn rosy pink in autumn, and violet berries.
↕ 3m (10ft) ↔ 2.5m (8ft)

SHRUB SMALL

Cryptomeria japonica

JAPANESE CEDAR This evergreen, dome-shaped, dwarf conifer has scaly, green leaves and green turning brown cones. Ideal for a gravel garden, use it to create a rounded shape between rocks, or plant in a large container as a focal plant.
↕↔ 1m (3ft)

SHRUB LARGE

Eriobotrya japonica

LOQUAT This architectural shrub produces large, glossy, dark green leaves, felted beneath, at the end of branched stems. It makes a dramatic focal point, but needs a sheltered site to produce its scented, white autumn and winter flowers.
↕↔ 8m (25ft)

SHRUB LARGE

Fatsia japonica ♈

JAPANESE ARALIA This evergreen shrub has large, hand-shaped, glossy, dark green leaves and small, spherical, white autumn flowerheads, followed by black fruits. Use as an attractive backdrop against a wall or fence. A variegated form is also available.
↕↔ 4m (12ft)

TREE LARGE

Ginkgo biloba

MAIDENHAIR TREE A large, deciduous, conical-shaped tree, with fan-shaped, green leaves that turn butter-yellow in autumn. Use it as a focal point in Japanese- and Chinese-style gardens. Plant dwarf cultivars in smaller spaces.
↕ 30m (100ft) ↔ 8m (25ft)

PERENNIAL SMALL

Hakonechloa macra 'Aureola' ♈

GOLDEN HAKONECHLOA This slow-growing grass has green-striped yellow leaves that age to reddish brown. Reddish brown flower spikes appear in early autumn and last into winter. Grow in a pot or gravel garden.
↕ 40cm (16in) ↔ 60cm (24in)

SHRUB MEDIUM

Hydrangea macrophylla

LACECAP HYDRANGEA This rounded shrub produces large, oval, light green leaves and, from midsummer to early autumn, lace-cap flowerheads formed of tiny, blue flowers, and larger pale blue (on acid soil), or pink petal-like florets.
↕ 2m (6ft) ↔ 2.5m (8ft)

SHRUB MEDIUM

Hydrangea quercifolia

OAK-LEAVED HYDRANGEA This mound-forming shrub has lobed, mid-green leaves that turn red and purple in autumn. From midsummer to autumn, cone-shaped, cream flowerheads appear, which age to pink-tinged white. Requires neutral to acid soil.
↕ 2m (6ft) ↔ 2.5m (8ft)

PERENNIAL MEDIUM

Iris laevigata

JAPANESE WATER IRIS This clump-forming perennial marginal has sword-shaped, mid-green leaves and dark purple early summer flowers, with gold marks on the lower petals. Plant it in aquatic baskets at the edge of a pond or pool.
↕ 90cm (36in) or more ↔ indefinite

PERENNIAL MEDIUM

Paeonia 'Bowl of Beauty'

PEONY This erect, clump-forming perennial has divided, green leaves and large, bowl-shaped, double flowers, with rose-pink outer petals and cream centres. Use it to contribute summer colour to a Japanese-style stroll garden.

↕↔ 1m (3ft)

SHRUB MEDIUM

Paeonia suffruticosa

TREE PEONY An upright, deciduous shrub, with deeply lobed, dark green leaves and, in late spring, bowl-shaped, white, pink, red, or purple flowers, sometimes scented. Use it to add height to a border in a Japanese-style gravel or stroll garden.

↕↔ 2m (6ft)

SHRUB LARGE

Phyllostachys bambusoides

JAPANESE TIMBER BAMBOO This large bamboo, with its green or yellow canes and lush foliage, lends height to a scheme. A perfect plant to add an Oriental note to a gravel garden. Alternatively, grow it as a screen.

↕ 8m (25ft) ↔ indefinite

SHRUB LARGE

Phyllostachys nigra

BLACK BAMBOO An evergreen, clump-forming bamboo, with grooved, greenish brown stems that turn black when mature, and long, narrow leaves. Grow it as a screen or as a focal point in a stroll or tea garden.

↕ 8m (25ft) ↔ indefinite

SHRUB LARGE

Pieris japonica

LILY OF THE VALLEY BUSH A compact, evergreen shrub, with leathery, dark green leaves, bright red when young, and pendent chains of urn-shaped, white, pink, or red flowers in spring. Grow it in a sheltered site, protected from cold wind.

↕ 4m (12ft) ↔ 3m (10ft)

SHRUB SMALL

Pinus mugo

DWARF MOUNTAIN PINE Pines are prized in Japan for their scaly bark and elegant forms. Set among boulders in a gravel garden, this compact, evergreen conifer will create a mound of needle-like, dark green leaves.

↕ 1m (3ft) ↔ 2m (6ft)

TREE LARGE

Prunus x *yedoensis*

YOSHINO CHERRY This spreading, deciduous tree often takes centre stage in Japanese gardens, and is revered for its abundance of almond-scented, blush white spring blossom. It also has dark red fruits in summer and yellow autumn leaf colour.

↕ 15m (50ft) ↔ 10m (30ft)

SHRUB SMALL

Sarcococca humilis

CHRISTMAS BOX An evergreen, clump-forming shrub, with tiny, fragrant, white winter flowers and glossy, dark green foliage, which can be clipped into a sphere and used in a gravel garden. Black berries appear after the blooms.

↕ 60cm (24in) ↔ 1m (3ft)

SHRUB LARGE

Semiarundinaria fastuosa

NARIHIRA BAMBOO This large, vigorous bamboo produces clumps of dark green canes and dark green, lance-shaped leaves, which are grey-green beneath. Plant it in a gravel garden in groves or in a line as a screen.

↕ 6m (20ft) ↔ indefinite

SHRUB MEDIUM

Viburnum davidii

This dome-shaped, evergreen shrub has deeply veined, dark green foliage and small, white spring flowers. Female plants bear decorative, metallic blue fruits if both sexes are grown. Use it for shady ground cover in a stroll or tea garden.

↕↔ 1.5m (5ft)

OTHER SUGGESTIONS

Perennials and grasses

Paeonia lactiflora

Shrubs, trees, and climbers

Abelia grandiflora • *Acer palmatum* 'Sango-kaku' • *Acer palmatum* var. *dissectum* 'Crimson Queen' • *Buxus macrophylla* • *Chamaecyparis obtusa* • *Camellia sinensis* • *Euonymus sieboldianus* • *Ilex crenata* • *Mahonia japonica* • *Nandina domestica* 'Firepower' • *Prunus* 'Kanzan' • *Punica granatum* • *Quercus dentata* • *Rhododendron dilatatum* • *Rhododendron indicum* • *Vaccinium bracteatum* • *Wisteria floribunda* • *Zelkova serrata*

Plants for cottage gardens

Fads and fashions come and go, yet the cottage garden endures as one of the most popular styles, loved for its nostalgic themes and romantic planting.

Traditional cottage gardens included herbs and vegetables, but today they tend to focus mainly on ornamentals. To create a cottage design, include wooden trellises and pergolas laced with roses, vines, and honeysuckle, and underplanted with cushions of lavender. Pack your beds with as many bulbs and perennials as you can squeeze in, planting in groups of three or more, and select a range that will flower consecutively from spring until autumn. Use pots of annuals to fill any gaps as they appear. You may also want to include a few roses, compact shrubs, and topiary balls or pyramids to add some structure to the mix.

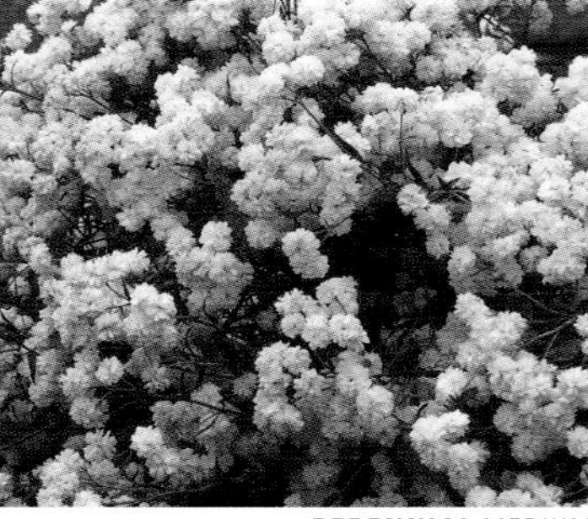

PERENNIAL MEDIUM

***Achillea ptarmica* The Pearl Group**

SNEEZEWORT An upright, spreading perennial, with linear, dark green leaves. In summer, it is covered with small, white, pompon flowers that resemble pearls. Use it to edge a pathway.

↕↔ 60cm (24in)

💧 ☼ ❄❄❄

PERENNIAL SMALL

Alchemilla mollis ⚘

LADY'S MANTLE This clump-forming perennial is ideal for the front of a border or shady corners. It has rounded, pale green leaves and sprays of tiny, greenish yellow flowers that last several weeks in summer. Trim back untidy leaves and spent blooms.

↕↔ 50cm (20in)

PERENNIAL LARGE

Anemone* x *hybrida

JAPANESE ANEMONE A slow spreading perennial, with divided, dark green leaves. From late summer to autumn, it bears saucer-shaped, pink or white flowers on upright stems. 'Honorine Jobert' AGM (above) has golden-eyed white flowers.

↕ 1.2m (4ft) ↔ 60cm (24in)

💧 ◑ ❄❄❄ ⓘ

PERENNIAL SMALL

Anthemis punctata* subsp. *cupaniana ⚘

SICILIAN CHAMOMILE This evergreen, drought-tolerant perennial makes a dome of silver foliage. In summer, it is covered with yellow-eyed white daisies. Use it at the front of a sunny border or plant in gravel.

↕↔ 30cm (12in)

💧 ☼ ❄❄

PERENNIAL MEDIUM

Aquilegia vulgaris

GRANNY'S BONNET This clump-forming perennial has fern-like foliage and nodding, bell-shaped late spring to early summer blooms, with spurred petals, on upright stems. Plant the pure white 'Nivea' AGM (above) in light shade with forget-me-nots.

↕ 90cm (36in) ↔ 45cm (18in)

PERENNIAL MEDIUM

Aquilegia vulgaris* var. *stellata

This upright perennial has round, divided leaves and, from late spring to early summer, pompon-like flowers on tall stems. 'Nora Barlow' (above) has greenish pink blooms, which become pink tipped with white as they age.

↕ 75cm (30in) ↔ 50cm (20in)

💧 ☼ ❄❄❄

PERENNIAL MEDIUM

Argyranthemum frutescens

MARGUERITE Plant this evergreen, woody-based, tender perennial in pots or raised beds. From late spring to autumn, its divided, green or grey-green leaves are teamed with copious, white, yellow, or pink daisy-like blooms, depending on the variety.

↕↔ 70cm (28in)

💧 ☼ ❄

PERENNIAL MEDIUM

Aster ericoides

This clump-forming, bushy perennial has small, lance-shaped, mid-green leaves and freely branching stems. From late summer to late autumn, it bears clusters of daisy-like, yellow-centred white flowerheads, sometimes shaded pink or blue.

↕ 1m (3ft) ↔ 30cm (12in)

💧 

PERENNIAL LARGE

Aster novae-angliae

NEW ENGLAND ASTER This perennial has slim, green, hairy leaves and, from late summer to autumn, daisy-like purple, pink, red, or white flowers, depending on the variety, on branched stems. Taller forms may require staking.

↕ up to 1.5m (5ft) ↔ 60cm (24in)

PERENNIAL SMALL

Astilbe x *arendsii*

FALSE GOAT'S BEARD A clump-forming perennial, with deeply divided, fern-like, dark green foliage. In summer, it bears cone-shaped, feathery, dark pink flowerheads on slim stems. It also grows well in full shade.

↕ 45cm (18in) ↔ 30cm (12in)

SHRUB MEDIUM

Ceanothus 'Dark Star' 🏆

CALIFORNIAN LILAC This wall shrub has arching branches, covered with small, oval, evergreen leaves and clusters of dark blue-purple flowers in late spring. Shelter plants from cold winds, and train on a sunny wall or fence.

↕ 2m (6ft) ↔ 3m (10ft)

PERENNIAL MEDIUM

Centaurea pulcherrima

This upright, drought-tolerant perennial is a favourite with butterflies. It forms clumps of divided, grey-green to silver foliage. In midsummer, large, cornflower-like, rose-pink flowers with pale yellow centres form. Try it with *Salvia* x *sylvestris* 'Mainacht' AGM.

↕ 75cm (30in) ↔ 60cm (24in)

TREE SMALL

Cercis canadensis 'Forest Pansy' 🏆

This deciduous, spreading tree bears pale pink, pea-like blooms on bare branches in spring, before the heart-shaped, rich purple leaves emerge. Use it as a specimen tree at the back of a cottage border.

↕↔ 5m (15ft)

ANNUAL/BIENNIAL MEDIUM

Cerinthe major

HONEYWORT This upright annual produces oval, grey-green leaves and, in summer, tubular, purple and yellow flowers on erect stems, which may need some support; plants may self-seed in gravel.

↕↔ 60cm (24in)

PERENNIAL SMALL

Chrysanthemum 'Grandchild' 🏆

KOREAN GROUP CHRYSANTHEMUM This compact perennial has lobed leaves and dense sprays of fully double, mauve-pink blooms, which are perfect for cutting. Grow it at the front of an autumn border.

↕ 45cm (18in) ↔ 40cm (16in)

CLIMBER MEDIUM

Clematis 'Jackmanii' 🏆

This late-flowering clematis blooms from midsummer to early autumn. The large, velvety, violet-purple flowers, with greenish cream stamens, look beautiful trained up a wall or fence as a border backdrop, or on a rustic wigwam.

↕ 3m (10ft)

CLIMBER MEDIUM

Clematis 'Madame Julia Correvon' 🏆

One of the Viticella Group of clematis, this vigorous climber bears abundant, wine-red blooms with a central tuft of cream stamens from mid- to late summer. Try growing it through a tree.

↕ 3.5m (11ft)

CLIMBER LARGE

Clematis montana

This vigorous, deciduous climber forms a mass of clambering stems, with divided, mid-green foliage. From late spring to early summer, it bears masses of white or pink flowers. It is ideal for clothing fences and pergolas, or growing through trees.

↕ 12m (40ft)

ANNUAL/BIENNIAL LARGE

Cosmos bipinnatus

This upright annual has feathery foliage and saucer-shaped blooms. It is easily grown from seeds and looks at home among perennial border flowers. Varieties include 'Sensation Mixed', with white, red, and pink flowers from summer to early autumn.

↕ up to 1.2m (4ft) ↔ 45cm (18in)

PERENNIAL LARGE

Crocosmia x *crocosmiiflora*

MONTBRETIA This clump-forming, bulbous perennial has long, narrow leaves and, from mid- to late summer, upright spikes of amber-yellow blooms. Darker flower buds provide an attractive highlight. Grow it front- to mid-border with purple-flowered plants.

↕↔ 1.2m (4ft)

BULB LARGE

Dahlia 'Bishop of Auckland'

This striking dahlia has divided, toothed, purple-black-flushed leaves and stems that show off the orange-centred red blooms from midsummer to early autumn. It works well with hot-hued, late-flowering perennials. Stake the stems.

↕ up to 1m (3ft)

BULB LARGE

Dahlia 'Franz Kafka'

A pompon dahlia, with divided, green leaves and, from midsummer to autumn, mauve-pink blooms, which are excellent for cutting. It is an ideal partner for pastel-shaded perennials, such as phlox and *Sidalcia*. Stems may need staking

↕ 80cm (32in)

PERENNIAL LARGE

Delphinium elatum

This bushy perennial has deeply-cut, dark green basal leaves. In summer, it produces tall spires of showy flowers in shades of blue and purple. Grow it at the back of a cottage border. Stems need to be supported.

↕ 2m (6ft) ↔ 90cm (36in)

PERENNIAL SMALL

Dianthus 'Dad's Favourite'

GARDEN PINK This compact, evergreen perennial forms a low mound of grey-green, grass-like foliage. Throughout summer, it bears clove-scented, semi-double, white flowers, with maroon markings, and is ideal for the front edge of a border.

↕ up to 45cm (18in) ↔ 30cm (12in)

PERENNIAL LARGE

Echinops ritro 🏆

GLOBE THISTLE This upright, branching perennial has prickly, divided, dark green leaves and distinctive round, spiky, steel-blue flowerheads in late summer. Plant it mid-border, and support the stems. 'Veitch's Blue' is popular.

↕ 1.2m (4ft) ↔ 75cm (30in)

PERENNIAL SMALL

Erigeron 'White Quakeress'

This clump-forming perennial, ideal for coastal sites, is grown for its white, daisy-like flowers, which appear over several weeks in summer. Plant it at the front of a border. Other varieties range in colours from pale blue and pink, to mauve.

↕↔ 50cm (20in)

PERENNIAL SMALL

Eryngium bourgatii

A drought-tolerant perennial, with divided, silver-veined leaves. From mid- to late summer, dark, silvery blue, cone-like flowerheads, encircled by a spiny "ruff", appear on upright, metallic blue stems. Grow it with achilleas or penstemons.

↕ 45cm (18in) ↔ 30cm (12in)

BULB LARGE

Fritillaria imperialis

CROWN IMPERIAL This tall perennial bulb forms a crown of blue-green leaves above large, pendent, bell-shaped, orange or yellow flowers, held on upright spikes in late spring. Plants die back after flowering.

↕1.5m (5ft)

SHRUB MEDIUM

Fuchsia magellanica var. *molinae*

An upright, deciduous shrub, with small, lance-shaped, green leaves and, from summer to early autumn, pendent, pale pink flowers on arching stems. Use it at the back of a border, or for screening.

↕↔2m (6ft)

SHRUB SMALL

Fuchsia 'Mrs Popple'

A deciduous, bushy shrub, with dark green leaves, speckled with small, pendent, bell-shaped, red and purple flowers in summer. Plant it mid-border or use to create an informal screen. It prefers a sheltered site, away from cold winds.

↕↔1.1m (3½ft)

BULB LARGE

Galtonia candicans

SUMMER HYACINTH This perennial bulb makes a clump of long, lance-shaped leaves. In late summer, fragrant, white, bell-shaped blooms appear on tall stems. A statuesque mid-border plant, it also thrives in gravel beds.

↕1.2m (4ft)

PERENNIAL MEDIUM

Geranium pratense

MEADOW CRANESBILL A clump-forming, compact perennial, with lobed, green leaves. In summer, it produces sprays of pale violet-blue flowers, with white veining. Plant it in groups in moist soil towards the front of a bed or border.

↕↔60cm (24in)

PERENNIAL MEDIUM

Geranium sylvaticum

WOOD CRANESBILL This clump-forming perennial has lobed, mid-green leaves and round, white-centred blue-purple flowers from late spring to early summer. Grow it in groups in a shady flower border. 'Album' AGM (above) has white blooms.

↕75cm (30in) ↔60cm (24in)

ANNUAL/BIENNIAL LARGE

Helianthus annuus

SUNFLOWER An iconic cottage plant, this annual has large, heart-shaped leaves and erect stems bearing yellow, orange, mahogany, or cream blooms, with a large, yellow or brown central disk. Varieties range from dwarfs to giants.

↕up to 3m (10ft) ↔45cm (18in)

PERENNIAL MEDIUM

Helleborus orientalis

LENTEN ROSE This evergreen, medium-sized, woodland perennial has leathery, dark green leaves, divided into slim leaflets. From midwinter to spring, it produces saucer-shaped, white or greenish cream flowers, which age to dark pink.

↕60cm (24in) ↔45cm (18in)

PERENNIAL MEDIUM

Hemerocallis 'Beauty to Behold'

DAYLILY A clump-forming perennial, with arching, strap-shaped leaves. Through summer, multiple buds open in succession to reveal double, trumpet-shaped, lemon-yellow blooms, each lasting a day.

↕60cm (24in) ↔75cm (30in)

PERENNIAL MEDIUM

Hyssopus officinalis

HYSSOP An aromatic, semi-evergreen, shrubby herb, with narrow leaves and spikes of small, violet-blue blooms from summer to early autumn. A front-of-border plant, combine it with sage and pinks. 'Roseus' is a pink variety.

↕60cm (24in) ↔1m (3ft)

SHRUB LARGE

Kalmia latifolia
MOUNTAIN LAUREL This is an evergreen, bushy, dense shrub. In early summer, large clusters of pink, saucer-shaped flowers open from distinctively crimped buds amid glossy, rich green foliage. Prefers acid soil in partial shade.
↕↔3m (10ft)

PERENNIAL LARGE

Kniphofia uvaria
RED-HOT POKER This perennial's strap-shaped, evergreen leaves and tall spikes of cone-shaped, red and yellow flowerheads create an eye-catching feature in late summer when planted in swathes through the centre of a border.
↕1.2m (4ft) ↔60cm (24in)

CLIMBER MEDIUM

Lathyrus odoratus
SWEET PEA This annual climbs by tendrils, and has divided, mid-green leaves and scented, pink, blue, purple, or white flowers from summer to early autumn. Plant it in large pots or in a border with a tripod support.
↕3m (10ft)

SHRUB SMALL

Lavandula angustifolia
A classic cottage garden evergreen shrub, with narrow, aromatic, silver-grey leaves and dense spikes of fragrant, deep purple flowers, held on wiry stems, from mid- to late summer. Other varieties have flowers in shades of white, pink, and mauve.
↕60cm (24in) ↔75cm (30in)

PERENNIAL MEDIUM

Leucanthemum x *superbum*
SHASTA DAISY Grow this upright perennial mid-border for its lance-shaped leaves and large, white, or sometimes yellow, daisies that appear from midsummer to autumn. 'Wirral Supreme' AGM (above) has striking double, yellow-centred white flowers.
↕1m (3ft) ↔60cm (24in)

BULB LARGE

Lilium longiflorum
EASTER LILY An upright, perennial bulb, with long, shiny, lance-shaped leaves. This classic lily bears sprays of fragrant, large, outward-facing, funnel-shaped, pure white summer flowers, with slightly reflexed petals. Flowers are excellent for cutting.
↕up to 1m (3ft)

BULB LARGE

Lilium regale
REGAL LILY This perennial bulb has tall, unbranched stems, topped with large, white, fragrant, trumpet-shaped flowers, yellow at the throat and pinkish purple on the outside. Plant it in pots or among tall delphiniums and shrub roses.
↕2m (6ft)

PERENNIAL LARGE

Lupinus 'The Chatelaine'
LUPIN This upright perennial has round leaves, divided into lance-shaped leaflets. Tall spikes of dark pink and white flowers appear in summer, adding colour to the middle of a border. Plants may need staking. The seeds are toxic.
↕1.2m (4ft) ↔45cm (18in)

ANNUAL/BIENNIAL MEDIUM

Malope trifida
ANNUAL MALLOW This fast-growing, upright annual, with lobed leaves, is a magnet for bees, and a good partner for soft fruits and vegetables. Trumpet-shaped, purple-pink blooms, with darker nectar guides, open from summer to autumn.
↕90cm (36in) ↔30cm (12in)

ANNUAL/BIENNIAL SMALL

Nemophila menziesii

BABY BLUE EYES This summer-flowering, low-growing, spreading annual has finely divided leaves and saucer-shaped, sky-blue blooms, with pale blue to white centres. Plant it *en masse* to edge a border in a sunny or lightly shaded spot.

↕20cm (8in) ↔15cm (6in)

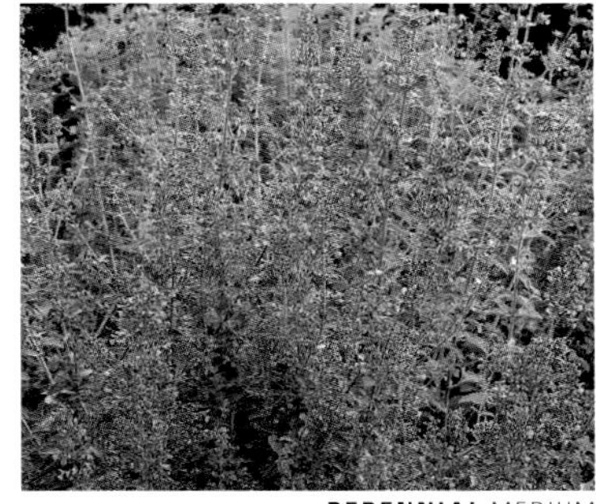

PERENNIAL MEDIUM

Nepeta 'Six Hills Giant'

CATMINT A vigorous, clump-forming perennial, with narrow, oval, toothed, aromatic, grey-green leaves. In summer, it is covered with spikes of tubular, lavender-blue flowers. Use it as border edging or in front of a hedge or wall.

↕1m (3ft) ↔1.2m (4ft)

ANNUAL/BIENNIAL SMALL

Nigella damascena

LOVE-IN-A-MIST Sow this ferny-leaved annual in gaps in borders, where it will produce light blue flowers, with delicate leafy ruffs, in summer. Ornamental pods follow the blooms. Persian Jewel Group has pastel flowers in various shades.

↕45cm (18in) ↔20cm (8in)

PERENNIAL MEDIUM

Paeonia lactiflora

PEONY This clump-forming perennial has divided, green leaves and, in early summer, large, bowl-shaped, single or double, white, pink, red, or mauve blooms. Varieties include 'Bowl of Beauty' AGM (above). Taller plants need support.

↕↔1m (3ft)

PERENNIAL MEDIUM

Paeonia mlokosewitschii

CAUCASIAN PEONY This clump-forming perennial, known as 'molly the witch', has pinkish shoots and bluish green leaves that complement the lemon, bowl-shaped late spring to early summer blooms. Try it front-of-border with *Geranium* 'Johnson's Blue'.

↕↔75cm (30in)

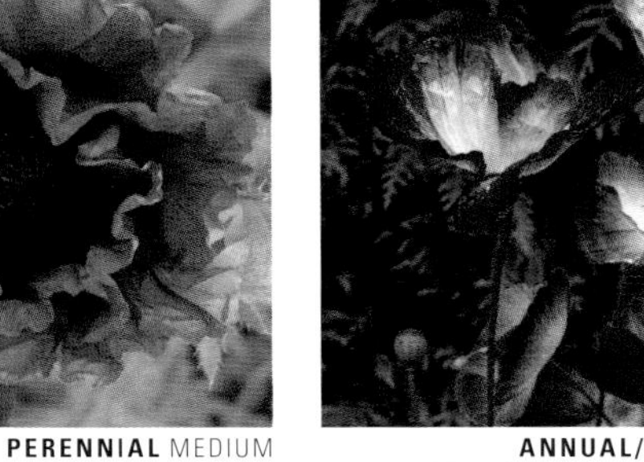

PERENNIAL MEDIUM

Papaver orientale

ORIENTAL POPPY This perennial has hairy, divided leaves and, in early summer, red, orange, pink, or white, bowl-shaped, ruffled blooms with prominent black centres. 'Patty's Plum' (above) has unusual purple-red flowers, effective with silvers.

↕↔90cm (36in)

ANNUAL/BIENNIAL MEDIUM

Papaver rhoeas

ANNUAL POPPY Bred originally from the field poppy, this annual has bowl-shaped blooms, with tissue-paper-like petals, in shades of white, pink, and red. All the coloured forms have white centres. Sow seeds directly in borders to fill gaps.

↕60cm (24in) ↔30cm (12in)

PERENNIAL MEDIUM

Penstemon 'Sour Grapes'

This semi-evergreen, upright perennial has lance-shaped, light green leaves and, from midsummer to autumn, tubular, bell-shaped, purple-blue flowers suffused with violet and white inside. Plant it mid-border in a sheltered garden.

↕60cm (24in) ↔45cm (18in)

PERENNIAL MEDIUM

Penstemon 'Stapleford Gem'

This bushy, semi-evergreen perennial has narrow spires of tubular, blue-flushed light purple flowers, with whitish throats, from late summer to autumn. Grow it mid-border against darker purples and magentas.

↕60cm (24in) ↔45cm (18in)

SHRUB SMALL

Potentilla fruticosa

CINQUEFOIL Grow this bushy shrub, with its summer to early autumn circular flowers in shades of white, pink, yellow, orange, or red, between perennials to brighten up borders. Varieties include the bright yellow 'Goldfinger' (above).

↕1m (3ft) ↔1.5m (5ft)

PERENNIAL SMALL

Primula japonica

PRIMROSE This perennial candelabra primula bears clusters of red, purple, orange, pink, or yellow flowers on upright stems in early summer. Grow it at the front of a damp border or pool side. 'Miller's Crimson' AGM (above) has rich pink blooms.

↕↔45cm (18in)

CLIMBER LARGE

Rosa 'Albertine' ♡

A vigorous rambler rose, with arching, thorny, reddish stems and glossy, dark green foliage. In summer, it bears clusters of scented, fully double, salmon-pink flowers in a single flush. Grow it through a tree or on a large pergola.

↕5m (15ft)

SHRUB MEDIUM

Rosa 'Ballerina' ♡

This shrub rose bears clusters of small, single, white-centred pale pink flowers in repeated flushes from midsummer to early autumn. Deadhead or pick regularly to prolong the display. Mulch annually in spring.

↕1.5m (5ft) ↔1.2m (4ft)

CLIMBER LARGE

Rosa banksiae 'Lutea' ♡

YELLOW BANKSIAN This vigorous, semi-evergreen rambling rose bears abundant clusters of unscented, soft yellow, double flowers in a single flush during late spring. The thornless stems clamber over fences and trees, draping them in early colour.

↕10m (30ft)

SHRUB SMALL

Rosa 'Buff Beauty' ♡

This mid-border shrub has long, arching stems, glossy leaves, and large clusters of fragrant, double blooms over a long period from summer to autumn. The apricot shading of the flowers lend an old-fashioned look to the garden.

↕↔1.2m (4ft)

SHRUB MEDIUM

Rosa 'Complicata'

This shrub rose flowers in a single flush in midsummer, producing open, mildly scented, single, white-centred mid-pink blooms. Grow it mid-border with late-flowering plants to give continued colour. Deadhead regularly, and mulch with compost in spring.

↕2.2m (7ft) ↔2.5m (8ft)

SHRUB MEDIUM

Rosa GERTRUDE JEKYLL ♡

Grown as a shrub or a small climber, this rose has large, richly scented, fully double, pink flowers from summer to early autumn. It is ideal for the back of a border or trained against a low fence. Deadhead regularly, and mulch with compost during spring.

↕2m (6ft) ↔1.2m (4ft)

SHRUB MEDIUM

Rosa 'Roseraie de l'Haÿ' ♡

This shrub produces prickly, upright stems, clothed in green, glossy, wrinkled foliage that provides a foil for the fragrant, double, purple-crimson flowers from summer to autumn. Tomato-like hips appear in autumn.

↕2.2m (7ft) ↔2m (6ft)

CLIMBER MEDIUM

Rosa SUMMER WINE ♡

This climbing rose has disease-resistant, dark green leaves and small clusters of fragrant, flat-faced, semi-double, coral-pink flowers from summer to autumn. It is ideal for growing on a wall, fence, arch, or pillar.

↕3m (10ft)

SHRUB MEDIUM

Rosa 'Zéphirine Drouhin'

Grow this rose as a shrub, or train it as a small climber. It bears double, bright pink flowers on thornless stems from summer to autumn. An ideal back-of-border plant, it tolerates shade. Deadhead often and mulch with compost in spring.

↕2.5m (8ft) ↔2m (6ft)

SHRUB SMALL

Salvia officinalis

SAGE An evergreen, shrubby perennial, with oval, aromatic, grey-green leaves that are used to flavour meat dishes and to make stuffing. It bears spikes of lilac-pink flowers in summer. 'Icterina' AGM (above) has variegated, yellow and green leaves.

↕80cm (32in) ↔1m (3ft)

PERENNIAL MEDIUM

Salvia pratensis

CLARY A cottage garden favourite, this perennial produces a clump of large, toothed leaves and, in early summer, upright stems of purple, two-lipped flowers that are irresistible to bees. Varieties include 'Indigo' AGM (above) with violet flowers.
↕90cm (36in) ↔30cm (12in)

SHRUB SMALL

Santolina pinnata

COTTON LAVENDER A bushy, evergreen, mound-forming shrub, with finely toothed, silver foliage that gives year-round colour. In summer, it bears pale lemon, pompon-like blooms. *S. pinnata* subsp. *neapolitana* 'Sulphurea' (above) is a popular variety.
↕75cm (30in) ↔1m (3ft)

PERENNIAL MEDIUM

Sidalcea malviflora

CHECKERBLOOM This upright, branching perennial has lobed, green leaves and, in summer, funnel-shaped, lilac or pale pink flowers on tall stems. Plant it in drifts in the middle of a border; stake taller plants as required.
↕90cm (36in) ↔45cm (18in)

SHRUB MEDIUM

Spiraea nipponica

TOSA SPIREA Use this shrub at the back of a border, or as an informal hedge. Small, oval leaves cover the wiry stems and rounded clusters of tiny, white blooms clothe the sideshoots in summer. 'Snowmound' AGM (above) is a vigorous variety.
↕↔2.5m (8ft)

PERENNIAL SMALL

Stokesia laevis

STOKES ASTER This evergreen perennial has narrow, mid-green leaves, and produces large, cornflower-like, lavender- or purple-blue flowers on short stems from midsummer to mid-autumn. 'Purple Parasols' (above) has violet-purple blooms.
↕↔45cm (18in)

SHRUB MEDIUM

Symphoricarpos x *doorenbosii*

SNOWBERRY This vigorous, deciduous shrub has small, round, dark green leaves and tiny, greenish white summer flowers, followed by showy clusters of round, white fruits that provide winter interest. Useful as a hedging plant or border backdrop.
↕2m (6ft) ↔indefinite

BULB MEDIUM

Tulipa 'China Pink'

This tall, late-spring-flowering bulb has greyish green leaves and slender, waisted, clear rose-pink flowers, with pointed, outward-curving petals. It looks good teamed with cream wallflowers or lime euphorbia.
↕55cm (22in)

BULB MEDIUM

Tulipa 'Spring Green'

This Viridiflora Group tulip bears lance-shaped, dark green leaves and cup-shaped, creamy white flowers, with green markings, in late spring. For a cool combination, try it with forget-me-nots and double bellis daisies.
↕38cm (15in)

PERENNIAL MEDIUM

Verbascum chaixii

MULLEIN This perennial, ideal for creating focal points in a border, makes a rosette of coarse leaves. Throughout summer, its upright stems are punctuated with long clusters of pale yellow flowers. 'Album' (above) has purple-eyed white blooms.
↕90cm (36in) ↔45cm (18in)

OTHER SUGGESTIONS

Annuals and biennials

Agrostemma githago 'Milas' • *Clarkia amoena* • *Myosotis sylvatica* 'Blue Ball'

Perennials and bulbs

Allium 'Mount Everest' • *Anemone* x *hybrida* 'September Charm' • *Aquilegia vulgaris* 'Black Barlow' • *Aster ericoides* 'Golden Spray' • *Aster novae-angliae* 'Andenken an Alma Potschke' • *Astilbe* x *crispa* 'Perkeo' • *Camassia quamash* • *Chrysanthemum* 'Ruby Mound' • *Dahlia* 'White Moonlight' • *Delphinium* 'Gillian Dallas' • *Dianthus* 'Gran's Favourite' • *Dianthus* 'Musgrave's Pink' • *Erigeron* 'Charity' • *Fritillaria persica* • *Kniphofia* 'Percy's Pride' • *Ornithogalum magnum* • *Tulipa* 'Menton' • *Veronica gentianoides*

Shrubs and trees

Ceanothus thyrsiflorus var. *repens* • *Rosa* 'Celeste' • *Rosa filipes* 'Kiftsgate'

Plants for play

Children love to explore – transform your garden into a jungle of secret dens and tunnels, and add plant curiosities to ensure they are never bored.

Weave narrow paths between tough shrubs, such as *Aucuba* and *Buddleja*, or tall bamboos that tolerate wear and tear. Also consider a living plant feature, such as a tunnel or tipi made from willow – you can buy the "withies" (young stems) in early spring from specialist suppliers, who will provide you with weaving instructions too. Plants that resemble animal features, such as the hare's-tail grass (*Lagurus*) and cute mouse plant (*Arisarum*), will keep children entertained, while easy-to-grow annuals help to stimulate their interest in gardening and nature – sunflowers and pot marigolds are good choices.

ANNUAL/BIENNIAL LARGE

Amaranthus caudatus

LOVE-LIES-BLEEDING This tall, bushy annual is fun to grow from seed and produces tall, sturdy stems from which hang long, pendulous, tassel-like, red flowers from summer to autumn. Children can also cut them for indoor displays.

↕ 1.2m (4ft) ↔ 45cm (18in)

ANNUAL/BIENNIAL SMALL

Antirrhinum majus

SNAP DRAGON Lending colour to summer pots and flowerbeds, this upright annual bears spikes of two-lipped, crimson, red, pink, burgundy, white, and yellow flowers. When pressed at the base, the blooms open up to reveal a dragon-like mouth.

↕↔ 45cm (18in)

PERENNIAL SMALL

Arisarum proboscideum

MOUSE PLANT This clump-forming perennial bears strange blooms that intrigue children. The maroon flowers have a white base and consist of a "hood" concealing tiny flowers, which is drawn out into a mouse-like shape.

↕ 15cm (6in) ↔ 30cm (12in)

SHRUB MEDIUM

Aucuba japonica

SPOTTED LAUREL This evergreen shrub has large, glossy, green leaves, decorated with small, yellow spots. This is a tough, resilient plant, and can be used to provide shelter and privacy near a play area. Female plants produce berries, which should not be eaten.

↕↔ 2m (6ft)

ANNUAL/BIENNIAL MEDIUM

Calendula officinalis

POT MARIGOLD Children will enjoy growing this bushy annual from seed. Its aromatic, pale green leaves are joined by single or double, daisy-like, edible flowers in shades of yellow and orange, from late spring to autumn.

↕↔ 60cm (24in)

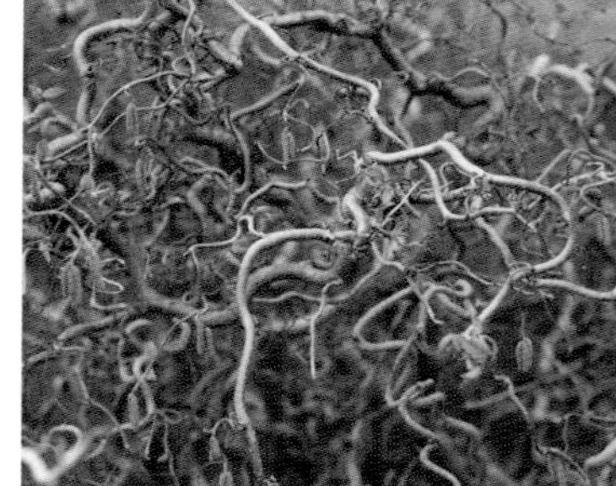

SHRUB LARGE

Corylus avellana

CORKSCREW HAZEL The twisted shoots of this deciduous shrub fascinate children, and give the garden a spooky look, especially when the green leaves fall in autumn. Grow it as an accent plant and use the twisted stems for indoor decorations.

↕↔ 5m (15ft)

PERENNIAL SMALL

Fragaria vesca

ALPINE STRAWBERRY This type of strawberry produces small, sweet berries all summer, which children love. Plant it in pots or directly in the soil. It is easy to look after and ideal for teaching children how to care for plants themselves.

↕ 30cm (12in) ↔ indefinite

ANNUAL/BIENNIAL MEDIUM

Helianthus annuus

SUNFLOWERS The edible seeds of this robust annual are easy for children to handle and sow. It bears classic daisy-like, orange or yellow sunflowers throughout summer. Dwarf varieties include 'Teddy Bear' (above), with fluffy, double blooms.

↕ up to 90cm (36in) ↔ 60cm (24in)

PERENNIAL SMALL

Lagurus ovatus

HARE'S-TAIL GRASS This clump-forming annual grass has green leaves and, in early summer, oval, furry, white flowerheads, which resemble soft rabbits' tails, and last into autumn. They also make beautiful indoor displays.

↕ 45cm (18in) ↔ 15cm (6in)

CLIMBER LARGE

Luffa cylindrica

SMOOTH LOOFAH Grow your own loofah sponge from seed with this climbing annual. It has maple-like leaves and yellow flowers, followed by elongated fruits that produce loofahs when dried. It needs a sunny sheltered site to thrive.

↕ 5m (15ft)

TREE SMALL

Malus domestica

APPLE TREE Plant a dessert apple tree for its delicious, sweet fruits and pretty spring blossom. Opt for one on a dwarfing rootstock, such as M9, for a small garden. Varieties include 'Cox's Orange Pippin' (above), with sweet, red apples.

↕↔ 8m (25ft)

SHRUB LARGE

Phyllostachys aureosulcata f. *spectabilis*

This bamboo has tall, upright canes, clothed in soft green leaves. It lends the garden a jungle-like appearance, and can be used to provide shelter and privacy to a play area, encouraging children to play outside.

↕ 6m (20ft) ↔ indefinite

PERENNIAL MEDIUM

Platycodon grandiflorus

BALLOON FLOWER The purple-blue flowers of this perennial open from balloon-like buds throughout early autumn, creating beautiful displays in pots filled with soil-based compost. Add a frill of violas around the edge to inject more colour.

↕ 60cm (24in) ↔ 45cm (18in)

TREE SMALL

Prunus x *subhirtella* 'Pendula Rubra'

This weeping cherry has pink buds that open into scented, white spring flowers. The spreading branches create a private play space for children. Its autumn colour also adds to its decorative value.

↕↔ 8m (25ft)

TREE SMALL

Pyrus salicifolia 'Pendula'

WEEPING SILVER PEAR This small tree creates a skirt of weeping stems, studded with narrow, silvery green leaves, beneath which children can hide. Creamy white spring flowers are followed by brown, inedible fruits.

↕ 8m (25ft) ↔ 6m (20ft)

TREE SMALL

Salix caprea

GOAT WILLOW Also known as pussy willow, this large, deciduous shrub or small tree produces arching stems that develop soft, furry catkins, which are loved by children, from late winter. The dark green leaves follow in spring.

↕↔ 2m (6ft)

PERENNIAL SMALL

Sempervivum arachnoideum

COBWEB HOUSELEEK This evergreen, low-growing perennial has rosettes of fleshy, green leaves covered with fine, white hairs that resemble cobwebs. Children can easily grow it in containers and raised beds.

↕ 12cm (5in) ↔ 10cm (4in) or more

ANNUAL/BIENNIAL SMALL

Solanum lycopersicum

BUSH TOMATO Tomato plants help to teach children how to care for plants, and reward even basic care with tasty summer fruits. Bush varieties that grow outside are the best choices, particularly dwarf types for patio containers.

↕↔ 45cm (18in)

PERENNIAL SMALL

Stachys byzantina

LAMBS' EARS This evergreen perennial produces downy, silver-grey foliage and spikes of small, mauve-pink flowers in summer. A tough plant, with soft, tactile leaves, it is perfect for edging a bed in a sunny family garden.

↕ 38cm (15in) ↔ 60cm (24in)

ANNUAL/BIENNIAL SMALL

Tropaeolum majus Alaska Series

NASTURTIUM A bushy annual with round, cream-splashed dark green leaves and trumpet-shaped, red and yellow summer to early autumn flowers. Grow it in tall pots for children to pick the edible leaves and flowers.

↕ 30cm (12in) ↔ 45cm (18in)

SHRUB MEDIUM

Vaccinium corymbosum

BLUEBERRY This easy-to-grow, soft fruit bush bears delicious sweet-tasting, blue-black fruits in summer. It is best planted in a large pot of ericaceous compost in a sunny spot. Children will love the berries, which are packed with healthy vitamins.

↕↔ up to 1.5m (5ft)

OTHER SUGGESTIONS

Annuals and biennials

Antirrhinum Bells Series • *Helianthus annuus* 'Giant'

Perennials and grasses

Cosmos atrosanguineus • *Geranium* 'Johnson's Blue' • *Stachys lanata*

Shrubs and trees

Buddleja davidii • *Cercidiphyllum japonicum* f. *pendulum* • *Choisya* 'Aztec Pearl' • *Choisya ternata* • *Corylus avellana* 'Red Majestic' • *Cotoneaster lacteus* • *Eleagnus pungens* • *Fargesia murielae* 'Simba' • *Lavandula angustifolia* • *Salix viminalis* 'Bowles Hybrid' • *Sedum spectabile*

Plants for wildlife gardens

Whatever your design style, bring your garden to life with a collection of berries, seeds, and nectar-rich plants to attract birds and beneficial insects.

A wildlife garden can be formal or informal, although the scope is limited in a tightly controlled design. You can create habitats for nesting birds with prickly hedging and trees to protect them from predators, and include shrubs that bear nuts and berries, such as hawthorn (*Crataegus*) and rowan (*Sorbus*), to feed your feathered friends over winter. Plants rich in pollen and nectar will attract beneficial insects, including butterflies and bees, as well as those whose larvae help control pest populations, such as hoverflies. A naturalistic pool surrounded by moisture-loving plants will also draw in birds and beasts to drink and bath.

TREE LARGE

Acer rubrum

RED MAPLE This large tree has maple-like, green leaves that turn yellow and orange in autumn. The red spring flowers provide pollen for bees, and winged fruits are food for birds and squirrels. 'Bowhall' has a columnar habit.

↕20m (70ft) ↔10m (30ft)

PERENNIAL LARGE

Ageratina altissima

WHITE SNAKEROOT This clump-forming perennial has oval, pointed, toothed-edged leaves. From summer to early autumn, upright stems are topped with clusters of tiny, white blooms that are a rich source of nectar for butterflies.

↕1.2m (4ft) ↔45cm (18in)

PERENNIAL LARGE

Alcea rosea

HOLLYHOCK A tall, upright, short-lived perennial or biennial that may self-seed, with rounded, lobed leaves. From mid- to late summer, towering stems bear purple, red, pink, white, or yellow, cup-shaped blooms that attract butterflies and bees.

↕2m (6ft) ↔60cm (24in)

BULB LARGE

Allium cernuum

LADY'S LEEK This upright, perennial bulb produces strap-shaped, grey-green leaves that fade before clusters of nectar-rich, drooping, purplish pink flowers appear on slim stems in early summer. Use it to naturalize in meadows.

↕70cm (28in)

PERENNIAL MEDIUM

Asclepias tuberosa

BUTTERFLY WEED This tap-rooted perennial produces flat heads of vivid orange or yellow flowers from late summer to autumn. Plant it in well-drained, sunny borders or wildflower meadows to attract butterflies. The milky sap is an irritant.

↕75cm (30in) ↔45cm (18in)

PERENNIAL MEDIUM

Aster x *frikartii*

A bushy, upright perennial, with yellow-centred lavender-blue, daisy flowers that bloom over a long period from midsummer to autumn. The flowers provide nectar for butterflies. Plant in bold swathes in a flower border or naturalistic schemes.

↕70cm (28in) ↔40cm (16in)

SHRUB LARGE

Berberis darwinii

DARWIN'S BARBERRY An evergreen shrub, with tiny, holly-like leaves, that is smothered in spring with pendent clusters of dark orange flowers that attract bees. Birds devour the blue-black berries that follow. Use as a border specimen or hedge.

↕↔3m (10ft)

PERENNIAL SMALL

Bergenia 'Silberlicht'

ELEPHANT'S EARS This neat, clump-forming, evergreen perennial has leathery, rounded leaves. In spring, reddish stems carry clusters of white, bell-shaped flowers that open from deep red buds, making an attractive contrast.

↕45cm (18in) ↔50cm (20in)

ANNUAL/BIENNIAL MEDIUM

Calendula officinalis

POT MARIGOLD The orange or yellow, daisy-like flowers of this fast-growing, bushy annual attract a wide range of insects. Choose single-flowered varieties for wildlife gardens, and sow seeds directly in flower borders or patio pots.

↕↔60cm (24in)

SHRUB SMALL

Calluna vulgaris

SCOTCH HEATHER This evergreen, ground-cover shrub has upright stems covered with scale-like green, gold, or silver leaves that often colour-up in autumn and winter. White, pink, red, or purple late summer to autumn flowers offer nectar.

↕60cm (24in) ↔45cm (18in)

PERENNIAL SMALL

Campanula rotundifolia

COMMON HAREBELL A perennial meadow plant, with small, rounded basal leaves and, in summer, bell-shaped, light blue flowers on slender, upright stems. Grow this dainty bee plant in grassy patches or rock gardens; avoid using fertilizer.

↕↔30cm (12in)

PERENNIAL SMALL

Cardamine pratensis

LADY'S SMOCK A neat, clump-forming perennial, with mid-green leaves, divided into leaflets. In spring, it bears spikes of single, lilac flowers, although double forms (above) are also available. It is ideal for boggy areas or pond perimeters.

↕45cm (18in) ↔30cm (12in)

PERENNIAL SMALL

Centaurea montana

PERENNIAL CORNFLOWER Plant this perennial in informal schemes and allow it to self-seed. Branched stems bear green leaves and, in early summer, thistle-like, red-purple-centred blue blooms that bees adore. Cut back for a second flowering.

↕50cm (20in) ↔60cm (24in)

PERENNIAL LARGE

Cephalaria gigantea

GIANT SCABIOUS Slot this clump-forming perennial at the back of a large border or naturalize in rough grass. In summer, pale yellow blooms that are attractive to butterflies appear above green, divided leaves on towering, branched stems.

↕2.5m (8ft) ↔60cm (24in)

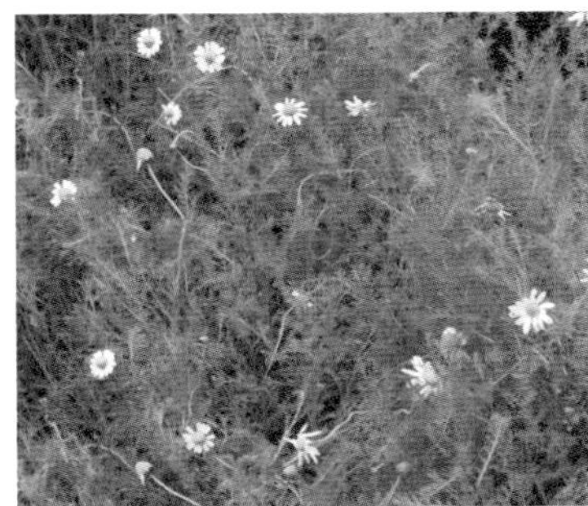

PERENNIAL SMALL

Chamaemelum nobile

CHAMOMILE An evergreen perennial that forms a carpet of feathery, aromatic leaves. From late spring to summer, it bears simple, white, daisy-like blooms. Plant it alongside wildflowers to create a naturalistic meadow effect.

↕10cm (4in) ↔45cm (18in)

CLIMBER LARGE

Clematis montana

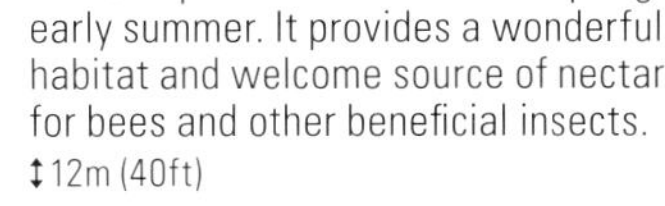

This vigorous, deciduous climber produces an abundance of small, yellow-centred white or pink blooms from late spring to early summer. It provides a wonderful habitat and welcome source of nectar for bees and other beneficial insects.

↕12m (40ft)

SHRUB LARGE

Cornus alba

DOGWOOD An upright, deciduous shrub, with bright scarlet shoots in winter. The creamy white early summer flowers are followed by bluish white fruits. Plant to provide cover around wildlife pools. Prune in late winter for more colourful stems.

↕↔3m (10ft)

SHRUB LARGE

Cornus mas

CORNELIAN CHERRY This useful deciduous shrub, or small tree, has reddish purple autumn leaves and edible red fruits, loved by birds. In late winter, it produces rounded clusters of tiny, nectar- and pollen-rich, yellow blooms on bare stems.

↕↔5m (15ft)

SHRUB LARGE

Corylus avellana

HAZEL A spreading deciduous shrub or small tree, this hedgerow plant bears round, dark green leaves, yellow in autumn, and dangling, yellow early spring catkins, sometimes followed by hazelnuts. Use it to create a small woodland garden.

↕↔ 5m (15ft)

SHRUB SMALL

Cotoneaster horizontalis

This deciduous shrub has small, glossy, dark green leaves and branches arranged in a distinctive "herringbone" pattern. In summer, it bears small, white blooms that are followed by vivid red berries, enjoyed by birds during autumn and early winter.

↕ 1m (3ft) ↔ 1.5m (5ft)

BULB SMALL

Crocus vernus

This dwarf perennial bulb, with its small, cup-shaped, lilac or purple blooms from spring to early summer, provides welcome nectar for early pollinating insects and can be naturalized in lawns. Plant the corms during autumn in groups or drifts.

↕ 12cm (5in)

BULB LARGE

Dahlia 'Bishop of Llandaff'

This perennial, grown from spring-planted tubers, has black-purple, glossy, divided leaves and semi-double, dark-eyed bright red blooms. Single and semi-double dahlias like these are important nectar sources for butterflies and bees.

↕ 1m (3ft)

PERENNIAL LARGE

Deschampsia cespitosa

A deciduous, tuft-forming, perennial grass, with clouds of tiny, golden-yellow flowers on long stems in summer. Provides a habitat for beneficial insects and nesting material and food for birds, as well as giving a naturalistic appearance to planting schemes.

↕ up to 2m (6ft) ↔ 50cm (20in)

BULB SMALL

Eranthis hyemalis

WINTER ACONITE This clump-forming tuber has lobed, green leaves and stalkless, cup-shaped, yellow flowers, with leaf-like ruffs, from late winter to early spring. Grow it in pockets between rocks, and combine with spring bulbs.

↕ 10cm (4in)

SHRUB SMALL

Erica carnea

WINTER HEATH This evergreen shrub, with needle-like leaves makes a good front-of-border or rock garden plant. The tiny pink, red, or white early winter to late spring flowers are a boon for insects emerging from hibernation.

↕ 30cm (12in) ↔ 45cm (18in) or more

PERENNIAL SMALL

Erythronium 'Pagoda'

DOG'S TOOTH VIOLET This tuberous-rooted perennial makes mounds of veined, mottled leaves. With its nodding, nectar-rich, pale yellow spring flowers, with swept-back petals, it is perfect for woodland-style plantings.

↕ 35cm (14in) ↔ 20cm (8in)

PERENNIAL LARGE

Filipendula rubra

QUEEN OF THE PRAIRIE An upright, moisture-loving perennial, with large, green, deeply cut leaves. From early- to midsummer, it produces fluffy pink flowerheads on tall, slender stems, which provide a source of pollen for bees.

↕ up to 2.5m (8ft) ↔ 1.2m (4ft)

BULB MEDIUM

Fritillaria meleagris

SNAKE'S-HEAD FRITILLARY A bulbous perennial, with narrow leaves and nodding, bell-shaped, rich purple or white spring flowers with a chequered pattern. Ideal for naturalizing in wildflower meadows, it also works well in cottage borders.
↕30cm (12in)

PERENNIAL SMALL

Galium odoratum

SWEET WOODRUFF Though steadily spreading, this carpeting perennial with emerald-green leaves charms its way into any semi-wild or woodland planting with its froth of starry, white blooms that appear from late spring to midsummer.
↕15cm (6in) ↔30cm (12in) or more

PERENNIAL MEDIUM

Geranium maculatum

WILD GERANIUM This bushy, upright perennial has deeply divided leaves and, in late spring, clusters of pale to dark pink or lilac blooms with white centres. Given sufficient soil moisture, this woodlander will naturalize. Attractive to butterflies.
↕75cm (30in) ↔45cm (18in)

PERENNIAL MEDIUM

Geranium pratense

MEADOW CRANESBILL A clump-forming perennial, with deeply divided, lobed, mid-green leaves that turn bronze in autumn. Saucer-shaped, violet-blue, veined flowers appear in summer. Plant it in meadows on heavy soils.
↕↔60cm (24in)

ANNUAL/BIENNIAL LARGE

Helianthus annuus

SUNFLOWER This classic cottage garden plant bears giant, daisy-like, yellow, orange, mahogany, or cream blooms throughout summer. Powerful bee attractors, the seedheads are used by hibernating insects and the seeds are eaten by birds.
↕up to 3m (10ft) ↔45cm (18in)

BULB MEDIUM

Hyacinthoides non-scripta

ENGLISH BLUEBELL This perennial bulb has narrow, strap-shaped leaves and blue, or occasionally white, bell-shaped, fragrant spring blooms, hanging from arching stems. It provides nectar for bees and butterflies. Plant *en masse* in light woodland.
↕40cm (16in)

TREE LARGE

Ilex aquifolium

ENGLISH HOLLY This evergreen, slow-growing tree has dark green, spiny leaves. Bright red winter berries form on female plants if grown near a male. Grow in wildlife borders or as a specimen tree. It offers habitat and food for insects and birds.
↕up to 20m (70ft) ↔6m (20ft)

SHRUB LARGE

Ilex 'Sparkleberry'

WINTERBERRY This deciduous, bushy shrub has pointed, oval leaves that develop bold autumn colour before falling. If a male plant is grown nearby, female plants bear small, bright red berries that last well into winter, or until eaten by birds.
↕5m (15ft) ↔4m (12ft)

PERENNIAL LARGE

Inula magnifica

GIANT INULA This imposing perennial makes a clump of large leaves and, in late summer, bears branched stems topped with daisy-like, yellow blooms, with thread-like petals. Grow it in damp meadows or next to a wildlife pond.
↕1.8m (6ft) ↔1m (3ft)

ANNUAL/BIENNIAL SMALL

Limnanthes douglasii ♀

POACHED-EGG FLOWER This low-growing, spreading annual has ferny foliage and, from summer to autumn, rounded, yellow-centred white blooms. A hoverfly magnet, plant it at the front of cottage borders, along path edges, and in gravel beds.
↕15cm (6in) ↔10cm (4in)

PERENNIAL MEDIUM

Knautia macedonica

MACEDONICAN SCABIOUS This upright perennial has lobed basal leaves and, in summer, a succession of button-like, crimson flowers on wiry stems. Combine this bee and butterfly attractor with ornamental grasses for a meadow effect.
↕75cm (30in) ↔60cm (24in)

PERENNIAL SMALL

Linaria alpina

ALPINE TOADFLAX This short-lived perennial bee attractor produces trailing stems, with a succession of snapdragon-like, orange-centred purple-violet flowers in summer. Allow it to trail over walls and raised beds, or spread through a gravel bed.
↕↔15cm (6in)

SHRUB SMALL

Lavandula angustifolia

LAVENDER An evergreen shrub, with linear, aromatic, silvery grey leaves. In midsummer, it produces a profusion of small, fragrant, violet-blue flowerheads that are irresistible to bees. Try it with self-seeded Californian poppies.
↕80cm (32in) ↔60cm (24in)

SHRUB MEDIUM

Leycesteria formosa

HIMALAYAN HONEYSUCKLE This deciduous shrub has large, heart-shaped leaves and pendent chains of white summer blooms that attract bees and insects. Black autumn berries provide food for birds. Plant it at the back of a border; it may self-seed.
↕↔2m (6ft)

CLIMBER LARGE

Lonicera periclymenum

DUTCH HONEYSUCKLE A large climber, with fragrant, nectar-rich, red and white summer flowers that attract pollinating insects and butterflies. Glossy, red berries appear from late summer to autumn, providing food for birds. Mature woody plants offer habitat.
↕7m (22ft)

SHRUB LARGE

Mahonia x *media*

This evergreen shrub flowers in winter, providing valuable nectar to early insects when few other plants are in bloom. The flowers are followed by chains of blue-black, edible berries, which are loved by birds. Plant it at the back of a large border.
↕5m (15ft) ↔4m (12ft)

PERENNIAL MEDIUM

Maianthemum racemosum ♀

FALSE SPIKENARD This upright perennial has prominently veined, oval, green leaves and, from mid- to late spring, scented, fluffy, cream plumes, sometimes followed by red berries. Plant it to provide a wildlife habitat; shelter from strong winds.
↕90cm (36in) ↔60cm (24in)

TREE MEDIUM

Malus x *moerlandsii* 'Profusion'

A deciduous tree, with pink spring blooms, which provide nectar for early insects, followed by crab apples that ripen to deep purple in autumn and are loved by birds. It provides habitat and nesting sites.
↕↔10m (30ft)

PERENNIAL SMALL

Meconopsis cambrica

WELSH POPPY This perennial has light green, divided leaves and lemon-yellow or orange, poppy-like blooms, which are borne in succession on slim stems from late spring to early summer. Plants colonize beside shaded walls and in gravel.
↕45cm (18in) ↔30cm (12in)

PERENNIAL MEDIUM

Monarda 'Cambridge Scarlet'

BERGAMOT This clump-forming, upright perennial has aromatic, dark green leaves and trumpet-shaped whorls of red summer blooms that attract bees and pollinating insects. Leave the spent stems standing for insect habitats and autumn interest.

↕1m (3ft) ↔45cm (18in)

PERENNIAL LARGE

Monarda fistulosa

WILD BERGAMOT A clump-forming perennial, with whorled heads of small, two-lipped, pale lilac blooms on tall stems in midsummer. The flowers attract bees and butterflies. Grow it at the edge of a woodland garden or in a prairie-style setting.

↕1.2m (4ft) ↔45cm (18in)

BULB MEDIUM

Muscari armeniacum 🏆

GRAPE HYACINTH A spring-flowering perennial bulb, with narrow, grass-like leaves and cone-shaped clusters of small, fragrant, bell-shaped, deep blue flowers, which attract early butterflies and bees. Allow it to naturalize in wild gardens.

↕20cm (8in)

PERENNIAL MEDIUM

Panicum virgatum

SWITCH GRASS A deciduous, clump-forming grass, with blue-green leaves and clouds of pink-tinged green flowerheads in summer. Gives a naturalistic appeal to gardens and habitat, food, and nesting material to birds, mammals, and insects.

↕1m (3ft) ↔75cm (30in)

BULB LARGE

Nectaroscordum siculum

HONEY GARLIC An upright bulb, with narrow leaves that fade as fountain-shaped clusters of pendent, bell-shaped, cream and purple flowers emerge in early summer on tall stems. Attractive seedheads follow the flowers. Try naturalizing it in meadows.

↕1.2m (4ft)

PERENNIAL SMALL

Nepeta x *faassenii* 🏆

CATMINT This bushy perennial has small, greyish green, aromatic leaves and spikes of lilac-purple flowers, irresistible to bees, from summer to autumn. Cut back between flushes for repeat flowering. Plant at the front of cottage borders or as path edging.

↕↔45cm (18in)

TREE LARGE

Nyssa sylvatica

BLACK GUM This graceful, broadly conical tree has insignificant greenish yellow late spring flowers, visited by honeybees. The deep blue fruits, which attract birds, ripen at the same time as when the oval leaves develop rich red or yellow autumnal shades.

↕20m (70ft) ↔10m (30ft)

ANNUAL/BIENNIAL MEDIUM

Papaver rhoeas

FIELD POPPY This annual wildflower produces divided leaves and bowl-shaped, black-centred scarlet blooms in summer. Sow seeds in spring or autumn, and turn over the soil annually to ensure repeat germination.

↕60cm (24in) ↔15cm (6in)

PERENNIAL SMALL

Primula veris 🏆

COWSLIP This evergreen or semi-evergreen perennial forms rosettes of corrugated, oval to lance-shaped leaves. Tight clusters of fragrant, nodding, tubular, butter-yellow flowers appear on stout stems in spring. It is attractive to bees and butterflies.

↕↔25cm (10in)

TREE MEDIUM

Prunus virginiana

CHOKECHERRY This suckering tree or shrub for wild gardens has short spikes of white flowers in spring, followed by red fruits, ripening to black, and attracts birds and other wildlife. Varieties include 'Schubert' (above), with dark purple, oval leaves.

↕10m (30ft) ↔8m (25ft)

PERENNIAL SMALL

Pulmonaria angustifolia

Forming a low clump of matt, mid-green leaves, this spreading perennial flowers in spring and bears blue, trumpet-shaped blooms. It provides nectar for early insects. The leaves often die back in summer; remove spent growth for tidiness.

↕23cm (9in) ↔30cm (12in) or more

TREE LARGE

Quercus rubra

RED OAK This large, forest tree with a spreading crown has leaves with pointed lobes that turn rich reddish brown in autumn, when the acorns ripen. Oaks feed native insects, including moths, and acorns are eaten by birds and mammals.

↕25m (80ft) ↔20m (70ft)

SHRUB LARGE

Rhododendron luteum

YELLOW AZALEA A deciduous, spring-flowering shrub, with oblong- to lance-shaped leaves, which develop bold autumn shades and fragrant, funnel-shaped, yellow flowers. It prefers a cool site and fertile soil.

↕↔4m (12ft)

SHRUB LARGE

Rubus 'Benenden'

This blackberry relative is a vigorous shrub with long, arching branches, peeling bark, and lobed leaves. The white, saucer-shaped blooms appear from spring to summer, and have central tufts of golden stamens, similar to wild roses.

↕↔3m (10ft)

SHRUB LARGE

Sambucus nigra

COMMON ELDER This large, bushy shrub or small tree has divided leaves and, in early summer, flat, rounded heads of creamy white flowers, which develop into bunches of black berries, much sought after by birds.

↕↔6m (20ft)

PERENNIAL MEDIUM

Scabiosa atropurpurea

PINCUSHION FLOWER This clump-forming perennial, grown as an annual, has narrow, toothed leaves and small, domed heads of purple summer flowers with creamy anthers. 'Chile Black' has dark maroon blooms. It attracts bees and butterflies.

↕up to 1m (3ft) ↔30cm (12in)

PERENNIAL SMALL

Scabiosa lucida

GLOSSY SCABIOUS Plant this neat scabious, with its basal clump of divided, grey-green leaves and pale-lilac, pincushion blooms from spring to autumn to attract bees and butterflies. Group plants at the front of a border or gravel bed.

↕20cm (8in) ↔15cm (6in)

BULB MEDIUM

Scilla siberica

SIBERIAN SQUILL This perennial bulb, with strap-shaped leaves and nodding, vivid blue, bell-shaped flowers, attracts bees in spring. Plant bulbs in drifts in autumn, with spring-blooming perennials like primrose, bugle, and wood anemone.

↕20cm (8in)

PERENNIAL SMALL

Sedum spectabile

This fleshy perennial blooms in late summer and produces large, flattened, pink flowerheads, attracting bees, pollinating insects, and butterflies. After flowering, leave the spent heads standing to provide interest during winter.

↕↔45cm (18in)

PERENNIAL LARGE

Selinum wallichianum

WALLICH MILK PARSLEY An upright, architectural perennial, with fern-like leaves and long-lasting, umbrella-shaped heads of starry, white flowers borne in summer. The blooms attract hoverflies; plant it for a meadow effect.

↕1.2m (4ft) ↔40cm (16in)

PERENNIAL MEDIUM

Silene dioica

RED CAMPION A semi-evergreen, clump-forming perennial, with clusters of small, vivid rose-pink flowers that appear from late spring to midsummer. Ideal for growing along hedgerows and in woodland areas. It attracts moths and butterflies.

↕80cm (32in) ↔45cm (18in)

ANNUAL/BIENNIAL SMALL

Tagetes patula

FRENCH MARIGOLD This annual has deeply divided, aromatic, dark green leaves and flowers in shades of yellow, orange, red, or mahogany, which appear from summer to early autumn. For hoverflies and butterflies, choose single or semi-double varieties.

↕↔30cm (12in)

PERENNIAL MEDIUM

Tellima grandiflora

FRINGE CUPS A semi-evergreen perennial, with bright green, heart-shaped, purple-tinged leaves and, from late spring to midsummer, long spikes of small, bell-shaped, cream flowers that provide nectar for insects. Grow in dry areas under trees.

↕↔60cm (24in)

PERENNIAL LARGE

Thalictrum aquilegiifolium

MEADOW RUE This perennial has clusters of fluffy, lilac-purple summer flowers and attractive seedheads, which appear above finely divided, grey-green leaves. Grow it in wild areas or cottage-style borders to attract beneficial insects.

↕1.2m (4ft) ↔45cm (18in)

PERENNIAL LARGE

Verbena bonariensis 🏆

PURPLETOP VERVAIN This tall perennial bears domed clusters of scented, violet-purple flowers from midsummer to autumn, which attract bees and butterflies. It is a versatile border plant, with "see-through" stems. It may self-seed.

↕1.5m (5ft) ↔60cm (24in)

PERENNIAL LARGE

Verbena hastata

AMERICAN BLUE VERVAIN This upright, bushy perennial spreads by self-seeding. It has narrow, toothed leaves and candelabra-like heads of violet-blue flowers that attract butterflies from midsummer to autumn. 'Rosea' (above) has lilac-pink flowers.

↕1.5m (5ft) ↔60cm (24in)

SHRUB LARGE

Viburnum prunifolium

BLACKHAW VIBURNUM A deciduous shrub, ideal for hedging, with glossy, oval leaves that turn red and purple-bronze in autumn. From late spring to early summer, it forms lacy, white flowers, attractive to butterflies. Birds love the blue-black fruits that follow.

↕5m (15ft) ↔4m (12ft)

OTHER SUGGESTIONS

Perennials and grasses

Anthriscus sylvestris • *Aster* x *frikartii* 'Mönch' 🏆 • *Centaurea cyanus* • *Chrysanthemum segetum* • *Convallaria majalis* 🏆 • *Digitalis purpurea* • *Dipsacus fullonum* • *Echinacea paradoxa* • *Echium vulgare* • *Leucanthemum vulgare* • *Lythrum salicaria* • *Phlox carolina* 'Bill Baker' 🏆 • *Polemonium foliosissimum* 'Scottish Garden' • *Ranunculus acris*

Shrubs

Buddleja davidii • *Calluna carnea* • *Cornus racemosa* • *Cotoneaster sternianus* 🏆 • *Euonymus europaeus* • *Lindera benzoin* • *Rosa rugosa* var. *alba* • *Ruscus aculeatus* • *Viburnum opulus*

Trees

Acer campestre 🏆 • *Crataegus monogyna* • *Koelreuteria paniculata* • *Malus* 'Evereste' 🏆 • *Sorbus aucuparia* 'Aspleniifolia'

PLANTS for SEASONAL INTEREST

By selecting plants that perform at different times of the year, you can ensure there is never a dull moment in your garden, even in the depths of winter. The key is to plan ahead and plant a few stars for each season, such as bulbs and blossoming trees in spring, flowering perennials for early and late summer colour, and berried shrubs and blazing foliage plants that perform in autumn. Also include evergreens and plants that bloom during the darkest days of the year to brighten up winter scenes.

Plants for spring borders

As the cold, dark days of winter fade, the new season bursts into life with a blaze of colourful flowers and fresh, green growth.

Hellebores are among the first to bloom in early spring, and look a treat together with bold groups of early daffodils and a sprinkling of anemones in front. As spring unfolds, more flowers join the party. Choose tulips, bergenias, and primulas in a range of bright or pastel shades to decorate borders, backed by shrubs and trees decked with blossom. Also include a few fragrant plants in your designs. Many viburnums have a delicious scent, as do daphnes and deciduous azaleas. However, large-flowered fritillaries have an unpleasant smell, so site them carefully with this in mind.

BULB MEDIUM

Allium neapolitanum
This upright, perennial bulb has strap-shaped, grey-green leaves and loose, spherical clusters of star-shaped, white flowers in spring. Plant the bulbs in autumn in groups towards the front of a spring border.
↕50cm (20in)

PERENNIAL MEDIUM

Amsonia tabernaemontana
EASTERN BLUESTAR An upright perennial with slim, tapering leaves that turn yellow before falling in autumn. Clusters of star-shaped, blue flowers appear from late spring to summer, offering a long season of colour.
↕1m (3ft) ↔30cm (12in)

PERENNIAL SMALL

Anemone blanda 🏆
WINTER WINDFLOWER Grow this tiny, perennial corm, with daisy-like, blue, pink, or white early spring flowers, where it can be seen at the front of a border or along a pathway. Plant the tubers in autumn.
↕10cm (4in) ↔15cm (6in)

PERENNIAL SMALL

***Anemone coronaria* De Caen Group**
This perennial has divided, green leaves and shallow, bowl-shaped early spring blooms in shades of red, blue-violet, and white, that inject borders with vibrant seasonal colour. Plant the tubers in autumn.
↕30cm (12in) ↔15cm (6in)

PERENNIAL MEDIUM

Aquilegia vulgaris
GRANNY'S BONNET An upright perennial, with green leaves, divided into rounded, lobed leaflets. From late spring to early summer, it produces pendent, single or double flowers in a wide range of colours, including the white 'Nivea' AGM (above).
↕90cm (36in) ↔45cm (18in)

PERENNIAL SMALL

Bergenia purpurascens 🏆
PURPLE BERGENIA Grow this evergreen perennial towards the front of a border, where its rounded, dark green leaves and spikes of dark pink spring flowers will contrast well with bulbs, such as daffodils and *Muscari*.
↕40cm (16in) ↔60cm (24in)

CLIMBER LARGE

Clematis montana
This vigorous, deciduous climber has divided, mid-green foliage and, in late spring and early summer, an abundance of single, white or pink flowers, with yellow centres. Plant the roots in shade, and stems in sun. It needs a large support.
↕12m (40ft)

SHRUB SMALL

Cytisus* x *praecox 🏆
BROOM This bushy, deciduous shrub has green stems, small leaves, and a profusion of pale creamy yellow, pea-like blooms, which lend an informal note to spring borders. Varieties include 'Allgold' AGM (above), with golden-yellow blooms.
↕1.2m (4ft) ↔1.5m (5ft)

SHRUB MEDIUM

Daphne* x *burkwoodii
Upright and semi-evergreen, this shrub has small, green leaves and highly fragrant, pink flowers, which complement tulips and aquilegias in a mixed, late spring border. Varieties include 'Somerset' AGM (above), with purple-pink blooms.
↕1.5m (5ft) ↔1m (3ft)

PERENNIAL SMALL

Dicentra formosa

BLEEDING HEART A spreading perennial, with finely divided, grey-green leaves and pendent, heart-shaped, dark pink flowers from late spring to early summer. Grow it towards the front of a border. 'Bacchanal' AGM (above) has dark red blooms.

↕45cm (18in) ↔30cm (12in)

PERENNIAL SMALL

Epimedium grandiflorum ♈

Grow this perennial, with its heart-shaped, green leaves, tinged bronze when young, in a shady border, where the nodding white, yellow, pink, or purple flowers will lend spring colour. Varieties include the pink-flowered 'Rose Queen' AGM (above).

↕↔30cm (12in)

BULB MEDIUM

Erythronium californicum ♈

FAWN LILY This clump-forming perennial has mottled, dark green leaves and faintly scented, nodding, creamy white spring flowers, with reflexed petals. Plant it at the front of a partly shaded border with dwarf daffodils and *Epimedium*.

↕35cm (14in)

PERENNIAL LARGE

Euphorbia characias subsp. *wulfenii*

MEDITERRANEAN SPURGE This evergreen, upright perennial, ideal for the back of a sunny, spring border, bears linear, grey-green leaves on tall stems and rounded clusters of yellow-green flowers.

↕↔1.2m (4ft)

SHRUB LARGE

Forsythia x *intermedia*

One of the first shrubs to bloom, it produces a bold display of bright yellow flowers on bare stems from late winter to early spring, before the leaves emerge. Plant it at the back of a border or as a screen, and prune after flowering.

↕↔3m (10ft)

BULB MEDIUM

Fritillaria camschatcensis

FRITILLARY An upright, perennial bulb, with dark green leaves and bell-shaped, dark purple flowers opening up along upright, slim stems in spring. It is ideal for a partly shaded border. Plant the bulbs in groups in autumn.

↕60cm (24in)

BULB LARGE

Fritillaria imperialis ♈

CROWN IMPERIAL Grow this tall, perennial bulb in a border, not too close to seating, as they emit an unpleasant, foxy scent. The bright orange-red, bell-shaped blooms, topped with a tuft of leaves, appear in late spring.

↕1.5m (5ft)

PERENNIAL MEDIUM

Geranium maculatum

Forming a low carpet of divided, green leaves, this perennial bears simple, pale pink to mauve flowers on short, branching stems in spring and early summer. It is ideal for cool, shady corners. Deadhead after flowering to prevent self-seeding.

↕75cm (30in) ↔45cm (18in)

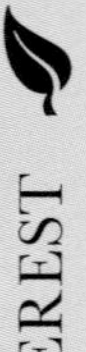

BULB MEDIUM

Muscari latifolium

GRAPE HYACINTH This dwarf bulb has strap-shaped, grey-green leaves and oval-shaped clusters of tiny, two-tone spring flowers, dark blue at the base and pale blue on top. Plant the bulbs in autumn in gritty soil to decorate the front of a spring border.

↕25cm (10in)

BULB MEDIUM

Narcissus cyclamineus

CYCLAMEN-FLOWERED DAFFODIL This tiny, perennial bulb, best planted at the front of a border, has narrow, green leaves and dainty, pendent, yellow flowers, with long, slim trumpets and swept-back petals. The flowers appear in early spring.

↕20cm (8in)

PERENNIAL SMALL

Helleborus* x *sternii

This evergreen perennial produces upright, branching heads of cup-shaped, purple-flushed, green spring flowers with prominent stamens, lasting several weeks. Its mottled, green foliage gives year-round interest. Deadhead to prevent self-seeding.

↕↔45cm (18in)

PERENNIAL MEDIUM

Mertensia virginica

VIRGINIA BLUEBELLS A compact perennial, with soft, blue-green leaves and clusters of nodding, funnel-shaped, blue flowers in spring. The plant dies down in summer, so grow it in a border with later-flowering perennials in front.

↕60cm (24in) ↔45cm (18in)

PERENNIAL SMALL

Phlox divaricata* subsp. *laphamii

This spreading, semi-evergreen perennial has small, lance-shaped, green leaves and, in late spring, small, pale blue, purple, white, or pink blooms, depending on the variety. Use it under taller plants.

↕30cm (12in) ↔20cm (8in)

PERENNIAL LARGE

Polygonatum* x *hybridum

SOLOMON'S SEAL An arching perennial, with oval, green leaves and, in late spring, clusters of small, tubular, green-tipped white flowers that hang like beads from the stems. Plant it in dappled shade in a mixed spring border.

↕up to 1.2m (4ft) ↔1m (3ft)

PERENNIAL SMALL

***Primula* Gold-laced Group**

POLYANTHUS Add drama to the front of a border with a group of these evergreen perennials. In spring, gold-centred mahogany flowers, with silver edges, appear on upright stems and complement dwarf daffodils and *Muscari*.

↕25cm (10in) ↔30cm (12in)

PERENNIAL SMALL

Primula japonica

JAPANESE PRIMROSE A perennial with basal rosettes of wrinkled, green leaves and, in late spring, clusters of white, pink, or red flowers held on upright stems. 'Postford White' AGM (above) has white flowers. Plant it in groups in moist soil.

↕↔45cm (18in)

PERENNIAL SMALL

Primula vulgaris ♡

PRIMROSE This perennial forms basal rosettes of oval, pale green leaves and primrose-yellow flowers, often scented, in early spring. Plant it in groups with spring bulbs at the front of a border in dappled shade.
↕20cm (8in) ↔35cm (14in)

PERENNIAL SMALL

Pulmonaria saccharata

BETHLEHEM SAGE Edge the front of a spring bed or border with this clump-forming, semi-evergreen perennial. Its white-spotted blue-green, bristly leaves and clusters of funnel-shaped, blue-purple flowers complement spring bulbs.
↕30cm (12in) ↔60cm (24in)

SHRUB MEDIUM

Ribes sanguineum

FLOWERING CURRANT Plant this deciduous shrub at the back of a border where its pendent clusters of crimson flowers add a splash of colour in spring, after which the lobed, green leaves lend a green backdrop in summer.
↕2m (6ft) ↔2.5m (8ft)

SHRUB MEDIUM

Spiraea x *vanhouttei*

BRIDAL WREATH This deciduous shrub forms a mound of small, green leaves and, from mid- to late spring, domed clusters of white flowers on arching stems. Good for cutting, plant it at the back of a border or as an informal screen.
↕2m (6ft) ↔1.5m (5ft)

PERENNIAL SMALL

Trillium grandiflorum ♡

AMERICAN WAKE-ROBIN Ideal for the front of a border in shade, this low-growing perennial has diamond-shaped, green leaves and elegant, three-petalled, white spring flowers. It looks best in groups beneath large shrubs or trees.
↕38cm (15in) ↔30cm (12in)

BULB MEDIUM

Tulipa 'Prinses Irene' ♡

TULIP Blooming in mid-spring, this award-winning tulip has bowl-shaped, orange flowers, with a purple flame on the outer petals, and grey-green leaves. Its sturdy stems hold up well in windy sites. Plant the bulbs in groups in autumn.
↕35cm (14in)

SHRUB MEDIUM

Viburnum carlesii

ARROW WOOD Plant this deciduous shrub at the back of a border, where its dark green leaves and clusters of highly fragrant, pink-budded white flowers will create a backdrop for spring bulbs. 'Aurora' AGM (above) has red-budded pink blooms.
↕2m (6ft) ↔2m (6ft)

OTHER SUGGESTIONS

Annuals

Myosotis sylvatica

Perennials and bulbs

Allium unifolium ♡ • *Anemone blanda* 'White Splendour' ♡ • *Bergenia* 'Beethoven' • *Dicentra* 'Spring Morning' • *Erythronium californicum* 'White Beauty' ♡ • *Fritillaria pontica* ♡ • *Helleborus* 'Bob's Best' • *Narcissus* 'Barnum' ♡ • *Narcissus poeticus* • *Primula* 'Belarina Series' • *Primula* 'Guinevere' ♡ • *Tulipa* 'Little Beauty' ♡ • *Tulipa* 'Orange Princess' ♡

Shrubs, trees, and climbers

Camellia japonica 'Adolphe Audusson' ♡ • *Clematis alpina* 'Pamela Jackman' ♡ • *Ornithogalum nutans* ♡ • *Rhododendron lutea* • *Salix hastata* 'Wehrhahnii' ♡ • *Salix lanata* ♡ • *Viburnum davidii* ♡

Trees for blossom

Even small gardens can accommodate a deciduous tree and the breathtaking display of blossom you will enjoy each spring is ample reward.

Trees also provide a structural counterpoint to low-growing bulbs, creating a canopy of flowers overhead followed by a carpet of petals as they fall. Before making a choice, consider the proportion as well as the decorative value of the tree. For tight spaces, choose upright or compact trees, such as *Prunus* 'Spire', *Amelanchier*, and flowering dogwoods (*Cornus*). Plant your tree at the back of a border in a sunny spot, or in the centre of a lawn as a feature, and underplant with spring bulbs, such as daffodils, *Muscari*, and anemones, leaving a space of at least 60cm (24in) around the trunk.

TREE LARGE

Aesculus x *carnea*

RED HORSE CHESTNUT This rounded, deciduous tree produces large, upright clusters of red flowers in late spring. It has dark green, divided leaves and autumn nuts. Varieties include 'Briotii' (above), with pink flowers.

↕20m (70ft) ↔15m (50ft)

SHRUB LARGE

Amelanchier canadensis

SNOWY MESPILUS This deciduous, upright shrub has oval, green leaves that develop fiery orange-red tones in autumn. From mid- to late spring, it produces starry, white flowers, followed by edible berries that ripen to red during summer.

↕6m (20ft) ↔3m (10ft)

TREE SMALL

Amelanchier laevis

ALLEGHENY SERVICEBERRY This small, deciduous tree or large shrub bears sprays of white flowers in spring, followed by red fruits. Its young bronze leaves turn dark green in summer, then red and orange in autumn.

↕↔8m (25ft)

SHRUB LARGE

Cercis canadensis

A deciduous, spreading shrub, grown for its pale pink, pea-like mid-spring flowers that emerge on bare stems. In autumn, its heart-shaped leaves turn vivid yellow, giving late-season colour. *C. canadensis* var. *alba* (above) is a white-flowered form.

↕↔10m (30ft)

TREE SMALL

Cercis chinensis

CHINESE REDBUD The branches of this large shrub or small tree are covered with pea-like, rosy pink flowers in late spring, before the foliage emerges, followed by bean-like seed pods. Its large, decorative green foliage is heart-shaped.

↕6m (20ft) ↔5m (15ft)

TREE MEDIUM

Cercis siliquastrum

JUDAS TREE This bushy, deciduous tree, perfect for small gardens, is a picture in spring when its bare stems are clothed with pink blooms. Attractive, heart-shaped, green leaves appear in summer, followed by purplish seed pods.

↕↔10m (30ft)

TREE SMALL

Cornus florida

FLOWERING DOGWOOD This conical, deciduous tree or shrub produces a bold display in late spring when covered with tiny, green flowers surrounded by showy white or pink bracts. The slightly twisted leaves turn red and purple in autumn.

↕6m (20ft) ↔8m (25ft)

TREE MEDIUM

Halesia monticola

SILVER BELL This slow-growing tree has an elegant habit and flowers freely in spring, bearing simple, white, bell-shaped blooms along its branches. It develops bright autumn colouring, making it a useful specimen tree for larger gardens.

↕ 12m (40ft) ↔ 8m (25ft)

TREE SMALL

Crataegus laevigata 'Paul's Scarlet' 🏆

A small, deciduous tree, with spiny stems of glossy, lobed, green leaves and, in late spring, a profusion of double, red flowers. It can be grown as a small specimen tree or planted as part of a wildlife hedge.

↕↔ 8m (25ft)

TREE LARGE

Davidia involucrata 🏆

DOVE TREE Provide this deciduous tree with space to spread its large branches, which, in late spring, are covered with tiny flowers, surrounded by white bracts, that resemble handkerchiefs, alongside oval, pointed green foliage.

↕ 15m (50ft) ↔ 10m (30ft)

TREE MEDIUM

Magnolia 'Heaven Scent' 🏆

This tree produces goblet-shaped, rosy pink flowers, with a magenta stripe on each petal, from mid-spring, creating a focal point in a lawn or border. It has oval, green leaves that appear at the same time as the flowers.

↕↔ 10m (30ft)

SHRUB LARGE

Magnolia stellata 🏆

STAR MAGNOLIA This deciduous, rounded shrub produces silky-haired buds on leafless stems that open to reveal fragrant, starry, white flowers. The narrow, green leaves appear just after the blooms. Protect the blooms from spring frosts.

↕↔ 10m (30ft)

TREE SMALL

Magnolia virginiana

SWEET BAY Grown as a large shrub or small tree, it has large, vanilla-scented, cup-shaped, creamy white flowers from early summer to early autumn, set against large, dark green leaves. Generally deciduous, it retains some leaves in milder areas.

↕ 10m (30ft) ↔ 6m (20ft)

TREE MEDIUM

Malus floribunda 🏆

JAPANESE CRAB APPLE This deciduous tree has showy, pale pink flowers that emerge from crimson buds in spring, after the oval leaves appear. Edible red and yellow fruits follow in autumn. Plant it in a border or lawn.

↕↔ 10m (30ft)

TREE MEDIUM

Malus hupehensis 🏆

HUPEH CRAB This spreading, deciduous tree is perfect for a lawn or border focal point. In spring, it produces masses of fragrant, white flowers, followed by cherry-like, red autumn fruits. Dark green, oval leaves appear before the blossom.

↕↔ 12m (40ft)

Trees for blossom

TREE MEDIUM

Malus x *moerlandsii* 'Profusion'

This deciduous tree bears masses of deep pink spring blooms set against purple-tinted leaves, which age to green and have bold autumn tints. Purple fruits follow the flowers, providing a long season of interest.

↕↔10m (30ft)

TREE SMALL

Mespilus germanica

MEDLAR Grown mainly for its edible fruits, this spreading, deciduous tree also bears attractive, single, white spring flowers, carried at the shoot tips. It is slow-growing and suitable for smaller gardens. Plant it in full sun in a sheltered position.

↕6m (20ft) ↔8m (25ft)

TREE SMALL

Prunus 'Accolade'

FLOWERING CHERRY Grow this compact, deciduous tree in a small garden or a courtyard. It produces semi-double, light pink spring flowers that open from dark pink buds, before the oval leaves appear. The foliage has good autumn colour.

↕↔8m (25ft)

TREE MEDIUM

Prunus 'Kanzan'

FLOWERING CHERRY Grow this vase-shaped, deciduous tree as a specimen in a lawn or border, where its double, pink spring flowers create a focal point. Bronze leaves, which follow the flowers, turn green in summer, then orange before falling.

↕↔10m (30ft)

TREE SMALL

Prunus pendula

WEEPING CHERRY Perfect as a focal point in a small garden, this deciduous tree has weeping branches that are covered with dark rose-pink blossom in early spring. The oval, pointed leaves turn orange and red in autumn.

↕↔6m (20ft)

SHRUB LARGE

Prunus incisa

FUJI CHERRY Use this deciduous, rounded shrub in a border or as hedging. Saucer-shaped, white or pale pink flowers appear in early spring, followed by purple fruits. The leaves add value in autumn when they turn orange-red.

↕↔8m (25ft)

TREE MEDIUM

Prunus jamasakura

HILL CHERRY This deciduous tree bursts into colour in spring, when clusters of pale pink buds open to reveal large, white flowers. The bronze young foliage, which turns red and yellow in autumn, adds to the effect.

↕↔12m (40ft)

TREE SMALL

Prunus 'Shirofugen'

FLOWERING CHERRY This compact, deciduous tree, ideal for a small garden, makes a splash in spring when masses of double, white flowers, fading to pink, appear. The coppery brown young leaves mature to green and turn orange in autumn.

↕8m (25ft) ↔10m (30ft)

TREE SMALL

Prunus 'Shirotae'

FLOWERING CHERRY This small, deciduous tree has a spreading crown. It will decorate a paved urban garden or lawn with its semi-double, fragrant, white spring flowers that appear with the green, serrated foliage; leaves also offer good autumn colour.

↕6m (20ft) ↔8m (25ft)

TREE SMALL

Prunus 'Shogetsu' ♡

FLOWERING CHERRY Grown for its double, pink spring flowers, which fade to white, this small deciduous tree creates a focal point in a border or lawn. The bronze young foliage matures to green and turns orange and red in autumn.

↕5m (15ft) ↔8m (25ft)

TREE MEDIUM

Prunus 'Spire' ♡

FLOWERING CHERRY This slim, deciduous tree, perfect for growing in a small urban garden, is covered with pale pink blossom in spring. The leaves, which follow the flowers, are bronze when young, and turn red and orange in autumn.

↕10m (30ft) ↔6m (20ft)

TREE SMALL

Prunus 'Ukon' ♡

FLOWERING CHERRY This spreading, deciduous tree, ideal for a lawn or border, has pink-budded, double, pink-tipped yellowish white flowers, which open in spring. The green foliage, bronze when young, turns reddish brown in autumn.

↕8m (25ft) ↔10m (30ft)

TREE LARGE

Prunus x *yedoensis*

YOSHINO CHERRY The stems of this spreading, deciduous tree are covered with almond-scented, blush white blossom in spring, when it will lend seasonal colour to a small garden. The dark green leaves turn yellow in autumn.

↕15m (50ft) ↔10m (30ft)

TREE MEDIUM

Robinia x *slavinii*

This deciduous tree produces pendent chains of bright pink, pea-like late spring flowers, which lead to knobbly, brown seed pods. Its lacy foliage produces only light shade, making it a good choice for small gardens. 'Hillieri' AGM (above) is popular.

↕↔10m (30ft)

TREE LARGE

Stewartia pseudocamellia ♡

DECIDUOUS CAMELLIA Grow this deciduous tree, prized for its large, cup-shaped, white flowers, as a feature plant in a summer garden. It also has attractive, flaking bark and good autumn leaf colour.

↕20m (70ft) ↔8m (25ft)

TREE MEDIUM

Styrax japonicus ♡

JAPANESE SNOWBELL This spreading, deciduous tree makes an elegant addition to a border or lawn. In early summer, it produces pendent clusters of fragrant, bell-shaped, white or pink-tinged flowers. The leaves provide good autumn colour.

↕10m (30ft) ↔8m (25ft)

OTHER SUGGESTIONS

Shrubs and trees

Cornus kousa • *Cornus* 'Norman Hadden' ♡ • *Crataegus flava* • *Magnolia denudata* 'Yellow River' • *Malus baccata* 'Street Parade' • *Malus* 'Jonagold' • *Malus* 'Snowcloud' • *Paulownia tomentosa* ♡ • *Prunus avium* 'Plena' ♡ • *Prunus padus* 'Colorata' ♡ • *Prunus padus* 'Watereri' ♡ • *Prunus* 'Pink Shell' • *Prunus* x *subhirtella* 'Pendula Rubra' • *Pyrus calleryana* 'Chanticleer'

Spring flowers for containers

Plan ahead and plant up spring containers in autumn. Many plants also provide a winter foliage display before the bulbs and bedding burst into life.

Plant bulbs from early- to late autumn; put tulips in at the end of the season to reduce the risk of tulip fire disease. Snowdrops, an exception, are best potted up in leaf when in flower in spring. Small bulbs, such as crocuses, grape hyacinths, and irises, are ideal for tiny pots, while tall daffodils and tulips look beautiful in large pots along with spring-flowering bedding, such as violas, wallflowers, and primroses. Plant these over the bulbs to provide interest through winter and a dramatic display in spring. Use large, containerized shrubs, such as camellias and rhododendrons, for their winter leaves, as well as spring blooms.

PERENNIAL SMALL

Aurinia saxatilis

GOLD DUST Masses of long-lasting, chrome-yellow flowers make this low-growing, evergreen perennial a good choice for spring pots. The blooms sit above oval, grey-green leaves. Varieties include 'Variegata', with cream-edged foliage.
↕23cm (9in) ↔30cm (12in)

PERENNIAL SMALL

Bellis perennis

DAISY This mound-forming perennial, grown as an annual, has mid-green leaves and, in spring, single, double, or pompon, pink, white, or red flowers. Plant in containers with multi-purpose compost, with spring bulbs.
↕↔20cm (8in)

PERENNIAL SMALL

Bergenia 'Silberlicht'

ELEPHANT'S EARS Grow this evergreen perennial with spring bulbs in containers of soil-based compost. It has large, oval, dark green leaves, which create a foil for the clusters of white flowers that appear on its red-tinged stems.
↕45cm (18in) ↔50cm (20in)

SHRUB LARGE

Camellia japonica

COMMON CAMELLIA Grow this upright, evergreen shrub in large containers of ericaceous compost in a sheltered site where cold winds and late frosts will not damage the large red, pink, or white early spring flowers.
↕3m (10ft) ↔2m (6ft)

SHRUB LARGE

Camellia x *williamsii*

An evergreen shrub with oval, pointed, glossy, green leaves and single or double, pink, white, or red blooms in spring. Varieties include 'Donation' AGM (above), with pink, semi-double flowers. Plant in ericaceous compost. Water well in summer.
↕5m (15ft) ↔2.5m (8ft)

SHRUB SMALL

Cassiope 'Edinburgh'

Plant this dwarf, evergreen shrub in a pot of gritty compost alongside short bulbs, such as species tulips. It has tiny, scale-like, dark green leaves and nodding, white, bell-shaped flowers, which appear from spring to summer.
↕↔20cm (8in)

CLIMBER MEDIUM

Clematis 'Frances Rivis'

Add height to a patio display with this deciduous, spring-flowering clematis. Plant it in a large pot of soil-based compost, with a trellis or pyramid to support the twining stems of divided foliage and bell-shaped, violet-blue flowers.
↕3m (10ft)

BULB SMALL

Crocus 'Snow Bunting'

Plant this perennial corm in small pots in autumn for an early spring display of fragrant, white flowers, with yellow centres and a purple blush on the outer petal surfaces. The grass-like leaves are green, with white lines.
↕7cm (3in)

BULB SMALL

Cyclamen coum ♀

EASTERN CYCLAMEN This perennial bulb forms a low mound of silvered leaves and, in early spring, produces dainty downward-facing, pink blooms with swept-back petals. Grow it in pots of soil-based compost. Many varieties are available.
↕↔ 10cm (4in)

SHRUB SMALL

Erica x *darleyensis*

DARLEY DALE HEATH This evergreen shrub has needle-like foliage and flowers in late winter and early spring, becoming speckled with small, nodding, urn-shaped, white or pink flowers. Ideal for early colour, plant it in containers in ericaceous compost.
↕ 25cm (10in) ↔ 50cm (20in)

ANNUAL/BIENNIAL SMALL

Erysimum cheiri

WALLFLOWER This biennial is planted in autumn, its green leaves providing some interest through winter in pots of multi-purpose compost. In mid-spring, it produces spikes of bright yellow, orange, red, or pink, scented flowers.
↕↔ 30cm (12in)

BULB MEDIUM

Galanthus nivalis ♀

SNOWDROP Either plant these perennial bulbs in small pots of soil-based compost or with hellebores and heucheras in mixed displays, where their nodding, white, late winter to early spring flowers and grass-like foliage add a delicate touch.
↕ 15cm (6in)

PERENNIAL SMALL

Heuchera villosa 'Palace Purple'

This evergreen perennial has large, lobed, glossy, purple foliage, offering year-round interest. In summer, small white flowers appear on wiry stems. Plant it in pots of soil-based compost.
↕↔ 45cm (18in)

BULB MEDIUM

Hyacinthus orientalis ♀

HYACINTH Introduce fragrance to a container collection with this early spring-flowering bulb. Its spikes of fragrant, bell-shaped, primrose-yellow flowers go well with evergreen perennials in pots of soil-based compost; plant in autumn.
↕ 25cm (10in)

BULB MEDIUM

Iris reticulata

DWARF IRIS Plant this perennial bulb with dwarf irises of other colours in autumn in small pots of gritty compost. The small reddish purple flowers, with yellow and white marks on the lower petals appear in early spring.
↕ 15cm (6in)

SHRUB LARGE

Itea virginica

Best for permanent planting in a large pot, this evergreen shrub bears arching stems of spiny, dark green leaves and flowers from midsummer to early autumn, with trailing, greenish white catkins. Grow in a sheltered spot, with ericaceous compost.
↕ 3m (10ft) ↔ 1.5m (5ft)

Spring flowers for containers

BULB MEDIUM

Muscari armeniacum

GRAPE HYACINTH Plant this spring bulb with daffodils and early tulips in autumn in pots of soil-based compost. The small, fragrant, deep blue flowers, held in cone-shaped clusters, are accompanied by grassy green leaves.

↕20cm (8in)

BULB MEDIUM

Narcissus 'Actaea'

DAFFODIL This late-spring-flowering perennial bulb, with its scented, white flowers and shallow, red-rimmed yellow cups, combines well with red tulips in large pots or half barrels. Plant the bulbs in autumn.

↕40cm (16in)

BULB MEDIUM

Narcissus 'Bridal Crown'

DAFFODIL This perennial bulb bears sweetly scented, fully double, creamy white blooms, with yellow-speckled centres, in spring. Plant in autumn with spring bedding in containers filled with soil-based compost.

↕40cm (16in)

BULB MEDIUM

Narcissus 'Canaliculatus'

DAFFODIL This bulb has slender, green leaves and slim stems bearing clusters of small, fragrant spring flowers, with reflexed white petals and yellow cups. Plant bulbs in autumn in pots of soil-based compost mixed with grit.

↕23cm (9in)

BULB MEDIUM

Narcissus 'Cheerfulness'

DAFFODIL A perennial bulb that mixes well with single-flowered, yellow daffodils. The fragrant, double, white flowers with yellow centres appear in mid-spring. Plant the bulbs in autumn in pots of soil-based compost.

↕40cm (16in)

BULB MEDIUM

Narcissus 'Jack Snipe'

DAFFODIL Try combining this early- to mid-spring-flowering perennial bulb with *Muscari* or polyanthus. It produces creamy white flowers with short, bright yellow cups and narrow, dark green leaves. Plant in autumn in soil-based compost.

↕23cm (9in)

BULB MEDIUM

Narcissus 'Tahiti'

DAFFODIL This double-flowered perennial bulb will brighten up mid-spring containers with its golden-yellow blooms and narrow, green leaves. Plant the bulbs during autumn in containers filled with soil-based compost.

↕45cm (18in)

PERENNIAL MEDIUM

Polygonatum odoratum

ANGLED SOLOMON'S SEAL This perennial has oval to lance-shaped, mid-green leaves and, in late spring to early summer, fragrant hanging, bell-shaped, green-tipped white flowers. Plant it in pots in autumn for spring colour, protect plants from sawflies.

↕60cm (24in) ↔30cm (12in)

PERENNIAL SMALL

Primula Crescendo Series

POLYANTHUS These evergreen perennials, grown as annuals, are ideal for spring pots and windowboxes. Their corrugated, dark green leaves and clusters of yellow-eyed, multicoloured flowers create a longlasting blaze of colour.

↕↔20cm (8in)

PERENNIAL SMALL

Primula Gold-laced Group

POLYANTHUS Add a touch of class to patio containers with these evergreen perennials. In spring, upright stems topped with gold-centred mahogany flowers with bright edging complement dwarf daffodils in pots of soil-based or multi-purpose compost.
↕25cm (10in) ↔30cm (12in)

PERENNIAL SMALL

Primula veris

COWSLIP This perennial forms a rosette of lance-shaped leaves and bears clusters of nodding, yellow, almond-scented blooms in spring. Deadhead to prolong the display. Plant in pots in autumn or spring, then plant out after flowering.
↕↔25cm (10in)

BULB MEDIUM

Scilla siberica

SIBERIAN SQUILL This perennial bulb flowers in spring, producing upright stems of nodding, bell-shaped, blue flowers. Plant groups of bulbs in small pots in autumn, then plant out into larger, mixed containers in spring.
↕20cm (8in)

BULB MEDIUM

Tulipa clusiana var. *chrysantha*

This tulip blooms from early spring and makes a dainty feature in pots or troughs of gritty, soil-based compost. It bears small, bowl-shaped, yellow blooms, tinged red on the outside. Plant the bulbs in autumn.
↕30cm (12in)

BULB MEDIUM

Tulipa 'Purissima'

TULIP This perennial bears bowl-shaped, creamy white flowers in mid-spring, set against purple-marked, grey-green leaves. Plant bulbs in late autumn in pots of gritty compost. The tulips can then be transferred to mixed container displays in spring.
↕40cm (16in)

PERENNIAL SMALL

Tiarella wherryi

FOAM FLOWER This leafy, shade-tolerant perennial, with deeply lobed, maroon-tinted green leaves, makes a great foil for spring bulbs, such as daffodils. In late spring or early summer, frothy white or pink flowerheads appear.
↕20cm (8in) ↔15cm (6in)

BULB MEDIUM

Tulipa 'Spring Green'

The late spring, cup-shaped blooms of this Viridiflora tulip are creamy white, with green streaks, and make elegant partners for hot-hued tulips and late-flowering daffodils. Plant bulbs in autumn in pots of gritty, soil-based compost.
↕38cm (15in)

PERENNIAL SMALL

Viola x *wittrockiana*

PANSY This valuable bedding plant produces oval leaves and late winter to spring flowers in almost every shade imaginable, providing a colourful frill around pots of spring bulbs. Plant it in soil-based or multi-purpose compost.
↕↔20cm (8in)

OTHER SUGGESTIONS

Biennials and perennials

Bellis perennis 'Pomponette' • *Erysimum* 'Apricot Twist' • *Primula* Charisma Series

Bulbs

Crocus tommasinianus • *Crocus sieberi* • *Hyacinthus* 'Dark Dimension' • *Iris* 'Pixie' • *Muscari* 'Album' • *Muscari neglectum* • *Narcissus* 'Dove Wings' • *Narcissus* 'Mount Hood' • *Scilla siberica* 'Alba' • *Tulipa* 'Angélique' • *Tulipa* 'Candela' • *Tulipa* 'Esperanto'

Shrubs and trees

Camellia 'Brushfield's Yellow' • *Camellia* 'Water Lily' • *Erica* x *darleyensis* 'Furzey' • *Rhododendron* 'Dopey' • *Rhododendron* 'Elizabeth' • *Rhododendron* 'Golden Torch' • *Rhododendron* 'Percy Wiseman'

Plants for summer borders

With a vast choice of flowers for summer borders, making a selection for your garden will be easier if you first decide on the look you want to create.

Borders filled with an assortment of roses and hardy perennials will produce a traditional cottage style, while bold groups of stiff-stemmed herbaceous plants convey a more contemporary design. Include perennials in groups of three or more to prevent a spotty, discordant appearance and choose a range of plants that flower consecutively to avoid a lull late in the season. *Agapanthus*, *Leucanthemum*, *Penstemon*, and *Perovskia* will all help to link summer with autumn to produce a continuous display, while fuchsias, hydrangeas, and repeat-flowering roses provide great value, blooming for many months.

PERENNIAL MEDIUM

Achillea 'Fanal'

YARROW Plant this perennial, with its flat heads of yellow-centred bright red flowers and green, ferny foliage, in the middle of a border for a splash of summer colour. It contrasts well with blue flower spikes.
↕75cm (30in) ↔60cm (24in)

PERENNIAL LARGE

Anemone x *hybrida*

JAPANESE ANEMONE Ideal for the back of a shady border, this upright perennial produces white or pink flowers on tall stems from late summer to autumn above divided leaves. 'Königin Charlotte' AGM (above) has semi-double, pale pink blooms.
↕1.2m (4ft) ↔ indefinite

PERENNIAL MEDIUM

Anthemis tinctoria

DYER'S CHAMOMILE Try this clump-forming perennial in a mixed border with blue or pink perennials that contrast with its daisy-like, white or yellow summer flowers, and feathery, aromatic leaves. 'E.C. Buxton' (above) has lemon-yellow flowers.
↕↔1m (3ft)

PERENNIAL SMALL

Calamintha nepeta

LESSER CALAMINT This perennial makes good edging for a wildlife summer border. Its spikes of tiny, lilac-pink late summer to early autumn flowers attract bees, and the small green leaves provide background colour.
↕45cm (18in) ↔75cm (30in)

PERENNIAL LARGE

Campanula lactiflora

MILKY BELLFLOWER This upright, branching perennial has tall, back-of-border varieties and shorter forms, such as the violet-blue 'Prichard's Variety' AGM (above) for the front. Nodding, bell-shaped, blue, pink, or white flowers appear in summer.
↕up to 1.2m (4ft) ↔60cm (24in)

PERENNIAL SMALL

Campanula punctata

BELLFLOWER A compact perennial that is perfect for the front of a moist border. Its light green leaves and pendent, tubular, bell-shaped, cream or dusky pink flowers make good companions for geraniums and phlox.
↕30cm (12in) ↔40cm (16in)

SHRUB MEDIUM

Ceanothus americanus

This is a dense, rounded, deciduous shrub that bears clusters of small, white, scented flowers from late spring to midsummer. It is best planted at the back of a border with plants in front that provide colour in summer.
↕↔2m (6ft)

SHRUB MEDIUM

Cistus x *dansereaui*

ROCK ROSE Provide a sunny, sheltered site for this evergreen shrub, with grey-green leaves and a succession of papery, bowl-shaped, white, pink, or purple flowers in summer. 'Decumbens' AGM (above) has white blooms, with purple central blotches.
↕60cm (24in) ↔1m (3ft)

CLIMBER MEDIUM

Clematis 'Abundance'

Use this deciduous clematis, with its twining stems of mid-green leaves and creamy yellow-centred wine-red flowers from midsummer to late autumn, to cover a large shrub at the back of a border.
↕3m (10ft)

CLIMBER LARGE

Clematis 'Fireworks'

Flowering from late spring to early summer, and again in late summer, this large-flowered, deciduous clematis, with its magenta-striped, blue-mauve blooms, provides a colourful back-of-border accent when threaded through trellis or a shrub.
↕4m (12ft)

CLIMBER LARGE

Clematis 'Huldine'

This deciduous climber, from mid- to late summer, bears abundant small, white flowers, mauve beneath. Climb it through large, spring-flowering shrubs to prolong their colour, or train it up freestanding obelisks. Keep the roots shaded.
↕4m (12ft)

PERENNIAL SMALL

Coreopsis verticillata

TICKSEED This bushy perennial is ideal for the front or middle of a border, and fills any gaps with its dense, ferny foliage and slender stems topped with daisy-like, yellow summer flowers. Varieties include the lemon-yellow 'Moonbeam' (above).
↕50cm (20in) ↔45cm (18in)

PERENNIAL MEDIUM

Crocosmia 'Lucifer'

MONTBRETIA Ideal for late-summer colour, this clump-forming perennial has upright, sword-shaped leaves. Vibrant red, tubular blooms appear on branched stems in late summer. Provide additional support; deadhead after flowering.
↕1m (3ft) ↔25cm (10in)

PERENNIAL LARGE

Delphinium elatum

An upright perennial, ideal for a mid-border spot, with deeply cut, green leaves and, in midsummer, tall spikes of bowl-shaped blooms in a wide range of colours. There are many tall and dwarf varieties to choose from; most forms require staking.
↕2m (6ft) ↔90cm (36in)

PERENNIAL LARGE

Echinops ritro

GLOBE THISTLE This upright perennial has prickly, divided, dark green leaves and globe-shaped, spiky, metallic blue flowerheads in late summer. Use it to plug the gaps left by tall alliums when they begin to fade earlier in the season.
↕1.2m (4ft) ↔75cm (30in)

SHRUB MEDIUM

Escallonia 'Apple Blossom'

A compact evergreen shrub that offers a green, leafy backdrop to perennials and bulbs for most of the year, and injects colour into the planting scheme from early- to midsummer, when it is covered with vase-shaped, pink flowers.
↕↔2.5m (8ft)

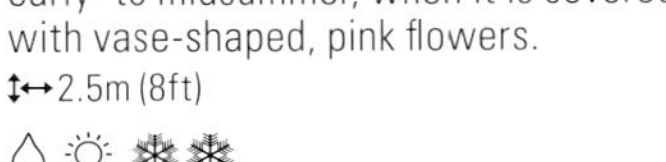

SHRUB SMALL

Fuchsia 'Lady Thumb'

This hardy fuchsia is a compact, deciduous shrub, with dark green leaves. It is ideal for the front of a summer border, where its semi-double, bell-shaped, carmine-pink and white flowers will provide interest for many months.
↕30cm (12in) ↔45cm (18in)

PERENNIAL MEDIUM

Gaillardia x *grandiflora*

BLANKET FLOWER Add this perennial to the front or middle of a border for a splash of colour. It produces slim, green leaves and rounded, red summer blooms, with a yellow ring at the outer edges. Varieties include 'Kobold' (above)
↕90cm (36in) ↔45cm (18in)

PERENNIAL LARGE

Gaura lindheimeri

WAND FLOWER An upright perennial that bears masses of simple, star-shaped, white flowers on wiry stems from midsummer to autumn. It has an open habit and can be planted mid-border with the plants behind remaining visible. Deadhead regularly.
↕1.5m (5ft) ↔90cm (36in)

PERENNIAL MEDIUM

Geranium pratense

MEADOW CRANESBILL Useful for sun or part-shade, this clump-forming perennial has divided leaves and saucer-shaped, violet-blue summer flowers. Ideal to use as a front-of-border plant. Varieties include the pinky grey 'Mrs Kendall Clark' AGM (above).

↕↔ 60cm (24in)

PERENNIAL MEDIUM

Geranium sylvaticum

WOOD CRANESBILL Ideal for the front of a shady border alongside pastel or hot-hued partners, this perennial produces clumps of lobed, mid-green leaves and blue-purple flowers from late spring to early summer. 'Album' AGM (above) has white blooms.

↕ 75cm (30in) ↔ 60cm (24in)

SHRUB SMALL

Hebe 'Great Orme'

This evergreen shrub, from midsummer to mid-autumn, bears slender spikes of pink blooms that fade with age. Useful at the back of a border, its deep purple shoots and glossy, dark green foliage provides a foil for neighbouring plants. Trim in early spring.

↕↔ 1.2m (4ft)

PERENNIAL MEDIUM

Hemerocallis 'Chicago Sunrise'

A clump-forming perennial, with slender, arching leaves and large, golden-yellow, trumpet-shaped summer flowers. Each flower lasts a day, but they are produced over several weeks. Water well in summer.

↕ 70cm (28in) ↔ 85cm (34in)

SHRUB LARGE

Hibiscus syriacus

ROSE MALLOW Plant this deciduous shrub for colour in late summer and early autumn, when it bears large, single or double, saucer-shaped, blue, violet, or white flowers. Grow in a sheltered site, at the back of a border. 'Oiseau Blue' (above) has blue flowers.

↕ 3m (10ft) ↔ 2m (6ft)

SHRUB LARGE

Hydrangea aspera Villosa Group

Ideal for providing shelter, or for the back of a border, this large deciduous shrub has large, velvety leaves. In late summer, it bears flattened heads of tiny, mauve florets, surrounded by larger white blooms.

↕↔ 3m (10ft)

SHRUB SMALL

Hydrangea macrophylla 'Mariesii Lilacina'

LACECAP HYDRANGEA This deciduous, compact shrub has broad, oval leaves and flattened clusters of tiny, mauve-pink or blue flowers surrounded by pink blooms. Use it in the middle of a mixed border.

↕↔ 1.2m (4ft)

SHRUB MEDIUM

Hydrangea quercifolia

OAK-LEAVED HYDRANGEA This mound-forming, deciduous shrub adds bulk to a shady border. From midsummer to autumn, its cone-shaped, cream flowerheads, which age to pink-tinged white, shine out from the gloom. Plants require neutral to acid soil.

↕ 2m (6ft) ↔ 2.5m (8ft)

PERENNIAL MEDIUM

Iris domestica

This iris is grown for its star-shaped, orange and yellow flowers, with speckled centres from mid- to late summer on upright stems. It is a rhizome- and clump-forming perennial, with sword-like, blue-green foliage, and is ideal for summer borders.

↕ 1m (3ft) ↔ 60cm (24in)

SHRUB SMALL

Kalmia angustifolia

SHEEP LAUREL This mound-forming, evergreen shrub provides year-round interest in a shady border, but peaks in early summer when it is covered with small, bowl-shaped, pink flowers. *K. angustifolia* f. *rubra* has dark red blooms.

↕ 60cm (24in) ↔ 1.5m (5ft)

PERENNIAL LARGE

Kniphofia uvaria

RED-HOT POKER With its strap-shaped, evergreen leaves, this perennial offers year-round interest, but it peaks in late summer, when tall spikes of cone-shaped, red and yellow flowerheads create a bold feature among either hot or cool colours.
↕1.2m (4ft) ↔60cm (24in)

PERENNIAL SMALL

Lychnis flos-jovis 🏆

FLOWER OF JOVE A dainty perennial, ideal for the front of a border. Its mat of grey, hairy leaves on erect stems makes a good edging, while the clusters of small, round flowers in red, pink, or white lend colour throughout summer.
↕↔45cm (18in)

PERENNIAL MEDIUM

Knautia macedonica

MACEDONIAN SCABIOUS The dainty, crimson summer flowers of this upright perennial add buttons of colour between larger blooms from the front to middle of a sunny border. The green leaves are lobed; stems may need support.
↕75cm (30in) ↔60cm (24in)

PERENNIAL MEDIUM

Leucanthemum x *superbum*

SHASTA DAISY This perennial's tall stems of golden-eyed white daisies create impact in a summer border. Plant it in swathes for the best effect, and use the blooms for cutting. Varieties include 'Beauté Nivelloise' (above).
↕up to 90cm (36in) ↔60cm (24in)

PERENNIAL MEDIUM

Monarda 'Cambridge Scarlet'

BERGAMOT Grow this upright, bushy perennial in soil that does not dry out during summer. The aromatic, green leaves and tufted red, pink, or white flowerheads provide colour and texture in summer.
↕1m (3ft) ↔45cm (18in)

PERENNIAL MEDIUM

Kniphofia 'Alcazar'

This upright, evergreen perennial gives a dramatic display in late summer when its spikes of bright orange, tubular flowers emerge from, and tower above, the strap-like leaves. Best in sun, use it to provide a vertical accent to mixed herbaceous borders.
↕1m (3ft) ↔45cm (18in)

PERENNIAL MEDIUM

Libertia grandiflora 🏆

NEW ZEALAND SATIN FLOWER Grown mainly for its fountains of grassy leaves, this evergreen perennial also produces slim stems of dainty, white, daisy-like flowers in summer. Try planting it at the front of a border.
↕75cm (30in) ↔60cm (24in)

PERENNIAL MEDIUM

Paeonia lactiflora

PEONY This perennial has divided, green leaves and large, bowl-shaped, single or double, white, pink, red, or mauve blooms in early summer. Try it with early-flowering geraniums and ceanothus. Varieties include 'Laura Dessert' AGM (above).
↕↔75cm (30in)

PERENNIAL LARGE

Lupinus 'The Chatelaine'

LUPIN This perennial produces cone-shaped spikes of pink and white, bicoloured flowers in shades of white, blue, pink, mauve, or yellow, and is best grown in summer borders with roses, delphiniums, and geraniums. The seeds are toxic.
↕1.2m (4ft) ↔45cm (18in)

PERENNIAL MEDIUM

Papaver orientale

ORIENTAL POPPY Grow this vibrant, early-summer-flowering perennial for its large, bright red flowers, with black marks and dark eyes, and decorative seedheads. Combine it with late-summer-flowering plants to plug the gaps left after it fades.
↕↔90cm (36in)

PERENNIAL MEDIUM

Penstemon 'Osprey' ♈

This perennial's spikes of tubular, pink-edged white summer to early autumn flowers and simple green leaves blend perfectly with *Perovskia* and pastel-coloured dahlias in a sunny, sheltered border. Deadheading prolongs the display.

↕1.1m (3½ft) ↔60cm (24in)

SHRUB SMALL

Perovskia 'Blue Spire' ♈

RUSSIAN SAGE An upright, deciduous shrub, with small, blue-purple flowers, held above silver-grey foliage from late summer to autumn. Plant it in a sunny, mixed border to provide late season colour. The spent flower stems will persist into winter.

↕1.2m (4ft) ↔1m (3ft)

PERENNIAL SMALL

Persicaria affinis 'Superba' ♈

This vigorous perennial forms a dense mat of oval, green leaves, which turn rich brown in autumn. From early- to late summer, it bears upright spikes of soft pink blooms that darken with age. Plant it at the front of a border and deadhead regularly.

↕25cm (10in) ↔60cm (24in)

PERENNIAL LARGE

Phlox paniculata

Producing large clusters of white, lilac, or pink flowers from summer to autumn, this perennial forms dense blocks of colour mid-border. Many colourful varieties are available, such as 'Blue Paradise' (above). Keep plants well watered in summer.

↕1.2m (4ft) ↔1m (3ft)

SHRUB SMALL

Phygelius aequalis

Best in a sheltered site, this shrubby perennial produces tall, upright stems that carry clusters of pendent, tubular, dusky pink or yellow blooms all summer. Plant it mid-border. 'Yellow Trumpet' AGM (above) has creamy yellow flowers.

↕↔1m (3ft)

SHRUB SMALL

Potentilla fruticosa

SHRUBBY CINQUEFOIL Grow this bushy, deciduous shrub, with circular, orange, red, yellow, white, or pink blooms, which appear from summer to early autumn, mid-border between perennials. 'Abbotswood' AGM (above) has white flowers.

↕1m (3ft) ↔1.5m (5ft)

PERENNIAL SMALL

Potentilla megalantha

CINQUEFOIL This compact, bushy perennial makes a neat edging in borders filled with hot-hued plants. Its divided, mid-green leaves and saucer-shaped, orange-centred bright yellow summer flowers provide a long season of interest.

↕20cm (8in) ↔15cm (6in)

PERENNIAL LARGE

Romneya coulteri ♈

TREE POPPY A back-of-border perennial, grown for its large, fragrant, papery, golden-centred white flowers, which appear in late summer on sturdy stems. Combine it with repeat-flowering shrub roses and mulleins in a sheltered spot.

↕↔2m (6ft)

SHRUB MEDIUM

Rosa 'Ballerina' ♈

This repeat-flowering shrub rose produces clusters of small, single, white-centred pale pink flowers from midsummer to early autumn, and creates a focal point in a border among dark pink, white, and purple flowered perennials.

↕1.5m (5ft) ↔1.2m (4ft)

SHRUB SMALL

Rosa Flower Carpet Series

Ideal for adding colour to large areas, this ground-cover rose forms a dense mound of disease-resistant foliage and bears cupped, semi-double, pink, red, yellow, or white flowers all summer. Deadhead regularly and mulch with compost in spring.

↕↔60cm (24in)

SHRUB MEDIUM

Rosa GERTRUDE JEKYLL ♈

This medium-sized shrub or short climber is perfect for the back of a mixed border, with its scrolled buds and large, scented, fully double, pink flowers providing colourful highlights throughout summer and early autumn. Mulch annually in spring.

↕2m (6ft) ↔1.2m (4ft)

SHRUB MEDIUM

Rosa 'Madame Hardy'

This erect, strong-growing damask rose, which makes a medium-sized, prickly shrub with mid-green foliage, offers a single flush of fragrant, fully double, white flowers that add an elegant note to mixed borders in early summer.

↕1.5m (5ft) ↔1.2m (4ft)

SHRUB MEDIUM

Rosa 'William Lobb'

Grow this moss rose on a trellis or pillar at the back of a border. Its arching, prickly stems, which hold cupped, fully double, scented, crimson-purple flowers and green leaves, create a colourful backdrop for pastel blooms in summer.

↕↔2m (6ft)

PERENNIAL MEDIUM

Rudbeckia fulgida var. *sullivantii*

CONEFLOWER This compact perennial, with its masses of daisy-like, golden flowers, injects bright colour into the front of a border from late summer to autumn. Leave faded flowerheads to stand over winter.

↕75cm (30in) ↔30cm (12in) or more

PERENNIAL MEDIUM

Salvia x *sylvestris*

WOOD SAGE Grow this clump-forming perennial, with its lance-shaped, wrinkled, green leaves and spikes of blue summer flowers, to provide a vertical accent at the front of a border. 'Blauhugel' AGM (above) has violet-blue flowers.

↕80cm (32in) ↔30cm (12in)

PERENNIAL SMALL

Scabiosa 'Butterfly Blue'

PINCUSHION FLOWER Grow this small, summer-flowering perennial, with its dark green leaves and wiry stems topped with small, rounded, lavender-blue flowers, between daisies and other flower spikes that bloom at the same time.

↕↔40cm (16in)

PERENNIAL MEDIUM

Sisyrinchium striatum

A good choice for the front of a dry, sunny border, this evergreen perennial's sword-shaped, grey-green leaves and spikes of small, pale yellow summer flowers combine well with bearded irises, which enjoy similar conditions.

↕60cm (24in) ↔30cm (12in)

PERENNIAL MEDIUM

Thermopsis villosa

CAROLINA LUPIN The spikes of small, pale yellow summer flowers, which appear above this perennial's divided, grey-green foliage, create a vertical accent behind geraniums and *Achillea* in a sunny border.

↕1m (3ft) or more ↔60cm (24in)

PERENNIAL MEDIUM

Verbascum chaixii

MULLEIN Use this perennial as a vertical accent towards the back of a sunny border. Spikes of pale yellow or white flowers appear on tall stems in summer above coarse leaves. Varieties include 'Album' (above), with purple-eyed white blooms.

↕90cm (36in) ↔45cm (18in)

PERENNIAL SMALL

Veronica spicata

SPEEDWELL Try combining this upright, clump-forming perennial's spikes of dark pink summer flowers and lance-shaped, green leaves with flat-headed *Achillea*, poppies, and daisy-like *Anthemis* towards the front of a sunny border.

↕↔50cm (20in)

OTHER SUGGESTIONS

Perennials and bulbs

Achillea 'Credo' • *Achillea* 'Summerwine' • *Agapanthus* 'Northern Star' • *Anemone hybrida* 'Serenade' • *Campanula punctata* 'Cherry Bells' • *Delphinium* 'Blue Nile' • *Delphinium* 'Sungleam' • *Echinops* 'Taplow Blue' • *Gaillardia* 'Oranges and Lemons' • *Geranium sylvaticum* 'Mayflower' • *Gladioli* 'Carine' • *Kniphofia* 'Percy's Pride' • *Paeonia mlokosewitschii* • *Penstemon* 'Raven' • *Scabiosa caucasica* 'Clive Greaves'

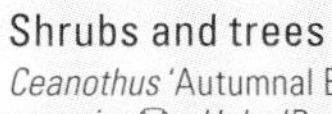

Shrubs and trees

Ceanothus 'Autumnal Blue' • *Cistus* x *cyprius* • *Hebe* 'Bowles's Variety' • *Hemerocallis* 'Red Precious' • *Hibiscus* 'Red Heart' • *Hydrangea aspera* 'Mauvette' • *Olearia macrodonta* • *Rosa* 'Southampton' • *Rosa* 'The Times Rose'

Summer flowers for containers

While relaxing on patios in summer, you observe containers more, so ensure that plants are in peak condition and provide season-long interest.

Annual bedding, such as pelargoniums, petunias, and nemesias, is invaluable in pots, providing continuous colour from early- to late summer. Maintain plants by keeping them well watered and removing faded blooms regularly, which stimulates plants to produce more flowers. As part of your summer display, include a few leafy specimens, such as hebes and hostas, to balance the flowery forms. If you include slow-release fertilizer granules when planting, you will not need to feed your plants again until late in the season, when they can be topped up with a liquid tomato fertilizer.

SHRUB LARGE

Abutilon 'Nabob'

FLOWERING MAPLE Grow this tall, evergreen shrub on a warm patio in a large pot of soil-based compost, with canes to support the stems. The maple-like foliage is joined by bowl-shaped, crimson flowers in summer. Protect it from frost.
↕↔ 3m (10ft)

ANNUAL/BIENNIAL SMALL

Ageratum houstonianum

FLOSS FLOWER This compact annual has abundant fluffy, pink, white, and blue flowers in summer, set against coarse, dark green leaves. Grow in pots, hanging baskets, and windowboxes in multi-purpose compost. Deadhead regularly, and water well.
↕↔ 30cm (12in)

PERENNIAL SMALL

Anthemis punctata subsp. *cupaniana*

SICILIAN CHAMOMILE This evergreen, spreading perennial has silvery leaves and masses of daisy-like summer blooms. Plant in soil-based compost with bedding in mixed planters. Deadhead regularly.
↕↔ 30cm (12in)

PERENNIAL SMALL

Antirrhinum majus Luminaire Series

This trailing plant weeps over the sides of containers and baskets, and produces short spikes of lightly scented, snapdragon flowers in many colours. Place pots out of strong winds; deadhead regularly.
↕↔ 50cm (20in)

PERENNIAL SMALL

Bidens ferulifolia

This annual's trailing stems of ferny green foliage and star-shaped, yellow flowers, are perfect for edging tall pots or baskets from summer to autumn. Grow it in multi-purpose compost with pelargoniums or dwarf zinnias.
↕ 30cm (12in) ↔ indefinite

ANNUAL/BIENNIAL SMALL

Brachyscome iberidifolia

SWAN RIVER DAISY Adding a decorative edge to pots and baskets, this bushy annual has feathery, green leaves and daisy-like, blue, pink, purple, or white flowers from summer to early autumn. Plant it in multi-purpose compost.
↕ 25cm (10in) ↔ 45cm (18in)

SHRUB LARGE

Brugmansia x *candida*

ANGELS' TRUMPETS Usually grown as a house plant, this shrub has huge, trumpet-shaped, scented, white, yellow, or pink blooms. It can spend summers on a warm patio in a pot of soil-based compost. 'Grand Marnier' AGM (above) has apricot blooms.
↕ 5m (15ft) ↔ 2.5m (8ft)

PERENNIAL SMALL

Calibrachoa Million Bells Series

This perennial, grown as an annual, produces trumpet-shaped, pink, yellow, red, white, or blue flowers from summer to early autumn. Use it for decorating containers. 'Cherry Pink' (above) is popular.
↕ 30cm (12in) ↔ 1m (3ft)

CLIMBER SMALL

Clematis SHIMMER

This compact, deciduous climber makes a beautiful patio feature and has large, deep lilac blooms, with a paler central bar on each petal, that appear throughout summer. Plant it in a large pot of soil-based compost, with a tripod support.
↕ 1.8m (6ft)

PERENNIAL SMALL

Convolvulus sabatius

BLUE ROCK BINDWEED This perennial, sold as summer bedding, produces trailing, leafy stems studded with funnel-shaped, purplish blue flowers that soften the edges of pots and windowboxes. Plant it in multi-purpose or soil-based compost.

↕20cm (8in) ↔30cm (12in)

ANNUAL/BIENNIAL LARGE

Cosmos bipinnatus

Large, daisy-like flowers in pink, red, and white, make this annual a favourite for summer patio pots. The blooms are set off by feathery, green foliage. Plant it in pots of multi-purpose compost, and deadhead to prolong flowering.

↕up to 1.2m (4ft) ↔45cm (18in)

ANNUAL/BIENNIAL MEDIUM

Cuphea ignea

CIGAR FLOWER This annual's lance-shaped leaves and small, tubular, black-tipped scarlet flowers make a decorative feature in windowboxes and pots of multi-purpose compost, together with upright plants, such as fuchsias or pelargoniums.

↕75cm (30in) ↔90cm (36in)

BULB MEDIUM

Dahlia 'Gallery Art Deco'

This decorative dahlia, with double flowers comprising burgundy-edged pale orange petals, offers a splash of colour in patio pots from midsummer to autumn. Grow it in multi-purpose compost, and protect the tubers from frost.

↕45cm (18in)

PERENNIAL SMALL

Diascia barberae

A mat-forming perennial, often grown as annual bedding, with wiry stems of apricot-pink blooms from summer to autumn. Use it to trail over the sides of pots and baskets. There are many varieties available. Feed and water well; deadhead regularly.

↕25cm (10in) ↔50cm (20in)

SHRUB SMALL

Fuchsia 'Thalia'

This unusual fuchsia, with its pendent, tubular, red flowers from summer to autumn and dark green leaves, maroon beneath, makes a beautiful centrepiece in a mixed display. Plant it in pots of soil-based compost; protect from frost.

↕↔60cm (24in)

SHRUB SMALL

Fuchsia 'Tom Thumb'

A good choice for windowboxes and baskets, this dwarf, upright shrub produces bell-shaped, pinky red and mauve-purple flowers throughout summer. Combine it with trailing lobelia and petunias, and plant in multi-purpose or soil-based compost.

↕↔50cm (20in)

ANNUAL/BIENNIAL SMALL

Gazania Talent Series

TREASURE FLOWER A colourful addition to patio pots, this dwarf annual produces daisy-like, yellow, orange, pink, or maroon flowers throughout summer. Plant it in containers or windowboxes with other annuals in multi-purpose compost.

↕↔25cm (10in)

PERENNIAL SMALL

Impatiens New Guinea Group

BUSY LIZZIE Add a frill of flowers to containers with this colourful short-lived perennial. The rounded pink, red, and white flowers provide interest on their own or in mixed displays in pots of multi-purpose compost.
↕35cm (14in) ↔30cm (12in)

PERENNIAL SMALL

Isotoma axillaris

STAR FLOWER The feathery, green foliage and star-shaped, lilac or blue flowers make this perennial, grown as an annual, a good choice for summer pots. Grow it singly or with fuchsias and dahlias in large pots of multi-purpose or soil-based compost.
↕↔30cm (12in)

PERENNIAL SMALL

Gerbera 'Mount Rushmore'

FLORIST GERBERA Use this pretty perennial, with lobed, green leaves and tall stems of daisy-like, black-eyed pink, yellow, or orange summer flowers, to decorate patios. Grow it in multi-purpose compost; protect from frost.
↕↔40cm (16in)

ANNUAL/BIENNIAL MEDIUM

Helianthus annuus

SUNFLOWER A summer-flowering annual, renowned for its large, yellow flowers. Dwarf forms, such as 'Teddy Bear' (above), are ideal for pots. Pinch out the stem tips for several small heads, or leave to form a larger, single bloom.
↕up to 90cm (36in) ↔60cm (24in)

SHRUB SMALL

Lavandula 'Willow Vale'

FRENCH LAVENDER A classic shrub to grow in terracotta pots, this compact, evergreen shrub has aromatic, grey-green foliage and deep purple blooms that are topped with wavy, reddish purple flower bracts from early- to midsummer.
↕↔70cm (28in)

ANNUAL/BIENNIAL SMALL

Lobelia erinus

TRAILING LOBELIA This annual has trailing stems of dark green leaves and tiny, blue-purple flowers that appear from summer to autumn if the plant is kept moist at all times. Plant in pots, windowboxes, and baskets in multi-purpose compost.
↕20cm (8in) ↔15cm (6in)

PERENNIAL SMALL

Nemesia 'KLM'

This perennial, commonly grown as an annual, forms a spreading mat of green leaves and wiry stems of small, scented, blue and white flowers all summer. Ideal for windowboxes, containers, and baskets. A number of varieties are available.
↕30cm (12in) ↔15cm (6in)

ANNUAL/BIENNIAL SMALL

Nicotiana Domino Series

TOBACCO PLANT This upright annual, with large, oval leaves and rounded, lightly scented summer flowers in shades of purple, pink, red, and white, combines well with trailing lobelia or swan river daisies in pots of multi-purpose compost.
↕45cm (18in) ↔40cm (16in)

PERENNIAL SMALL

Osteospermum jucundum

AFRICAN DAISY Long-flowering, daisy-like, pale pink blooms, which appear above this evergreen perennial's grey-green foliage from summer to early autumn, are perfect for summer pots of soil-based compost. Move it to a sheltered site in winter.
↕↔30cm (12in)

CLIMBER MEDIUM

Rhodochiton atrosanguineus

PURPLE BELL VINE Add height to patio displays by planting this annual climber in a large pot of multi-purpose compost. It has heart-shaped leaves and dangling summer flowers, with pinky red "hats" and maroon tubes. Stems need a tripod support.
↕3m (10ft)

CLIMBER MEDIUM

Thunbergia alata

BLACK-EYED SUSAN Ideal for large pots, grow this annual climber in a sunny spot up a wigwam of canes, which it will clothe with heart-shaped leaves and, bright orange or yellow dark-eyed flowers. Water and feed well; remove spent blooms.
↕3m (10ft)

PERENNIAL MEDIUM

Rudbeckia hirta

This perennial, usually grown as an annual, bears large, daisy-like, brown-centred golden flowers from summer to autumn. There are many varieties to choose from, such as 'Becky Mixed' (above). Ideal for pots; water well, and deadhead regularly.
↕up to 90cm (36in) ↔45cm (18in)

ANNUAL/BIENNIAL SMALL

Verbena Novalis Series

This annual is a popular container plant, grown for its grey-green leaves and rounded red, pink, orange, purple, or white flowerheads, which appear all summer. Use it to fill gaps between larger flowers in pots of multi-purpose compost.
↕↔25cm (10in)

PERENNIAL SMALL

Pelargonium 'Lord Bute'

This evergreen perennial, usually grown as an annual, has deep purple-red blooms throughout summer, set against rounded, hairy leaves. There are many varieties to grow and all are ideal for windowboxes and pots; plant in multi-purpose compost.
↕45cm (18in) ↔30cm (12in)

ANNUAL/BIENNIAL SMALL

Salvia splendens Cleopatra Series

Bearing short, upright spikes of tubular, red or purple flowers, this annual makes a vibrant addition to a planting scheme. Plant it in pots of multi-purpose compost, water and feed well, and deadhead often.
↕↔30cm (12in)

ANNUAL/BIENNIAL SMALL

Zinnia Thumbelina Series

Perfect for pots, this dwarf annual has dark green leaves and semi-double flowers in shades of red, yellow, maroon, and pink, from summer to early autumn. Grow it in pots, windowboxes, and baskets in multi-purpose compost.
↕15cm (6in) ↔ 21cm (8in)

ANNUAL/BIENNIAL SMALL

Petunia PHANTOM

Dress up summer pots and baskets with this eye-catching bedding plant. It produces a mound of trumpet-shaped, almost black flowers, with a golden central star. It combines well with the biennial *Rudbeckia* in containers of multi-purpose compost.
↕↔30cm (12in)

ANNUAL/BIENNIAL SMALL

Sutera cordata

This spreading perennial, grown as an annual, is used to trail over the sides of containers, baskets, and windowboxes, and forms a curtain of small, white or pale pink flowers and green leaves. It flowers all summer, if watered and fed well.
↕10cm (4in) ↔indefinite

OTHER SUGGESTIONS

Annuals and biennials

Convolvulus tricolor • *Rudbeckia* 'Toto Gold' • *Verbena* 'Aztec Red' • *Verbena* 'Diamond Merci'

Perennials

Begonia Illumination Series • *Bidens* 'Gold Star' • *Diascia* 'Sydney Olympics' • *Eustoma grandiflorum* • *Isotoma* 'Avant Garde' • *Osteospermum* 'Serena' • *Osteospermum* Sunny Series • *Petunia* Tumbelina Series • *Sutera* 'Snowflake'

Shrubs and climbers

Anthemis punctata • *Argyranthemum* 'Yellow Empire' • *Clematis* 'Josephine' • *Fuchsia* 'Firecracker' • *Lavandula* 'Fathead' • *Rosa* 'Regensburg'

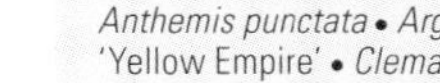

Plants for autumn borders

Often forgotten when planting in early summer, these late performers inject a new lease of life into tired borders, just when they need it most.

Boost colour and interest at this time of year with a range of key plants, including asters and rudbeckias, which are available in a rainbow of colours to suit your style, from hot reds and yellows, to pastels and whites. Flowers are not the only features to consider – ornamental grasses produce decorative seedheads at this time of the year and make perfect partners for many autumn blooms. The leaves of some deciduous shrubs also fire up now, while others are studded with decorative fruits. Combine a variety of plant forms to create a scene of contrasting colours, shapes, and textures.

TREE SMALL

Acer palmatum 'Sango-kaku' 🏆

CORAL-BARK MAPLE Perfect for a lightly shaded border in autumn, this tree puts on a spectacular display of coral-red stems and yellow, lobed leaves, which are pinkish yellow in spring and green in summer.
↕8m (25ft) ↔4m (12ft)

TREE LARGE

Acer rubrum

RED MAPLE Ideal as a focal point at the back of a large border, the rounded, lobed foliage of this sizeable tree is dark green in summer and turns bright yellow or red in autumn. It requires neutral to acid soil.
↕20m (70ft) ↔10m (30ft)

PERENNIAL LARGE

Actaea simplex

This upright perennial forms a clump of attractive divided foliage, above which rise tall, wiry spikes of small, fluffy, off-white flowers from early- to mid-autumn. Plant it at the back of a border or as a backdrop to a pond or water feature.
↕1.5m (5ft) ↔60cm (24in)

PERENNIAL MEDIUM

Amsonia tabernaemontana

EASTERN BLUESTAR An upright perennial, with star-shaped, blue flower clusters from late spring to summer. Its autumn interest comes from its foliage, which develops rich seasonal shades before falling. Grow it mid-border with autumn blooms.
↕1m (3ft) ↔30cm (12in)

PERENNIAL LARGE

Anemone hupehensis

This perennial is grown for its large, simple, white, pink, red, or mauve autumn flowers, which are held on upright stems above dark green, divided leaves. 'Hadspen Abundance' AGM (above) has pale pink blooms. Provide support for tall varieties.
↕up to 1.5m (5ft) ↔60cm (24in)

PERENNIAL LARGE

Aster cordifolius

An upright, bushy perennial, with dark green, oval, toothed-edged leaves. Sprays of small, pale blue, daisy-like flowers appear on arching stems from late summer to autumn. Plant it near the back of a border; provide additional support.
↕1.2m (4ft) ↔1m (3ft)

SHRUB LARGE

Arbutus unedo

STRAWBERRY TREE A large, evergreen shrub, with dark green, leathery, oval leaves. In autumn, clusters of white, urn-shaped flowers appear at the same time as clusters of round, green and amber strawberry-like fruits ripen to red.
↕↔8m (25ft)

PERENNIAL SMALL

Aster amellus

ITALIAN ASTER Use this clump-forming perennial at the front of a sunny border, where its yellow-eyed lilac-blue flowers will gleam in early autumn, against the lance-shaped, green leaves. 'King George' AGM (above) has violet-blue blooms.
↕↔50cm (20in)

PERENNIAL MEDIUM

Aster 'Coombe Fishacre' 🏆

Use this autumn-flowering perennial as a contrast to the bright yellows and reds that abound at this time of the year. Its sprays of small, pale lilac-pink flowers are set against dark foliage, adding interest to the middle of a border.
↕90cm (36in) ↔35cm (14in)

PERENNIAL MEDIUM

Aster x *frikartii*

Use this perennial to cool hot colours or blend with pastels. From late summer to early autumn, it bears daisy-like, orange-centred violet blue flowers on top of slim stems. Varieties include the lavender-blue 'Wunder von Stäfa' AGM (above).

↕70cm (28in) ↔40cm (16in)

PERENNIAL LARGE

Aster novae-angliae

With slim green leaves and daisy-like purple, pink, red, or white flowers from late summer to autumn, this tall perennial is a cottage garden staple. It is more resistant to mildew disease than *A. novi-belgii*. 'Purple Dome' (above) has purple flowers.

↕up to 1.5m (5ft) ↔60cm (24in)

SHRUB SMALL

Caryopteris x *clandonensis*

This deciduous, bushy shrub forms an upright, compact mass of lance-shaped, grey-green leaves. It bears dense clusters of tubular, blue to purplish blue flowers from late summer to autumn. 'Worcester Gold' (above) has contrasting golden leaves.

↕1m (3ft) ↔1.5m (5ft)

SHRUB SMALL

Ceratostigma willmottianum 🏆

BLUE-FLOWERED LEADWORT This deciduous shrub is prized for its bright red autumn foliage and small sky-blue flowers, which appear at the same time. It dies back in winter and re-shoots in spring.

↕1m (3ft) ↔1.5m (5ft)

PERENNIAL MEDIUM

Chelone obliqua

TURTLE HEAD On reliably moist soil, this upright perennial will produce spikes of two-lipped, dark pink or purple flowers from late summer to early autumn, above toothed-edged, green leaves. Try combining it with *Eupatorium*.

↕1m (3ft) ↔50cm (20in)

PERENNIAL SMALL

Chrysanthemum 'Grandchild' 🏆

KOREAN GROUP CHRYSANTHEMUM

This perennial offers colour to the autumn garden, with sprays of single or double, rounded, red, pink, white, or orange blooms, which appear above lobed foliage.

↕45cm (18in) ↔40cm (16in)

PERENNIAL LARGE

Gaura lindheimeri 🏆

WAND FLOWER From summer to autumn, this upright perennial bears starry, white or pink flowers on long, wiry stems that move freely in the wind. Plant it mid-border as a see-through plant. 'Karalee White' (above) is a popular modern variety.

↕1.5m (5ft) ↔90cm (36in)

PERENNIAL MEDIUM

Helenium 'Moerheim Beauty' 🏆

SNEEZEWEED This free-flowering perennial bears copper-red, daisy-like flowers from midsummer to late summer. As the blooms fade, the centres remain, providing interest into autumn. Plant it mid-border.

↕1m (3ft) ↔60cm (24in)

PERENNIAL LARGE

Eupatorium perfoliatum

Forming a dense clump of upright stems and coarse, dark green leaves, this perennial bears branching heads of small, pincushion-shaped, white flowers from midsummer to early autumn. It needs moist soil and is ideal for bog gardens.

↕2m (6ft) ↔1.2m (4ft)

PERENNIAL LARGE

Eupatorium purpureum

JOE PYE WEED An upright perennial, it forms a dense clump of tough stems and large, coarse, dark green leaves. From late summer to early autumn, purple-pink, fluffy blooms appear. Best for larger, naturalistic gardens with moist soil.

↕2.2m (7ft) ↔1m (3ft)

PERENNIAL LARGE

Helianthus 'Lemon Queen'

Add a splash of colour to the back of an autumn border with this upright perennial, which produces dark green leaves and large, daisy-like, pale yellow flowers on branched stems. The tall stems may need staking.

↕1.5m (5ft) ↔60cm (24in) or more

PERENNIAL LARGE

Leucanthemella serotina

AUTUMN OX-EYE Ideal for a naturalistic or wild garden, this tall perennial produces yellow-eyed white daisies on top of leafy stems in autumn. Use it along with asters and *Eupatorium* to brighten up the back of a border.

↕1.5m (5ft) ↔90cm (36in)

PERENNIAL SMALL

Liriope muscari

LILYTURF Use this evergreen perennial as edging for a shady border, where its grass-like, glossy, dark green leaves will be joined in autumn by spikes of tiny, lavender or purple-blue flowers, followed by black berries. It prefers acid soil.

↕30cm (12in) ↔45cm (18in)

PERENNIAL MEDIUM

Panicum virgatum

SWITCH GRASS Use this upright, deciduous grass as a leafy foil for autumn-flowering-plants. It produces blue-green foliage, which turns yellow in autumn, and clouds of pink-tinged-green summer flowerheads, which fade to beige.

↕1m (3ft) ↔75cm (30in)

PERENNIAL MEDIUM

Lysimachia clethroides

GOOSENECK LOOSESTRIFE This spreading perennial has grey-green foliage and, from late summer to autumn, white, long, pointed flowerheads. Can be invasive, but is ideal for large or wildlife gardens. Support the stems; remove unwanted growth in autumn.

↕↔1m (3ft)

PERENNIAL MEDIUM

Pennisetum alopecuroides

DWARF FOUNTAIN GRASS This deciduous grass produces arching, green leaves that turn bronze and spikes of bristly, purple-tinged flowerheads in late summer, followed by autumn seedheads. Try it with dahlias, rudbeckias, and sedums.

↕1m (3ft) ↔45cm (18in)

PERENNIAL MEDIUM

Rudbeckia fulgida

This upright perennial has narrow, mid-green leaves. From late summer to autumn, it is awash with daisy-like, golden flowers with brown centres, which give winter interest when the blooms fade. Plant mid-border in a sunny site.

↕90cm (36in) ↔45cm (18in)

BULB MEDIUM

Nerine bowdenii

BOWDEN CORNISH LILY This spring-planted perennial bulb sleeps in summer, but bursts to life in autumn when spidery, pink flowers on upright stems emerge. Strap-shaped leaves unfurl later. Grow it at the front of a sunny border.

↕60cm (24in)

PERENNIAL LARGE

Rudbeckia laciniata

CONEFLOWER A towering perennial that adds interest to the back of an early autumn border. The daisy-like, bright yellow flowers that grow atop tall stems, combine well with asters. It performs best in soil that does not dry out.

↕2m (6ft) ↔75cm (30in)

PERENNIAL LARGE

Salvia guaranitica

This tall perennial is at its best in autumn when the tall stems are dotted with two-lipped, deep blue flowers above mid-green leaves. Varieties include the rich blue 'Blue Enigma' AGM (above). Apply a thick mulch in winter.

↕1.5m (5ft) ↔60cm (24in)

PERENNIAL LARGE

Salvia involucrata ♀

ROSY-LEAF SAGE This tall perennial has heart-shaped leaves and is best combined with grasses and anemones at the back of a border, where its spikes of small, tubular, pink flowers will bloom until mid-autumn. Protect it from hard frosts.

↕1.5m (5ft) ↔1m (3ft)

PERENNIAL MEDIUM

Salvia nemorosa

BALKAN CLARY A compact, upright perennial that blooms until early autumn, when spikes of violet-blue flowers with pink bracts accompany lance-shaped, mid-green leaves. Try planting it with sedums, grasses, and *Echinacea*.

↕up to 75cm (30in) ↔60cm (24in)

PERENNIAL MEDIUM

Salvia patens ♀

A perennial, sometimes grown as an annual, that bears upright stems of vivid blue, two-lipped flowers, held above hairy, green leaves from midsummer to autumn. Plant it at the front of a sunny border; deadhead regularly.

↕60cm (24in) ↔45cm (18in)

PERENNIAL MEDIUM

Schizostylis coccinea

CRIMSON FLAG LILY The sword-shaped leaves of this perennial are joined by spikes of salmon-pink, starry flowers throughout autumn. Try it with blue asters and sedums in moist but well-drained soil, and provide a thick mulch in winter.

↕60cm (24in) ↔30cm (12in)

PERENNIAL SMALL

Sedum erythrostictum 'Mediovariegatum'

STONECROP Allow this perennial's stems of oval, fleshy, cream and green variegated leaves and tiny, star-shaped, greenish white early autumn flowers to spread across the front of a sunny border.

↕30cm (12in) ↔60cm (24in)

PERENNIAL SMALL

Sedum spectabile ♀

ICE PLANT Ideal for the front of a sunny autumn border, this clump-forming perennial produces flat heads of rose-pink flowers, followed by brown seedheads that persist over winter. The oval, fleshy leaves are grey-green.

↕↔45cm (18in)

PERENNIAL LARGE

Solidago 'Goldkind'

GOLDENROD A reliable perennial for the back of a border, it has flattened clusters of tiny, golden flowers, held on tall, leafy stems in autumn, and lance-shaped, dark green foliage. A good choice for dry soils, try combining it with grasses.

↕1.2m (4ft) ↔1m (3ft)

PERENNIAL LARGE

Thalictrum delavayi ♀

This clump-forming perennial has ferny, green leaves and, from late summer to autumn, large, airy panicles of tiny, lavender blooms. Varieties include 'Hewitt's Double' AGM (above), with dainty double flowers. Grow on moist soil.

↕1.5m (5ft) ↔60cm (24in) or more

PERENNIAL LARGE

Veronicastrum virginicum

CULVER'S ROOT Tall spires of white, blue, or pink, star-shaped flowers, on top of stems bearing lance-shaped, dark green foliage, make this an ideal back-of-border plant for late summer and early autumn interest. 'Album' (above) has white flowers.

↕1.2m (4ft) ↔45cm (18in)

OTHER SUGGESTIONS

Perennials and bulbs

Anemone 'Praecox' • *Aster* 'Coombe Fishacre' ♀ • *Aster* 'Photograph' ♀ • *Chrysanthemum* 'Sea Urchin' ♀ • *Dahlia* 'Arabian Night' • *Echinacea* 'Bradford Strain' • *Echinacea paradoxa* • *Echinacea* 'Sundown' • *Echinacea* 'White Lustre' • *Eupatorium maculatum* 'Riesenschirm' ♀ • *Leucanthemum* 'Freak' • *Rudbeckia* 'Goldquelle' • *Rudbeckia maxima* • *Rudbeckia triloba* ♀ • *Salvia nemorosa* 'Amethyst' ♀ • *Sedum erythrostictum* • *Tulbaghia violacea* ♀

Grasses

Miscanthus sinensis • *Pennisetum setaceum* 'Rubrum'

Shrubs and trees

Acer palmatum 'Osakazuki' ♀ • *Berberis thunbergii* f. *atropurpurea* 'Rose Glow' ♀ • *Hypericum* 'Magical Beauty'

Plants for autumn leaves

While bulbs lift the spirits in spring, vibrant foliage colour provides the perfect denouement to the drama in the garden come autumn.

The stars of the autumn show are maples (*Acer*), which sport sculptural leaves that turn startling shades of crimson and orange before falling. Place them in pride of place in a lawn or border to create a seasonal focal point. House walls and garden fences can also be set ablaze with the scarlet foliage of *Parthenocissus* and vines (*Vitis*), which make a beautiful backdrop to container displays on a patio or seasonal borders. Contrast these bright reds with the warm, buttery tones of birch and hydrangea leaves and burnished bronze hornbeam and beech.

TREE SMALL

Acer henryi

This small, often multi-stemmed, tree has dark green leaves, with three elliptical leaflets, which develop glowing red and orange tints in early autumn. Plant in fertile soil in sun or light, dappled shade in a border or lawn sheltered from strong winds.

↕8m (25ft) ↔10m (30ft)

TREE MEDIUM

Acer japonicum 'Vitifolium' 🏆

A bushy, deciduous shrub or tree for the back of a border, it produces drooping, red flower clusters in spring and large, maple-shaped leaves. In autumn, the leaves turn purple, crimson-red, and orange.

↕↔10m (30ft)

TREE SMALL

Acer palmatum 'Atropurpureum'

JAPANESE MAPLE A dainty tree, with lobed, purple-red tinged summer foliage. Shelter, dappled shade, and moisture-retentive soil encourage the fiery red autumn display and prevent leaf scorch.

↕↔8m (25ft)

TREE SMALL

Acer palmatum 'Nicholsonii'

JAPANESE MAPLE This is an elegant, arching tree, with long, finely pointed, deeply lobed leaves. The foliage is purple-red in spring and turns green in summer. In autumn, it makes a dazzling show of yellow, orange, and red.

↕↔8m (25ft)

TREE SMALL

Acer palmatum 'Osakazuki' 🏆

This rounded tree makes a fine specimen towards the back of a sheltered border. It has deeply lobed, green leaves and drooping, red flower clusters and winged seeds. Given adequate moisture, the orange and scarlet-red autumn foliage is striking.

↕↔6m (20ft)

TREE SMALL

Acer palmatum 'Shinobuga-oka'

A small, sparsely branched, upright tree, with cascading leaves that have long, very narrow, bright green leaflets. In autumn, the leaf colour transforms to a translucent yellow and gold.

↕↔8m (25ft)

TREE LARGE

Acer rubrum 'October Glory' ♡

This form of the red maple makes a fine specimen tree, with glossy, dark green, lobed leaves. In spring, bare branches carry red flower clusters. Given neutral to acid soil, the autumn leaves glow bright crimson.
↕20m (70ft) ↔10m (30ft)

TREE MEDIUM

Betula alleghaniensis

YELLOW BIRCH This medium-sized, upright tree is ideal for a woodland glade, with smooth, light bronze bark, peeling in horizontal slivers. The toothed-edged, oval, pointed leaves turn butter-yellow in autumn.
↕12m (40ft) or more ↔3m (10ft)

TREE LARGE

Betula lenta

CHERRY BIRCH This large tree has glossy, browny grey bark with prominent horizontal markings, often developing black vertical fissures as the tree matures. The broad, oval leaves develop beautiful golden-yellow tints in autumn.
↕15m (50ft) ↔12m (40ft)

TREE LARGE

Carya ovata

SHAG-BARK HICKORY This walnut relative forms a large tree, with leaves divided into pairs, and bears edible nuts in autumn. As it matures, the bark peels away, hanging in rough strips. In autumn, the foliage turns yellow and golden-brown.
↕25m (80ft) ↔15m (50ft)

TREE LARGE

Cercidiphyllum japonicum ♡

KATSURA This large tree, ideal for a sheltered woodland setting, is grown for its paired, heart-shaped leaves, bronze-tinted when young. In autumn, the leaves develop beautiful orange, pink, and yellow tints, and a toffee fragrance.
↕20m (70ft) ↔15m (50ft)

SHRUB LARGE

Cornus alba 'Sibirica' ♡

DOGWOOD This deciduous shrub displays bright red bare stems in winter and bears flattened heads of white flowers in early summer, followed by bluish white fruits. In autumn, the leaves flush with rich shades of yellow, orange, and red before falling.
↕↔3m (10ft)

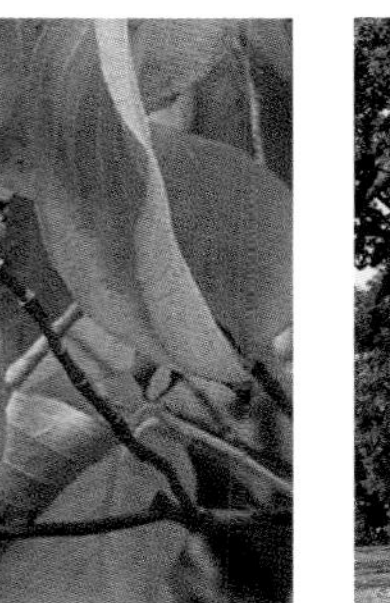

TREE SMALL

Cornus florida

FLOWERING DOGWOOD This deciduous shrub or tree has slightly curled, green leaves, and is covered with white or pink, flower-like bracts in late spring. In autumn, its leaves develop bold orange and red shades before falling.
↕6m (20ft) ↔8m (25ft)

TREE SMALL

Cornus kousa var. *chinensis*

Although this large shrub or small tree is grown for its tiny, late spring flowers surrounded by cream-coloured, petal-like bracts, its autumn display is equally fine. The oval leaves, which hang in tiers, colour rich red and orange.
↕7m (22ft) ↔5m (15ft)

SHRUB LARGE

Cotinus 'Grace'

SMOKE BUSH The broadly oval leaves of this shrub gradually turn from purple to scarlet-red and flame-orange in autumn. Whether purple- or green-leaved, all smoke bush varieties provide an eye-catching autumn display.
↕6m (20ft) ↔5m (15ft)

SHRUB LARGE

Disanthus cercidifolius 🏆
This bushy shrub, which thrives in moist soil and shade has large, green, rounded to heart-shaped leaves that colour crimson, purple, and orange in autumn, and are borne alongside small, scattered, starry, maroon blooms.
↕↔3m (10ft)

SHRUB SMALL

Fothergilla gardenii
Suitable for the middle of a border, this dwarf shrub produces fragrant, fluffy, spring flowers followed by deep blue-green, oval leaves. In autumn, the foliage lights up with fiery red, orange, and yellow tints.
↕↔1m (3ft)

SHRUB MEDIUM

Hydrangea quercifolia
OAK-LEAVED HYDRANGEA The lobed leaves of this shrub make a fine backdrop for the cream, cone-shaped clusters of midsummer to autumn blooms. In autumn, after developing maroon tints, the leaves take on striking red and purple hues.
↕2m (6ft) ↔2.5m (8ft)

SHRUB MEDIUM

Fothergilla major 🏆
The large, rounded leaves of this medium-sized, spring-flowering shrub with a dense, bushy habit, light up with a brilliant scarlet display in autumn, making a striking contrast when placed amongst dark evergreen plants.
↕2.5m (8ft) ↔2m (6ft)

PERENNIAL SMALL

Geranium x *magnificum* 🏆
Although this ground-cover perennial is mainly grown for its violet-blue, early summer blooms, in autumn the deeply cut leaves take on pale yellow, pink, or, sometimes, deep crimson tones as they age.
↕45cm (18in) ↔60cm (24in)

TREE LARGE

Liquidambar styraciflua
SWEET GUM Only suitable for large gardens, this deciduous tree is magnificent in autumn when its lobed, dark green leaves turn orange, red, and purple. Select one of the compact cultivars, such as 'Gum Ball', for a smaller garden.
↕25m (80ft) ↔12m (40ft)

TREE MEDIUM

Nyssa sinensis
CHINESE TUPELO This spreading tree or large shrub is grown for its spectacular autumn colour, when the narrow, oval leaves turn amber and scarlet. It performs best in moist soil in a sheltered site.
↕↔10m (30ft)

TREE SMALL

Parrotia persica
PERSIAN IRONWOOD This widespreading specimen tree or large shrub, sometimes multi-stemmed, has peeling bark and is transformed in autumn when the broad, oval leaves turn red, purple, and amber, highlighting the veins.
↕8m (25ft) ↔10m (30ft)

CLIMBER LARGE

Parthenocissus tricuspidata

BOSTON IVY A vigorous, deciduous, self-clinging climber, this large plant has glossy, lobed, green leaves that display spectacular crimson and red colours in autumn. Grow on a large house wall or fence behind an autumn border.
↕20m (70ft)

TREE LARGE

Quercus alba

AMERICAN WHITE OAK This large tree, has a broad, spreading crown at maturity, and produces deeply lobed, oak-shaped autumn leaves that turn a spectacular wine red, with some brown and yellow tints, the brown persisting into winter.
↕↔30m (100ft)

TREE LARGE

Quercus palustris 🏆

PIN OAK This tall, pyramid-shaped tree has deeply lobed, oak-shaped leaves that are green in summer, but in autumn turn vibrant reddish brown or crimson red, making a magnificent focal point in a large garden.
↕20m (70ft) ↔12m (40ft)

TREE LARGE

Prunus sargentii

Like many flowering cherries, this large tree has smouldering red and amber autumn leaf colour. 'Rancho' (above) is columnar, with foliage that opens red-brown and large, single, pink flowers that appear in spring.
↕20m (70ft) ↔15m (50ft)

TREE LARGE

Quercus rubra

RED OAK This large tree with a spreading to rounded crown, tolerates urban pollution and, in autumn, provides a treat when its oak-shaped leaves with pointed lobes turn a rich reddish brown or crimson.
↕25m (80ft) ↔20m (70ft)

SHRUB LARGE

Rhus typhina 🏆

STAGHORN SUMAC A small, deciduous tree or suckering shrub, with large, dissected, bright green leaves, which turn yellow, red, and orange in autumn and are a perfect foil for the cone-shaped, maroon fruit clusters. 'Dissecta' AGM (above) has finely cut leaves.
↕↔3m (10ft)

CLIMBER LARGE

Vitis coignetiae 🏆

CRIMSON GLORY VINE The impressively large, heart-shaped leaves of this tendril climbing vine are green in summer and, given a warm sheltered aspect, develop spectacular purple, red, orange, and yellow shades through autumn.
↕15m (50ft)

TREE LARGE

Zelkova serrata 🏆

This large tree with a spreading to vase-shaped crown has alternately arranged, long, pointed leaves with prominent veins and serrated margins. The autumn foliage is rich yellow-orange to red-brown. Best in large gardens.
↕30m (100ft) ↔25m (80ft)

OTHER SUGGESTIONS

Shrubs and climbers

Cotinus 'Flame' 🏆 • *Euonymus alatus* • *Euonymus oxyphyllus* 🏆 • *Parthenocissus henryana* 🏆 • *Vaccinium corymbosum* 'Pioneer' • *Vaccinium parvifolium* • *Vitis* 'Brant' 🏆

Trees

Acer laxiflorum • *Acer palmatum* 'Beni Maiko' 🏆 • *Acer palmatum* 'Orange Dream' 🏆 • *Acer pensylvanicum* • *Acer platanoides* 'Palmatifidum' • *Acer saccharum* subsp. *nigrum* • *Carpinus betulus* 🏆 • *Fagus sylvatica* 🏆 • *Picrasma quassioides* • *Prunus avium* • *Rhus glabra* • *Rhus trichocarpa* • *Sorbus sargentiana* 🏆

Plants for ornamental fruits

While flower choices are more limited in autumn, there are fruits aplenty, adding colour and texture to designs, as well as food for wildlife.

Decorate your garden with ornamental and edible fruits, such as apples and pears, for both taste and colour. The classic red fruits of crab apples (*Malus*), roses, and cotoneasters introduce bright beads of colour, and combine well with berries in other shades. Try the violet berries of a *Callicarpa*, the baby pink fruits of *Sorbus vilmorinii*, and the purple and pink, flower-like berries of *Clerodendron*. Although many fruits are borne on large trees and shrubs, skimmias are ideal for a small garden, and climbers, such as *Ampelopsis*, will scale boundary, fences, and house walls while taking up very little ground space.

PERENNIAL MEDIUM

Actaea pachypoda

WHITE BANEBERRY This clump-forming perennial has divided, bright green leaves. In midsummer, it produces upright spikes of fluffy, white flowers, followed by oval, black-tipped white berries on stiff, bright red stalks in autumn.
↕1m (3ft) ↔50cm (20in)

SHRUB LARGE

Arbutus unedo

STRAWBERRY TREE This large, evergreen shrub has dark green, leathery leaves and urn-shaped, white flowers, which appear at the same time as the strawberry-like fruits, which turn red in autumn. Use it for seasonal interest at the back of a border.
↕↔8m (25ft)

CLIMBER MEDIUM

Billardiera longiflora

CLIMBING BLUEBERRY This evergreen, twining climber has narrow, green leaves and small, bell-shaped, greenish yellow flowers. In autumn, unusual purple-blue fruits with a metallic sheen develop. Plant on warm sheltered walls.
↕2m (6ft)

SHRUB LARGE

Aronia x *prunifolia*

PURPLE CHOKEBERRY An upright shrub, with dark green, oval leaves, this North American native has white spring flowers, which are followed by purple-black berries that make a stunning contrast to the glowing red autumn foliage.
↕3m (10ft) ↔2.5m (8ft)

SHRUB LARGE

Callicarpa bodinieri var. *giraldii*

BEAUTY BERRY A bushy, deciduous shrub, with oval, purple-tinted leaves and tiny, lilac flowers, followed, in autumn, by dense clusters of small, spherical, violet berries, set at intervals along the stems.
↕3m (10ft) ↔2.5m (8ft)

SHRUB SMALL

Callicarpa dichotoma

This small, rounded shrub has arching stems, which, in summer, bear clusters of lilac flowers in the leaf axils. By early autumn, these develop into clusters of violet berries, which persist as the oval, green leaves turn yellow.
↕↔1.2m (4ft)

SHRUB LARGE

Clerodendrum trichotomum var. *fargesii* 🏆

This small tree or large, upright shrub has broad, heart-shaped leaves and, from late summer to mid-autumn, fragrant, white flowers that lead to kingfisher-blue berries, surmounted by maroon bracts.
↕↔ 6m (20ft)

TREE MEDIUM

Cornus capitata

This evergreen tree or large shrub favours a warm, sheltered site. In early summer, the branches are smothered with flowers that have showy creamy petal-like bracts. Edible, strawberry-like fruits, ripening in autumn, follow the blooms.
↕↔ 12m (40ft)

TREE SMALL

Cornus kousa

This early summer-flowering, slow-growing tree or large shrub has glossy, oval leaves and showy petal-like, cream-coloured bracts that develop pink tints. Hanging clusters of knobbly, pinkish red, spherical fruits follow in autumn.
↕ 7m (22ft) ↔ 5m (15ft)

SHRUB SMALL

Cotoneaster horizontalis

Best trained against a wall at the back of a narrow border, this spreading, deciduous shrub produces stems in a "herringbone" pattern. These are dotted with red berries in autumn and early winter, before they fall or are eaten by birds.
↕ 1m (3ft) ↔ 1.5m (5ft)

TREE SMALL

Crataegus pedicellata

syn. ***Crataegus coccinea*** This spreading, deciduous tree has broad leaves, with jagged, toothed edges. In spring, white flowers appear, followed in autumn by gleaming scarlet haws that are attractive to wildlife.
↕↔ 6m (20ft)

SHRUB LARGE

Decaisnea fargesii

BLUE SAUSAGE A deciduous, upright shrub, with large, divided leaves and hanging clusters of clematis-like, yellow-green early summer flowers that lead to autumn-ripening, pendulous, metallic-blue bean pods. Protect from severe frosts.
↕↔ 6m (20ft)

SHRUB LARGE

Euonymus hamiltonianus

CHINESE SPINDLE TREE A large, deciduous shrub, with lance-shaped leaves that have fiery autumn colours. In autumn, it produces unusual ornamental fruits, which are flesh-pink and split to reveal glossy, orange-red seeds.
↕↔ 8m (25ft)

SHRUB SMALL

Gaultheria mucronata

This evergreen shrub with dark, prickly leaves and tiny, white spring blooms, is chiefly grown for its spherical berries in shades of white, pink, or crimson. Females fruit provided a male is planted close by.

↕↔ 1.2m (4ft)

SHRUB LARGE

Ilex 'Sparkleberry'

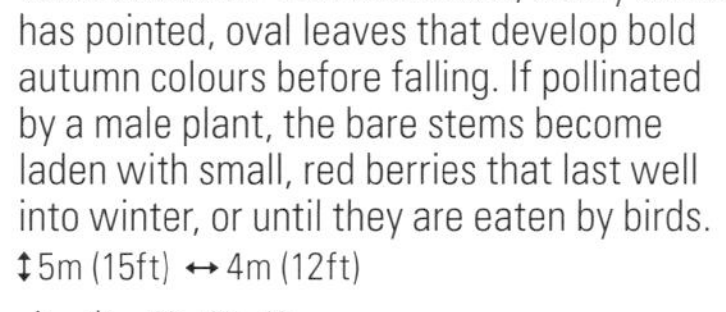

WINTERBERRY This deciduous, bushy shrub has pointed, oval leaves that develop bold autumn colours before falling. If pollinated by a male plant, the bare stems become laden with small, red berries that last well into winter, or until they are eaten by birds.

↕ 5m (15ft) ↔ 4m (12ft)

PERENNIAL MEDIUM

Iris foetidissima

STINKING IRIS This evergreen, shade- and drought-tolerant perennial has sword-shaped leaves and yellow-tinged-purple blooms from early- to midsummer. In autumn, its long seed pods split to reveal orange-red berries.

↕ 90cm (36in) ↔ indefinite

TREE MEDIUM

Malus floribunda

JAPANESE CRAB APPLE Plant this deciduous tree in a border or lawn. Its showy pale pink spring flowers emerge from crimson buds, after the oval leaves appear. Edible red and yellow fruits follow in autumn.

↕↔ 10m (30ft)

SHRUB LARGE

Pyracantha 'Mohave'

FIRETHORN This evergreen shrub has prickly stems and clusters of white blooms in summer, followed by orange-red berries. Other varieties are also available, with fruits that ripen in shades of orange, red, and yellow. All are ideal for wall-training.

↕↔ 4m (12ft)

TREE MEDIUM

Malus x *robusta* 'Red Sentinel'

This deciduous tree is smothered in single white flowers in spring and, in autumn, bunches of glossy, deep red, cherry-like fruits form. The fruits last beyond leaf fall, decorating the bare winter branches.

↕ 12m (40ft) ↔ 10m (30ft)

PERENNIAL MEDIUM

Physalis alkekengi

A spreading perennial with upright stems and small, cream flowers in midsummer, followed by highly ornamental, orange "Chinese lanterns", concealing berry-like fruits. Cut early for indoor decoration.

↕ 75cm (30in) ↔ 90cm (36in)

SHRUB SMALL

Rosa 'Fru Dagmar Hastrup'

This shrub rose has healthy crinkled foliage and produces single, fragrant, soft pink blooms in flushes from summer to early autumn, and large, deep red, tomato-like hips alongside later flowers. The hips are attractive to birds.

↕ 1m (3ft) ↔ 1.2m (4ft)

SHRUB LARGE

Rosa moyesii

This large, arching shrub rose with single, orange-red summer flowers, each with a central tuft of golden stamens, is noted for its hanging clusters of glossy, red, flask-shaped fruits, particularly fine in the variety 'Geranium' AGM.

↕4m (12ft) ↔3m (10ft)

SHRUB MEDIUM

Skimmia japonica

This neat, evergreen shrub with dark green, glossy leaves has dense heads of tiny, fragrant, pink- or red-budded, white flowers in spring. Female plants develop red berries, which persist into winter, if a male plant is nearby.

↕↔1.5m (5ft)

SHRUB SMALL

Skimmia japonica 'Fructo Albo'

This dwarf, evergreen shrub has neat, dark green foliage and crops of white berries that develop from white spring flower clusters. To help fruiting, plant with a male such as the red-budded 'Rubella' AGM.

↕60cm (24in) ↔1m (3ft)

SHRUB LARGE

Viburnum prunifolium

BLACKHAW VIBURNUM A deciduous shrub, ideal for hedging, with glossy, oval leaves that turn red and purple-bronze in autumn. From late spring to early summer, it forms lacy, white flowers, attractive to butterflies. Birds love the blue-black fruits that follow.

↕5m (15ft) ↔4m (12ft)

SHRUB LARGE

Zanthoxylum simulans

A large shrub with glossy, divided, aromatic leaves, long thorns and in late spring, small, greenish yellow flowers. Reddish brown fruits, splitting to reveal black seeds, form in autumn if both sexes are planted together.

↕6m (20ft) ↔5m (15ft)

SHRUB MEDIUM

Symphoricarpos x *doorenbosii*

SNOWBERRY Use this vigorous, deciduous shrub as an informal hedge. The small, round, dark green leaves and tiny, greenish white summer flowers create a useful border backdrop, and showy round, white, fruit clusters provide winter interest.

↕2m (6ft) ↔ indefinite

OTHER SUGGESTIONS

Shrubs and climbers

Actinidia deliciosa • *Ampelopsis brevipedunculata* • *Aronia arbutifolia* • *Berberidopsis corralina* • *Berberis* x *carminea* 'Pirate King' • *Billardiera longiflora* 🏆 • *Cotoneaster* 'Cornubia' 🏆 • *Cotoneaster lacteus* 🏆 • *Cotoneaster salicifolius* • *Euonymus myrianthus* • *Euonymus oxyphyllus* 🏆 • *Euonymus* 'Red Cascade' 🏆 • *Pyracantha* 'Golden Charmer' • *Rosa rugosa* • *Skimmia japonica* 'Wakehurst White' • *Viburnum opulus* 'Compactum' 🏆 • *Viburnum plicatum*

Trees

Crataegus monogyna • *Ilex aquifolium* 'Amber' 🏆 • *Malus* 'John Downie' • *Sambucus nigra* • *Sorbus aucuparia* • *Sorbus cashmiriana* 🏆 • *Sorbus commixta* • *Sorbus forrestii* • *Sorbus vilmorinii* 🏆

Autumn and winter flowers for containers

As summer bedding begins to fade, rejuvenate tired container displays with a range of autumn- and winter-flowering plants.

The choice of autumn and winter flowers is not huge, but the colour range equals that of summer schemes, with hot-hued chrysanthemums, primulas, and dahlias, or cool-coloured Chinese asters (*Callistephus*), dainty cyclamen, and nerines. When buying bedding plants in early autumn, check the labels for hardiness if you want your plants to soldier on after the first frost. Also remember that winter-flowering pansies do not actually bloom that well during very cold weather – small-headed violas are more reliable, and the best displays can be achieved by sheltering pots on windowsills or close to house walls.

PERENNIAL SMALL

Aster novi-belgii 'Professor Anton Kippenberg'

This perennial, with its pastel-coloured, daisy-like flowers, is perfect for autumn pots filled with soil-based compost, or in moist soil at the front of a border. Water well to prevent powdery mildew disease.
↕35cm (14in) ↔45cm (18in)

PERENNIAL SMALL

Bellis perennis

DAISY This dwarf perennial, grown as an annual, produces double or single, pink, white, or red flowers from late winter in sheltered sites, above green foliage. Plant it in groups in multi-purpose compost with evergreen shrubs and violas.
↕↔20cm (8in)

ANNUAL/BIENNIAL MEDIUM

Callistephus chinensis

CHINA ASTER This bushy, annual bedding plant has oval, lobed, green leaves and autumn flowers in shades of pink, purple, white, and yellow that add colour to pots of soil-based compost. Deadhead to prolong the flowering display.
↕60cm (24in) ↔45cm (18in)

PERENNIAL SMALL

Ceratostigma plumbaginoides

BLUE-FLOWERED LEADWORT Plant this bushy perennial in a windowbox or pot of soil-based compost. Small, blue late season flowers appear annually. Its oval leaves turn from green to red in autumn.
↕45cm (18in) ↔20cm (8in)

PERENNIAL SMALL

Chrysanthemum 'Grandchild'

KOREAN GROUP CHRYSANTHEMUM
Plant this perennial in soil-based compost in a pot on a sheltered patio. Sprays of double, rounded, pink blooms appear above lobed, green foliage.
↕45cm (18in) ↔40cm (16in)

BULB MEDIUM

Colchicum byzantinum

AUTUMN CROCUS Use this dainty perennial corm in small pots or windowboxes for an autumn display of open funnel-shaped, soft lilac flowers. The leaves appear later. Plant the corms in summer in soil-based compost.
↕20cm (8in)

BULB MEDIUM

Colchicum 'Waterlily'

AUTUMN CROCUS This perennial, grown from corms, blooms in autumn, when fully double, pinkish lilac flowers develop, followed by strap-shaped leaves soon after. Plant the corms in summer in small pots of soil-based compost.
↕15cm (6in)

PERENNIAL MEDIUM

Cosmos atrosanguineus

CHOCOLATE COSMOS This perennial's bowl-shaped, chocolate-scented, dark maroon flowers are held on tall, slim stems, from late summer to early autumn, above divided, green leaves. Plant it in large pots of gritty, soil-based compost.
↕60cm (24in) or more ↔45cm (18in)

BULB SMALL

Cyclamen coum

EASTERN CYCLAMEN Enjoy the late winter colour afforded by this dwarf perennial bulb's silver-marbled leaves and deep pink flowers, with twisted, swept-back petals. Plant it in small pots or windowboxes in soil-based compost mixed with grit.
↕10cm (4in)

BULB SMALL

Cyclamen hederifolium

IVY-LEAVED CYCLAMEN Add autumn colour to small containers filled with gritty, soil-based compost with this dwarf perennial bulb. Its pink blooms bear swept-back petals and the patterned silvery-green leaves appear after the flowers.
↕10cm (4in)

BULB MEDIUM

Dahlia 'Yellow Hammer' ♡

This dwarf bedding dahlia blooms up to the first frosts in autumn, with single, bright yellow flowers above dark bronze foliage. Grow it in pots of gritty, soil-based compost. Lift tubers and store inside after plants are blackened by frost.
↕60cm (24in) ↔45cm (18in)

BULB MEDIUM

Galanthus nivalis ♡

SNOWDROP In late winter, this perennial bulb produces small, nodding, white flowers, with green markings on some varieties, among clumps of grey-green, grassy leaves. Plant it in flower in pots of soil-based compost.
↕15cm (6in)

PERENNIAL SMALL

Helleborus orientalis

LENTEN ROSE This evergreen perennial is among the first to flower each year, and bears saucer-shaped flowers in shades of white or greenish cream, ageing to dark pink, in midwinter on sturdy stems. Plant in pots of soil-based compost.
↕60cm (24in) ↔45cm (18in)

SHRUB SMALL

Juniperus communis 'Compressa' ♡

Forming a slim cone of blue-grey foliage, this dwarf, slow-growing, evergreen conifer makes a good partner for small shrubs, flowers, and bulbs in autumn and winter in containers of soil-based compost.
↕80cm (32in) ↔45cm (18in)

BULB LARGE

Nerine bowdenii ♡

BOWDEN CORNISH LILY This perennial bulb, with its spidery, pink flowers, followed by strap-shaped leaves, decorates autumn borders and brightens up late-season containers. Plant the bulbs in spring in soil-based compost mixed with grit.
↕60cm (24in) ↔15cm (6in)

PERENNIAL MEDIUM

Platycodon grandiflorus ♡

BALLOON FLOWER The purple-blue flowers of this perennial open from balloon-like buds throughout early autumn, creating beautiful displays in pots filled with soil-based compost. Add a frill of violas around the edge to inject more colour.
↕60cm (24in) ↔45cm (18in)

PERENNIAL SMALL

Primula Crescendo Series

POLYANTHUS This evergreen perennial, grown as an annual, is ideal for late winter pots and windowboxes. It features dark green, corrugated leaves and clusters of yellow-eyed flowers in shades of red, yellow, white, and purple.
↕↔20cm (8in)

PERENNIAL SMALL

Sedum erythrostictum 'Frosty Morn'

This perennial's variegated white and grey-green fleshy foliage and flat-headed clusters of pale pink flowers in early autumn, add colour to container displays. Plant it in gritty, soil-based compost.
↕30cm (12in) ↔45cm (18in)

PERENNIAL SMALL

Sedum 'Ruby Glow' ♡

STONECROP With spreading, dark red stems of oval, purplish green leaves and star-shaped, ruby-red flowers, this perennial makes a decorative edging to pots of soil-based compost and grit from late summer to early autumn.
↕20cm (8in) ↔40cm (16in)

PERENNIAL SMALL

Senecio cineraria

CINERARIA Although this evergreen perennial, grown as an annual, is not fully hardy, when planted in pots of soil-based compost mixed with grit, its decorative, deeply cut, silver leaves continue to provide interest for most of autumn.
↕↔30cm (12in)

PERENNIAL SMALL

Viola x *wittrockiana*

PANSY This decorative bedding plant forms the mainstay of cold-season containers. Plant the colourful, rounded flowers in autumn and winter pots and baskets in multi-purpose compost. It flowers best in winter in sheltered sites.
↕23cm (9in) ↔30cm (12in)

OTHER SUGGESTIONS

Annuals and biennials

Callistephus chinensis 'Milady Series' • *Callistephus chinensis* 'Ostrich Plume'

Perennials and bulbs

Chrysanthemum 'Breitner' ♡ • *Chrysanthemum carinatum* 'Court Jesters' • *Chrysanthemum* 'Clara Curtis' • *Cyclamen mirabile* ♡ • *Dahlia* 'Art Pablo' • *Dahlia* 'Fairy' • *Nerine sarniensis* ♡ • *Nerine undulata* • *Sedum erythrostictum* • *Senecio pulcher* • *Viola* 'Antique Shades' • *Viola* 'Ochre Trailing'

Shrubs

Buddleja 'Blue Chip' • *Fuchsia magellanica* ♡ • *Hebe* 'Great Orme' ♡ • *Skimmia japonica* 'Rubella' ♡

Plants for winter borders

Days may be cold and dark, but with a selection of winter flowers, berries, stems, and leaves, gardens can continue to look stunning.

Evergreens are the mainstay of the winter garden and include shades of red (*Leucothöe* and *Nandina*) and splashes of gold (*Euonymus*), as well as green. Flowers are at a premium at this time of year, but this makes those that do face the cold all the more charismatic. Some, including *Mahonia japonica*, sweet box (*Sarcococca*), and witch hazel (*Hamamelis*) also offer scent into the bargain. Combine these beauties with the seedheads of summer flowers and grasses, and plants with colourful stems and dangling catkins, such as *Garrya elliptica*.

SHRUB MEDIUM

Abeliophyllum distichum

WHITE FORSYTHIA Plant this deciduous shrub in a sheltered spot where hard frosts will not damage the fragrant, star-shaped, pink-tinged white flowers, which appear on bare stems in late winter. The dark green leaves appear later in spring.

↕↔1.5m (5ft)

PERENNIAL SMALL

Ajuga reptans

BUGLE The purple-leaved or variegated forms of this evergreen perennial create mats of foliage interest in winter borders. The short-spiked, blue flowers appear later in spring. Varieties include 'Atropurpurea' (above) with bronze-tinted leaves.

↕15cm (6in) ↔90cm (36in)

PERENNIAL SMALL

Arum italicum

LORDS AND LADIES The arrow-shaped, dark green leaves of this perennial appear in late autumn, persisting over winter and joined in spring by white or yellow petal-like spathes. 'Marmoratum' has cream-patterned leaves that offer the best winter interest.

↕30cm (12in) ↔15cm (6in)

TREE LARGE

Betula utilis var. *jacquemontii*

HIMALAYAN BIRCH Brightening up winter borders with its white trunk and stems, this deciduous tree can be used to create small woodlands. The diamond-shaped, dark green leaves, which turn yellow in autumn, unfurl in spring after yellow catkins appear.

↕18m (60ft) ↔10m (30ft)

PERENNIAL MEDIUM

Carex flagellifera

This evergreen sedge produces tufts of reddish brown leaves, with late summer spikes of light brown flowers, followed by red-brown seedheads. Plant it at the front of a border, and in free-draining soil in areas that experience hard frosts.

↕80cm (32in) ↔60cm (24in) or more

PERENNIAL MEDIUM

Carex 'Ice Dance'

JAPANESE SEDGE Plant this decorative, evergreen, perennial sedge at the front of a sunny or part-shaded border. It produces mounds of grass-like, creamy edged green leaves and small, white flowers in spring. It may become invasive.

↕60cm (24in) ↔75cm (30in)

SHRUB LARGE

Chimonanthus praecox

WINTERSWEET This deciduous shrub, with lance-shaped, mid-green leaves, bears sulphur-yellow flowers in winter. 'Grandiflorus' AGM (above) is a popular form, with fragrant, maroon-centred yellow flowers.

↕4m (12ft) ↔3m (10ft)

CLIMBER MEDIUM

Clematis cirrhosa

Plant this evergreen climber with toothed, green leaves on a sheltered surface to protect its bell-shaped, cream flowers, spotted red inside, which appear in late winter and early spring. Shade the roots. 'Freckles' has speckled, creamy pink flowers.

↕3m (10ft)

CLIMBER LARGE

Clematis tangutica

The lantern-shaped, yellow flowers of this vigorous, deciduous clematis are followed by fluffy, silvery seedheads that provide winter interest at the back of a border. Cut stems to the ground in late winter; plant with the roots in shade.

↕6m (20ft)

SHRUB LARGE

Cornus alba

RED-BARKED DOGWOOD Grow this deciduous shrub in a border where its colourful display of young, bright red shoots will be most effective. For the best colour, cut down old stems in late winter to spur new growth, and plant in moist soil.

↕↔3m (10ft)

SHRUB LARGE

Cornus mas

CORNELIAN CHERRY Grow this deciduous shrub or small tree at the back of a border, where it will transform a drab, late winter scene with its rounded clusters of tiny, yellow blooms on bare stems. Oval, green leaves and edible, red fruits follow.

↕↔ 5m (15ft)

SHRUB LARGE

Cornus sanguinea

Grown for its reddish green stems, the foliage of this deciduous shrub also has good autumn colour. 'Winter Beauty' (above) has bright orange-yellow and red winter shoots. Prune old stems to the ground in late winter.

↕3m (10ft) ↔2.5m (8ft)

SHRUB MEDIUM

Cornus sericea 🏆

This deciduous shrub is grown for its colourful red winter stems, or yellow-green in varieties such as 'Flaviramea' AGM (above). Cut old stems to the ground in late winter to encourage colourful new growth. It also has fiery autumn leaves.

↕2m (6ft) ↔4m (12ft)

SHRUB SMALL

Cotoneaster horizontalis

Best trained against a wall at the back of a narrow border, this spreading, deciduous shrub produces stems in a "herringbone" pattern, which are dotted with red berries in autumn and early winter, before they fall or are eaten by birds.

↕1m (3ft) ↔1.5m (5ft)

SHRUB LARGE

Daphne bholua

NEPALESE PAPER PLANT Ideal for the back of a border, this evergreen shrub bears oval, leathery, green leaves and, in late winter, fragrant, white flowers, flushed purple-pink, followed by black berries. 'Jacqueline Postill' AGM (above) has pink flowers.

↕3m (10ft) ↔1.5m (5ft)

PERENNIAL MEDIUM

Dryopteris affinis 🏆

GOLDEN MALE FERN This evergreen fern's "shuttlecock" of tall, lance-shaped, divided, pale green fronds, which mature to dark green, make a decorative addition to winter borders beneath trees and shrubs. Cut off old growth in early spring.

↕↔ 90cm (36in)

SHRUB MEDIUM

Edgeworthia chrysantha

PAPER BUSH Given a sheltered site, this deciduous shrub will produce a beautiful display of fragrant, tubular, yellow flowers, covered in silky white hairs, from late winter to early spring. The oval, dark green leaves follow in spring.

↕↔ 1.5m (5ft)

SHRUB SMALL

Erica carnea

WINTER HEATH Plant this small, evergreen shrub at the front of a border or in a pot of ericaceous compost on a patio where it will add colour from winter to spring with its needle-like, dark green leaves and spikes of tiny pink, red, or white flowers.

↕30cm (12in) ↔45cm (18in) or more

BULB MEDIUM

Galanthus nivalis 🏆

SNOWDROP Allow this perennial bulb to spread through borders where its small, nodding, white flowers, with green markings on some varieties, will appear among grey-green, grassy leaves in late winter. Plant it in groups after flowering in spring.

↕15cm (6in)

SHRUB LARGE

Garrya elliptica

SILK-TASSEL BUSH This dense, evergreen shrub has leathery, wavy-edged, grey-green foliage and grey-green catkins from midwinter to early spring. Plant it in a sheltered area against a wall or fence; prune when the catkins fade.

↕↔ 4m (12ft)

SHRUB LARGE

Hamamelis x *intermedia*

A vase-shaped shrub, grown for its lightly scented, spidery flowers, which appear on bare stems from early- to midwinter. The broadly oval leaves appear later in spring. Varieties include 'Jelena' AGM (above), with large, coppery orange flowers.

↕↔4m (12ft)

PERENNIAL MEDIUM

Helleborus foetidus

STINKING HELLEBORE This evergreen perennial brightens up late winter borders with its cup-shaped, nodding, pale green flowers with red margins, held on red stems. The dark green leaves are comprised of narrow, toothed leaflets.

↕↔60cm (24in)

PERENNIAL SMALL

Helleborus niger

CHRISTMAS ROSE This semi-evergreen, clump-forming perennial has leathery leaflets and, from winter to early spring, cup-shaped, nodding, white or pink-flushed flowers, with a touch of green at the base of the tuft of stamens.

↕↔30cm (12in)

PERENNIAL SMALL

x *Heucherella* 'Tapestry'

CORAL BELLS In sheltered areas, this evergreen perennial retains its deeply lobed, green foliage, with purple centres and veins. Ideal for edging borders or in containers, it forms sprays of small, pink flowers in early summer.

↕↔30cm (12in)

SHRUB LARGE

Hydrangea paniculata

PANICULATE HYDRANGEA Although this shrub is deciduous, it provides winter interest after the dark green leaves have fallen, when conical clusters of white or pink flowerheads fade to form dried bronze sculptural features on skeletal stems.

↕↔3m (10ft)

TREE LARGE

Ilex aquifolium

ENGLISH HOLLY An iconic winter evergreen shrub, or tree, with wavy-edged, spiny, dark green leaves and, on pollinated female plants, scarlet berries. Choose a variegated cultivar for additional cold-season colour, and plant at the back of a border.

↕up to 20m (70ft) ↔6m (20ft)

SHRUB LARGE

Jasminum nudiflorum

WINTER JASMINE This deciduous, arching shrub has oval, dark green leaves, followed by starry, bright yellow flowers on bare, green shoots from winter to early spring. Use as a backdrop to a border by training the stems against a wall or fence.

↕↔3m (10ft)

SHRUB MEDIUM

Leucothöe fontanesiana

DOG HOBBLE Decorate a winter garden with this evergreen, mid- to back-of-border shrub, with arching, red-tinged stems, green leaves, and cream, bell-shaped spring flowers. 'Scarletta' (above) has bronze-tinted winter leaves.

↕up to 1.5m (5ft) ↔2m (6ft)

SHRUB SMALL

Mahonia aquifolium

OREGON GRAPE An evergreen shrub, with leathery, spine-edged, dark green leaves that turn purplish in winter, providing interest before dense, yellow flower clusters appear in spring. Round, black berries follow the blooms. It thrives in shady borders.

↕1m (3ft) ↔1.5m (5ft)

SHRUB MEDIUM

Mahonia japonica

JAPANESE MAHONIA The leathery, spine-edged, dark green leaves of this evergreen shrub add colour to winter borders when they become tinted with purple. Dense clusters of fragrant, yellow flowers open through winter, followed by black berries.

↕2m (6ft) ↔3m (10ft)

PERENNIAL MEDIUM

Ophiopogon jaburan

Use this evergreen perennial in a sheltered winter border to create an edge of dark green, grass-like leaves. Other features include the white, bell-shaped summer flowers and violet-blue fruits. 'Vittatus' (above) is cream- and green-striped.

↕60cm (24in) ↔30cm (12in)

PERENNIAL MEDIUM

Panicum virgatum

SWITCH GRASS An upright, deciduous grass that forms clumps of blue-green leaves, which turn yellow in autumn. Clouds of pink-tinged green summer flowerheads, held on tall stems, fade to beige in autumn and persist through winter.

↕1m (3ft) ↔75cm (30in)

SHRUB LARGE

Phyllostachys aureosulcata f. *aureocaulis*

GOLDEN GROOVE BAMBOO This stately, clump-forming bamboo, with tall, yellow canes, sometimes crooked at the base, can be used as a screen or focal point in a border in a winter garden.

↕6m (20ft) ↔indefinite

SHRUB LARGE

Pittosporum tenuifolium

Plant this evergreen shrub, with its wavy-edged, grey-green leaves held on dark stems, at the back of a sheltered border to inject winter colour. 'Silver Queen' AGM (above) is a variegated form, and will thrive in the shelter of a south-facing wall.

↕10m (30ft) ↔5m (15ft)

SHRUB LARGE

Pyracantha 'Mohave'

FIRETHORN Train this evergreen shrub against a wall or fence at the back of a winter border, where its prickly stems of glossy, dark green foliage provide a foil for plants in front, while the orange-red berries add splashes of colour early in the season.

↕↔4m (12ft)

SHRUB MEDIUM

Rubus cockburnianus

A dense, deciduous shrub, with arching prickly shoots, coloured brilliant white in winter, creates a ghostly effect in cold-season borders. In spring, diamond-shaped, dark green foliage appears; inedible black fruits follow the purple summer flowers.

↕↔2.5m (8ft)

SHRUB MEDIUM

Sarcococca hookeriana var. *digyna*

SWEET BOX Grow this evergreen shrub in a winter border to enjoy the fragrance of its tiny white flowers, which appear among narrow, lance-shaped, green leaves. Spherical black fruits follow the blooms.

↕1.5m (5ft) ↔2m (6ft)

SHRUB LARGE

Stachyurus praecox

The pale greenish yellow, bell-shaped flowers of this deciduous shrub hang in clusters from bare stems from late winter to early spring, creating a focal point at the back of a border filled with winter foliage. Dark green leaves unfurl later in spring.

↕4m (12ft) ↔3m (10ft)

PERENNIAL MEDIUM

Stipa calamagrostis

This deciduous or semi-evergreen, perennial grass forms tufts of long, blue-green leaves in summer, which turn vibrant yellow in autumn. It produces tall, feathery flowerheads in summer, which dry *in situ* and provide interest well into winter.

↕↔80cm (32in)

SHRUB LARGE

Viburnum x *bodnantense*

A deciduous shrub, with richly fragrant, pink or white late autumn to early spring flower clusters on bare stems. A good back-of-border plant, grow it with evergreens in a winter border. 'Dawn' AGM (above) has deep pink buds and pink flowers.

↕3m (10ft) ↔2m (6ft)

SHRUB LARGE

Viburnum farreri

Plant this deciduous shrub at the back of a border close to a south-facing wall, and it will reward you with fragrant, white or pale pink flowers on bare stems throughout winter. The dark green foliage is bronze when it appears in spring.

↕3m (10ft) ↔2.5m (8ft)

OTHER SUGGESTIONS

Perennials and bulbs

Ajuga reptans 'Burgundy Glow' • *Galanthus caucasius* • *Sedum spectabile* 'Autumn Joy'

Grasses

Miscanthus sinensis 'Morning Light' • *Miscanthus sinensis* 'Gracillimus'

Shrubs and climbers

Cornus mas 'Golden Glory' • *Coronilla valentina* subsp. Glauca • *Erica carnea* 'Ann Sparkes' • *Erica carnea* 'Irish Dusk' • *Erica umbellata* • *Euonymus japonicus* 'Microphyllus variegatus' • *Hamamelis* x *intermedia* 'Barmstedt Gold' • *Hypericum* 'Magical Red Star' • *Ilex aquifolium* 'J.C. van Tol' • *Nandina domestica* 'Fire Power' • *Phyllostachys nigra* • *Prunus lusitanica* 'Myrtifolia' • *Pyracantha* 'Golden Dome'

Plants for evergreen effects

Providing a permanent stage for seasonal stars, evergreens have special value in small gardens, and help create continuity during lulls.

Although many are quiet and unassuming, in combination, evergreens with contrasting textures, shapes, and colours can make beautiful displays of their own. For example, large, hand-like *Fatsia* leaves, variegated holly foliage, and feathery ferns make an elegant combination for a shady garden. When creating a new planting scheme, place your evergreens first to provide a permanent framework for your design. Also check heights and spreads to ensure that they do not block seasonal plants and flowers or, if they are small and compact like hebes, adjacent perennials won't swamp them in summer.

TREE SMALL

Abies nordmanniana 'Golden Spreader' ♀

Use this slow-growing conifer to brighten up borders. Combine it with dark green- and blue-green-leaved evergreens for year-round colour. It has spreading branches of bright golden-yellow, needle-like leaves.
↕1m (3ft) ↔1.5m (5ft)

SHRUB LARGE

Arbutus unedo

STRAWBERRY TREE Plant this evergreen shrub in a sheltered garden in fertile soil for the best foliage effects through the year. It has dark green, leathery leaves and white, urn-shaped flowers that appear in autumn alongside red, strawberry-like fruits.
↕↔8m (25ft)

SHRUB SMALL

Arctostaphylos uva-ursi

BEARBERRY Grow this low-growing, evergreen shrub at the front of a border in acid soil. The oval, dark green leaves, pink-tinted white summer flowers, and round, scarlet autumn fruits provide year-round interest.
↕10cm (4in) ↔50cm (20in)

PERENNIAL SMALL

Bergenia purpurascens ♀

PURPLE BERGENIA This clump-forming, evergreen perennial has large, rounded, glossy, green leaves and, in spring, spikes of dark pink blooms. Plant near the front of a border, and remove tired or damaged leaves to keep it tidy.
↕40cm (16in) ↔60cm (24in) or more

PERENNIAL MEDIUM

Blechnum spicant ♀

HARD FERN Perfect for evergreen colour and texture, this fern forms low, arching hummocks of divided, stiff, leathery leaves, with narrow leaflets, and thrives in deep shade. The upright, spore-bearing fronds resemble fish bones.
↕75cm (30in) ↔45cm (18in)

SHRUB MEDIUM

Buxus sempervirens 'Elegantissima' ♀

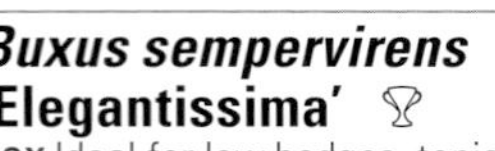

BOX Ideal for low hedges, topiary shapes, or dots of colour in a border, this compact, rounded, evergreen shrub has oval, white-margined green leaves. Clip topiary and hedges in early and late summer.
↕↔1.5m (5ft)

SHRUB SMALL

Calluna vulgaris

SCOTS HEATHER The tiny, grey, yellow, or bright green leaves of this bushy, evergreen shrub liven up border edges, given acid soil and full sun. The bell-shaped flowers appear from midsummer to late autumn in shades of pink, white, and purple.
↕30cm (12in) ↔35cm (14in)

TREE MEDIUM

Chamaecyparis pisifera 'Filifera Aurea' ♀

SAWARA CYPRESS Use this slow-growing, evergreen conifer as a colourful, textured specimen with blue-leaved conifers in a mixed display. It makes a mound of weeping stems, with golden-yellow, scaly leaves.
↕12m (40ft) ↔5m (15ft)

SHRUB SMALL

Cryptomeria japonica 'Globosa Nana' ♡

JAPANESE CEDAR Thriving in a sheltered spot, this rounded, evergreen conifer's rich green foliage gives year-round interest in a border or gravel bed. Short, cord-like, arching stems carry the scale-like leaves.

↕↔ 1m (3ft)

SHRUB MEDIUM

Daphne odora

A rounded, evergreen shrub, grown for its dark green, glossy leaves and small clusters of richly scented, carmine-edged white flowers, which appear from winter to early spring. 'Aureomarginata' AGM (above) has yellow-edged foliage.

↕↔ 1.5m (5ft)

PERENNIAL MEDIUM

Dryopteris affinis ♡

GOLDEN MALE FERN Evergreen in sheltered sites, this fern produces a "shuttlecock" of tall, lance-shaped, divided, pale green fronds, which mature to dark green, with scaly, golden-brown midribs. Use it to decorate partly shaded borders.

↕↔ 90cm (36in)

SHRUB MEDIUM

Erica arborea

The dark green, needle-like leaves of this upright, evergreen shrub lend year-round colour. In spring, pyramid-shaped clusters of bell-shaped, honey-scented, greyish white flowers appear. *E. arborea* var. *alpina* AGM is more compact than the species.

↕ up to 2m (6ft) ↔ 85cm (34in)

SHRUB LARGE

Fargesia murielae ♡

UMBRELLA BAMBOO This large, clump-forming, evergreen bamboo bears arching, yellow-green canes and lance-shaped, bright green leaves. It makes a decorative year-round screen or accent plant in a border or gravel bed. Best in fertile soil.

↕ 4m (12ft) ↔ indefinite

SHRUB LARGE

Fatsia japonica ♡

JAPANESE ARALIA The large, deeply lobed, evergreen leaves of this shrub inject colour and texture into containers and borders, and the round clusters of tiny, white autumn flowers and black fruits add to the effect. 'Variegata' AGM has white-edged leaves.

↕↔ 4m (12ft)

SHRUB SMALL

Gaultheria mucronata

This evergreen shrub has prickly, dark green leaves that create a year-round backdrop to flowers in a border. Its tiny, white late spring flowers are followed, on female plants, by colourful berries. 'Wintertime' (above) has large, white berries.

↕↔ 1.2m (4ft)

SHRUB SMALL

Hebe 'Red Edge' ♡

The red-edged blue-green leaves of this compact evergreen shrub lend year-round interest to a sheltered, sunny site. Clusters of pale mauve to white flowers add a colourful note in summer. Trim lightly after flowering, if necessary.

↕ 45cm (18in) ↔ 60cm (24in)

PERENNIAL MEDIUM

Helleborus argutifolius ♡

HOLLY-LEAVED HELLEBORE Grown for its large, spiny-edged, dark green leaves, this evergreen, architectural perennial has, from late winter to spring, long-lasting clusters of nodding, pale green, bowl-shaped blooms. Cut spent leaves and flowers in autumn.

↕ 60cm (24in) ↔ 45cm (18in)

PERENNIAL SMALL

x *Heucherella* 'Tapestry'

CORAL BELLS This perennial, evergreen in sheltered sites, produces a carpet of deeply lobed, green foliage, with purple centres and veins. It is ideal for edging borders or in containers. In early summer, sprays of small, pink flowers form on upright stems.

↕↔ 30cm (12in)

SHRUB SMALL

Hypericum calycinum

ROSE OF SHARON This vigorous, evergreen shrub makes good ground cover. Its dark green foliage provides year-round colour. Large, open, bright yellow flowers, with fluffy stamens in the centre, appear from midsummer to mid-autumn. Can be invasive.

↕60cm (24in) ↔indefinite

SHRUB MEDIUM

Leucothöe fontanesiana

DOG HOBBLE An evergreen shrub, with arching, red-tinged stems, clothed in lance-shaped, green leaves, red-tinted or mottled in some varieties. Cream flowers hang from the stems in spring. 'Rainbow' (above) has mottled cream- and pink-variegated leaves.

↕up to 1.5m (5ft) ↔2m (6ft)

PERENNIAL MEDIUM

Luzula nivea

SNOWY WOODRUSH This evergreen perennial forms loose clumps of dark green, grass-like leaves, ideal for edging a border. From early- to midsummer, tight clusters of tiny, white flowers appear on slim stems. Try it in a shady spot in a wildlife garden.

↕↔60cm (24in)

TREE LARGE

Magnolia grandiflora

Creating an eye-catching focal point in a sheltered garden or beside a south-facing wall, this evergreen tree produces a rounded head of large, glossy, dark green leaves. Large, fragrant, white flowers appear from midsummer to early autumn.

↕18m (60ft) ↔15m (50ft)

SHRUB SMALL

Mitchella repens

PARTRIDGE BERRY This evergreen subshrub, with trailing stems of small, oval, white-striped green leaves, is ideal for draping over the side of a raised bed. In early summer, tiny, fragrant, white flowers appear, followed by red berries.

↕5cm (2in) ↔indefinite

SHRUB MEDIUM

Mahonia x *media*

An architectural, evergreen shrub, with long, dark green leaves, divided into paired, spiny leaflets. Throughout winter, long clusters of scented, yellow flowers appear. 'Buckland' AGM has longer flower spikes and red-tinted winter leaves.

↕1.8m (6ft) ↔4m (12ft)

SHRUB SMALL

Microbiota decussata 🏆

Grow this low-growing, evergreen, coniferous shrub for its sprays of tiny, scale-like, bright green leaves, bronze-purple in winter. The flowers and cones are inconspicuous. Use it as a foil for shrubs with colourful foliage and flowers.

↕1m (3ft) ↔indefinite

PERENNIAL SMALL

Ophiopogon planiscapus 'Nigrescens' 🏆

BLACK MONDO GRASS Perfect for border edging or pots, this perennial's grass-like, shiny, black foliage contrasts well with bright gravel. Black autumn berries follow the clusters of purple-pink summer blooms.

↕23cm (9in) ↔30cm (12in)

SHRUB LARGE

Osmanthus heterophyllus

FALSE HOLLY The holly-like, glossy, bright green leaves of this evergreen shrub make a decorative feature or screen in sheltered gardens. It produces fragrant, tiny, white flowers in autumn. 'Aureomarginatus' (above) has gold-edged leaves.

↕↔5m (15ft)

PERENNIAL LARGE

Phormium tenax

NEW ZEALAND FLAX An evergreen, clump-forming perennial with long, upright, then arching, sword-shaped, grey-green leaves that create a dramatic effect at the back of a border. During hot summers, dark red or yellow flowers may form on tall stems.
↕3m (10ft) ↔2m (6ft)

TREE LARGE

Picea breweriana 🏆

BREWER'S SPRUCE Make a year-round statement in a large garden with this cone-shaped, evergreen conifer. Its drooping stems, covered with dark grey-green needles, create curtains of colour. Grow in neutral to acid soil in a sheltered site.
↕15m (50ft) ↔4m (12ft)

SHRUB SMALL

Picea pungens 'Montgomery'

DWARF BLUE SPRUCE This compact, mound-forming, slow-growing evergreen conifer, with needle-like, grey-blue leaves, lends a splash of year-round colour to a gravel bed or rock garden. Plant with contrasting green foliage plants. Needs neutral to acid soil.
↕↔60cm (24in)

SHRUB LARGE

Pieris japonica

LILY OF THE VALLEY BUSH This decorative evergreen shrub has leathery, dark green leaves, bright red when young, and, in spring, clusters of urn-shaped, white, pink, or red flowers appear. Grow it in acid soil in a sheltered, sunny site.
↕4m (12ft) ↔3m (10ft)

SHRUB SMALL

Pinus mugo

DWARF MOUNTAIN PINE This compact conifer forms a mound of needle-like, evergreen foliage and makes a good foil for spring bulbs, perennials, and small shrubs. Select a dwarf form like 'Mops' AGM (above) for containers.
↕1m (3ft) ↔2m (6ft)

SHRUB LARGE

Pittosporum tenuifolium 'Silver Queen' 🏆

The wavy-edged, grey-green leaves of this evergreen shrub create a year-round, textured effect in sheltered sites. Use as a backdrop to a sunny border, and select variegated forms for additional interest.
↕10m (30ft) ↔5m (15ft)

SHRUB LARGE

Pittosporum tobira 🏆

JAPANESE PITTOSPORUM Grow this evergreen shrub in a sheltered site, where it will produce a neat dome of long, oval, dark green leaves. Clusters of sweetly scented, starry, white flowers, which age to creamy yellow, add to the effect.
↕10m (30ft) ↔3m (10ft)

SHRUB LARGE

Pyracantha 'Mohave'

FIRETHORN Ideal for training against a wall or fence, the dark green leaves of this evergreen shrub provide a foil for plants growing in front. In late summer, it bears orange-red berries that persist into winter and are attractive to birds.
↕↔4m (12ft)

SHRUB MEDIUM

Rosmarinus officinalis

ROSEMARY This evergreen, culinary, shrubby herb retains its aromatic, needle-like, dark green foliage year-round, given a sheltered site. Do not harvest the leaves in winter. The small, blue flowers appear in spring. Cut back old growth in early summer.
↕↔1.5m (5ft)

OTHER SUGGESTIONS

Perennials

Bergenia 'Bressingham Ruby' • *Bergenia* 'Silberlicht' 🏆 • *Dryopteris erythrosora* 🏆 • *Polystichum setiferum* Divisilobum Group 🏆

Shrubs and climbers

Chamaecyparis pisifera 'Snow' • *Choisya* x *dewitteana* 'Aztec Pearl' • *Cryptomeria japonica* 'Cistata' • *Daphne laureola* subsp. *philippi* • *Elaeagnus pungens* 'Maculata' • *Enkianthus perulatus* 🏆 • *Erica vulgaris* 'Boeley Gold' • *Euonymus fortunei* • *Gaultheria mucronata* 'Mulberry Wine' 🏆 • *Hedera helix* • *Ilex aquifolium* 'Argentea Marginata' 🏆 • *Leucothöe fontainesiana* 'Carinella' • *Mahonia nervosa* • *Nandina domestica* • *Pyracantha* 'Teton' 🏆 • *Ruscus aculeatus* • *Skimmia japonica* 'Robert Fortune' • *Taxus cuspidata* 'Aurescens'

PLANTS for COLOUR and SCENT

Scent and colour affect our moods and perceptions, and they can be used to great effect in the garden. Try planting lavenders close to seating, where the fragrance will help soothe frayed nerves, or use pastel colours, which produce a similar calming effect. Conversely, bright reds and oranges are stimulants, and help to energize planting schemes, and act as focal points. You can also mimic wider landscapes, where colours pale as they extend towards the horizon, giving an illusion of distance.

Plants for cool colours

Ever since Vita Sackville West created her famous white garden in the 1930s, designers have raved about the beauty of monochrome planting schemes.

To achieve a similar effect, plant a succession of white flowers, starting in spring with hellebores and snowdrops, peaking in summer with peonies, roses, and shasta daisies, and concluding in autumn with asters and Japanese anemones. If you find an all-white scheme too restrictive, include a few blues, which will blend well without jarring the overall effect. You can also create a cool, neutral backdrop for pastel or brightly coloured flowers with blue and grey foliage, or use silver-leaved plants, such as *Artemisia, Convolvulus,* and *Tanacetum,* to reflect the sun and twinkle from beds and borders like spotlights.

TREE LARGE

Abies concolor

WHITE FIR An evergreen conifer, with an attractive "Christmas tree" shape, it makes a good lawn specimen or can feature in a mixed boundary scheme. The silver-grey needles are its main feature. Prune carefully, if required.
↕40m (130ft) ↔7m (22ft)

PERENNIAL MEDIUM

Aconitum 'Stainless Steel'

MONKSHOOD This perennial makes a substantial clump of deeply divided leaves. Its mid- to late summer spires of unusual silvery blue, hooded blooms create an elegant impression within a shaded border. All parts are poisonous.
↕1m (3ft) ↔60cm (24in)

PERENNIAL MEDIUM

Agapanthus africanus

AFRICAN LILY This evergreen perennial has strap-shaped leaves and spherical heads of deep blue, trumpet-shaped flowers on sturdy, unbranched stems during summer. Suitable for mixed or single plantings. Provide protection from frost.
↕1m (3ft) ↔50cm (20in)

PERENNIAL LARGE

Anemone x *hybrida*

JAPANESE ANEMONE An upright, clump-forming, slowly spreading perennial, with lobed leaves. With its spherical buds and saucer-shaped, single or semi-double, pink or white blooms, it makes a lovely feature in late summer borders.
↕1.2m (4ft) ↔60cm (24in)

PERENNIAL SMALL

Anthemis punctata subsp. *cupaniana*

SICILIAN CHAMOMILE This carpeting, evergreen perennial has filigree silver foliage, green in winter. It bears a mass of yellow-centred white daisies, with a single bloom on each short stem, in early summer.
↕↔30cm (12in)

PERENNIAL MEDIUM

Aquilegia vulgaris

GRANNY'S BONNET This upright, clump-forming perennial has ferny foliage and, from late spring to early summer, nodding flowers with spurred petals in mainly pastel shades. 'Nivea' AGM (above) has white blooms.
↕90cm (36in) ↔45cm (18in)

PERENNIAL MEDIUM

Artemisia ludoviciana 'Silver Queen'

WESTERN MUGWORT Grown as a foil for flowers, this spreading perennial has narrow, silver-grey leaves on upright stems. Remove the small, yellow summer flowers to maintain the foliage effect.
↕↔75cm (30in)

SHRUB SMALL

Artemisia 'Powis Castle'

WORMWOOD A semi-evergreen subshrub, with fern-like, silvery grey leaves that provide a foil for green-leaved plants, including roses and sun-loving border perennials. Snip off the yellow, pompon flowers to enhance the effect.
↕↔1m (3ft)

PERENNIAL MEDIUM

Aster ericoides 'White Heather'

Plant this clump-forming perennial mid-border with small-leaved, autumn-interest plants. Sprays of tiny, white daisies are borne for weeks and make an excellent foil for larger, more solid blooms and leaves.
↕1m (3ft) ↔30cm (12in)

PERENNIAL SMALL

Athyrium niponicum var. *pictum* 🏆

JAPANESE PAINTED FERN This deciduous fern has deeply cut, grey-green leaves, with metallic purple and silver highlights. Use between blocks of green foliage in shady borders. Protect from hard frost.

↕30cm (12in) ↔ indefinite

SHRUB SMALL

Ballota pseudodictamnus 🏆

This low-growing, compact, evergreen shrub has vertical stems with evenly spaced pairs of small, round, grey-green leaves that are covered with woolly, white hairs. Clip over after the whorls of tiny, pink blooms fade.

↕60cm (24in) ↔ 90cm (36in)

PERENNIAL SMALL

Bergenia 'Silberlicht' 🏆

ELEPHANT'S EARS An evergreen, compact, clump-forming perennial, with leathery, round leaves and bell-shaped, white spring flowers held on red-tinged stems. The blooms, which sometimes become pink-tinged, have contrasting deep red bud cases.

↕45cm (18in) ↔ 50cm (20in)

PERENNIAL SMALL

Brunnera macrophylla

A ground-cover perennial, it is useful for adding sparkle under trees or in a shady, urban setting. It has broad, oval to heart-shaped leaves, often overlaid or edged with white. Sprays of blue flowers appear in spring.

↕45cm (18in) ↔ 60cm (24in)

PERENNIAL SMALL

Calamintha nepeta

LESSER CALAMINT Grown for its upright spikes of tiny, lilac-pink flowers, which appear from late summer to early autumn, this perennial also has herbal-scented, green leaves. Grow near a path where it can be brushed against to release the aroma.

↕45cm (18in) ↔ 75cm (30in)

PERENNIAL SMALL

Campanula carpatica 🏆

TUSSOCK BELLFLOWER This low-growing perennial forms a dense mat of heart-shaped leaves. Throughout summer, it is smothered in bell-shaped, violet-blue or white flowers. Use it as ground cover or to fill gaps in walls and paving.

↕10cm (4in) ↔ 30cm (12in)

BULB LARGE

Cardiocrinum giganteum

GIANT LILY A giant perennial bulb, with heart-shaped leaves. In summer, white, fragrant, trumpet-shaped blooms, with purple throats, open at the top of sturdy, unbranched stems. Plant it in a sheltered site and rich soil.

↕3m (10ft)

SHRUB MEDIUM

Caryopteris incana

BLUEBEARD This domed shrub has slender stems that bear grey-green, aromatic, toothed-edged leaves. In early autumn, tiny, blue flowers form in dense clusters in the leaf axils. It is attractive to bees and butterflies.

↕1.2m (4ft) ↔ 1.5m (5ft)

SHRUB MEDIUM

Ceanothus americanus

NEW JERSEY TEA This dense, rounded, deciduous shrub tolerates drought but needs a sheltered site. From late spring to midsummer, it produces thimble-shaped heads of tiny, white or blue flowers.

↕↔ up to 2m (6ft)

SHRUB MEDIUM

Ceanothus x *delileanus* 'Gloire de Versailles' 🏆

CALIFORNIAN LILAC This bushy, deciduous shrub has small, oval leaves and clusters of tiny, fragrant, powder-blue flowers in loose, thimble-shaped heads from summer to autumn. It makes a cooling foil for roses.

↕↔ 1.5m (5ft)

TREE LARGE

Cedrus atlantica Glauca Group

BLUE ATLAS CEDAR This evergreen conifer makes a conical-shaped tree when young, and a broad crown when mature. It bears striking, silvery blue-green needles. Plant it as a lawn specimen in a large garden.

↕40m (130ft) ↔10m (30ft)

PERENNIAL SMALL

Centaurea montana

PERENNIAL CORNFLOWER This vigorous, self-seeding perennial has tapered leaves, with hairy undersides. In early summer, it produces open, thistle-like, purple, blue, white, or pink flowerheads, with red-purple centres. Cut back to promote re-flowering.

↕50cm (20in) ↔60cm (24in)

SHRUB SMALL

Convolvulus cneorum

This low, spreading, evergreen shrub is grown for its narrow, oblong, metallic silver leaves. In summer, wide, funnel-shaped, white flowers open from pink buds. Grow it at the front of a sheltered border or in gravel.

↕60cm (24in) ↔1m (3ft)

BULB LARGE

Crinum x *powellii* 'Album'

This richly fragrant perennial bulb has broad, arching, strap-shaped leaves. In late summer and autumn, it produces large, white, funnel-shaped blooms on sturdy stems. Provide protection in winter in cold regions.

↕1m (3ft)

PERENNIAL LARGE

Delphinium elatum

This tall perennial has deeply cut, green leaves and is famous for its striking columns of midsummer flowers. Pale blue or white varieties provide a cooling contrast for hot shades. The stems need sturdy stakes.

↕2m (6ft) ↔90cm (36in)

SHRUB MEDIUM

Deutzia x *magnifica*

This early summer-flowering, deciduous shrub makes a lovely arching specimen or back-of-border feature. Small, oval leaves appear along slender, branched stems, and snowy, double flowers open in profusion in spring.

↕2.5m (8ft) ↔2m (6ft)

PERENNIAL MEDIUM

Echinacea purpurea 'White Swan'

This perennial has a neat, upright habit and oval leaves. From summer to autumn, it produces large, daisy-like flowers, with drooping, white petals surrounding prominent, gingery brown domes.

↕60cm (24in) ↔45cm (18in)

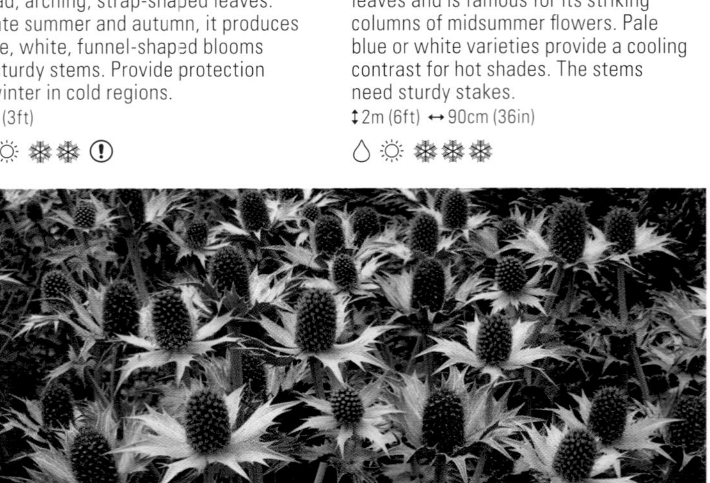

PERENNIAL LARGE

Eryngium giganteum

SEA HOLLY This short-lived, self-seeding perennial produces a basal clump of marbled, heart-shaped, grey-green leaves. Tall, upright stems hold the cone-like, silvery blue summer flowers, which are surrounded by silver, leaf-like bracts.

↕90cm (36in) ↔30cm (12in)

TREE LARGE

Eucalyptus gunnii

CIDER GUM This evergreen, slender tree has rounded, aromatic, blue-grey young foliage, which, as the tree matures, gives way to leathery, sickle-shaped leaves. Cut it back periodically to produce more young growth and colourful foliage.

↕up to 25m (80ft) ↔15m (50ft)

TREE MEDIUM

Eucryphia glutinosa

This upright, deciduous or semi-evergreen tree or shrub has green foliage, divided into glossy, toothed-edged leaflets. In summer, white, bowl-shaped, single or double blooms, with a central tuft of stamens, appear.

↕10m (30ft) ↔6m (20ft)

PERENNIAL LARGE

Euphorbia characias subsp. *wulfenii*

MEDITERRANEAN SPURGE This evergreen perennial has multiple upright stems, with whorls of slender, grey to blue-green leaves. In the second spring, long-lasting yellow-green flowerheads appear. Sap is an irritant.

↕↔1.2m (4ft)

PERENNIAL SMALL

Euphorbia myrsinites

SPURGE This evergreen, ground-hugging perennial has woody stems clothed in scale-like, fleshy, grey leaves. Stems are tipped with vivid yellow-green flowerheads in spring. Remove the spent blooms. The sap can cause skin irritation.

↕8cm (3in) ↔20cm (8in)

SHRUB MEDIUM

Exochorda x *macrantha*

PEARL BUSH This deciduous shrub makes a mound of dense stems, covered with small, dark green leaves that develop bold autumn tints. From late spring to early summer, small, saucer-shaped, white blooms smother the arching branches.

↕2m (6ft) ↔3m (10ft)

PERENNIAL SMALL

Festuca glauca

BLUE FESCUE This evergreen grass is grown for its arching tufts of fine, steel-blue, thread-like leaves. Short flower spikes appear in summer, but are best removed to maintain the plant's uniform shape and colour.

↕↔50cm (20in)

BULB MEDIUM

Galanthus 'S. Arnott'

SNOWDROP This perennial bulb, effective when planted in drifts, produces grass-like, grey-green leaves and, in late winter, small, fragrant, nodding, white flowers, with green "V" marks on the inner segments.

↕15cm (6in)

PERENNIAL MEDIUM

Geranium clarkei 'Kashmir White'

This spreading perennial cranesbill has elegantly divided leaves and white, saucer-shaped late summer blooms, with pink nectar guides on the petals. Use it to underplant roses or at the front of a border.

↕↔60cm (24in)

PERENNIAL MEDIUM

Helictotrichon sempervirens

BLUE OAT GRASS This evergreen perennial grass has long, silvery blue leaves, forming an airy mound, and light blue flowers from early- to midsummer on arching stems, with long-lasting, straw-coloured seedheads.

↕1m (3ft) ↔60cm (24in)

PERENNIAL SMALL

Helleborus foetidus

STINKING HELLEBORE An evergreen perennial with handsome dark green, leaves, divided into toothed leaflets, and, from late winter to early spring, clusters of cup-shaped, pale green flowers, with red-edged petals.

↕↔60cm (24in)

PERENNIAL SMALL

Helleborus niger

CHRISTMAS ROSE Perfect for cold-season gardens, this semi-evergreen perennial forms clumps of lobed, dark green leaves and white, saucer-shaped blooms, ageing to pink, from winter to early spring. 'Potter's Wheel' (above) has pure white blooms.

↕↔30cm (12in)

SHRUB MEDIUM

Hydrangea arborescens

This rounded shrub has oval, mid-green leaves and, in midsummer, flattened, dome-shaped clusters of long-lasting blooms that open from lime-green buds and age to creamy white, and produce papery winter skeletons.

↕↔1.5m (5ft)

Plants for cool colours

SHRUB MEDIUM

Hydrangea macrophylla

LACECAP HYDRANGEA This broad-leaved shrub, from midsummer to early autumn, bears flat heads of tiny, blue flowers, surrounded by petal-like florets. Other varieties may bear pure white, purple, pink, or, on acid soils, blue blooms.

↕2m (6ft) ↔2.5m (8ft)

PERENNIAL MEDIUM

Iris confusa ♀

This spreading, evergreen perennial produces tufts of strap-shaped leaves on upright stems. In mid-spring, the white or pale blue flowers, with yellow- or purple-speckled yellow crests appear on branched stems.

↕1m (3ft) ↔indefinite

PERENNIAL LARGE

Iris sibirica

SIBERIAN IRIS A vertical, clump-forming perennial, with broad, grass-like leaves and blue, purple, or white blooms opening from tapered buds, from early- to midsummer. The base of the lower petals is blotched and streaked.

↕up to 1.2m (4ft) ↔50cm (20in)

SHRUB SMALL

Lavandula 'Willow Vale' ♀

FRENCH LAVENDER This evergreen, compact shrub has narrow, aromatic, grey-green leaves. From early- to midsummer, upright stems bear short spikes of dark flowers that are topped with tufts of showy, violet-purple, wavy bracts.

↕↔70cm (28in)

SHRUB SMALL

Juniperus squamata 'Blue Star' ♀

FLAKY JUNIPER This dense, evergreen conifer forms a low mound of short, stiff, tightly packed, blue-grey leaves. It slowly spreads, making an irregular and undulating carpet; ideal in gravel or amongst rocks.

↕40cm (16in) ↔1m (3ft)

PERENNIAL MEDIUM

Leucanthemum x *superbum*

SHASTA DAISY An upright perennial, with spreading clumps of dark green, lance-shaped leaves. In summer, unbranched stems carry large, single or double, white daisies. 'Beauté Nivelloise' (above) has thread-like petals.

↕up to 90cm (36in) ↔60cm (24in)

PERENNIAL MEDIUM

Lamprocapnos spectabilis 'Alba' ♀

syn. ***Dicentra spectabilis*** A clump-forming perennial, with fern-like, deeply cut, light green leaves and heart-shaped, white blooms from late spring to early summer. It dies back in late summer.

↕75cm (30in) ↔60cm (24in)

BULB LARGE

Lilium longiflorum ♀

EASTER LILY This perennial bulb, ideal for planting mid-border, has shiny, lance-shaped leaves and, from mid- to late summer, large, highly fragrant, flared, funnel-shaped, pure white, outward-facing blooms on tall stems.

↕1m (3ft)

BULB LARGE

Lilium regale ♀

REGAL LILY This bulbous perennial has upright stems, clothed in narrow leaves and topped, in summer, by numerous large, white, fragrant, outward-facing trumpets. The petals are yellow at the throat and pinkish purple outside.

↕2m (6ft)

SHRUB LARGE

Magnolia stellata

STAR MAGNOLIA The fragrant silk-budded, star-shaped, white flowers make this deciduous shrub a favourite for early spring borders. Slim, green leaves appear after the blooms. Plant it in a sheltered site to protect the blooms from frost.

↕↔ 10m (30ft)

PERENNIAL LARGE

Meconopsis betonicifolia

HIMALAYAN BLUE POPPY This short-lived perennial forms a rosette of heart-shaped, rusty haired leaves. In late spring, leafy stems bear poppy-like, yellow-centred startling blue blooms. Plant it in cool, moist conditions.

↕ 1.2m (4ft) ↔ 45cm (18in)

BULB MEDIUM

Narcissus 'Ice Follies'

DAFFODIL Flowering in mid-spring, this perennial bulb makes a cool partnership with purple honesty or blue grape hyacinth. Its creamy white flowers have frilled, primrose cups that fade to white. Plant the bulbs in drifts in autumn.

↕ 40cm (16in)

PERENNIAL SMALL

Nepeta x *faassenii*

CATMINT This bushy, edging perennial, has small, greyish green, aromatic leaves. From summer to autumn, masses of soft purple-blue flower spikes appear. Cut back after the first main flush to keep the plant compact.

↕↔ 45cm (18in)

SHRUB LARGE

Osmanthus x *burkwoodii*

This large, rounded, slow-growing, evergreen shrub has small, oval, dark green leaves and, in spring, clusters of tiny, funnel-shaped, highly fragrant, white flowers. Black berries may follow. Grow in borders or as hedging.

↕↔ 3m (10ft)

SHRUB SMALL

Perovskia atriplicifolia

RUSSIAN SAGE A woody-based perennial, with erect, white stems clothed in deeply cut, grey-green, aromatic leaves. Airy heads of lavender-blue flowers appear from late summer to autumn. The bare, white winter stems are a feature.

↕ 1.2m (4ft) ↔ 1m (3ft)

SHRUB MEDIUM

Philadelphus 'Beauclerk'

MOCK ORANGE This deciduous shrub has slightly arching stems that bear oval, deeply veined leaves. In early summer, highly fragrant, single, white blooms, with a faint purple staining at the centre, open. Grow at the back of a border.

↕↔ 2.5m (8ft)

SHRUB SMALL

Phlomis fruticosa

JERUSALEM SAGE A spreading, evergreen shrub, with round whorls of yellow, trumpet-shaped flowers on upright stems in summer and grey-green leaves, which release a musky, herbal scent when brushed. The seedheads provide interest over winter.

↕ 1m (3ft) ↔ 1.5m (5ft)

PERENNIAL LARGE

Phlox paniculata 'Mount Fuji'

This clump-forming perennial produces tall stems of lance-shaped leaves, topped with large domed heads of white, fragrant blooms from midsummer to early autumn, provided the soil is moist.

↕ 1.2m (4ft) ↔ 60cm (24in)

SHRUB MEDIUM

Phygelius x *rectus*

This semi-evergreen, upright shrub has lance-shaped, green leaves and clusters of long, tubular flowers in summer. It may die back in winter when young. Varieties include 'Moonraker' (above), with custard-yellow blooms.

↕ 1m (3ft) ↔ 1.2m (4ft)

TREE LARGE

Picea pungens 'Koster'

COLORADO SPRUCE This evergreen conifer is cone-shaped and bears stiff, bright silvery blue, needle-like foliage. Upright cones form towards the ends of the branches. Grow it in a sheltered site in neutral to acid soil.

↕ 15m (50ft) ↔ 5m (15ft)

SHRUB SMALL

***Potentilla fruticosa* 'Abbotswood'**

SHRUBBY CINQUEFOIL This tough, reliable, deciduous shrub has a bushy, rounded or domed habit and white, round blooms for many weeks from summer to autumn. The foliage is divided into small leaflets.

↕1m (3ft) ↔1.5m (5ft)

BULB MEDIUM

Puschkinia scilloides* var. *libanotica

STRIPED SQUILL A perennial bulb, with strap-shaped leaves and, in spring, clusters of bell-shaped, white blooms, with a blue stripe on each petal. Grow in drifts in lawns, or with spring bulbs and alpines.

↕15cm (6in)

TREE SMALL

***Pyrus salicifolia* 'Pendula'**

WEEPING SILVER PEAR A small tree, with branches trailing to the ground and cream spring flowers that appear before the narrow, silver-grey leaves are fully open. Its size and spread may be controlled by trimming into a bell-domed standard.

↕8m (25ft) ↔6m (20ft)

PERENNIAL LARGE

Romneya coulteri

TREE POPPY This late summer-flowering, back-of-border, shrubby perennial has elegant, deeply divided, grey-green leaves and large, white, fragrant, bowl-shaped flowers, with a central dome of golden stamens.

↕↔2m (6ft)

CLIMBER LARGE

***Rosa* 'Félicité Perpétue'**

This vigorous, rambling rose has semi-evergreen, glossy leaves. Grow it up a large tree, or over a wall or pergola where the spectacular summer flush of blush pink to white, fully double, fragrant blooms can be appreciated.

↕5m (15ft)

SHRUB SMALL

Ruta graveolens

COMMON RUE A rounded, evergreen shrub for herb gardens, with blue-green, aromatic, ferny foliage, divided into oval leaflets. Remove the yellow flowers to promote a stronger leaf effect. Toxic, if ingested; sap may blister skin in sunlight.

↕1m (3ft) ↔75cm (30in)

SHRUB SMALL

***Salvia officinalis* 'Purpurascens'**

PURPLE SAGE An evergreen, woody-based shrub, with aromatic, oval leaves, purple when young and dusty purple-green when mature, and lilac summer flowers. Use as edging or at the front of mixed borders.

↕up to 80cm (32in) ↔1m (3ft)

PERENNIAL MEDIUM

Salvia patens

This herbaceous perennial has oval to triangular, hairy leaves that clothe the upright stems. From midsummer to autumn, the stems are topped with open heads of two-lipped, vivid blue flowers. 'Cambridge Blue' AGM has paler blooms.

↕60cm (24in) ↔45cm (18in)

PERENNIAL MEDIUM

Santolina chamaecyparissus

COTTON LAVENDER A dwarf, evergreen perennial, often grown as low hedging or edging, with dense foliage made up of narrow, dissected, silver-grey, aromatic leaves. Remove the flowers before they open to maintain the silvery effect.

↕75cm (30in) ↔1m (3ft)

PERENNIAL SMALL

Senecio cineraria

CINERARIA An evergreen, woody-based perennial, often grown as an annual in bedding schemes, but in mild gardens and sheltered sites the finely cut, felted, silver leaves last well into autumn. Cut back to prevent flowering.

↕↔30cm (12in)

SHRUB MEDIUM

Spiraea* x *vanhouttei

BRIDAL WREATH A deciduous shrub, with a mound-forming habit. It has small leaves and arching stems of white flowerheads from mid- to late spring. Grow as an informal flowering hedge, a back-of-border plant, or as a specimen.

↕2m (6ft) ↔1.5m (5ft)

PERENNIAL SMALL

Stachys byzantina

LAMBS' EARS An evergreen, mat-forming perennial, with soft-to-the-touch, oval, silvery grey leaves. In summer, spikes of tiny, clustered, mauve-pink flowers form on felted stems. It overwinters best in sunny, sheltered areas.
↕38cm (15in) ↔60cm (24in)

TREE MEDIUM

Styrax japonicus

JAPANESE SNOWBELL A deciduous tree, with pointed, oval leaves and a spreading habit, it makes a choice specimen for a sheltered border or lawn. Clusters of white, fragrant bell-shaped blooms hang from the branches in early summer.
↕10m (30ft) ↔8m (25ft)

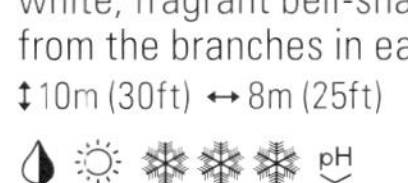

SHRUB LARGE

Syringa vulgaris 'Madame Lemoine'

LILAC A large, deciduous shrub or small tree, with heart-shaped leaves and, in spring, trusses of cone-shaped heads of double, strongly perfumed flowers that open from creamy buds.
↕↔7m (22ft)

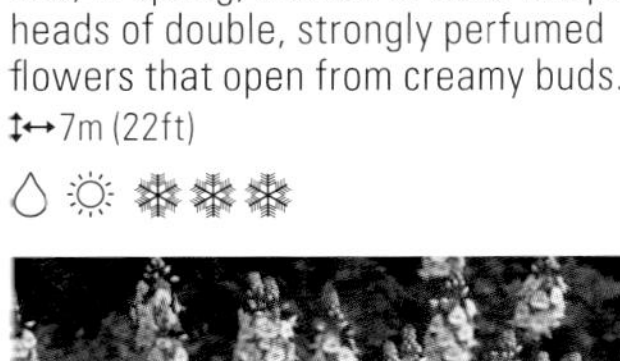

PERENNIAL SMALL

Trillium ovatum

This woodland perennial, slowly spreading by rhizomes, produces single stems, each with three, equally spaced, heart-shaped leaves, above which sit three-petalled, white flowers in early spring. Leave plants undisturbed to slowly naturalize.
↕38cm (15in) ↔20cm (8in)

ANNUAL/BIENNIAL LARGE

Verbascum olympicum

OLYMPIAN MULLEIN This biennial, during its first year, forms a rosette of large, grey, felted leaves, which may die back in winter. The next spring, it bears tall, upright, silver stems and golden, saucer-shaped flowers from mid- to late summer. May need staking.
↕2m (6ft) ↔1m (3ft)

PERENNIAL SMALL

Veronica gentianoides

A perennial with mats of evergreen, glossy, oval leaves and slender spires of very pale blue blooms in early summer. It is ideal for border edging or rock gardens. Remove spent flower spikes, and water in dry spells.
↕↔45cm (18in)

SHRUB MEDIUM

Viburnum acerifolium

MAPLE LEAF VIBURNUM This deciduous, upright shrub thrives in partially shaded borders, adding highlights with its green, lobed leaves and early summer heads of fluffy, creamy white flowers. Autumn colour and fruits are a bonus.
↕2m (6ft) ↔1.2m (4ft)

SHRUB LARGE

Viburnum x *carlcephalum*

FRAGRANT SNOWBALL A bushy, rounded, deciduous shrub, with dense ball-shaped clusters of small, sweetly fragrant, white blooms that open from pink buds from mid- to late spring. Red-turning-black summer fruits sometimes follow.
↕↔3m (10ft)

OTHER SUGGESTIONS

Annuals and biennials

Onopordum acanthium

Perennials and bulbs

Allium 'Mount Everest' • *Artemisia absinthium* 'Lambrook Silver' • *Astelia chathamica* • *Brunnera* 'Jack Frost' • *Delphinium* 'Blue Butterfly' • *Dianthus* 'Milky Way' • *Hosta* 'Blue Moon' • *Puschkinia scilloides* 'Alba' • *Tanacetum ptarmiciflorum* 'Silver Feather'

Shrubs

Camellia japonica 'Alba Plena' • *Choisya* x *dewitteana* 'Aztec Pearl' • *Cistus battandieri* • *Cornus kousa* 'Milky Way' • *Halimium umbellatum* • *Hebe* 'White Gem' • *Hydrangea* 'Annabelle' • *Osmanthus delavayi* • *Picea pungens* 'Hoopsii' • *Rhododendron* 'Polar Bear' • *Syringa vulgaris* 'Madame Florent Stepman'

Plants for pastel colours

Romantic and feminine, a planting scheme of pinks, purples, peaches, and custard-yellows produces a calming effect in the garden.

These easy-on-the-eye shades are much sought-after and you will find a wide variety of plants to suit such a scheme. The only danger is that designs made up exclusively of soft shades can look rather bland, so to balance the pale hues add a few high- and lowlights, such as bright pinks and dark burgundy. A sprinkling of small, white blooms will also lighten subdued pastel schemes. In addition, lift the focus to eye level and above by planting climbers, such as a pink climbing rose or an elegant purple clematis, and use foliage plants like *Cornus florida*, *Solenostemon*, and *Spiraea* for long-lasting colour.

SHRUB LARGE

Abutilon x *suntense*

INDIAN MALLOW The grey-green foliage of this upright, deciduous shrub sets off the pale to deep purple, bowl-shaped flowers that appear from late spring to early summer. Grow it in a sheltered site as a backdrop to a pastel colour-themed border.

↕4m (12ft) ↔2.5m (8ft)

TREE LARGE

Acer negundo

This large tree has lobed, reddish brown young leaves that turn mid-green in summer. Grow the variety 'Flamingo' (above) for its white- and pink-variegated foliage, but remember that the leaves of all forms turn bright yellow in autumn.

↕15m (50ft) ↔10m (30ft)

PERENNIAL MEDIUM

Achillea 'Taygetea'

The combination of this perennial's feathery, grey-green foliage and flat heads of pale yellow summer flowers provide a subtle highlight in a pastel-themed bed or border. Plant it with purple and dark pink flower spikes in free-draining soil and full sun.

↕60cm (24in) ↔50cm (20in)

PERENNIAL MEDIUM

Agastache 'Black Adder'

GIANT HYSSOP Decorate sunny beds and borders with this perennial's spikes of fluffy, violet-purple flowers, which appear from late summer to mid-autumn above slim, green leaves. Use it with other pastels or to contrast with white or fiery shades.

↕60cm (24in) ↔45cm (18in)

BULB MEDIUM

Allium cristophii 🏆

STAR OF PERSIA Prized for its spherical, large heads of star-shaped, violet flowers in early summer, this bulb also forms decorative seedheads, providing a long season of interest. The narrow, grey-green leaves fade as the flowers appear.

↕40cm (16in)

BULB LARGE

Allium 'Purple Sensation' 🏆

This perennial bulb offers globes of small, purple flowers, held aloft on sturdy stems in early summer, followed by decorative buff-coloured seedheads. Disguise the strap-shaped leaves, which fade as the flowers appear.

↕80cm (32in)

BULB MEDIUM

Allium senescens

This early summer-flowering bulb has short stems, topped with spherical, purple-pink flowerheads that bloom for many weeks, and blue-green foliage. Plant it in autumn. The most popular form is *A. senescens* subsp. *glaucum*, with pale lilac flowers.

↕60cm (24in)

PERENNIAL SMALL

Androsace lanuginosa

WOOLLY ROCK JASMINE This trailing, evergreen perennial forms a low mat of silky, grey-green, oval, tapering leaves and makes a beautiful border edging or rock garden specimen. It bears clusters of small, lilac-pink flowers in midsummer.

↕4cm (1½in) ↔18cm (7in)

PERENNIAL SMALL

Anemone coronaria De Caen Group

Plant this blue-purple perennial corm on gravel beds and border edges, but avoid the variant with red flowers. The shallow bowl-shaped blooms are held above divided, green leaves from early spring.

↕30cm (12in) ↔15cm (6in)

PERENNIAL MEDIUM

Aster 'Coombe Fishacre'

This dainty perennial is perfect for pastel-themed borders; its dark foliage provides a foil for dense sprays of small, pale lilac-pink flowers from late summer to mid-autumn. Try it mid-border with white Japanese anemones.

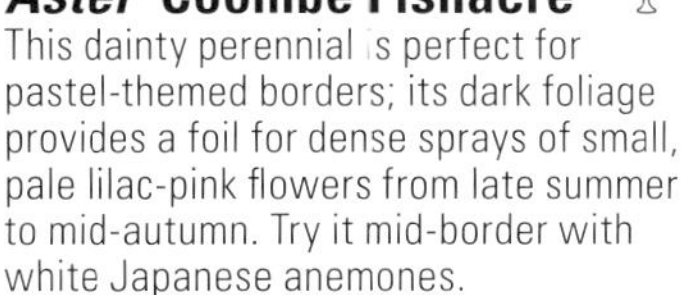

↕90cm (36in) ↔35cm (14in)

PERENNIAL MEDIUM

Aster x *frikartii*

Add this long-flowering perennial to the middle of a sunny border, where it will produce yellow-centred lavender-blue or lilac, daisy-like flowers from midsummer to early autumn. Try it with coneflowers and *Anemone* x *hybrida*.

↕70cm (28in) ↔40cm (16in)

PERENNIAL SMALL

Aster novi-belgii

This dwarf perennial is perfect for pots and borders. The daisy-like red, pink, white, blue, and purple blooms are held on leafy stems. Keep the soil moist to prevent powdery mildew disease. 'Professor Anton Kippenberg' has blue blooms.

↕35cm (14in) ↔45cm (18in)

CLIMBER MEDIUM

Billardiera longiflora

CLIMBING BLUEBERRY This evergreen, twining climber works well as a backdrop to a sheltered border. It produces slender, bell-shaped, green-yellow summer flowers, followed by purple-blue autumn fruits. Provide winter protection in cold climes.

↕2m (6ft)

SHRUB LARGE

Camellia x *williamsii*

The flowers of the williamsii camellias come in a range of pastel pinks, and appear on this evergreen shrub in spring among glossy, dark green leaves. It tolerates full shade and prefers a sheltered, north- or west-facing site.

↕5m (15ft) ↔2.5m (8ft)

PERENNIAL LARGE

Campanula lactiflora

MILKY BELLFLOWER Both the tall varieties of this perennial and shorter forms, such as 'Prichard's Variety' AGM (above), bear bell-shaped, nodding, violet-blue, occasionally pink or white, flowers all summer. Plant in groups among other pastel-coloured blooms.

↕up to 1.2m (4ft) ↔60cm (24in)

PERENNIAL SMALL

Campanula punctata

BELLFLOWER The dusky pink, pendent blooms of this perennial lend an elegant touch to the front of a border. Partner it with cool colours or purple blooms, such as geraniums, but avoid bright, hot colours that may eclipse its delicate hue.

↕30cm (12in) ↔40cm (16in)

PERENNIAL MEDIUM

Catananche caerulea

Plant this hardy perennial in groups in a wild or naturalistic garden, where its dainty blue or white, star-like blooms will wave in the breeze on tall, wiry stems above grey-green foliage. 'Bicolor' (above) is a white, semi-double form.

↕60cm (24in) ↔30cm (12in)

SHRUB MEDIUM

Chaenomeles speciosa

Choose a pink- or peach-flowered form of this deciduous, bushy shrub to dress up a pastel scheme in spring. The thorny stems are ideal for training up a sunny or part-shaded wall. 'Moerloosei' AGM (above) is a popular variety.

↕2.5m (8ft) ↔5m (15ft)

CLIMBER MEDIUM

Clematis alpina

ALPINE CLEMATIS This deciduous clematis, with divided, mid-green leaves and lantern-shaped, blue, pink, or white flowers, makes a beautiful feature when grown through a spring-flowering shrub or up an arch. Fluffy seedheads follow the flowers.

↕3m (10ft)

CLIMBER MEDIUM

Clematis 'Bees Jubilee'

Grow this compact, large-flowered, deciduous clematis in partial shade against a wall or fence, or up a rose arch to complement pastel plantings. The rose-pink early summer flowers feature a central, darker pink stripe on each petal.

↕2.5m (8ft)

TREE SMALL

Cornus florida

FLOWERING DOGWOOD White or pink, flower-like bracts cover this deciduous tree or shrub in late spring and make a focal point in a pastel-themed setting. The curled red and purple autumn leaves provide a colourful backdrop to late-season borders.

↕6m (20ft) ↔8m (25ft)

CLIMBER MEDIUM

Clematis 'Guernsey Cream'

The large, cream flowers, with a green bar down the centre of each petal, appear on this compact evergreen clematis in early summer and again in late summer. Use them to decorate an arch, or a wall or fence, at the back of a border.

↕2.5m (8ft)

SHRUB MEDIUM

Clethra alnifolia

SWEET PEPPER BUSH This bushy shrub bears slender spires of small, fragrant, bell-shaped, white blooms in late summer. It offers a complementary backdrop to pastel pinks and purples in a damp, shady spot. The foliage turns yellow in autumn.

↕↔2.5m (8ft)

TREE SMALL

Cornus florida 'Welchii'

This conical-shaped tree or large, deciduous shrub has tiny, green flowers surrounded by white or pink, petal-like bracts in late spring and variegated leaves, which feature pink autumn leaf tints.

↕6m (20ft) ↔8m (25ft)

SHRUB MEDIUM

Cornus mas 'Aurea' 

Use this deciduous shrub as a backdrop to a border of dark pinks and purples. Its pale golden-yellow spring leaves turn lime-green in summer and clusters of small, bright yellow flowers appear on bare branches in early spring.

↕2m (6ft) ↔3m (10ft)

BULB SMALL

Crocus vernus

This dwarf, spring to early summer-flowering bulb produces small, cup-shaped, lilac or purple flowers and can be naturalized in lawns, or combined with other pastel flowers in troughs or pots filled with gritty, soil-based compost.
↕12cm (5in)

BULB SMALL

Cyclamen coum Pewter Group ♡

This perennial, grown from autumn-planted tubers, has rounded, dark green leaves, often with silver patterning. The small, pink flowers with swept-back petals appear in winter or early spring with the leaves.
↕8cm (3in)

BULB LARGE

Dahlia 'Franz Kafka'

Grown from spring-planted tubers, this perennial bulb bears pink pompon flowers with a honeycomb texture above the foliage, from mid- to late summer and early autumn. There are many pastel-coloured dahlia varieties to choose from.
↕80cm (32in)

PERENNIAL LARGE

Delphinium elatum

Available in a range of pastel blues, pinks, and purples, this upright perennial adds spikes of colour to a border when its flowers appear in midsummer above deeply cut, green leaves. Provide support for the stems.
↕2m (6ft) ↔90cm (36in)

PERENNIAL SMALL

Diascia barberae 'Blackthorn Apricot' ♡

Perfect for pots or the front of a sunny border, this perennial, sold as annual bedding, produces spikes of apricot-pink flowers above heart-shaped, tapering leaves from summer to autumn.
↕25cm (10in) ↔50cm (20in)

SHRUB SMALL

Deutzia x *elegantissima*

Plant this deciduous, rounded shrub where the scent of its flowers can be enjoyed. Most varieties are in shades of pastel pink, lending colour and fragrance to late spring and early summer beds. 'Rosealind' AGM (above) has pink-flushed white blooms.
↕1.2m (4ft) ↔1.5m (5ft)

PERENNIAL MEDIUM

Dianthus caryophyllus

BORDER CARNATION This perennial, with grey-green leaves and tall, slender stems of rounded flowers in a range of pastel shades, is suited to a cottage or informal garden. Dwarf forms, ideal for pots or the front of a border, are also available in pastel shades.
↕80cm (32in) ↔23cm (9in)

BULB LARGE

Dierama pulcherrimum

ANGEL'S FISHING ROD The elegant, arching stems of this summer-flowering corm bear dangling, funnel-shaped, deep pink flowers, and emerge from a clump of narrow, strap-shaped, evergreen leaves. Use it to edge a border with cream and purple blooms.
↕1.5m (5ft)

SHRUB LARGE

Dipelta floribunda ♡

ROSY DIPELTA Unassuming for much of the year, this deciduous shrub, with lance-shaped, green leaves, comes to life from late spring to early summer, when clusters of funnel-shaped, creamy white flowers, with orange and pink throats, appear.
↕↔4m (12ft)

PERENNIAL LARGE

Eremurus robustus ♡

FOXTAIL LILY This perennial produces tall, slim spikes of pale pink blooms and looks most dramatic when planted in groups where the sun can bake the tubers in late summer. Add grit to the soil before planting tubers in spring; stems may need staking.
↕3m (10ft) ↔1.2m (4ft)

SHRUB SMALL

Erica x *darleyensis*

DARLEY DALE HEATH Plant drifts of this dwarf, evergreen shrub to form a carpet of urn-shaped, white or pink flowers from late winter to early spring. The lance-shaped foliage is dark green. A popular variety is f. *albiflora* 'White Glow' (above).

↕25cm (10in) ↔50cm (20in)

PERENNIAL LARGE

Eryngium x *tripartitum*

This perennial produces a rosette of spiny, grey-green leaves and tall, branched, violet-blue stems, topped with small, cone-like, purple flowerheads, surrounded by narrow, dark blue bracts. Grow in a border with peach and white neighbours.

↕1.2m (4ft) ↔50cm (20in)

PERENNIAL LARGE

Gaura lindheimeri

This upright perennial has wand-like stems dotted with starry, butterfly-shaped, pink-budded white flowers throughout summer, which appear above small, lance-shaped, green leaves. Use its "see-through" stems at the front of a border or beside a path.

↕1.5m (5ft) ↔90cm (36in)

PERENNIAL MEDIUM

Geranium sylvaticum

WOOD CRANESBILL Producing a carpet of saucer-shaped, blue-purple flowers with white centres over lobed leaves from late spring to early summer, this clump-forming perennial makes excellent ground cover for a pastel border in sun or part shade.

↕75cm (30in) ↔60cm (24in)

SHRUB MEDIUM

Escallonia 'Apple Blossom'

Use this bushy, evergreen shrub as hedging or a backdrop to a border. Its glossy, spine-edged, dark green leaves are joined by small, tubular, pink and white flowers from summer to early autumn, which complement cool and pastel colours.

↕↔2.5m (8ft)

SHRUB MEDIUM

Fuchsia magellanica var. *molinae*

Suitable for an informal hedge or the back of a partly shaded border, this tall, hardy, deciduous shrub bears pendent, bell-shaped, pale pink flowers from summer to early autumn, and small, lance-shaped leaves.

↕↔2m (6ft)

BULB LARGE

Gladiolus 'Passos'

This bulb's deep purple- and red-centred pale purple flower spikes create a focal point in summer borders, while the sword-like foliage also adds interest. Plant bulbs at two-week intervals from early spring to summer to lengthen the flowering period.

↕1.2m (4ft)

PERENNIAL SMALL

Helleborus purpurascens

LENTEN ROSE Use this clump-forming perennial in shady borders with pale yellow or white daffodils that also bloom in early spring, when small, nodding, cup-shaped, deep purple flowers with cream stamens appear above dark green, divided leaves.

↕↔30cm (12in)

PERENNIAL SMALL

Hemerocallis 'Arctic Snow' 🏆

DAYLILY A deciduous perennial, with arching, strap-like, green leaves that provide a foil for other flowers before sturdy stems of large, cream flowers, with pale yellow margins and throats, appear from summer to early autumn, each bloom lasting a day.
↕55cm (22in) ↔50cm (20in)

PERENNIAL MEDIUM

Hyssopus officinalis

HYSSOP This upright perennial is a decorative herb that can also be used in mixed sunny borders. It has aromatic leaves that smell and taste of aniseed, and spikes of fluffy, lavender-blue summer flowers that attract bees and butterflies.
↕60cm (24in) ↔1m (3ft)

PERENNIAL MEDIUM

Hemerocallis 'Dan Mahony'

DAYLILY This perennial forms a clump of upright, arching, strap-like foliage. Large, trumpet-shaped, pink flowers, with darker centres appear over several weeks in summer, and each bloom lasts a day. It resents root disturbance.
↕65cm (26in) ↔11cm (4½in)

PERENNIAL SMALL

Heuchera 'Plum Pudding'

This leafy, evergreen perennial is grown for its rounded, lobed and veined, maroon-purple leaves, which make a great match for pink, cream, and pale yellow flowers. It produces small, white flowers in summer. Remove spent blooms and old foliage.
↕50cm (20in) ↔30cm (12in)

SHRUB LARGE

Hydrangea aspera Villosa Group 🏆

This upright, deciduous shrub makes a good partner for pale-hued perennials. It has large, velvety leaves and small, mauve late summer flower clusters, surrounded by pale blooms.
↕↔3m (10ft)

PERENNIAL MEDIUM

Iris 'Harriette Halloway'

BEARDED IRIS This bearded iris's large, scented, pale blue summer flowers, with yellow centres, and sword-like foliage lend elegance to the front of a sunny border or path edge. Plant the rhizomes just above the soil surface to allow them to bake in the sun.
↕70cm (28in) ↔30cm (12in)

PERENNIAL MEDIUM

Lamprocapnos spectabilis 🏆

syn. *Dicentra spectabilis* Use this perennial in a late spring border, when its fern-like, mid-green foliage is joined by arching stems that bear heart-shaped, rose-red and white flowers. Try it in a border with early-flowering geraniums.
↕↔1m (3ft)

CLIMBER MEDIUM

Lathyrus odoratus

SWEET PEA Add height to a pastel scheme by training this annual tendril climber on a tripod support. The pink, blue, purple, or white flowers bloom from summer to early autumn when deadheaded regularly. Sow the seeds indoors in autumn or early spring.
↕3m (10ft)

PERENNIAL LARGE

Liatris spicata

GAYFEATHER This upright perennial produces poker-like spikes of fluffy, pinky purple or white flowers from late summer to early autumn above clumps of linear leaves. 'Kobold' (above) has vivid purple-pink blooms.
↕1.5m (5ft) ↔45cm (18in)

PERENNIAL MEDIUM

Malva moschata
MUSK MALLOW This perennial produces heart-shaped basal leaves and a profusion of saucer-shaped, pale pink flowers from summer to early autumn. Its loose habit makes it ideal for a wildflower or naturalistic scheme.
↕1m (3ft) ↔60cm (24in)

BULB MEDIUM

***Narcissus* 'Canaliculatus'**
DAFFODIL A good partner for pale lemon daffodils and purple hellebores, this bulb has clusters of small, fragrant flowers, with reflexed white petals and yellow cups, that appear from early to mid-spring among linear, green leaves. Plant bulbs in autumn.
↕23cm (9in)

ANNUAL/BIENNIAL SMALL

Nigella damascena
LOVE-IN-A-MIST Sow this annual in a sunny border for its feathery foliage effect and white, lavender, purple, rose, and blue, saucer-shaped flowers. The unusual blooms sport a "ruff" of spidery leaves and are followed by large, decorative seed pods.
↕45cm (18in) ↔20cm (8in)

ANNUAL/BIENNIAL SMALL

Nemophila menziesii
BABY BLUE EYES Use this annual, with its serrated, grey-green leaves and saucer-shaped, white-centred, blue summer flowers, to edge patio pots or sunny borders. Combine it with taller plants that have pastel or hot-hued blooms.
↕20cm (8in) ↔15cm (6in)

BULB MEDIUM

Nerine bowdenii
BOWDEN CORNISH LILY Decorate the front of autumn borders with this spring-planted perennial bulb, which produces slim stems topped with spidery pink flowers before the strap-shaped leaves unfurl. Try it with the pastel-coloured *Aster amellus*.
↕60cm (24in)

PERENNIAL SMALL

Oenothera speciosa
PINK EVENING PRIMROSE Ideal for the front of a free-draining, sunny border, this perennial has divided, green leaves and cup-shaped, fragrant, pink-flushed white flowers from late spring to early summer. Try it alongside *Nemophila* and geraniums.
↕↔30cm (12in)

PERENNIAL MEDIUM

Paeonia lactiflora
PEONY A reliable perennial for a pastel-colour-themed border, with red, mottled stems of dark green, divided leaves and, from early- to midsummer, single or double, bowl-shaped, fragrant, white to pale pink flowers, with yellow centres.
↕↔70cm (28in)

SHRUB MEDIUM

Paeonia suffruticosa
TREE PEONY With many pastel-flowered varieties, this shrub bears single or double blooms in late spring against lobed, dark green leaves. Use it to add height and architectural foliage interest to a border. 'Reine Elisabeth' has double, pink flowers.
↕↔2m (6ft)

PERENNIAL MEDIUM

***Papaver orientale* 'Patty's Plum'**
ORIENTAL POPPY Use this perennial, with its large, ruffled, dusky purple summer flowers, which are followed by decorative seedheads, as a contrast to cream and pale yellow blooms. Stake the stems.
↕90cm (36in) ↔60cm (24in)

PERENNIAL SMALL

Pelargonium 'Lady Plymouth'

SCENTED-LEAVED PELARGONIUM This perennial, usually grown as an annual, bears eucalyptus-scented, lobed, silver-margined green leaves and clusters of lavender-pink summer flowers.

↕40cm (16in) ↔20cm (8in)

PERENNIAL SMALL

Penstemon 'Evelyn'

This compact, bushy, semi-evergreen perennial, with its narrow foliage and slim panicles of tubular, rose-pink flowers that appear from summer to early autumn, makes a colourful addition to summer borders. Protect with a dry mulch in winter.

↕↔45cm (18in)

PERENNIAL MEDIUM

Persicaria bistorta

In sun or part-shade, this clump-forming perennial produces a mat of oval foliage below spikes of soft pink flowers, which bloom from early to late summer. Plant it with purple, white, and dark pink partners in moist soil.

↕75cm (30in) ↔60cm (24in)

PERENNIAL SMALL

Petunia 'Pastel Mixed'

This perennial, grown as an annual, bears trumpet-shaped blooms from summer to early autumn, and provides a rich source of pastels for colour-themed containers and raised beds. Deadhead the flowers regularly to ensure a long display.

↕25cm (10in) ↔45cm (18in)

PERENNIAL MEDIUM

Phlomis russeliana

Offering a long season of interest, this perennial has large, rough-textured, heart-shaped leaves, and unusual hooded, butter-yellow summer flowers, set at intervals up the tall stems. The seedheads provide interest over winter.

↕1m (3ft) ↔60cm (24in) or more

PERENNIAL MEDIUM

Phormium 'Evening Glow'

With forms available in a range of pinks and purples, this leafy perennial produces fountains of evergreen foliage creating focal points in borders or large containers filled with soil-based compost, and complementing pastel-coloured flowers.

↕↔75cm (30in)

SHRUB SMALL

Phygelius aequalis

In a sunny, sheltered site, this shrub bears tall stems of pendent, tubular, dusky pink or yellow summer blooms. If hit by winter frosts, cut stems to the ground in spring and it should reshoot. 'Yellow Trumpet' (above) has creamy yellow flowers.

↕↔1m (3ft)

PERENNIAL SMALL

Primula denticulata

This perennial produces a rosette of toothed, green leaves and, from mid-spring to summer, spherical clusters of yellow-eyed purple or white flowers on stout stems. Plant it at pond edges in damp soil. *P. denticulata* var. *alba* has white flowers.

↕↔45cm (18in)

PERENNIAL MEDIUM

Primula vialii

ORCHID PRIMROSE Perfect for areas with damp soil, this clump-forming perennial bears unusual tapering, two-tone cones of bluish purple and dark red flowers on sturdy stems in late spring. The blooms appear above a rosette of oval leaves.

↕60cm (24in) ↔30cm (12in)

PERENNIAL SMALL

Pulmonaria saccharata

BETHLEHEM SAGE Use this clump-forming, semi-evergreen perennial with daffodils and Dutch irises in pastel-themed beds. Its white-spotted blue-green leaves make great edging and the funnel-shaped, blue-purple spring flower clusters attract bees.
↕30cm (12in) ↔60cm (24in)

CLIMBER LARGE

Rosa 'Albertine'

This rambler rose's stems of dark green leaves and scented, fully double, salmon-pink flowers, which appear in a single flush in summer, make a beautiful backdrop to a border when trained on a large wall or fence, or over a pergola.
↕5m (15ft)

SHRUB SMALL

Rosa 'Buff Beauty'

A vigorous shrub, with long, arching stems of glossy foliage and clusters of small, fragrant, semi-double, apricot-yellow and buff flowers from summer to autumn, which contrast well with purples and dark pinks.
↕↔1.2m (4ft)

CLIMBER LARGE

Rosa 'Climbing Lady Hillingdon'

Train this vigorous hybrid tea rose on a wall or fence behind a pastel-themed border, where its fragrant, apricot summer flowers and dark green foliage, coppery mahogany when young, will form a colourful backdrop.
↕5m (15ft)

PERENNIAL MEDIUM

Sidalcea malviflora

CHECKERBLOOM The informal style of this upright perennial lends itself to wildlife and cottage schemes. Tall stems of pink or lilac-pink, funnel-shaped flowers appear above kidney-shaped basal leaves in summer. Combine it with geraniums and achilleas.
↕90cm (36in) ↔45cm (18in)

SHRUB MEDIUM

Rosa glauca

BLUE-LEAVED ROSE The single, rose-pink summer flowers, with pale centres, and red stems of greyish purple leaves make this rose a perfect addition to pastel-themed plantings. Use it at the back of a border or as an informal hedge; prune in late winter.
↕2m (6ft) ↔1.5m (5ft)

PERENNIAL MEDIUM

Salvia farinacea

MEALY SAGE Use this upright perennial, grown as an annual, as border edging or in pots with pink, lavender, and cream blooms. Its spikes of small, purple-blue flowers appear over a long period from summer to autumn, if deadheaded often.
↕60cm (24in) ↔30cm (12in)

PERENNIAL SMALL

Sedum erythrostictum

STONECROP This clump-forming perennial bears fleshy, grey-green leaves and starry, greenish white flowers in early autumn. Varieties include 'Mediovariegatum' (above), with cream-splashed variegated leaves. Grow it in sun in gritty soil.
↕30cm (12in) ↔60cm (24in)

SHRUB LARGE

Stachyurus praecox

From late winter to early spring, the pale greenish yellow, bell-shaped flowers of this deciduous shrub hang in clusters from bare stems, creating an exciting feature. Slim, tapering, dark green leaves unfurl in spring.
↕4m (12ft) ↔3m (10ft)

PERENNIAL SMALL

Stokesia laevis

STOKES ASTER Ideal for cottage or informal borders, this evergreen perennial has mid-green leaves and lavender- or purple-blue, cornflower-like flowers on short stems, from midsummer to mid-autumn. 'Purple Parasols' (above) has violet-purple blooms.
↕↔45cm (18in)

PERENNIAL LARGE

Thalictrum aquilegiifolium

MEADOW RUE This useful perennial for shady borders and moist soil has clusters of fluffy, lilac-purple summer flowers that are ideal for pastel-coloured plantings. Its finely divided, grey-green leaves and attractive autumn seedheads add to its appeal.

↕1.2m (4ft) ↔45cm (18in)

SHRUB MEDIUM

Syringa x *persica* ♡

PERSIAN LILAC This compact, bushy shrub, suited to the back of a border in a small garden, infuses the air with the scent of its purple flowers, which appear in small, dense clusters in late spring. The lance-shaped foliage is dark green.

↕↔2m (6ft)

SHRUB LARGE

Syringa vulgaris

COMMON LILAC Grown in borders or as a boundary screen, this vigorous, deciduous shrub offers scent and colour when the large, conical clusters of pastel-coloured flowers appear in late spring, after the heart-shaped foliage unfurls.

↕↔7m (22ft)

BULB MEDIUM

Tulipa 'China Pink' ♡

This lily-flowered form bears rosy pink, lily-like blooms in late spring and has broad, grey-green foliage. Other varieties offer a spectrum of pastel-coloured flowers from mid- to late spring. Plant the bulbs in groups in autumn.

↕55cm (22in)

BULB MEDIUM

Tulipa 'Negrita'

Combine this dark purple-pink tulip with pale pink and cream varieties that flower at the same time in mid-spring and use it in a pastel border or to fill a formal parterre. Plant the bulbs in groups in a sunny site in autumn.

↕45cm (18in)

PERENNIAL LARGE

Veronica spicata

A mat of slim, lance-shaped, green foliage sits beneath this perennial's upright spikes of blue, pink, white, or purple, star-shaped summer flowers. Select a purple form, such as 'Romiley Purple' (above), or a pink variety for a pastel scheme.

↕1.2m (4ft) ↔60cm (24in)

PERENNIAL LARGE

Veronicastrum virginicum

CULVER'S ROOT This upright perennial bears slender spikes of small, star-shaped, white, blue, or pink flowers on tall, leafy stems. Plant it at the back of a border and provide support. 'Album' (above) has cool white flowers.

↕1.2m (4ft) ↔45cm (18in)

OTHER SUGGESTIONS

Perennials and bulbs

Agapanthus 'Purple Cloud' • *Anthyrium niponicum* var. *pictum* 'Burgundy Lace' • *Crocus corsicus* • *Dahlia* 'Sascha' ♡ • *Delphinium* 'Gillian Dallas' • *Gaura* 'Crimson Butterflies' • *Gentiana asclepiadea* ♡ • *Geranium sylvaticum* 'Amy Doncaster' • *Hemerocallis* 'Summer Wine' • *Lilium* 'Mona Lisa' • *Paeonia cambessedesii* ♡ • *Pelargonium* 'Mr Henry Cox' ♡ • *Primula* 'Guinevere' ♡ • *Salvia patens* ♡ • *Sidalcea* 'Oberon' • *Tulipa* 'Menton' ♡ • *Verbascum* 'Cotswold Beauty' • *Veronicastrum virginicum* 'Fascination'

Shrubs and climbers

Abutilon x *suntense* 'Violetta' • *Fuchsia* 'Delta's Sarah' • *Hebe* 'Purple Queen' • *Kolwitzia ambalis* 'Pink Cloud' • *Lavatera* x *clementii* 'Rosea' ♡ • *Rhododendron* 'Purple Spendour' • *Rosa* 'Many Happy Returns' ♡

Plants for hot and dark colours

Hot-hued flowers create drama in the garden, especially when teamed with smouldering dark purple or blue-black foliage plants.

Brightly coloured blooms also produce striking combinations with golden-leaved plants, such as *Choisya* SUNDANCE and *Sambucus racemosa*, and look effective when cooled with a selection of neutral greens. Purple foliage complements almost any colour, but take care when matching up golds with pastel shades of pink or peach to avoid unsightly clashes. Borders aglow with sizzling reds, oranges, and yellows catch the eye, and add highlights to more subtle colour schemes, while compact plants, such as bedding dahlias and varieties of *Salvia splendens*, make arresting focal points in container displays.

SHRUB MEDIUM

Abutilon megapotamicum ♡

TRAILING ABUTILON In a sheltered, sunny site, this evergreen shrub will brighten schemes with its bell-shaped, yellow and red summer flowers, which contrast with its heart-shaped, dark green leaves. Train the stems up a wall and protect against frost.

↕↔2m (6ft)

SHRUB MEDIUM

Acer palmatum var. *dissectum*

JAPANESE MAPLE The dark purple foliage of this deciduous shrub complements bright hot shades. It has a domed habit and bears lobed leaves, with finely divided, tapering "fingers". Plant in a sheltered site.

↕1.5m (5ft) ↔1m (3ft)

TREE SMALL

Acer shirasawanum 'Aureum' ♡

This deciduous shrub or small tree, with lobed, bright yellow foliage, complements a hot-hued border or gravel garden. It bears small, crimson spring flowers and bright yellow leaves that turn scarlet in autumn.

↕↔6m (20ft)

PERENNIAL MEDIUM

Achillea 'Moonshine' ♡

This short-lived perennial is perfect for the front of a sunny, hot colour-themed border. Its flat-topped, yellow flowerheads, which appear over a long period in summer, are held on slim stems above silvery, ferny foliage. It is resistant to flopping.

↕60cm (24in) ↔50cm (20in)

PERENNIAL MEDIUM

Achillea 'Paprika'

YARROW This semi-evergreen perennial's flat heads of tiny, scarlet flowers, which fade to buff as they mature, combine well with other scarlet blooms and contrasting golden-yellow plants. Its feathery leaves are silvery green.

↕80cm (32in) or more ↔40cm (16in)

PERENNIAL LARGE

Actaea simplex

BUGBANE Use the dark purple, divided foliage of this perennial as a foil for bright yellows and reds, or to complement moodier shades. From early- to mid-autumn, it bears bottlebrush, off-white tinged purple blooms. 'Brunette' (above) is a popular form.

↕1.2m (4ft) ↔60cm (24in)

CLIMBER LARGE

Actinidia kolomikta ♡

This vigorous, deciduous climber is grown for its large, heart-shaped leaves, many of which are marked with bold silver or pink patches. Planted on a wall or fence, it makes a useful backdrop to brightly coloured plants. Best in a sheltered site.

↕4m (12ft)

TREE SMALL

Aesculus pavia

RED BUCKEYE This relative of the horse chestnut lights up in spring when upright panicles of orange-red, tubular flowers appear. The large, divided, dark green leaves create an architectural feature. The autumn seeds are poisonous.

↕5m (15ft) ↔3m (10ft)

BULB MEDIUM

Anemone pavonina

To create a dazzling early spring display in raised beds and borders, combine this diminutive bulb, with its dark-eyed, bright red, pink, or purple flowers, with yellow daffodils and irises. The flowers sit above lobed, dissected leaves.

↕25cm (10in)

PERENNIAL MEDIUM

Aquilegia vulgaris

GRANNY'S BONNETS A few forms of this perennial are in darker shades, including 'William Guinness' (above), which features intricate, white-centred purple-black flowers from late spring to early summer, above fern-like, mid-green leaves.

↕90cm (36in) ↔45cm (18in)

PERENNIAL MEDIUM

Begonia sutherlandii

Create an eye-catching display in a basket or pot with this perennial, grown from spring-planted tubers. Its trailing stems of lobed, green leaves feature clusters of small, bright orange summer flowers. Treat it as an annual, or lift and protect tubers.

↕80cm (32in) ↔indefinite

ANNUAL/BIENNIAL SMALL

Brassica oleracea

ORNAMENTAL CABBAGE This annual offers decorative, colourful leaves to brighten up autumn and winter containers, and bedding displays. With loose rosettes of plain or frilly-edged, white, red, or pink leaves, it combines well with violas.

↕↔45cm (18in)

ANNUAL/BIENNIAL MEDIUM

Calendula officinalis

POT MARIGOLD The orange or yellow, daisy-like flowers of this bushy annual inject punchy hot colours into summer schemes. Sow seeds in flower borders or patio pots in spring, and use in an informal garden or formal parterres.

↕↔60cm (24in)

SHRUB SMALL

Calluna vulgaris 'Gold Haze'

HEATHER Use this gold-leaved, small evergreen shrub to carpet the front of a border with colour, or plant it with dwarf conifers in a large pot of ericaceous compost. Small, white flowers appear from late summer to autumn. It needs acid soil.

↕45cm (18in) ↔60cm (24in)

SHRUB LARGE

Camellia japonica

COMMON CAMELLIA The glossy, evergreen leaves of this shrub lend a neutral backdrop to hot-hued borders year-round, punctuated by early spring blooms. Choose a richly hued form, such as the salmon-red 'Blood of China' (above), for hot schemes.

↕3m (10ft) ↔2m (6ft)

PERENNIAL LARGE

Canna 'Durban'

The large, paddle-shaped foliage of this perennial produces a spectacular display of orange- and dark green-striped leaves from late spring, followed by ruffled, orange flowers on tall stems in late summer, offering a complete hot-hued package.

↕1.6m (5½ft) ↔50cm (20in)

PERENNIAL SMALL

Carex comans

NEW ZEALAND SEDGE This perennial forms clumps of arching, hair-like, green or bronze foliage. Inconspicuous flowerheads appear on long stems in summer. Use it to cascade over the sides of pots or try it in a raised bed as a foil for orange and red flowers.

↕35cm (14in) ↔75cm (30in)

PERENNIAL SMALL

Carex elata

SEDGE Ideal for lively bog garden schemes, this evergreen perennial forms tufts of long, arching, golden-yellow leaves. Thin, grass-like, blackish brown flower spikes appear in summer. Try it with scarlet lobelias and golden marsh marigolds.

↕40cm (16in) ↔15cm (6in)

TREE MEDIUM

Catalpa bignonioides 'Aurea'

INDIAN BEAN TREE This deciduous tree, with its large, golden-yellow leaves, makes a bright backdrop to a hot-hued border. The foliage turns light yellow-green in summer, when showy white flowers appear.

↕↔10m (30ft)

PERENNIAL SMALL

***Celosia argentea* 'Fairy Fountains'**

This perennial, sold as annual summer bedding, brightens up borders and pots with its large, plume-like, fluffy, hot-shaded flowerheads that appear all summer above the pale green, lance-shaped leaves.
↕↔30cm (12in)

SHRUB SMALL

Chaenomeles* x *superba

JAPANESE QUINCE Heat up the colour of a late-spring garden with this spiny, deciduous shrub. Its cup-shaped, orange, white, pink, or red flowers clothe the stems before the leaves unfurl. For a burst of orange-red, choose 'Crimson and Gold' AGM (above).
↕1m (3ft) ↔2m (6ft)

PERENNIAL MEDIUM

***Crocosmia* 'Lucifer'**

MONTBRETIA A clump-forming perennial, with tubular, red late summer flowers that inject fiery colour into beds and borders. Other varieties have flowers in bright oranges, yellows, or reds. The architectural, sword-shaped leaves are a bonus.
↕1m (3ft) ↔25cm (10in)

TREE SMALL

***Cercis canadensis* 'Forest Pansy'**

A deciduous, spreading, small tree or shrub, grown primarily for its heart-shaped, purple leaves, which turn yellow in autumn and make a fiery feature in the garden. Pale pink, pea-like flowers appear in mid-spring.
↕↔5m (15ft)

TREE SMALL

Cordyline australis

CABBAGE PALM This evergreen, palm-like tree makes a fountain of sword-shaped leaves in a range of green and purple. Use it as an accent plant in a scheme of brightly coloured blooms. 'Purpurea' is a good purple-leaved form.
↕3m (10ft) ↔1m (3ft)

ANNUAL/BIENNIAL MEDIUM

Cuphea ignea

CIGAR FLOWER This annual's slim, black-tipped red flowers shine all summer from among the green, lance-shaped leaves, making a decorative border edging or feature in pots of multi-purpose compost. It thrives in both full sun and part shade.
↕75cm (30in) ↔90cm (36in)

SHRUB LARGE

***Cotinus* 'Grace'**

SMOKE BUSH Plant this large, deciduous shrub at the back of a border, where it will create a blaze of fiery red, oval leaves in autumn. This variety has rich purple leaves in summer; other forms produce green or pale yellow summer foliage.
↕6m (20ft) ↔5m (15ft)

BULB LARGE

***Dahlia* 'Bishop of Llandaff'**

Few flowers can outshine the semi-double, clear red flowers of this popular tuber. The smouldering dark foliage provides the perfect foil for the fiery blooms. Plant the tubers outside after the frosts; stems may need staking.
↕1m (3ft)

PERENNIAL MEDIUM

Crocosmia* x *crocosmiiflora

MONTBRETIA This upright perennial, with its arching, sword-shaped, green leaves and trumpet-shaped summer to autumn flowers, is perfect for hot-hued borders. Try the golden form 'Star of the East' AGM (above) with orange-yellow blooms.
↕70cm (28in) ↔8cm (3in)

BULB LARGE

***Dahlia* 'Fire pot'**

The flowers of this decorative dahlia, grown from spring-planted tubers, sport a blend of sizzling pink and orange, with a dash of yellow in the centre. Use it to brighten up the back of a border, or in a large patio pot of soil-based compost.
↕90cm (36in)

BULB MEDIUM

Dahlia 'Yellow Hammer' ♡

The single, orange-streaked bright yellow flowers of this dwarf bedder dahlia will draw the eye in hot-coloured borders, while the dark bronze foliage provides a dramatic contrast. Lift and store tubers after the plants are blackened by frost.

↕60cm (24in)

PERENNIAL SMALL

Dicentra formosa

BLEEDING HEART Use this perennial in a brightly coloured border, where its pendent, heart-shaped, pink or red flowers will combine well with orange-tinted tulips from late spring to early summer. Its divided, grey-green foliage provides a leafy foil.

↕45cm (18in) ↔30cm (12in)

ANNUAL/BIENNIAL SMALL

Eschscholzia californica

CALIFORNIA POPPY Sprinkle the seeds of this clump-forming annual in spring at the front of a sunny, well-drained border to be rewarded with masses of bowl-shaped, orange, red, and yellow flowers in summer. It may self-seed after its first season.

↕30cm (12in) ↔15cm (6in)

PERENNIAL MEDIUM

Euphorbia griffithii

Ideal for injecting colour into a partly shaded border in early summer, this perennial offers red stems of narrow, copper-tinged dark green leaves, topped with domed clusters of orange-red flowers. Best in moist, fertile soil. The sap is a skin irritant.

↕75cm (30in) ↔50cm (20in)

SHRUB SMALL

Euryops acraeus ♡

MOUNTAIN EURYOPS Brighten up a rock or gravel garden with this evergreen, dome-shaped shrub, with its bright yellow, daisy-like flowers from late spring to early summer, and silvery blue leaves. Try with purple aubretia and late-flowering tulips.

↕↔30cm (12in)

SHRUB LARGE

Forsythia x *intermedia* 'Lynwood Variety' ♡

Few plants outshine this vigorous, upright, deciduous shrub from late winter to spring, when its long, bare stems are clothed with small, star-shaped, bright yellow flowers, before small, bright green leaves appear.

↕↔3m (10ft)

BULB LARGE

Fritillaria imperialis ♡

CROWN IMPERIAL Combine this perennial bulb with hot-hued tulips for a dazzling late spring display. It forms a crown of blue-green leaves above large, orange or yellow, bell-shaped flowers, held on sturdy stems. Plants die back after flowering.

↕1.5m (5ft)

SHRUB SMALL

Fuchsia 'Genii' ♡

This deciduous, bushy shrub offers a fiery combination of gold-green young foliage, fading to lime-green in summer, and a profusion of pendent, cerise and purple blooms from summer to autumn. Plant it in borders or in a pot of soil-based compost.

↕↔75cm (30in)

ANNUAL/BIENNIAL SMALL

Gazania Kiss Series

TREASURE FLOWER Grow this annual in a shallow pot of multi-purpose compost or at the front of a sunny border to enjoy its pink, bronze, gold, or white, daisy-like blooms, with contrasting eyes, set against dark green leaves from summer to early autumn.

↕30cm (12in) ↔25cm (10in)

PERENNIAL SMALL

Geum coccineum

This clump-forming perennial has cheerful red or orange flowers, with prominent yellow stamens, and adds a vibrant highlight to the front of a sunny border in summer. Best in damp soils. 'Cooky' (above) is a popular orange variety.

↕↔30cm (12in)

PERENNIAL SMALL

Helianthemum 'Fire Dragon'

ROCK ROSE Use the trailing stems of soft, grey-green leaves and vivid, fiery orange-red, rounded summer blooms of this evergreen perennial to add sparkle to the edge of a raised bed or sunny border.
↕30cm (12in) ↔45cm (18in)

TREE MEDIUM

Gleditsia triacanthos 'Sunburst'

HONEY LOCUST This deciduous tree is prized for its large, golden-yellow leaves, divided into leaflets, which turn green in summer and bright yellow in autumn. Pendent, long seed pods appear in autumn.
↕12m (40ft) ↔10m (30ft)

SHRUB LARGE

Hamamelis x *intermedia*

WITCH HAZEL This exceptional vase-shaped shrub blooms in the depths of winter, producing lightly scented, spidery flowers on bare stems. The leaves appear in spring. 'Jelena' AGM (above) has large, coppery orange flowers.
↕↔4m (12ft)

PERENNIAL MEDIUM

Helenium 'Wyndley'

SNEEZEWEED Sprays of daisy-like, orange-brown flowers with brown central disks top the sturdy stems of this perennial from midsummer to early autumn. The foliage is lance-shaped and dark green. Plant it mid-border with dahlias and grasses.
↕80cm (32in) ↔50cm (20in)

ANNUAL/BIENNIAL LARGE

Helianthus annuus

SUNFLOWER A classic cottage plant, this easy-to-grow annual includes dwarf and tall forms, with daisy-like, yellow, orange, mahogany, or cream summer blooms. The seedheads that follow provide food for birds. Protect young plants against slugs.
↕up to 3m (10ft) ↔45cm (18in)

PERENNIAL MEDIUM

Hemerocallis 'All American Chief'

DAYLILY This perennial forms an upright clump of arching, strap-like leaves and trumpet-shaped, deep red summer blooms, with yellow throats, each flower lasting just a day. Plant in a sunny border.
↕80cm (32in) ↔75cm (30in)

PERENNIAL SMALL

Heuchera 'Amber Waves'

This evergreen, clump-forming perennial has lobed, ruffled, orange-yellow leaves, pale burgundy underneath, and sprays of small, bell-shaped, pink summer flowers. Use it to edge pots or borders. It works well with dark-leaved forms.
↕30cm (12in) ↔50cm (20in)

PERENNIAL SMALL

Heuchera 'Plum Pudding'

The dark maroon-purple, rounded, lobed leaves of this leafy, evergreen perennial provide a contrasting foil for hot pink, yellow, or red flowers. Wiry stems of small, insignificant, white flowers appear in summer.
↕50cm (20in) ↔30cm (12in)

PERENNIAL SMALL

Hosta 'Fire and Ice'

PLANTAIN LILY A clump-forming perennial, with large, heart-shaped, ribbed, green leaves, which are boldly marked with prominent white splashes. It contrasts well with brightly coloured blooms, and can be used to underplant larger shrubs.
↕40cm (16in) ↔indefinite

PERENNIAL SMALL

Ipomoea batatas 'Blackie'

SWEET POTATO VINE This trailing, evergreen perennial, grown as an annual, has dramatic, lobed, ivy-shaped, almost black leaves that blend beautifully with hot-coloured flowers. Use it to edge containers of multi-purpose compost or a raised bed.
↕25cm (10in) ↔60cm (24in)

CLIMBER MEDIUM

Ipomoea coccinea

RED MORNING GLORY An exotic, twining, annual climber, with heart-shaped, green leaves and small, fragrant, tubular, yellow-throated scarlet flowers. Grow it up a tripod to produce an accent of colour in a border or large pot from summer to early autumn.
↕3m (10ft)

BULB LARGE

Lilium Golden Splendor Group ♀

YELLOW TRUMPET LILY Use this upright, tall bulb, with scented, trumpet-shaped, golden summer flowers to create a dramatic highlight at the back of a border. Plant the bulbs in spring; stake the stems.

↕2m (6ft)

BULB MEDIUM

Lilium lancifolium

LANCE-LEAVED LILY This upright, perennial bulb has long, narrow, lance-shaped leaves and, from summer to early autumn, nodding, turks-cap, pink- to red-orange flowers. 'Splendens' AGM (above) has larger, black-spotted bright red-orange blooms.

↕60cm (24in or more)

PERENNIAL MEDIUM

Monarda 'Cambridge Scarlet'

BERGAMOT Plant this clump-forming, upright perennial mid-border in moist soil to enjoy the aromatic, dark green leaves and spidery, red blooms in summer. Try it with hot pink and orange flowers for a sizzling display. Stems may need staking.

↕1m (3ft) ↔45cm (18in)

SHRUB MEDIUM

Lonicera nitida 'Baggesen's Gold' ♀

SHRUBBY HONEYSUCKLE Use this evergreen shrub as a backdrop to a fiery border of red, yellow, and orange blooms. Its arching stems of yellow-green leaves make for a dense screen.

↕2m (6ft) ↔3m (10ft)

PERENNIAL MEDIUM

Lychnis coronaria ♀

This perennial or biennial, with its profusion of small, round, bright pink summer flowers, is ideal for the front of a border. The silver-grey, slightly downy foliage adds to its charm. Allow it to self-seed to create a naturalistic effect.

↕60cm (24in) ↔45cm (18in)

PERENNIAL SMALL

Meconopsis cambrica

WELSH POPPY This perennial is perfect for naturalistic gardens. Use its lemon-yellow or orange, poppy-like blooms to brighten borders and gravel beds from late spring to summer. The divided leaves are light green. It will self seed readily in free-draining soil.

↕45cm (18in) ↔30cm (12in)

BULB MEDIUM

Narcissus 'Pipit'

JONQUIL PIPIT Plant this dwarf perennial bulb *en masse* in borders or large containers to create a sea of fragrant, lemon-yellow flowers with creamy white cups from mid- to late spring. Plant the bulbs in autumn with tulips that flower at the same time.

↕25cm (10in)

SHRUB MEDIUM

Paeonia delavayi

TREE PEONY This upright, deciduous shrub has tall stems, with large, deeply lobed, dark green leaves, blue-green beneath, and nodding, cup-shaped, dark red late spring flowers. Plant it at the back of a hot colour-themed border or along a boundary.

↕↔2m (6ft)

PERENNIAL MEDIUM

Paeonia mlokosewitschii 

CAUCASIAN PEONY The showy, bowl-shaped, lemon-yellow flowers of this clump-forming perennial look dramatic from late spring to early summer in a sunny or partly shaded border. Pinkish leaf buds unfurl to reveal oval, bluish green foliage.

↕↔75cm (30in)

PERENNIAL MEDIUM

Papaver orientale

ORIENTAL POPPY Grow this early-summer-flowering perennial for its large, bright red blooms, with black marks and dark eyes, and decorative seedheads. Combine it with peonies and late-summer-flowering plants to plug the gaps left after it fades.

↕↔ 90cm (36in)

SHRUB MEDIUM

Phygelius x *rectus*

CAPE FUCHSIA This semi-evergreen, upright subshrub produces long stems dripping with pendent, tubular, yellow or red flowers, above lance-shaped, green leaves. Varieties include 'African Queen' AGM (above), with orange-red blooms.

↕ 1m (3ft) ↔ 1.2m (4ft)

PERENNIAL MEDIUM

Penstemon digitalis 'Husker Red'

The lance-shaped, dark red young leaves and stems of this bushy perennial create a smouldering foil for orange and red flowers at a border front. From late spring to midsummer, white, tubular flowers appear.

↕ 75cm (30in) ↔ 30cm (12in)

SHRUB MEDIUM

Philadelphus coronarius 'Aureus' 🏆

This upright, deciduous shrub offers a package of bright yellow young leaves, which age to lime green and, in early summer, scented, creamy white blooms. Use it as backdrop to a hot-hued border.

↕ 2.5m (8ft) ↔ 1.5m (5ft)

SHRUB LARGE

Physocarpus opulifolius

This deciduous shrub has lobed, green, yellow, or purple-red leaves, which can be used as a backdrop to hot-hued schemes. Domed clusters of pale pink summer blooms are followed by brown fruits. 'Lady in Red' AGM (above) has bold red foliage.

↕ up to 3m (10ft) ↔ 1.5m (5ft)

SHRUB MEDIUM

Rosa CRAZY FOR YOU 🏆

This floribunda-type rose has dark green, glossy leaves and, from summer to autumn, cupped, semi-double, cherry-flecked cream flowers, with a light, fruity perfume. The foliage is dark green and disease-resistant.

↕ 1.5m (5ft) ↔ 1.1m (3½ft)

CLIMBER MEDIUM

Rosa SUMMER WINE 🏆

This climbing rose, from summer to autumn, bears fragrant, semi-double, coral flowers, which add vibrancy to a bed of red- or yellow-flowered plants. A disease-resistant variety, it has glossy, dark green leaves. Deadhead regularly and mulch in spring.

↕ 3m (10ft)

PERENNIAL MEDIUM

Rudbeckia hirta

Grown as an annual, this short-lived perennial adds a wealth of large, daisy-like, purple-centred golden-yellow flowers to the front of a sunny border from summer to early autumn. The mid-green leaves offer a cool-coloured foil.

↕ up to 90cm (36in) ↔ 45cm (18in)

ANNUAL/BIENNIAL SMALL

Salvia splendens

SCARLET SAGE Grow this annual bedder at the front of a border or in pots of multi-purpose compost for its tubular, pink, red, or purple flower spikes, which appear all summer above spear-shaped, dark green leaves. Deadhead faded flower spikes.

↕ 25cm (10in) ↔ 35cm (14in)

SHRUB LARGE

Sambucus nigra f. *porphyrophylla* 'Eva' ♡

The divided, fern-like, almost black foliage of this rounded, deciduous shrub offers a dark foil for brightly coloured flowers. In late spring, it bears pale pink flowerheads, followed by blackish red berries in autumn.
↕↔ 6m (20ft)

PERENNIAL SMALL

Solenostemon scutellarioides

FLAME NETTLE A small, bushy perennial, grown as an annual, with spear-shaped, pink, red, green, and yellow leaves, which add fiery hues to path and border edges, and pots and containers filled with multi-purpose compost.
↕ 45cm (18in) ↔ 30cm (12in) or more

ANNUAL/BIENNIAL SMALL

Tropaeolum Jewel Series

NASTURTIUM This fast-growing annual will quickly smother the front of a border with its combination of round, green leaves and funnel-shaped flowers in bright shades of red, yellow, or orange from summer to early autumn. Also use to edge tall pots.
↕ 30cm (12in) ↔ 45cm (18in)

ANNUAL/BIENNIAL SMALL

Tagetes patula

FRENCH MARIGOLD With its divided, dark green, aromatic leaves and yellow, orange, red, or mahogany flowers from summer to early autumn, this annual will brighten up the edges of fiery borders. Sow seeds indoors in early spring; protect from frost.
↕↔ 30cm (12in)

CLIMBER MEDIUM

Tropaeolum speciosum ♡

FLAME CREEPER Let the twining stems of this perennial climber scramble through a large shrub at the back of a border, where its scarlet flowers, set against lobed, blue-green leaves, will offer a focal point. Bright blue fruits follow in autumn.
↕ 3m (10ft)

SHRUB LARGE

Sambucus racemosa

RED-BERRIED ELDER Prized for its bronze, serrated leaves, ageing to golden-yellow in early summer, this deciduous, upright shrub is ideal for the back of a hot-hued, partly shaded border. The scarlet fruits that follow yellow spring flowers add to the effect.
↕↔ 3m (10ft)

BULB MEDIUM

Tulipa 'Prinses Irene' ♡

This mid-spring-flowering tulip bears bowl-shaped, orange blooms, with contrasting purple markings on the outer petals. Use it to create hot, fiery displays with other red or yellow varieties. Plant the bulbs in groups in autumn in borders or pots.
↕ 35cm (14in)

PERENNIAL MEDIUM

Sedum telephium Atropurpureum Group

ORPINE This perennial's dark purple stems, fleshy leaves, and pinkish white flower clusters from late summer to autumn offer a dramatic contrast to hot-hued flowers and green-leaved perennials.
↕ 60cm (24in) ↔ 30cm (12in)

CLIMBER MEDIUM

Thunbergia alata

BLACK-EYED SUSAN Grow this annual, twining climber up a tripod in a border to add a pyramid of toothed, heart-shaped to oval leaves and rounded, flat-faced, dark-eyed, orange, yellow or peach flowers from early summer to early autumn.
↕ 3m (10ft)

OTHER SUGGESTIONS

Annuals and biennials

Salvia splendens Vista Series • *Tagetes* 'Tangerine Gem'

Perennials and bulbs

Aster novae-angliae 'Andenken an Alma Pötschke' • *Helenium* 'Wesergold' ♡ • *Hemerocallis* 'Golden Chimes' • *Heuchera* 'Circus' • *Imperata cylindrical* 'Rubra' • *Pelargonium* 'Ardens' ♡ • *Penstemon* 'Chester Scarlet' ♡ • *Phormium* 'Platt's Black' • *Rudbeckia* 'Prairie Sun' • *Tulipa* 'Black Hero'

Shrubs and trees

Acer japonicum 'Aureum' • *Buddleja davidii* 'Black Knight' ♡ • *Calluna vulgaris* 'Golden Feather' • *Camellia japonica* 'Bob's Tinsie' ♡ • *Chaenomeles speciosa* 'Simonii' • *Chamaecyparis obtusa* • *Choisya* SUNDANCE ♡ • *Genista lydia* ♡ • *Prunus cerasifera* 'Nigra' ♡ • *Rosa* 'Golden Showers' • *Spiraea japonica* 'Goldflame'

Plants for colourful stems

As temperatures plummet and deciduous plants lose their plumage, a few species choose to put on their best performance of the year.

Dogwoods decorate winter gardens with spiky bouquets of scarlet, gold, and lime green stems, while the stark white trunks of silver birches make ghostly silhouettes in barren landscapes. Other plants produce intricately patterned bark to make your garden sparkle. The stems of the snake bark maples (*Acer capillipes* and *A. davidii*) resemble that beautiful reptile's skin, and the plane tree's (*Platinus* x *acerifolia*) multicoloured stems look like designer wallpaper – only select this large tree if you have plenty of space. Bamboos work well in modern, minimalist planting schemes – their graphic stems injecting a hint of gold, blue, or apple green.

TREE MEDIUM

Acer capillipes

SNAKE-BARK MAPLE This stunning, deciduous tree features decorative, green- and grey-striped bark, and has an elegant spreading habit. The three-lobed, dark green leaves turn spectacular shades of orange and red in autumn.

↕↔10m (30ft)

TREE MEDIUM

Acer griseum 🏆

PAPER-BARK MAPLE This deciduous, spreading tree is grown for its distinctive bronze flaking bark that can be seen most clearly in winter after the leaves have fallen. The lobed, dark green, decorative foliage turns bright orange and red in autumn.

↕↔10m (30ft)

TREE SMALL

Acer palmatum 'Sango-kaku' 🏆

CORAL-BARK MAPLE The coral-red stems of this deciduous tree look great when combined with the yellow autumn leaves. The lobed foliage unfurls pinkish yellow in spring, and turns green in summer.

↕8m (25ft) ↔4m (12ft)

TREE MEDIUM

Acer pensylvanicum

SNAKE-BARK MAPLE An upright, deciduous tree, with green- and white-striped bark, which looks beautiful year-round, and rounded, green leaves, which turn yellow in autumn. 'Erythrocladum' (above) has reddish brown and white-striped bark.

↕12m (40ft) ↔10m (30ft)

ANNUAL/BIENNIAL SMALL

Beta vulgaris

SWISS CHARD Grown as an edible crop, this annual has brightly coloured stems in shades of red, orange, and yellow, and can be treated as an ornamental. Sow seeds from spring to summer for colour all year. Plant it at the front of a border or in pots.

↕↔45cm (18in)

TREE LARGE

Betula nigra

RIVER BIRCH The reddish brown, peeling bark of this upright, deciduous tree makes a focal point in winter gardens, and looks good all year, especially when contrasted with the yellow spring catkins. Its diamond-shaped, green leaves turn yellow in autumn.

↕15m (50ft) ↔10m (30ft)

TREE LARGE

Betula utilis var. *jacquemontii*

HIMALAYAN BIRCH Grown for its white trunk and stems, this deciduous tree looks stunning from winter to spring, when yellow catkins appear. The diamond-shaped leaves turn yellow in autumn. Multi-stemmed forms are ideal for small gardens.

↕18m (60ft) ↔10m (30ft)

SHRUB LARGE

Chusquea culeou 🏆

CHILEAN BAMBOO Best on free-draining soils, this clump-forming bamboo produces glossy, yellow-green to olive-green canes, with long, papery-white leaf sheaths, giving young canes a striped appearance. It has mid-green, linear leaves. Ideal for screening.

↕5m (15ft) ↔2.5m (8ft) or more

SHRUB LARGE

Cornus alba

DOGWOOD Grown for its bright red winter stems, this deciduous shrub looks good year-round. In early summer, it produces flattened heads of white flowers. Its leaves turn yellow, orange, and red in autumn. Cut the stems to the ground in late winter.

↕↔3m (10ft)

SHRUB LARGE

Cornus sanguinea

The young stems of this deciduous shrub turn reddish green in winter, and the green foliage that follows in spring develops a fiery autumn colour. 'Winter Beauty' has bright orange-yellow and red winter shoots. Prune old stems to the ground in late winter.

↕3m (10ft) ↔2.5m (8ft)

SHRUB MEDIUM

Cornus sericea

A deciduous shrub grown for its stunning red winter stems, or bright yellow-green in varieties such as 'Flaviramea' AGM (above). Cut old stems to the ground in late winter to encourage new growth. Its fiery autumn leaves are an added attraction.

↕2m (6ft) ↔4m (12ft)

TREE LARGE

Eucalyptus gunnii ♈

CIDER GUM This evergreen, slender tree has smooth, whitish green bark that peels in late summer to reveal pink-flushed greyish green young bark, creating a two-tone effect. The rounded, blue-grey young foliage leads to sickle-shaped, green leaves.

↕25m (80ft) ↔15m (50ft)

SHRUB LARGE

Fargesia murielae ♈

UMBRELLA BAMBOO This large clump-forming bamboo produces arching, yellow-green canes and lance-shaped, bright green leaves, and makes a beautiful screen. 'Simba' (above) is a compact form, ideal as a focal plant in a small family garden.

↕4m (12ft) ↔ indefinite

SHRUB LARGE

Phyllostachys aureosulcata f. *aureocaulis*

GOLDEN GROOVE BAMBOO A tall, upright, clump-forming bamboo, with yellow canes, striped at the base, and lance-shaped, mid-green leaves. Ideal as a screen. Remove the roots with a sharp spade to limit its spread.

↕6m (20ft) ↔ indefinite

SHRUB LARGE

Phyllostachys nigra ♈

BLACK BAMBOO This large, evergreen, clump-forming bamboo produces grooved, greenish brown canes, which turn black in their second season and look stunning against a light-coloured backdrop. The green foliage is lance-shaped.

↕8m (25ft) ↔ indefinite

SHRUB LARGE

Phyllostachys viridiglaucescens

An evergreen, clump-forming bamboo, with greenish brown canes, ageing to yellow-green, and bright green leaves. Try it in blocks in a geometric bed in a modernist scheme. Remove old canes in spring.

↕8m (25ft) ↔ indefinite

TREE MEDIUM

Prunus serrula

A deciduous tree, prized for its gleaming, coppery red, peeling bark, which looks particularly effective on multi-stemmed specimens. Oval, tapering, dark green leaves, which turn yellow in autumn, are joined by small, white flowers in late spring.

↕↔10m (30ft)

SHRUB MEDIUM

Rubus cockburnianus

The arching, prickly shoots of this deciduous shrub are bright white in winter, creating a dramatic, ghostly effect. Diamond-shaped, dark green foliage follows in spring; inedible black fruits follow the purple summer flowers.

↕↔2.5m (8ft)

OTHER SUGGESTIONS

Perennials

Bergenia purpurascens ♈

Trees and shrubs

Acer cappadocicum 'Rubrum' ♈ • *Acer davidii* 'Ernest Wilson' • *Acer grosseri* • *Betula albosinensis* • *Cornus stolonifera* • *Eucalyptus pauciflora* subsp. *niphophila* ♈ • *Luma apiculata* 'Glanleam Gold' • *Phyllostachys violascens* • *Platanus* x *acerfolia* • *Salix alba* var. *vitellina* 'Britzensis' • *Semiarundinaria yashadake* f. *kimmei* • *Thamnocalamus crassinodus* 'Kew Beauty' ♈

Plants for fragrant blooms

Guaranteed to lure visitors into the garden, scented plants are a welcome asset, and add a new dimension to any garden design.

Site fragrant flowers close to seating areas, doorways, and paths, and ensure that plants that require nudging to release their perfume, such as lavender and scented-leaved pelargoniums, are brushed past to achieve this effect. Plants with a heady perfume like regal lilies may be too overpowering for beds close to dining areas, so test them out before including any. To keep your spirits raised year-round, include roses that bloom in autumn, *Mahonia* and witch hazel for winter gardens, and a selection of the vast legion of fragrant plants that flower in spring and summer.

SHRUB LARGE

Abelia x *grandiflora*

GLOSSY ABELIA This arching, semi-evergreen shrub, with small, dark green leaves, has pale pink to white, bell-shaped blooms with a light fragrance, which appear for weeks in midsummer. It is ideal for a sheltered spot next to a sunny patio.
↕3m (10ft) ↔4m (12ft)

PERENNIAL MEDIUM

Agastache 'Tangerine Dreams'

GIANT HYSSOP This tender perennial has grey-green, minty-aniseed-scented leaves and, from summer to autumn, long-lasting spikes of small, tubular, pale orange-pink, nectar-rich blooms. Overwinter as cuttings.
↕↔60cm (24in)

SHRUB SMALL

Arctostaphylos uva-ursi

BEARBERRY This creeping, evergreen shrub has glossy, oval, green leaves, with a silver reverse, and small, urn-shaped, pink or white, fragrant flowers that bloom from spring to early summer. The flowers are followed by red berries.
↕10cm (4in) ↔50cm (20in)

SHRUB LARGE

Brugmansia x *candida*

ANGELS' TRUMPETS This large shrub, usually overwintered in a conservatory, will enjoy summer on a warm patio. The huge, hanging trumpets of night-scented blooms are white, yellow, or pink-apricot in 'Grand Marnier' (above). It is highly toxic.
↕5m (15ft) ↔2.5m (8ft)

SHRUB LARGE

Chimonanthus praecox

WINTERSWEET This deciduous shrub, with lance-shaped leaves, produces sulphur-yellow winter flowers, which produce a heady fragrance, on bare stems. 'Grandiflorus' AGM (above) has maroon centres.
↕4m (12ft) ↔3m (10ft)

SHRUB MEDIUM

Choisya ternata

MEXICAN ORANGE BLOSSOM This rounded, evergreen shrub has whorls of aromatic, glossy, green leaves. It has sweetly perfumed late spring flower clusters, which repeat in autumn. 'Aztec Pearl', with narrow leaves, is particularly free-flowering.
↕↔2.5m (8ft)

CLIMBER MEDIUM

Clematis recta

This herbaceous perennial climber scrambles up over other shrubs. Large heads of small, white, fragrant blooms appear from summer to early autumn, followed by fluffy seedheads. 'Purpurea' has coppery young foliage.
↕2m (6ft)

SHRUB MEDIUM

Clethra alnifolia

SWEET PEPPER BUSH This bushy shrub, with neat, serrated leaves that turn yellow in autumn, bears candle-like spires in late summer made up of tiny, white bells with fluffy stamens and a rich floral perfume. Plant close to pathways and entrances.

↕↔2.5m (8ft)

PERENNIAL MEDIUM

Cosmos atrosanguineus

CHOCOLATE COSMOS The added pleasure of growing this maroon-flowered, slightly tender perennial is that its dish-shaped blooms, appearing from late summer to early autumn on slender stems, are chocolate-scented. Grow in patio pots.

↕60cm (24in) or more ↔45cm (18in)

BULB LARGE

Crinum x *powellii* 'Album'

This evergreen, bulb-forming perennial produces arching, strap-shaped leaves and large, white, funnel-shaped, richly perfumed flowers from late summer to autumn on sturdy, erect stems. Protect plants in cold regions.

↕1m (3ft) ↔60cm (24in)

SHRUB LARGE

Daphne bholua

NEPALESE PAPER PLANT This evergreen shrub with oval, dark green leaves is grown for its late winter clusters of very sweetly fragrant, pinkish-purple-flushed white flowers, ideal near doorways. 'Jacqueline Postill' AGM (above) is popular.

↕3m (10ft) ↔1.5m (5ft)

SHRUB MEDIUM

Daphne x *burkwoodii*

This is a neat, semi-evergreen shrub, with small, green leaves and pink, highly fragrant late spring flower clusters, which are followed by red fruits. Varieties like 'Somerset' (above) are perfect for narrow patio borders.

↕1.5m (5ft) ↔1m (3ft)

SHRUB SMALL

Deutzia gracilis

This deciduous, upright to arching shrub, with small, bright green leaves, is smothered in clusters of starry, white blooms from late spring to early summer. The mass of blooms produces an attractive fragrance.

↕↔1m (3ft)

ANNUAL/BIENNIAL MEDIUM

Dianthus barbatus

SWEET WILLIAM This bushy, upright biennial is a mainstay of cottage garden borders with its lance-shaped leaves and domed heads of sweet-scented, pink, red, burgundy, white, or bicoloured blooms in early summer.

↕70cm (28in) ↔30cm (12in)

PERENNIAL MEDIUM

Dianthus caryophyllus

BORDER CARNATION This traditional perennial, ideal for a cottage or cutting garden, has narrow, grey-green leaves and tall, slender stems of sweetly scented blooms in a range of pastel shades. Dwarf forms are also available.

↕80cm (32in) ↔23cm (9in)

PERENNIAL SMALL

Dianthus 'Dad's Favourite'

GARDEN PINK This dwarf, evergreen perennial, ideal as a border edging, has short, grey-green, grassy leaves and a succession of semi-double, white flowers with maroon markings all summer. The blooms have a rich clove scent.

↕up to 45cm (18in) ↔30cm (12in)

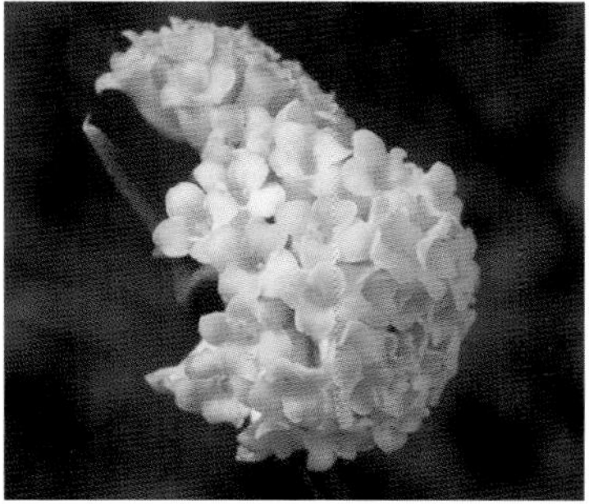

SHRUB MEDIUM

Edgeworthia chrysantha

PAPER BUSH This late winter- to early spring-flowering shrub thrives in sheltered borders, and have a spicy, clove fragrance. The rounded clusters of yellow blooms are silky haired in bud, creating a frosted look.

↕↔1.5m (5ft)

ANNUAL/BIENNIAL SMALL

Erysimum cheiri

WALLFLOWER A cottage garden favourite, this powerfully fragrant biennial, planted in autumn, produces mid-spring heads of bright yellow, orange, red, or pink blooms, attracting bees and butterflies. 'Blood Red' (above) is popular.

↕↔ 30cm (12in)

SHRUB MEDIUM

Euphorbia mellifera

HONEY SPURGE This dome-shaped, evergreen shrub gives a Mediterranean effect when planted in gravel. In late spring, rounded heads of honey-scented, brownish flowers form at the ends of leafy shoots.

↕ 2m (6ft) ↔ 2.5m (8ft)

SHRUB LARGE

Hamamelis x *intermedia*

WITHCH HAZEL This spreading shrub is grown for its shredded citrus peel-like, yellow, copper, or mahogany scented blooms on bare stems from early- to midwinter, in. 'Jelena' AGM (above) has coper-coloured blooms.

↕↔ 4m (12ft)

SHRUB LARGE

Hamamelis mollis

CHINESE WITCH HAZEL A large, spreading shrub with broad leaves, turning yellow in autumn, this late winter-flowering plant has fragrant, spidery, yellow blooms that sprout from the bare branches. Plant close to a path or driveway to enjoy the scent.

↕↔ 4m (12ft) or more

SHRUB SMALL

Heliotropium arborescens

CHERRY PIE This bushy, evergreen shrub, grown as an annual for pots and bedding displays, has stems of oval, dark green, crinkled leaves, topped with clusters of purple or white, spicy, vanilla-scented blooms all summer long.

↕↔ 45cm (18in)

BULB MEDIUM

Hyacinthus orientalis

HYACINTH This early spring-flowering perennial bulb has broad spikes of heavily fragrant, waxy, bell-shaped blooms in colours including blue, purple, pink, red, orange, yellow, and white. 'City of Haarlem' AGM (above) has pale yellow blooms.

↕ 25cm (10in)

CLIMBER LARGE

Jasminum officinale

COMMON JASMINE This vigorous, semi-evergreen, twining climber, with paired leaflets produces clusters of white flowers from summer to autumn. Headily perfumed in the evening, it's a great choice for a pergola over a patio.

↕ 12m (40ft)

CLIMBER MEDIUM

Lathyrus odoratus

SWEET PEA An annual, tendril climber, with oval leaflets and richly perfumed flowers in shades of pink, blue, purple, or white from summer to early autumn, if regularly picked. Choose "old-fashioned" mixes for a stronger scent.

↕ 3m (10ft)

SHRUB SMALL

Lavandula angustifolia

ENGLISH LAVENDER This evergreen shrub, with narrow, aromatic, silver-grey leaves has short spikes of fragrant, blue-purple, pink, or white flowers from mid- to late summer. Clip after blooming. 'Hidcote' AGM (above) is a dwarf form.

↕ 60cm (24in) ↔ 75cm (30in)

BULB LARGE

Lilium 'Red Hot'

ORIENTAL LILY The exotic perfume of lilies is unmistakable. This large bulb produces arching stems that bear slender leaves and are topped with white-edged red blooms in summer. The blooms are particularly fragrant.
↕1m (3ft)

TREE MEDIUM

Magnolia virginiana

SWEET BAY This moisture-loving deciduous or semi-evergreen shrub or tree has large, glossy leaves, bluish white beneath, with lemon-vanilla scented, cup-shaped, cream flowers in early summer, and sporadically till autumn.
↕10m (30ft) ↔6m (20ft)

BULB LARGE

Lilium regale 🏆

REGAL LILY A cottage garden classic, this richly perfumed bulb, in summer, produces several large, white trumpets on top of upright stems. The throats are yellow, with pinkish purple-marked outer petals. Stake and watch for lily beetle.
↕2m (6ft)

CLIMBER LARGE

Lonicera x *heckrottii*

GOLDFLAME HONEYSUCKLE This large, deciduous or semi-evergreen, twining climber has oval, dark green leaves. Its summer flowers are richly perfumed, pink with orange throats, the scent especially strong in the evenings and early mornings.
↕5m (15ft)

CLIMBER LARGE

Lonicera periclymenum

COMMON HONEYSUCKLE This deciduous, twining climber forms clusters of flared, tubular summer blooms that are a combination of creamy white, yellow, and pink, depending on the variety. The perfume is stronger at twilight.
↕7m (22ft)

SHRUB SMALL

Mahonia aquifolium

OREGON GRAPE This spreading, evergreen shrub has spine-edged, leathery leaves, which are flushed purple-bronze in winter, providing a foil for the sweetly scented, yellow flower clusters in spring. Round black berries follow. Tolerates dry shade.
↕1m (3ft) ↔1.5m (5ft)

SHRUB MEDIUM

Mahonia japonica 🏆

JAPANESE MAHONIA This architectural evergreen shrub, with prickly edged, leathery leaflets becomes purple tinged in winter. From late autumn to early spring, drooping, pale yellow flower tassels fill the air with lily-of-the-valley scent.
↕2m (6ft) ↔3m (10ft)

SHRUB MEDIUM

Mahonia x *media*

This hybrid adds architectural interest to the winter garden with its large, glossy, divided leaves and stems topped with clusters of fragrant, yellow flower spikes. The blooms of 'Buckland' AGM (above) are striking.
↕1.8m (6ft) ↔4m (12ft)

ANNUAL/BIENNIAL SMALL

Matthiola incana

STOCK Grown as a half-hardy annual, the large, pastel-shaded flowerheads of grey-green-leaved stocks, such as Cinderella Series (above), fill the cottage garden border with sweet, clove fragrance from late spring to summer.
↕↔25cm (10in)

BULB MEDIUM

Narcissus 'February Gold'

DAFFODIL This variety is one of the first daffodils to bloom in spring, bearing single rich yellow flowers, with a peppery, sweet scent. Plant in groups in autumn in lawns and borders, and leave to naturalize. It is also ideal for spring containers.
↕30cm (12in)

BULB MEDIUM

Narcissus 'Pipit'

JONQUIL PIPIT Jonquils, noted for their powerful perfume, are perfect for pots and planters around doorways or for planting *en masse* in raised beds. The lemon-yellow, cream-cupped, multi-headed flowers bloom in mid-spring.
↕25cm (10in)

ANNUAL/BIENNIAL MEDIUM

Nicotiana 'Lime Green'

TOBACCO PLANT An upright annual, grown for summer bedding, with spoon-shaped leaves that are sticky to touch. In late summer and autumn, it produces open, trumpet-shaped, twilight-scented, greenish yellow flowers.
↕60cm (24in) ↔25cm (10in)

PERENNIAL LARGE

Nicotiana sylvestris

FLOWERING TOBACCO This perennial, grown as a summer annual, has very large, sticky leaves and tall stems topped with heads of tubular, white flowers, flared at the ends. Plant in pots or patio borders to enjoy the evening scent.
↕1.5m (5ft) ↔75cm (30in)

PERENNIAL SMALL

Oenothera fruticosa

EVENING PRIMROSE An upright perennial, good for large rock gardens, this day-flowering form of evening primrose has cup-shaped, golden, scented summer flowers. The compact 'Fyrverkeri' (above) has bronze-tinted leaves on reddish purple stems.
↕↔38cm (15in)

PERENNIAL SMALL

Petunia Prism Series

This compact, large-flowered perennial, grown as an annual, has trumpet-shaped blooms all summer in a wide range of shades. Plant in baskets and windowboxes around doorways. The sweet scent is strongest in the evening.
↕35cm (14in) ↔50cm (20in)

SHRUB LARGE

Philadelphus 'Virginal'

MOCK ORANGE This vigorous, upright shrub with dark green, oval leaves is festooned with clusters of richly perfumed, white, double or semi-double blooms, that fill large areas of the garden with its scent from early- to midsummer.
↕3m (10ft) ↔2.5m (8ft)

SHRUB SMALL

Rosa 'Buff Beauty'

This romantic-looking musk rose is a strong grower, with long, arching stems, glossy leaves, and clusters of small, apricot-yellow and buff, semi-double blooms, which add fragrance from summer to autumn.
↕↔1.2m (4ft)

SHRUB MEDIUM

Rosa GERTRUDE JEKYLL

This upright shrub rose or short climber has richly perfumed, fully double, mid-pink blooms, which open flat, from summer to early autumn. Grow around an arbour or against the house wall near a doorway.
↕2m (6ft) ↔1.2m (4ft)

CLIMBER LARGE

Rosa 'New Dawn'

This vigorous, disease-resistant climbing rose has dark, glossy leaves and heads of lightly fragrant, pale pink, double blooms, with elegant buds, which are produced through summer into autumn. It is useful for a north wall.
↕5m (15ft)

SHRUB MEDIUM

Syringa meyeri 'Palibin'

This delightful dwarf lilac is slow-growing, bushy, and deciduous, with upright heads of fragrant, lilac-pink flowers in late spring and early summer. An ideal shrub for courtyard gardens and narrow patio borders.

↕↔ 1.5m (5ft)

SHRUB SMALL

Sarcococca confusa

SWEET BOX This bushy, evergreen shrub, with dark green, glossy, lance-shaped leaves is grown for the sweet winter fragrance emanating from the tiny, white flowers. A must for walkways through trees or shady city courtyards.

↕↔ 1m (3ft)

SHRUB MEDIUM

Sarcococca hookeriana var. *digyna*

SWEET BOX This evergreen, clump-forming shrub has narrow, lance-shaped leaves and tiny, fragrant, white flowers in winter. Plant it near a doorway or drive. 'Purple Stem' has pink-flushed flowers.

↕ 1.5m (5ft) ↔ 2m (6ft)

SHRUB LARGE

Syringa vulgaris

LILAC This vigorous, deciduous shrub is renowned for its fragrant flowers, which appear in late spring in large, cone-shaped clusters in shades of purple, pink, yellow, and white. Choose compact varieties for smaller plots.

↕↔ 7m (22ft)

CLIMBER LARGE

Trachelospermum jasminoides

STAR JASMINE This evergreen climber has twining stems and small, oval leaves that turn bronze with colder nights. Clusters of small, pin wheel shaped, white flowers form in midsummer, releasing a rich scent.

↕ 9m (28ft)

SHRUB LARGE

Viburnum x *bodnantense*

A deciduous, upright shrub that produces sweet, almond-scented, pink to white flowers in rounded clusters along bare branches from late autumn to early spring. Varieties of this shrub include 'Dawn' AGM (above).

↕ 3m (10ft) ↔ 2m (6ft)

SHRUB MEDIUM

Viburnum carlesii

In late spring, this deciduous shrub produces domed heads of pink buds, opening to white, trumpet-shaped blooms, with a sweet, spiced fragrance, followed by red and black autumn berries. Plant it at the back of a sunny border.

↕↔ 2m (6ft)

PERENNIAL SMALL

Viola odorata

SWEET VIOLET This creeping, semi-evergreen perennial, with heart-shaped leaves flowers from late winter to early spring, bearing violet or, sometimes, white flowers with a sweet but delicate scent. Plant beneath shrubs or in paving cracks.

↕ 20cm (8in) ↔ 30cm (12in)

PERENNIAL SMALL

Viola x *wittrockiana*

PANSY These bedding plants bloom at any time, depending on the variety and time of sowing, producing flat-faced flowers in a wide range of colours and shades, which have a sweet scent, most noticeable in bold groups.

↕↔ 20cm (8in)

OTHER SUGGESTIONS

Annuals and biennials

Alyssum maritima • *Matthiola bicornis* • *Tagetes erecta*

Perennials and bulbs

Convallaria majalis • *Dianthus* 'Mystic Star' • *Hyacinthoides non-scripta* • *Lilium* 'Casablanca' • *Lilium* 'Tiger Woods' • *Lilium* 'Pink Perfection' • *Mirabilis jalapa* • *Petunia* 'Blue Daddy' • *Sternbergia candida*

Shrubs and climbers

Brugmansia suaveolens • *Chimonanthus praecox* 'Luteus' • *Clematis montana* 'Elizabeth' • *Clerodendrum trichotomum* • *Coronilla valentina* subsp. *glauca* • *Daphne odora* • *Jasminum humile* 'Revolutum' • *Lavandula stoechas* • *Magnolia denudata* • *Rhododendron arborescens* • *Rhododendron* 'Edgeworthii' • *Rhododendron lutea* • *Rosa* 'Frau Dagmar Hastrup' • *Viburnum* x *burwoodii* • *Wisteria sinensis*

PLANTS for SHAPE and TEXTURE

While colourful flowers enjoy celebrity status in the garden, designers often focus on plants with striking shapes and textures before considering blooms, since these features offer more enduring interest. Choose a selection of foliage plants and those with attractive stems to create a decorative backdrop to flowers, or use them as focal features to embolden your schemes. Trees with peeling bark, such as *Betula nigra*, make a statement, while architectural leaves, including those of a *Fatsia* or *Phormium*, produce dramatic effect.

Stems for shape and texture

While bright colours shout "look at me", contrasting shapes and textures can be used as the *corps de ballet*, supporting and defining a planting design scheme.

Although most stem textures are quite subtle, creating intrigue as you get up close, a few also compete for centre stage. The cinnamon-coloured peeling bark of *Acer griseum* and shaggy orange and white stems of the river birch (*Betula nigra*) are hard to miss and make beautiful focal points. Twisted stems also provide impact, particularly in winter. If you have space, the wiggly stems of the twisted willow make a dramatic display; in small gardens, include the corkscrew hazel, with its mop of curly stems and dangling catkins, which appear in spring.

TREE MEDIUM

Acer griseum ♡

PAPER-BARK MAPLE One of the best specimens for textural interest, this deciduous, spreading tree has cinnamon, flaking bark. Remove lower stems to show off the trunk. It also has lobed leaves that turn bright orange and red in autumn.

↕↔10m (30ft)

SHRUB LARGE

Chusquea culeou ♡

CHILEAN BAMBOO A clump-forming bamboo, with glossy, yellow-green canes that have long, papery white leaf sheaths. The resulting striped effect is a feature. Remove weak canes and the bottom third of side branches to enhance the look.

↕5m (15ft) ↔2.5m (8ft) or more

TREE LARGE

Betula nigra

RIVER BIRCH The peeling bark of this upright, deciduous tree creates eye-catching texture, and looks especially effective when contrasted with the yellow spring catkins. Its diamond-shaped, green leaves turn yellow in autumn.

↕15m (50ft) ↔10m (30ft)

TREE LARGE

Betula papyrifera

PAPER BIRCH This tall, deciduous birch tree is narrow and sparsely branched when young, with a smooth, white bark that peels in broad, scrolling pieces to reveal its pale orange reverse. Perfect as a multi-stemmed specimen.

↕20m (70ft) ↔10m (30ft)

TREE LARGE

Cedrus atlantica Glauca Group ♡

BLUE ATLAS CEDAR This evergreen conifer has a spreading, tiered crown, but as a young, loosely conical tree, the slender, dipping branches, covered in silvery blue-green needles, create an attractive tracery.

↕40m (130ft) ↔10m (30ft)

SHRUB LARGE

Corylus avellana

CORKSCREW HAZEL This deciduous shrub, with corkscrew stems, is most attractive from winter to early spring when the sculptural branches are decked with catkins. The broad, oval leaves are distorted like the shoots. Plant as a winter accent.

↕↔5m (15ft)

PERENNIAL LARGE

Dicksonia antarctica ♡

TREE FERN An evergreen or, in colder areas, deciduous fern for a large pot, this perennial features a fibrous "trunk" made up of old leaf bases, creating a chevron pattern. On top is a "shuttlecock" of much-divided leaves and unfurling fronds.

↕up to 6m (20ft) or more ↔4m (12ft)

TREE SMALL

Parrotia persica

PERSIAN IRONWOOD In winter, this wide spreading tree or large shrub, shows a pattern of peeling bark on its multiple stems. Brown, outer layers drop, revealing a patchwork of greys, greens, and whites. The autumn leaf colour is stunning.

↕8m (25ft) ↔10m (30ft)

SHRUB LARGE

Phyllostachys aureosulcata f. *aureocaulis*

GOLDEN GROOVE BAMBOO A clump-forming bamboo, the new upright, occasionally zig-zagged, canes are pale yellow. These darken with age and some have vertical green stripes at the base.
↕6m (20ft) ↔ indefinite

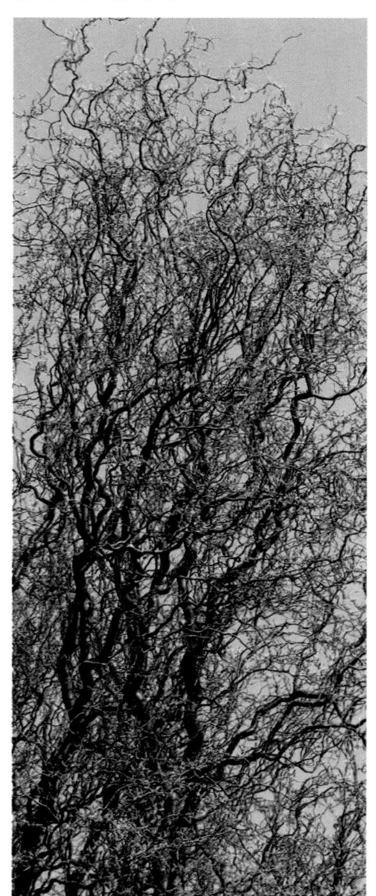

TREE MEDIUM

Salix babylonica var. *pekinensis* 'Tortuosa'

CORKSCREW WILLOW This deciduous tree is grown for its corkscrewed shoots and narrow, wavy leaves, which partially disguise the stems in summer. Its size may be controlled by pollarding.
↕↔12m (40ft)

SHRUB LARGE

Phyllostachys nigra

BLACK BAMBOO A clump-forming bamboo, with gently arching, greenish brown canes, which turn glossy black when mature. Thin out weak stems and prune off lower-level side branches to show off the stems.
↕8m (25ft) ↔ indefinite

TREE LARGE

Pinus pinea

STONE PINE This pine has a characteristic parasol-shaped crown when grown as a single-stemmed tree. The bark is patterned, forming smooth, irregularly-shaped, rusty coloured, or chestnut plates, outlined by dark fissures.
↕20m (70ft) ↔12m (40ft)

TREE SMALL

Salix caprea 'Kilmarnock'

KILMARNOCK WILLOW This weeping pussy willow is a small, mushroom-headed tree with long, trailing branches. In late winter, it bears soft, silver-coloured, furry catkins. Prune to thin the head and maintain an airy framework.
↕↔2m (6ft)

TREE LARGE

Pinus nigra

BLACK PINE This exposure-tolerant pine forms a large, domed-headed evergreen tree, with long needles and yellow-brown cones. The ornamental bark, split by dark fissures, shows irregular patches of caramel and grey.
↕30m (100ft) ↔8m (25ft)

TREE MEDIUM

Prunus serrula

TIBETAN CHERRY This deciduous tree is grown for its gleaming coppery red bark that peels as it matures, leaving trunks and stems looking like they have been bound in satin ribbon. Yellow autumn leaf colour adds to the effect.
↕↔10m (30ft)

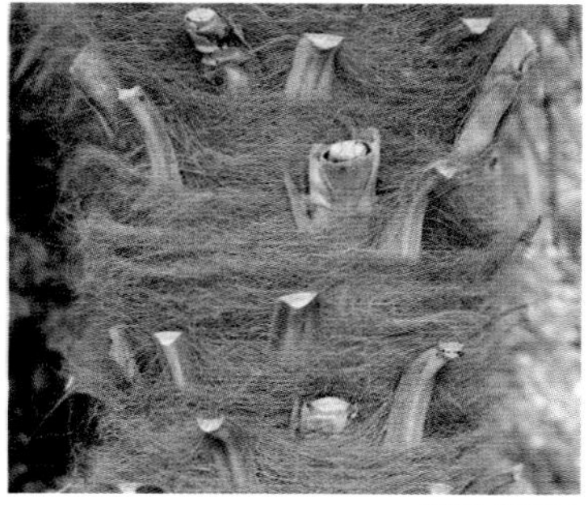

TREE SMALL

Trachycarpus fortunei

CHUSAN PALM This evergreen palm has large, fan-like leaves, held on long, flexible stems that sprout from a central trunk. Young plants have a bushy habit and take several years to develop the woody trunk with its fibrous coating.
↕up to 2m (6ft) ↔2.5m (8ft)

SHRUB MEDIUM

Rubus cockburnianus

This deciduous shrub, in winter, forms strands of prickly, arching shoots that have a ghostly white "bloom". In summer, its stems are concealed by dark green leaves and purple blooms. Inedible, black fruits follow the flowers.
↕↔2.5m (8ft)

OTHER SUGGESTIONS

Shrubs

Hydrangea aspera subsp. *sargentiana* • *Rosa sericea* subsp. *omeiensis* f. *pteracantha* • *Physocarpus opulifolius* 'Diabolo'

Trees

Acer capillipes • *Acer rufinerve* • *Betula ermanii* • *Castanea sativa* • *Eucalyptus glaucescens* • *Juglans regia* • *Pinus nigra* subsp. *pallasiana* • *Pinus radiata* • *Pinus sylvestris* • *Populus serotina* • *Prunus rufa* • *Quercus robur* • *Quercus suber* • *Salix fragilis* • *Stewartia pseudocamellia*

Plants for textured foliage

Plants with textured foliage make ideal specimens, creating impact when used on their own as focal points, or as a contrast to their neighbours.

One of the main tools of a garden designer's trade, textured foliage offers year-round interest, and looks most effective when teamed with contrasting smooth or shiny leaves. Pines, such as *Abies* and *Picea*, produce stems of evergreen, brush-like leaves that combine well with box topiary or an underskirt of heather for a year-round textural display. As part of a seasonal scheme at the front of a sunny border, try the downy leaves of lamb's ears (*Stachys*) with the blue, spiky foliage of a sea holly (*Eryngium*), or in a shady spot bring together a collection of ferns, hostas, and frilly, burgundy-leaved heucheras for a feast of textures.

TREE SMALL

Acer palmatum 'Sango-kaku'

CORAL-BARK MAPLE This deciduous tree's divided foliage, borne on coral-red stems, offers three seasons of interest, unfurling pinkish yellow in spring, turning green in summer, and butter-yellow in autumn.

↕8m (25ft) ↔4m (12ft)

PERENNIAL SMALL

Begonia 'Escargot'

This tender, evergreen perennial offers large, attractive, spiral-shaped, purple-tinted green leaves, with silver swirly markings. It produces small, pink flowers in autumn. Grow it as a house plant in winter and outdoors in a pot in summer.

↕25cm (10in) ↔50cm (20in)

TREE LARGE

Betula ermanii

This decorative, deciduous tree, prized for its peeling, cream bark on the trunk and papery, brown bark on the branches, also features textured, toothed, distinctly veined oval foliage, which turns yellow in autumn. Catkins appear in spring.

↕20m (70ft) ↔12m (40ft)

PERENNIAL SMALL

Adiantum aleuticum

MAIDENHAIR FERN This deciduous or semi-evergreen fern has small, branching, pale green fronds, clothed with oblong leaflets, which create an exciting textural effect. New fronds may be tinged pink. Try it at the front of a shady border with dwarf daffodils.

↕↔45cm (18in)

PERENNIAL SMALL

Brunnera macrophylla

Grown for its oval, tapering leaves, with soft hairs, often speckled or edged with silvery white (above), this perennial is ideal for use as ground cover in partly-shaded gardens. Sprays of small, bright blue flowers appear in spring.

↕45cm (18in) ↔60cm (24in)

TREE SMALL

Acacia baileyana

This small, evergreen tree has decorative, finely divided, silvery grey leaves, which contrast well with plants bearing smooth, green foliage. From late winter to spring, finger-like clusters of fluffy, yellow flowers appear. Requires neutral to acid soil.

↕↔6m (20ft)

pH

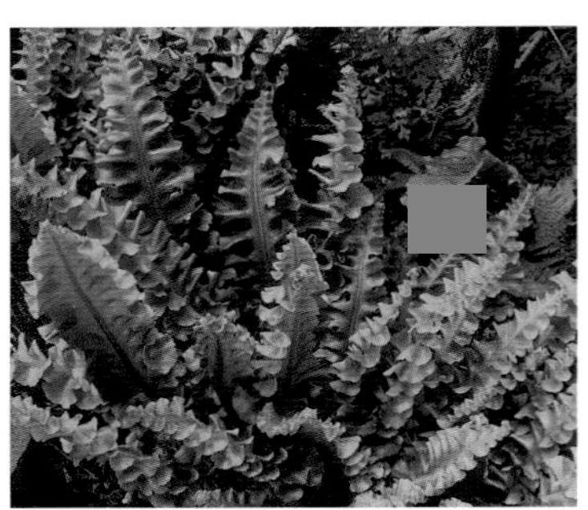

PERENNIAL MEDIUM

Asplenium scolopendrium

HART'S TONGUE FERN This rosette-forming, evergreen fern has leathery, bright green, tongue-shaped fronds. The spores beneath the fronds create a striped effect on upper surfaces. Forms such as Crispum Group AGM (above) have crinkled-edged leaves.

↕70cm (28in) ↔60cm (24in)

TREE SMALL

Carpinus japonica ♡

JAPANESE HORNBEAM This deciduous tree produces serrated, textured, prominently veined foliage, which contrasts well with plants bearing smooth leaves. Green catkins, which mature to brown, appear in spring.

↕↔ 8m (25ft)

PERENNIAL LARGE

Crocosmia masoniorum ♡

MONTBRETIA Grown primarily for its arching stems of orange-red flowers in late summer, this clump-forming perennial also offers sword-shaped, dark green leaves, which are pleated lengthways, and offer dramatic shape and texture.

↕ 1.4m (4½ft) ↔ 45cm (18in)

PERENNIAL LARGE

Cynara cardunculus ♡

CARDOON This tall, upright perennial is grown for its ornamental, spiny, silver-grey foliage, which can be used as a focal point in a sunny border. Large, thistle-like, purple flowers appear from summer to autumn. Provide support in exposed sites.

↕ 2m (6ft) ↔ 1m (3ft)

PERENNIAL LARGE

Ensete ventricosum ♡

ABYSSINIAN BANANA This evergreen perennial has large, paddle-shaped, green leaves, red beneath, with cream midribs, and marked horizontal veins that create a ridged effect. Grow it in a large pot in cold regions, and bring under cover in winter.

↕↔ 3m (10ft) in a pot

PERENNIAL SMALL

Eryngium bourgatii

This upright perennial offers a wealth of textures. Its silver-veined leaves have spiny edges, creating a prickly appearance, while the cone-like summer flowers sport spiny ruffs. The seedheads retain this spiky look. Combines well with grasses.

↕ 45cm (18in) ↔ 30cm (12in)

TREE LARGE

Eucalyptus gunnii ♡

CIDER GUM The smooth, blue-grey young foliage of this tall, evergreen tree makes a good contrast with the rough-textured leaves. The leaves become grey-green and more leathery as they mature. Its colourful, peeling bark also offers textural interest.

↕ up to 25m (80ft) ↔ 15m (50ft)

TREE LARGE

Fagus sylvatica ♡

COMMON BEECH This large, deciduous tree, often clipped to create hedging, has prominently veined, textured, wavy-edged, green leaves, which turn coppery brown in autumn. It retains its autumn foliage through winter when used as hedging.

↕ 25m (80ft) ↔ 15m (50ft)

PERENNIAL LARGE

Foeniculum vulgare

SWEET FENNEL A tall, perennial herb, with a cloud of ferny leaves on slender, upright stems. In summer, it bears clusters of tiny, yellow flowers held in large, flat heads. Use it in borders to contrast with plants that produce larger, smooth-textured foliage.

↕ 1.8m (6ft) ↔ 45cm (18in)

TREE LARGE

Ginkgo biloba

MAIDENHAIR TREE This deciduous, conical tree produces fan-shaped, pale green, frilly-edged leaves that create a textural effect from spring to summer, and inject colour in autumn when they turn yellow. Use it as a specimen in a lawn or border.

↕ 30m (100ft) ↔ 8m (25ft)

Plants for textured foliage

PERENNIAL SMALL

Heuchera villosa 'Palace Purple'

Prized for its rounded, lobed and veined, metallic bronze-purple leaves, combine this evergreen perennial with small-leaved plants, such as *Buxus*. Wiry stems of small, white flowers appear in summer.
↕↔ 45cm (18in)

PERENNIAL SMALL

Hosta 'Halcyon' 🏆

The long, heart-shaped, slug-resistant, blue-grey leaves of this compact perennial have a ribbed, thick-textured appearance, and make a good foil for dark green foliage plants. The plant provides both colour and texture in shady borders.
↕ 40cm (16in) ↔ 70cm (28in)

SHRUB MEDIUM

Hydrangea quercifolia

OAK-LEAVED HYDRANGEA This mound-forming shrub has rough-textured, lobed leaves, which develop maroon tints, and take on striking red and purple hues in autumn. Cone-shaped flower clusters appear from midsummer to autumn.
↕ 2m (6ft) ↔ 2.5m (8ft)

PERENNIAL MEDIUM

Matteuccia struthiopteris 🏆

OSTRICH-FEATHER FERN Most ferns offer interesting textures, but this deciduous, shuttlecock-shaped species is among the best. It produces tall, pale green fronds, divided into toothed segments, and is ideal for damp soils in a woodland setting.
↕ 1m (3ft) ↔ 45cm (18in)

TREE LARGE

Ilex aquifolium 🏆

ENGLISH HOLLY A slow-growing, evergreen tree, known for its dark green, spiny leaves. Offering a prickly year-round textural effect, it makes a good foil for smooth-leaved plants. Female forms produce red winter berries; variegated varieties are available.
↕ up to 20m (70ft) ↔ 6m (20ft)

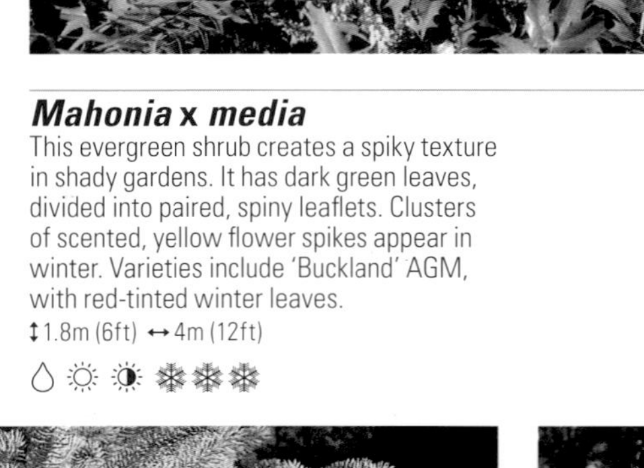

SHRUB MEDIUM

Mahonia x *media*

This evergreen shrub creates a spiky texture in shady gardens. It has dark green leaves, divided into paired, spiny leaflets. Clusters of scented, yellow flower spikes appear in winter. Varieties include 'Buckland' AGM, with red-tinted winter leaves.
↕ 1.8m (6ft) ↔ 4m (12ft)

SHRUB SMALL

Picea pungens 'Montgomery'

DWARF BLUE SPRUCE The mound of grey-blue, needle-like leaves produced by this compact, slow-growing, evergreen conifer creates a bristly, textured effect. Combine it with smooth, green foliage plants in a sunny site. Requires neutral to acid soil.
↕↔ 60cm (24in)

TREE LARGE

Pinus radiata

MONTEREY PINE A vigorous, evergreen conifer, suitable for large gardens, with a broad crown of dense, dark green, needle-like foliage, which will create a textured feature in a lawn. The tree also produces large, conical to oval-shaped cones.
↕ 40m (130ft) ↔ 12m (40ft)

SHRUB SMALL

Plectranthus argentatus ♡

SILVER SPURFLOWER Use this spreading evergreen subshrub, with its silver stems and grey-green foliage, to create textural contrasts with leathery- and glossy-leaved plants. It bears bluish white summer flower spikes. Makes a decorative edging for pots.

↕↔ up to 1m (3ft)

PERENNIAL LARGE

Rheum palmatum

The huge leaves of this upright perennial provide texture and drama. The foliage is dark green, purple-red beneath, jaggedly lobed, and features distinctive veining. In early summer, large plumes of fluffy, cream to red flowers appear.

↕↔ 2m (6ft)

PERENNIAL LARGE

Rodgersia podophylla

This perennial produces large, divided leaves made up of jagged-edged leaflets, with a distinctive veined texture, which are bronze-purple when young, maturing to green, and turning bronze-red in autumn. Creamy flower panicles appear in summer.

↕ 1.5m (5ft) ↔ 1.8m (6ft)

PERENNIAL MEDIUM

Salvia argentea

SILVER SAGE This short-lived perennial forms a rosette of silvery leaves with a soft, woolly texture that combine well with green foliage plants in sunny or lightly shaded borders. Spikes of blush white flowers appear in late summer.

↕ 90cm (36in) ↔ 60cm (24in)

PERENNIAL SMALL

Sempervivum giuseppii ♡

HOUSELEEK Use this evergreen perennial to inject textural interest into rock gardens, small pots, and cracks in walls. Its rosettes of spiky, green leaves with pointed, purple tips are joined by upright stems topped with star-shaped, pink or red flowers in summer.

↕ in flower 10cm (4in) ↔ 10cm (4in)

PERENNIAL SMALL

Stachys byzantina

LAMBS' EARS An evergreen, mat-forming perennial, with oval, silvery grey leaves that have a soft, downy texture. In summer, spikes of tiny, clustered, mauve-pink blooms form on felted stems. The foliage remains attractive in winter in mild areas.

↕ 38cm (15in) ↔ 60cm (24in)

PERENNIAL MEDIUM

Stipa tenuissima

MEXICAN FEATHER GRASS A deciduous, perennial grass with fine, green leaves, and silvery green summer flowers that turn beige as seeds form, creating a hazy, hair-like texture. Use it as a front-of-border plant with flowering perennials.

↕ 60cm (24in) ↔ 40cm (16in)

TREE SMALL

Trachycarpus fortunei ♡

CHUSAN PALM This evergreen palm adds texture to exotic-style schemes with its large, glossy, green, fan-shaped leaves, which are held on long, flexible stems. Plant it in a sheltered spot, or in a large container of soil-based compost when young.

↕ up to 2m (6ft) ↔ 2.5m (8ft)

ANNUAL/BIENNIAL LARGE

Verbascum olympicum

OLYMPIAN MULLEIN This biennial forms a rosette of large, grey leaves, with a soft, tactile, felted texture. These may die back in their first winter. Saucer-shaped, bright golden flowers appear the next year from mid- to late summer. It may need staking.

↕ 2m (6ft) ↔ 1m (3ft)

OTHER SUGGESTIONS

Annuals and biennials

Onopordum acanthium

Perennials and grasses

Aloe polyphylla ♡ • *Aspelenium bulbiferum* • *Echeveria elegans* ♡ • *Helichrysum* 'Ruby Cluster' • *Helleborus* x *sternii* • *Hosta* 'Krossa Regal' ♡ • *Mentha spicata* • *Musa basjoo* ♡ • *Salvia corrugata* • *Sempervivum arachnoideum* ♡ • *Veratrum nigrum* ♡ • *Verbascum bombyciferum*

Shrubs and trees

Abies procera 'Glauca' • *Acacia dealbata* ♡ • *Carpinus betulus* ♡ • *Carpinus cordata* • *Cornus avellana* • *Hoheria angustifolia* • *Ilex aquifolium* 'Ferox Argentea' ♡ • *Sorbus aria* 'Lutescens' ♡ • *Thymus pseudolanuginosus* • *Ulmus glabra* 'Lutescens'

Plants for architectural foliage

The large, shapely leaves of a tree or shrub create features that can equal the dramatic impact of the brightest, most colourful flowers.

Use plants with sculptural foliage as a focal point at the end of a path or as sentries to guard an entrance way. Many big-leaved species prefer shady locations, where a group sporting foliage of different sizes, shapes, and textures will produce an elegant scheme that sings out from the gloom. In sunny sites, plant spikes and sword-shaped leaves to draw the eye – their dramatic outlines create an effective contrast to rounded and small, tear-shaped foliage. Some trees and shrubs, including *Paulownia*, should be pruned hard in late winter to encourage young stems that bear the largest leaves.

PERENNIAL LARGE

Acanthus mollis

BEAR'S BREECHES This semi-evergreen, upright perennial lends an architectural note to sunny and shaded gardens. The long, dark green, glossy leaves are deeply cut. In summer, tall spikes of funnel-shaped, white and purple flowerheads appear.

↕1.2m (4ft) ↔60cm (24in)

PERENNIAL LARGE

Actaea simplex

BUGBANE An upright perennial, grown for its sculptural, dark green, divided foliage. From early- to mid-autumn, it produces arching stems of bottlebrush, off-white-tinged purple flowers. 'Brunette' AGM (above) is a dark purple-leaved variety.

↕1.2m (4ft) ↔60cm (24in)

PERENNIAL MEDIUM

Aeonium arboreum

Grow this tender, evergreen perennial in a pot and stand it outside in summer to enjoy its rosettes of fleshy, spoon-shaped, pale green leaves, purple-black in some forms, which appear on branching stems. Protect it from frost in winter by bringing under cover.

↕↔60cm (24in)

TREE LARGE

Araucaria araucana

MONKEY PUZZLE This evergreen conifer, suitable for large gardens, has an unusual, branching, open habit. Its stems are densely clothed in scale-like, prickly, green leaves, giving it a sculptural quality. Slow-growing at first, it grows rapidly after a few years.

↕25m (80ft) ↔10m (30ft)

PERENNIAL LARGE

Beschorneria yuccoides 🏆

Ideal for a large pot of gritty, soil-based compost, this evergreen perennial produces pointed, strap-shaped, fleshy, grey-green leaves, creating a sculptural focal point. In summer, yellow-green tubular flowers, with red bracts, may appear. Protect from frost.

↕↔1.5m (5ft)

PERENNIAL LARGE

Calamagrostis acutiflora

FEATHER REED GRASS The fountain of long, arching, green leaves produced by this deciduous, clump-forming grass provides architectural interest for most of the year. The tall, bronze, summer flowerheads fade to buff. Cut back in early spring.

↕1.8m (6ft) ↔1.2m (4ft)

PERENNIAL LARGE

Canna 'Durban'

Grown for its large, paddle-shaped foliage, this perennial offers a spectacular display of orange, and dark green, striped leaves, creating a sculptural focal point. In late summer, tall stems topped with ruffled, orange flowers appear.

↕1.6m (5½ft) ↔50cm (20in)

TREE LARGE

Catalpa bignonioides 🏆

INDIAN BEAN TREE This deciduous, spreading tree has distinctive, large, broadly oval, pale green leaves. It can be pollarded to enhance the foliage. White flowers appear in summer, followed by bean-like seedpods.

↕↔up to 15m (50ft)

SHRUB LARGE

Chamaerops humilis 🏆

DWARF FAN PALM This evergreen palm has architectural, fan-shaped, divided, green leaves, ideal for foliage schemes. Mature plants produce clusters of small, yellow summer flowers. It is more likely to survive winters if grown in free-draining soil.

↕3m (10ft) ↔2m (6ft)

TREE SMALL

Cordyline australis

CABBAGE PALM This evergreen, palm-like tree produces a sculptural fountain of sword-shaped leaves in a range of green and purple shades. Grow in a large pot and use it as an accent plant in a border or patio display. Protect young specimens from frost.

↕3m (10ft) ↔1m (3ft)

PERENNIAL LARGE

Cynara cardunculus

CARDOON Grown for its ornamental spiny, silver-grey foliage, this tall, upright perennial can be used as a sculptural focal point in a sunny border. Large, thistle-like, purple flowers appear from summer to autumn. Provide support in exposed sites.

↕2m (6ft) ↔1m (3ft)

SHRUB LARGE

Daphniphyllum macropodum

The large, slender, oval leaves of this hardy, yet exotic-looking, evergreen shrub are borne in clusters on branching stems, creating an architectural effect in sheltered gardens. Small, purple-pink male or green female flowers appear in early summer.

↕↔6m (20ft)

PERENNIAL LARGE

Darmera peltata

UMBRELLA PLANT This perennial is prized for its large, round, deeply veined, green leaves, which can grow larger than dinner plates, and turn red in autumn. In spring, it bears clusters of white or pale pink flowers on hairy stems, before the foliage appears.

↕1.2m (4ft) ↔60cm (24in)

PERENNIAL MEDIUM

Dryopteris affinis

GOLDEN MALE FERN This evergreen fern's "shuttlecock" of tall, lance-shaped, divided fronds, which are pale green when young and mature to dark green create a sculptural effect in shady gardens. Cut off old growth in early spring.

↕↔90cm (36in)

PERENNIAL LARGE

Ensete ventricosum

ABYSSINIAN BANANA This evergreen perennial is ideal for foliage displays. It has tropical-style, large, paddle-shaped, green leaves, red beneath. 'Maurelii' AGM has red-tinted leaves. Grow in a large pot in cold regions, and bring under cover in winter.

↕2m (6ft) ↔1m (3ft)

SHRUB LARGE

Eriobotrya japonica

LOQUAT This architectural shrub produces large, glossy, dark green leaves, felted beneath, at the end of branched stems, and makes a dramatic focal point. Grow it in a sheltered, sunny spot to promote the scented, white autumn and winter flowers.

↕↔8m (25ft)

PERENNIAL MEDIUM

Farfugium japonica

The large, kidney-shaped, green leaves of this evergreen perennial lend sculptural interest to shaded gardens. 'Argentea' (above) has variegated foliage. Daisy-like, bright yellow flowers appear in late autumn. Apply a deep mulch in winter.

↕↔60cm (24in)

SHRUB MEDIUM

x *Fatshedera lizei*

TREE IVY The dark green leaves of this upright, evergreen shrub are deeply lobed and glossy. When trained as a climber up a pillar or against a wall, it creates tall columns of decorative foliage. Sprays of small, white flowers appear in autumn.

↕2m (6ft) ↔3m (10ft)

TREE SMALL

Ficus carica
FIG Although in warm climes, this small tree or deciduous shrub is grown for its fruits, it lends sculptural interest to ornamental gardens in cooler climes, with its rounded, lobed, grey-green foliage. Grow it in a sheltered, sunny or lightly shaded site.
↕3m (10ft) ↔4m (12ft)

TREE LARGE

Gleditsia triacanthos
HONEY LOCUST A spreading, deciduous tree, grown for its glossy, dark green leaves, divided into many slim, oval leaflets, creating a dramatic effect. Long, twisted seed pods in autumn add to its charms. 'Rubylace' has dark bronze-red young leaves.
↕15m (50ft) ↔5m (15ft)

PERENNIAL MEDIUM

Helleborus argutifolius 🏆
HOLLY-LEAVED HELLEBORE The dark green, spiny-edged leaves of this architectural perennial provide interest year-round. From late winter to spring, clusters of nodding, bowl-shaped, pale green flowers appear. Cut old leaves and flowerheads in autumn.
↕60cm (24in) ↔45cm (18in)

PERENNIAL MEDIUM

Hosta sieboldiana 🏆
The large, rounded leaves of this perennial create a spectacular effect in a shady border, where their rich blue-green colour and puckered texture catch the eye. Spikes of lilac-tinged white flowers appear in early summer. Protect from slugs.
↕1m (3ft) ↔1.2m (4ft)

TREE LARGE

Liriodendron tulipifera 'Aureomarginatum' 🏆
A deciduous tree, ideal for big gardens, it is notoriously slow to flower, but gives interest with its unusual, square-shaped, yellow-edged green foliage. Pale green, tulip-like summer blooms form on old specimens.
↕30m (100ft) ↔15m (50ft)

TREE LARGE

Magnolia grandiflora
Although its large, fragrant, white summer flowers are a definite attraction, this evergreen tree's rounded head of large, glossy, dark green leaves also make an eye-catching feature in a sheltered garden, or beside a south-facing wall.
↕18m (60ft) ↔15m (50ft)

SHRUB MEDIUM

Mahonia x *media*
Grown for its large leaves, comprised of holly-like, dark green leaflets, this evergreen shrub makes an exciting architectural feature in shady sites. The long, finger-like spikes of small, scented, yellow winter flowers attract bees.
↕1.8m (6ft) ↔4m (12ft)

SHRUB LARGE

Melianthus major 🏆
HONEYBUSH Grown for its striking sculptural foliage, this evergreen subshrub bears blue-grey leaves, divided into toothed leaflets. Small, brownish red flowers appear from late spring. Shelter it from cold winds and provide a dry mulch in winter.
↕↔up to 3m (10ft)

PERENNIAL LARGE

Osmunda regalis 🏆
ROYAL FERN This deciduous, clump-forming fern, with divided, green fronds, pinkish when young and red-brown in autumn makes a focal point in pondside plantings. Mature plants bear tassel-like spikes of rust-brown spores at the ends of tall fronds.
↕2m (6ft) ↔1m (3ft)

PERENNIAL LARGE

Rheum palmatum

The huge leaves of this upright perennial create an architectural feature in gardens with moist soil. The dark green, jaggedly lobed foliage makes a statement even before the plumes of fluffy, cream to red flowers appear in early summer.
↕↔2m (6ft)

SHRUB MEDIUM

Paeonia delavayi var. *delavayi* f. *lutea*

TREE PEONY Lobed, green leaves, blue-green beneath, extend this shrub's appeal, giving interest even after the cup-shaped late spring flowers have faded. Grow at the back of a border, or beside a fence or wall.
↕2m (6ft) ↔1.2m (4ft)

TREE SMALL

Rhus typhina 'Dissecta' AGM

STAG'S HORN This small, deciduous tree or suckering shrub produces large, dramatic, finely cut, bright green leaves on velvety stems, which turn yellow, red and orange in autumn, when cone-shaped, maroon fruits also appear.
↕↔3m (10ft)

PERENNIAL SMALL

Podophyllum peltatum

MAY APPLE The large, glossy leaves of this perennial will lend a sculptural quality to the front of a shady border. Single, white to pale pink, slightly scented flowers appear beneath the leaves from spring to summer, followed by edible fruits.
↕45cm (18in) ↔30cm (12in)

PERENNIAL LARGE

Rodgersia pinnata

This clump-forming perennial is grown for its large, corrugated, divided, dark green leaves. Conical spires of small, pink, red, or yellow-white flowers appear in summer. 'Superba' AGM (above) has bronze-tinged leaves and bright pink flowers.
↕1.2m (4ft) ↔75cm (30in)

TREE LARGE

Sciadopitys verticillata AGM

JAPANESE UMBRELLA PINE This slow-growing, evergreen conifer bears long, linear, dark green leaves, which produce an architectural effect. It has textured red-brown bark and oval cones. It makes a good focal feature in a large lawn.
↕20m (70ft) ↔8m (25ft)

TREE SMALL

Trachycarpus fortunei AGM

CHUSAN PALM The architectural, deeply divided, fan-like leaves of this evergreen palm are held on an upright stem that becomes fibrous with age. It makes a dramatic focal point in a sheltered position, away from cold, drying winds.
↕up to 2m (6ft) ↔2.5m (8ft)

TREE MEDIUM

Trochodendron aralioides

This broadly columnar tree or large shrub produces large, oval, tapering, glossy leaves. From late spring to early summer, clusters of unusual, green, spider-shaped flowers appear. Shelter from cold, drying winds. It suits tropical-style schemes.
↕10m (30ft) ↔8m (25ft)

SHRUB MEDIUM

Yucca filamentosa AGM

The spiky, sword-shaped, green leaves of this clump-forming, evergreen shrub make it a useful accent plant in a sunny border or large pot. Pendulous, tulip-like flowers appear from mid- to late summer. 'Bright Edge' AGM (above) is a variegated form.
↕2m (6ft) ↔1.5m (5ft)

OTHER SUGGESTIONS

Perennials and bulbs

Arundo donax • *Astelia nervosa* • *Dryopteris filix-mas* • *Gunnera manicata* AGM • *Gunnera tinctoria* AGM • *Hedychium forrestii* • *Rheum palmatum* var. *tanguticum* • *Ricinus communis* • *Rodgersia aesculifolia* AGM

Grasses

Miscanthus sinensis 'Malepartus'

Shrubs and trees

Acer pseudoplatanus 'Brilliantissimum' AGM • *Albizia julibrissin* f. *rosea* • *Magnolia delavayi* • *Magnolia macrophylla* • *Paulownia tomentosa* AGM • *Platanus orientalis* • *Pseudopanax chathamica* • *Sorbus aucuparia* • *Tilia cordata* • *Trachycarpus wagnerianus*

Leaves for containers

Foliage is an essential ingredient of a well-balanced container display. Use leaves as foils for flowers and to add shape and form to a design.

Ferns, heucheras, and ivy leaves add rich texture to an autumn or winter display of violas, while bronze sedges set against dwarf narcissi will liven up a spring pot. To create a tropical look in summer, pair dark-leaved *Solenostemon* or the broody *Ipomoea batatus* 'Blackie' with scarlet and cerise dahlias, verbenas, and salvias. Some leafy plants need no flowery partners to make a statement. Striped and spiked agaves are best planted solo; a single large-leaved hosta will produce a beautiful display of colour and form; and *Fatsia japonica's* handsome glossy foliage sets off a glazed pot to perfection.

PERENNIAL SMALL

Acorus gramineus

SLENDER SWEET FLAG A semi-evergreen grown for its grassy foliage that arches over the sides of containers. Plant as a specimen or in a mixed display in permanently wet soil-based compost. 'Ogon' (above) has creamy yellow-striped foliage.

↕25cm (10in) ↔15cm (6in)

PERENNIAL SMALL

Adiantum aleuticum

MAIDENHAIR FERN Ideal for containers in shade, this semi-evergreen fern has small, pale green, branching fronds, clothed with tiny leaflets. Best grown as a specimen or in a mixed display with other ferns. Water well and protect from dry winds.

↕↔45cm (18in)

PERENNIAL MEDIUM

Aeonium 'Zwartkop'

This tender, evergreen succulent has upright stems, clothed with glossy, purple-black leaves. Grow it in containers of well-drained compost in a sunny, sheltered site. Protect it from frost during winter by bringing it under cover.

↕60cm (24in) ↔1m (3ft)

PERENNIAL LARGE

Agave americana

CENTURY PLANT A tender, perennial succulent, grown for its fleshy, lance-shaped, sharply pointed leaves. Grow in a large container of free-draining compost in full sun. Protect from frost. 'Variegata' AGM (above) has cream-edged leaves.

↕↔1.5m (5ft)

PERENNIAL MEDIUM

Asparagus densiflorus 'Myersii'

FOXTAIL FERN Grown for fine, lacy leaves, this tender perennial has a dense, plume-like appearance. It is ideal for containers and windowboxes in light shade. Move it under cover for winter protection.

↕1m (3ft) ↔50cm (20in)

PERENNIAL MEDIUM

Aspidistra elatior

This slow-growing, evergreen perennial has upright, glossy, dark green leaves, and looks attractive in a large container. It needs a sheltered site, away from cold winds. Varieties include 'Variegata' AGM (above), with cream-striped leaves.

↕60cm (24in) ↔45cm (18in)

PERENNIAL SMALL

Athyrium niponicum var. *pictum*

JAPANESE PAINTED FERN This deciduous fern has deeply cut, grey-green leaves, with metallic purple and silver highlights. Grow in a large container. Water well in summer and protect from hard frost.

↕30cm (12in) ↔indefinite

PERENNIAL SMALL

Begonia 'Escargot'

This tender, evergreen perennial is grown for its silver-edged green-tinted purple, heart-shaped leaves that have a distinctive coiled appearance. Stand it outside in summer in a sheltered, shady spot, and move it under cover for winter.

↕25cm (10in) ↔50cm (20in)

PERENNIAL LARGE

Canna 'Durban'

Ideal for large containers, this perennial is grown for its large paddle-shaped leaves, with vivid orange veins. It flowers in late summer, producing tall stems topped with orange blooms. Water well, remove spent flowers, and protect during winter.

↕1.6m (5½ft) ↔50cm (20in)

PERENNIAL MEDIUM

Carex buchananii

SEDGE This evergreen perennial has narrow, grass-like, upward, then arching, copper-coloured leaves that are red towards the base. Suitable for baskets and windowboxes for summer and winter displays; the leaves will trail over the sides.

↕60cm (24in) ↔20cm (8in)

PERENNIAL SMALL

Carex comans

NEW ZEALAND SEDGE This is an evergreen, grass-like perennial, grown for its arching, hair-like, copper-brown leaves that form an airy tuft. Ideal for containers, it cascades over the sides; it is also suitable for winter interest. Remove dead growth in spring.

↕35cm (14in) ↔75cm (30in)

PERENNIAL SMALL

Dichondra argentea 'Silver Falls'

SILVER PONYFOOT This evergreen perennial has trailing, silver stems, with rounded, shiny, silvery green leaves. Plant in baskets and windowboxes. Quick-growing, it can swamp neighbouring plants.

↕50cm (20in) ↔indefinite

PERENNIAL SMALL

Carex oshimensis

ORNAMENTAL SEDGE Useful for summer and winter container displays, this perennial has glossy, arching, dark green leaves, with cream stripes. 'Evergold' AGM (above) is yellow-striped and brightens up shady areas. Water well; shelter from cold winds.

↕↔20cm (8in)

BULB MEDIUM

Dahlia 'Roxy'

Dwarf dahlias are perfect for pots and are grown for their flowers and foliage. This variety has dark purple-green leaves, which provide a useful backdrop to other flowering plants in mixed containers. Protect from frost.

↕45cm (18in)

SHRUB LARGE

Dasylirion acrotrichum

This slow-growing, evergreen shrub has slender, stiff, spear-like leaves that radiate out from a central crown, forming a leafy globe. Mature plants develop a trunk, topped with leaves. Best in a large container of free-draining compost; protect in winter.

↕6m (20ft) ↔2.2m (7ft)

SHRUB LARGE

Fatsia japonica

JAPANESE ARALIA Suited to large containers, this evergreen shrub has large, hand-shaped, glossy, dark green leaves. Plant it in soil-based compost in a sheltered position for year-round interest. Use young plants for winter bedding in windowboxes.

↕↔4m (12ft)

PERENNIAL MEDIUM

Geranium maculatum

This perennial bears simple, pink flowers from spring to early summer, but is also grown for its divided, green leaves, which form a low mound and provide contrast in a mixed planter. Ideal for cool, shady corners; remove the spent flowers to keep it tidy.

↕75cm (30in) ↔45cm (18in)

PERENNIAL SMALL

Hakonechloa macra

GOLDEN HAKONECHLOA This slow-growing, deciduous grass is grown for its vivid green and yellow-striped foliage and elegant, trailing habit. Grow it in a tall container. Red-brown flower spikes appear in early autumn and last into winter.

↕40cm (16in) ↔60cm (24in)

SHRUB SMALL

Helichrysum petiolare

LIQUORICE PLANT Grown as an annual, this shrub has downy, grey-green leaves carried on trailing stems. Ideal for mixed containers and windowboxes, it can be overwintered under cover. 'Limelight' AGM (above) has pale lime-green foliage.

↕15cm (6in) ↔30cm (12in)

PERENNIAL SMALL

Heuchera 'Amber Waves'

Suitable for containers, this evergreen, clump-forming perennial has vivid, orange-yellow, lobed leaves providing colour and texture. It also flowers in summer, bearing sprays of tiny, white blooms. Other forms have green, silver, or purple foliage.
↕30cm (12in) ↔50cm (20in)

PERENNIAL SMALL

Hosta 'June'

Grown for its large, blue-green leaves that are marked with pale green flashes and form a dense, leafy crown, this perennial bears short-lived flowers in summer. Use it as a foliage specimen in a large container of soil-based compost. Protect from slugs.
↕38cm (15in) ↔70cm (28in)

PERENNIAL SMALL

Hosta 'Revolution' ♡

Grown as a foliage plant, this perennial has cream-splashed, dark green leaves that will add colour to sites in dappled shade. Best grown in a pot of soil-based compost; remove the short-lived, lavender flowers as soon as they fade. Protect from slugs.
↕50cm (20in) ↔1.1m (3½ft)

PERENNIAL SMALL

Hosta 'So Sweet'

This clump-forming perennial has dark green, textured leaves, with irregular, pale yellow edges. Best grown as a specimen plant in a large pot of soil-based compost, it can be sited in the shade of taller plants. Water well; protect from slugs.
↕35cm (14in) ↔55cm (22in)

PERENNIAL SMALL

Ipomoea batatas 'Blackie'

SWEET POTATO VINE Grown as an annual, this tender perennial has trailing stems of deeply lobed, ivy-shaped, blackish leaves. Ideal for containers, windowboxes, and baskets, it can used to provide contrast to brightly coloured plants in mixed displays.
↕25cm (10in) ↔60cm (24in)

PERENNIAL MEDIUM

Lamium galeobdolon

DEAD NETTLE This evergreen perennial bears two-lipped, yellow summer flowers, but is mainly grown for its nettle-like, silver-variegated leaves. Use the popular 'Hermann's Pride' (above) to give colour to summer and winter container displays.
↕60cm (24in) ↔indefinite

PERENNIAL LARGE

Leymus arenarius

LYME GRASS A vigorous, upright perennial grass, with long, slender, steel-blue leaves, forming an airy clump in summer, with tall flower stems that persist into autumn. Plant in a large pot of gritty soil-based compost. Deadhead to prevent self-seeding.
↕1.5m (5ft) ↔indefinite

PERENNIAL SMALL

Ophiopogon planiscapus 'Nigrescens' ♡

BLACK MONDO GRASS Ideal for bedding and long-term container planting, this evergreen perennial forms a mound of black leaves. It bears purple-pink flower clusters in summer. Lends contrast to brightly coloured plants.
↕23cm (9in) ↔30cm (12in)

PERENNIAL MEDIUM

Panicum virgatum

SWITCH GRASS An upright, clump-forming, deciduous grass, it gives a long season of interest in a container, with its green, grassy leaves, which turn yellow in autumn, and clouds of pink-tinged summer flowerheads. The dried flowers persist into winter.
↕1m (3ft) ↔75cm (30in)

PERENNIAL SMALL

Pelargonium 'Lady Plymouth' ♡

SCENTED-LEAVED PELARGONIUM A bushy, evergreen perennial, perfect for pots and windowboxes, with scented, lobed, silver-margined green leaves and lavender-pink summer blooms. Overwinter under cover.
↕40cm (16in) ↔20cm (8in)

PERENNIAL SMALL

Pelargonium 'Royal Oak' ♡

This bushy, evergreen perennial is grown for its lobed, crinkled, dark green leaves that develop crimson central flashes as they age. Simple, pink flowers form sparingly in summer. Grow it in containers and windowboxes as a foil for bright flowers.

↕38cm (15in) ↔30cm (12in)

PERENNIAL MEDIUM

Persicaria microcephala 'Red Dragon'

Ideal as a specimen plant in large containers of moist compost, this spreading perennial bears heart-shaped, reddish green leaves, with bold silver and bronze markings. Tiny, white flowers form in midsummer.

↕70cm (28in) ↔1m (3ft)

PERENNIAL MEDIUM

Phormium 'Bronze Baby'

NEW ZEALAND FLAX This evergreen, compact perennial forms an upright fountain of dark bronze leaves that give year-round interest. Use it as a specimen plant in a large container of soil-based compost. Protect from frost in colder sites.

↕↔60cm (24in)

SHRUB SMALL

Plectranthus madagascariensis

MINTLEAF Grown as an annual, this trailing, evergreen shrub has lobed, green leaves. Use the variegated form 'Variegated Mintleaf' AGM (above) in mixed plantings in pots. Pinch out long stems for bushy growth.

↕30cm (12in) ↔indefinite

CLIMBER MEDIUM

Senecio macroglossus 'Variegatus' ♡

A tender, twining climber, grown for its angular, ivy-like, yellow-edged dark green leaves. Train it up a wigwam in a large pot, or let the stems trail. It flowers sparingly in summer. Overwinter under cover.

↕up to 3m (10ft)

PERENNIAL SMALL

Solenostemon 'Black Prince'

FLAME NETTLE Grown as an annual, this bushy perennial has large, toothed, dark purple leaves. It can be planted singly or in mixed summer containers. Pinch out the tips regularly to encourage bushiness and to deter flowering. Water well.

↕↔50cm (20in)

PERENNIAL SMALL

Solenostemon scutellarioides

FLAME NETTLE A bushy perennial, grown as an annual, with spear-shaped, pink, red, green, and yellow leaves. Best in large containers, it makes a bold specimen plant. Pinch out the tips to promote bushy growth and to deter flowering. Water well.

↕45cm (18in) ↔30cm (12in) or more

PERENNIAL MEDIUM

Stipa tenuissima

MEXICAN FEATHER GRASS This clump-forming perennial grass gives interest to pots from summer to autumn, with its fine green leaves and airy summer flowerheads that dry and persist until winter. Grow it as a specimen plant in soil-based compost

↕60cm (24in) ↔40cm (16in)

PERENNIAL SMALL

Tiarella cordifolia var. *collina*

FOAM FLOWER This evergreen, spreading perennial has lobed, green leaves, with veins that turn bronze-red in winter. In late spring, spikes of fluffy, white flowers form. Use it to lend interest in winter and spring, then plant it out in a border in summer.

↕20cm (8in) ↔30cm (12in) or more

PERENNIAL MEDIUM

Tolmiea menziesii

PICK-A-BACK PLANT A semi-evergreen perennial, with textured, green, ivy-shaped leaves that provide contrast in mixed container displays and hanging baskets. 'Taff's Gold' AGM (above) has cream- and green-variegated foliage.

↕60cm (24in) ↔2m (6ft)

OTHER SUGGESTIONS

Annuals and biennials

Perilla frutescens

Perennials and bulbs

Adiantum capillus-veneris • *Begonia* 'Connie Boswell' • *Begonia* 'Fire Flush' • *Begonia rex* • *Chlorophytum comosum* 'Variegatum' ♡ • *Dahlia* 'Bishop of Llandaff' ♡ • *Heuchera* 'Palace Purple' • *Heuchera* 'Chocolate Ruffles' • *Solenostemon* 'Midnight' • *Solenostemon* 'Winter Sun' • *Strobilanthes dyeriana* ♡ • *Uncinia rubra*

Grasses

Carex comans 'Frosted Curls'

Shrubs, trees, and climbers

Cordyline 'Torbay Red' ♡ • *Hedera helix* • *Vinca minor* 'Illumination'

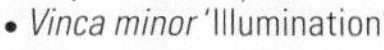

Trees for small gardens

Large plants can actually make small gardens look bigger by blurring the boundaries, and creating the impression that the space extends further.

When selecting trees for a small garden, choose compact varieties that won't loom over the whole plot, and site taller specimens on the north and east side of your property, where they will create the least shade. Opt for slim or vase-shaped trees to fit into corners, and a spreading species to create a shady feature in the centre of your garden. Also consider the seasonal interest when making your choices. Evergreens are tempting as they afford year-round colour and privacy, but deciduous trees that produce spring blossom, fruits, and autumn tints offer great value too.

TREE SMALL

Acer palmatum 'Atropurpureum'

JAPANESE MAPLE A small, bushy-headed tree, with purple-red summer foliage, turning fiery red in autumn. In winter, older specimens form a pleasing tracery of branches. Plant in shelter to prevent scorch.

↕↔ 8m (25ft)

TREE SMALL

Amelanchier laevis

ALLEGHENY SERVICEBERRY An upright shrub or small, spreading tree, with bronze-tinted young leaves, maturing to green, and finally becoming orange-red in autumn. White spring blossom is followed by round, red to purple, edible fruits.

↕↔ 8m (25ft)

TREE LARGE

Betula nigra

RIVER BIRCH Grow this as a multi-stemmed tree to make more of the cinnamon, peeling bark on young wood. The delicate branches carry yellow spring catkins; the small, diamond-shaped leaves turn yellow in autumn.

↕ 15m (50ft) ↔ 10m (30ft)

TREE SMALL

Betula pendula 'Youngii'

YOUNG'S WEEPING BIRCH This weeping form of the silver birch has grey-white bark and a wide-spreading crown of cascading branches that touch the ground. Brownish yellow catkins appear in spring, and the foliage turns yellow in autumn.

↕ 8m (25ft) ↔ 10m (30ft)

TREE LARGE

Betula utilis var. *jacquemontii*

HIMALAYAN BIRCH Grow this eye-catching, white-stemmed birch as a multi-stemmed tree to enhance its effect as a lawn or border specimen. The dangling, yellow catkins and yellow autumn foliage colour are added attractions.

↕ 18m (60ft) ↔ 10m (30ft)

TREE LARGE

Carpinus betulus 'Fastigiata' ♈

HORNBEAM This large tree, with its young, columnar form and flame-shaped mature crown, is perfect for limited spaces. The oval, toothed leaves have a pleated look and turn yellow in autumn.

↕ 25m (80ft) ↔ 20m (70ft)

TREE SMALL

Cercis chinensis

CHINESE REDBUD Grow this small, multi-branched tree in a sunny space to encourage good crops of the vivid, pea-like flowers. The flowers are unusual in that clusters sprout direct from the wood, before the heart-shaped leaves appear.

↕ 6m (20ft) ↔ 5m (15ft)

TREE SMALL

Cornus alternifolia 'Argentea'

SILVER PAGODA DOGWOOD This aptly named, large, spreading shrub or small, deciduous tree has beautiful, horizontally tiered branches, covered with oval, white-edged green leaves, turning red in autumn.

↕↔ 6m (20ft)

TREE MEDIUM

Cercis siliquastrum

JUDAS TREE A compact and bushy, deciduous tree, it will suit a sunny courtyard. In mid-spring, clusters of rosy pink, pea-like blooms sprout from bare branches, followed by heart-shaped leaves and purplish seed pods.

↕↔ 10m (30ft)

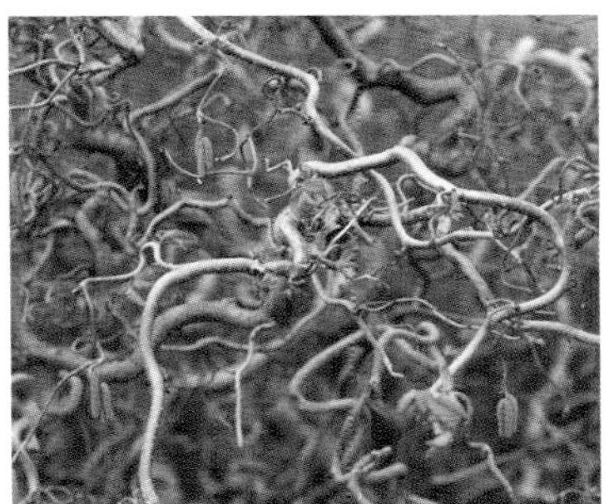

SHRUB LARGE

Corylus avellana 'Contorta'

CORKSCREW HAZEL This deciduous, bushy shrub, with corkscrewed shoots, makes an interesting focal point, especially in a small, courtyard garden. In late winter, pale yellow catkins form on bare stems. Broad, crinkled leaves follow.

↕↔ 5m (15ft)

TREE SMALL

Chionanthus virginicus

FRINGE TREE Perfect where space is at a premium, this unusual tree or large shrub has a spreading crown, decked with long-petalled, fragrant, white summer flowers. Once the oval leaves fall, its peeling bark is revealed.

↕↔ 3m (10ft)

TREE LARGE

Gleditsia triacanthos 'Rubylace'

HONEY LOCUST A spreading, deciduous tree, grown for its large, glossy leaves, divided into ferny leaflets. The deep red new leaves darken to bronze-green, making it a handsome tree for a small plot.

↕ 15m (50ft) ↔ 5m (15ft)

TREE MEDIUM

Gleditsia triacanthos 'Sunburst'

HONEY LOCUST This deciduous, golden-yellow tree has an airy appearance, with the foliage divided into small, bright, ferny, leaflets, which darken through summer. Pendent seed pods appear in autumn.

↕ 12m (40ft) ↔ 10m (30ft)

TREE LARGE

Ilex aquifolium

ENGLISH HOLLY A slow-growing, columnar to conical, evergreen tree that can be clipped to shape. It has dark green, glossy, spiny leaves and red winter berries on female plants. Choose one of the the variegated varieties to light up shade.

↕ up to 20m (70ft) ↔ 6m (20ft)

TREE SMALL

Juniperus scopulorum

ROCKY MOUNTAIN JUNIPER The narrow forms of this evergreen conifer with scale-like leaves include 'Skyrocket' (above), with blue-grey foliage, and the steely-blue 'Blue Arrow' AGM. Both make bold columns, but don't eat up valuable space.
↕6m (20ft) ↔75cm (30in)

TREE SMALL

Magnolia x *loebneri* 'Merrill' 🏆

This small, bushy, deciduous tree has an upright habit and oblong leaves that open after the flowers appear. The sta-ry, goblet-shaped, many-petalled, white blooms are produced in abundance in mid-spring.
↕8m (25ft) ↔7m (22ft)

TREE SMALL

Magnolia x *soulangeana*

SAUCER MAGNOLIA A deciduous, spreading tree, branching close to the ground, with goblet-shaped, white, pink, or purple mid- to late spring flowers. It makes a lovely lawn or border specimen with carpeting bulbs. Protect buds from late frost.
↕↔6m (20ft)

TREE MEDIUM

Malus floribunda 🏆

JAPANESE CRAB APPLE This small, deciduous tree, perfect for a small space, has a rounded head, with pale pink flowers that open from crimson buds in spring. The edible, red and yellow fruits are followed by bold autumn foliage.
↕↔up to 10m (30ft)

TREE MEDIUM

Malus x *moerlandsii* 'Profusion'

CRAB APPLE Ideal for limited space, this tree offers a long season of interest. Purple fruits and purple-tinted leaves follow the masses of deep pink blossom in spring. It develops bold autumn foliage colour.
↕↔10m (30ft)

TREE SMALL

Malus 'Royalty'

CRAB APPLE With a spreading head, this small tree has dark purple foliage, which turns red in autumn. This tree also gives a spectacular show in spring, when purple-red blossom smothers the branches. Small, dark purple fruits form in late summer.
↕↔8m (25ft)

TREE SMALL

Mespilus germanica

MEDLAR Though this tree develops a wide-spreading crown and often has a multi-stemmed habit, it can be trimmed to a more formal shape. The large, white, open flowers in spring are ornamental and develop into edible fruits.
↕6m (20ft) ↔8m (25ft)

TREE MEDIUM

Prunus 'Kanzan' 🏆

JAPANESE FLOWERING CHERRY
Sumptuous in bloom, this deciduous cherry has a vase-shaped head and double, pink spring blossoms opening just as the bronze-tinged foliage unfurls. Green in summer, the leaves turn orange in autumn.
↕↔10m (30ft)

TREE MEDIUM

Prunus 'Okame'

This large shrub or medium-sized, bushy tree is an early spring-flowering ornamental cherry, with delicate, fragrant, single, pink blooms opening from reddish buds. The leaves colour in autumn, with fiery oranges and reds.

↕10m (30ft) ↔8m (25ft)

TREE MEDIUM

Prunus 'Spire' 🏆

FLOWERING CHERRY This columnar cherry is ideal for adding height in a bijou plot. In spring, pale pink blossom covers the upright branches. The new leaves are bronze tinted, turning green in summer and colouring orange and red in autumn.

↕10m (30ft) ↔6m (20ft)

TREE SMALL

Ptelea trifoliata

HOP TREE Related to a citrus plant, this is a rounded shrub or small tree, with aromatic, three part leaves and fragrant, but small, greenish white flowers opening in early summer, which are followed by papery, hanging fruit clusters.

↕↔7m (22ft)

TREE SMALL

Sophora SUN KING 🏆

In cold areas, this bushy evergreen tree is best planted against a south-facing wall. Its narrow leaves are made up of small, oval leaflets. In late winter, large clusters of golden-yellow, bell-shaped flowers appear.

↕↔ up to 3m (10ft)

TREE SMALL

Rhus typhina

STAG'S HORN SUMACH This large, suckering shrub or small, spreading tree, with forked, felted stems, has large, divided leaves that turn red, orange, and yellow in autumn. Deep red, cone-shaped fruiting heads form on female plants.

↕5m (15ft) ↔6m (20ft)

TREE SMALL

Ulmus glabra 'Camperdownii'

CAMPERDOWN ELM A wide spreading, weeping, deciduous tree, this grafted form has reddish green spring flowers and later, papery seed cases, which appear among the opening leaves. The lumpy, layered texture of the tree in leaf is attractive.

↕↔8m (25ft)

TREE SMALL

Salix caprea 'Kilmarnock'

KILMARNOCK WILLOW This small, mushroom-headed tree has long, trailing branches. In late winter, silver-coloured, "pussy willow" catkins appear. Thin out select stems and prune away dead wood to keep the head airy.

↕↔2m (6ft)

OTHER SUGGESTIONS

Trees

Acer palmatum 'Sango kaku' 🏆 • *Amelanchier* 'Ballerina' • *Betula utilis* 'Long Trunk' • *Cercis siliquastrum* 'Alba' • *Chamaecyparis lawsoniana* 'Stardust' 🏆 • *Corylus avellana* 'Zellernus' • *Crataegus laevigata* 'Paul's Scarlet' • *Crataegus* x *prunifolia* • *Ginkgo biloba* 'Nana' • *Ilex castaneifolia* • *Koelreuteria paniculata* • *Ligustrum lucidum* 'Variegata' • *Malus* 'John Downie' • *Malus* 'Profusion' • *Morus alba* 'Platanifolia' • *Olea europaea* • *Prunus cerasifera* 'Nigra' • *Prunus* 'Pandora' 🏆 • *Prunus* 'Pink Perfection' 🏆 • *Sorbus aria* 'Lutescens' 🏆 • *Sorbus aucuparia* 'Joseph Rock' • *Sorbus aucuparia* 'Vilmorinii' • *Sorbus cashmiriana* • *Sorbus commixta* 'Embley' 🏆 • *Tamarix aestivalis* • *Tilia mongolica*

Plants for focal points and topiary

Some plants are natural-born stars, upstaging others with their radiant colour, graphic shape, or detailed texture to produce striking features.

While evergreens provide focal points all year round, deciduous plants often put on a show for just a brief spell and should be planted where they can be seen most clearly at those key times. For example, a cherry tree is best placed where its blossom will be bathed in spring sunshine, with an underskirt of bulbs to complement the flowery canopy. Unlike focal plants, topiary specimens generally have neither colour nor form to set them apart. However, their small, insignificant leaves are exactly what's needed to mould them into amazing features guaranteed to draw the crowds.

TREE SMALL

Acer palmatum 'Bloodgood' ♡

JAPANESE MAPLE A shapely tree, with delicate, lobed, purple leaves that turn brilliant red in autumn, it is especially eye-catching set against evergreens. The skeletal winter framework is also pleasing.
↕↔ 5m (15ft)

TREE LARGE

Araucaria araucana

MONKEY PUZZLE This evergreen conifer has a striking branching habit and stems densely clothed with small, scale-like, prickly, green leaves. It is slow-growing at first but grows rapidly after a few years, so allow plenty of space.
↕ 25m (80ft) ↔ 10m (30ft)

TREE LARGE

Betula utilis var. *jacquemontii*

HIMALAYAN BIRCH Planted singly as a multi-stemmed tree, or grouped to form a small copse, this white-stemmed birch makes a striking statement. The dangling, yellow catkins and yellow autumn foliage colour are an extra bonus.
↕ 18m (60ft) ↔ 10m (30ft)

SHRUB LARGE

Buxus sempervirens

COMMON BOX This traditional evergreen hedging and topiary shrub has small, oval, glossy leaves. When close-clipped, it will form intricate shapes and green sculptures, such as spirals. Prune between early and mid- to late summer.
↕↔ 5m (15ft)

TREE LARGE

Catalpa speciosa ♡

This exotic-looking flowering tree has heart-shaped leaves that bear clusters of white, orchid-like blooms, with purple and yellow spots on the throats, in late summer. Long, thin, green-turning-dark brown seed pods follow.
↕↔ 15m (50ft)

TREE LARGE

Cedrus deodara ♡

DEODAR CEDAR This magnificent cedar needs plenty of space to accommodate its broad, spreading canopy of tiered branches covered with short, blue-green needles. The shape begins as conical, changing as the tree reaches maturity.
↕ 40m (130ft) ↔ 10m (30ft)

TREE SMALL

Cordyline australis ♡

CABBAGE PALM An evergreen accent plant, it makes a sculptural fountain of sword-shaped leaves, eventually growing into a lightly branched, palm-like tree. Foliage varies from green, through bronze and purple, to variegated.
↕ 3m (10ft) ↔ 1m (3ft)

SHRUB LARGE

Cornus alba 'Sibirica' ♡

DOGWOOD This deciduous, winter interest shrub is grown for its bright red, strongly upright, winter stems, particularly showy when grown in a swathe against an evergreen backdrop. Hard prune in late winter for brighter stems.
↕↔ 3m (10ft)

TREE SMALL

Cornus kousa

DOGWOOD This early summer-flowering tree draws attention with its cream, petal-like bracts and hanging clusters of knobbly, pinky red fruit clusters. In autumn, the oval leaves turn into a striking reddish purple colour.

↕7m (22ft) ↔5m (15ft)

PERENNIAL LARGE

Cortaderia selloana 'Silver Comet'

PAMPAS GRASS This variegated grass forms an arching tussock of evergreen white-edged leaves. It looks most splendid in summer when tall stems carry creamy white flower plumes. Remove seedheads.

↕1.5m (5ft) ↔1m (3ft)

TREE LARGE

Cornus controversa

A large, deciduous tree, with an elegant, tiered habit and glossy, elliptical leaves that hang down from the branches. In summer, flat heads of tiny, white flowers appear, and in autumn, the leaves turn purple.

↕↔15m (50ft)

TREE SMALL

Cornus florida

FLOWERING DOGWOOD This deciduous tree or shrub is transformed by the showy, white or pink flower-like bracts that smother the branches in spring. In autumn, the curled green leaves develop rich red and purple colouring.

↕6m (20ft) ↔8m (25ft)

BULB MEDIUM

Eucomis bicolor

PINEAPPLE LILY In patio pots, this late-summer bulb draws attention when the large head, reminiscent of a pineapple, packed with greenish white flowers, and topped with leafy bracts, rises from the rosette of bold leaves.

↕50cm (20in)

TREE MEDIUM

Eucryphia glutinosa

In flower, this upright, semi-evergreen tree turns heads. The pure white, bowl-shaped blooms, with a tuft of central stamens, appear in summer against the dark green, glossy leaves, which later show rich autumn tints.

↕10m (30ft) ↔6m (20ft)

PERENNIAL LARGE

Euphorbia characias subsp. *wulfenii*

MEDITERRANEAN SPURGE Plant this evergreen subshrub in gravel to enhance its architectural form. In spring, the upright stems, covered with grey-green leaves, are topped with tiny, yellow-green flowerheads.

↕↔1.2m (4ft)

PERENNIAL SMALL

Hakonechloa macra

GOLDEN HAKONECHLOA Use this yellow-striped, deciduous grass singly in a pot, or grouped at the front of a border, or alongside a path. The glowing colour and arching form of the ribbon-like leaves can really light up a shady spot.

↕40cm (16in) ↔60cm (24in)

TREE MEDIUM

Laurus nobilis

BAY LAUREL This large, evergreen tree, with sweetly aromatic, dark green leaves, can be shaped into a number of topiary forms including cones and columns, or ball-headed standards for patio pots, given winter protection in cold areas.
↕12m (40ft) ↔10m (30ft)

TREE MEDIUM

Malus x moerlandsii 'Profusion'

CRAB APPLE Spectacular when in full bloom and smothered in deep pink spring blossom, this tree offers a long season of interest, with purple-tinged new growth, small, purple fruits, and rich autumn colour.
↕↔10m (30ft)

SHRUB SMALL

Hebe rakaiensis

This naturally dome-shaped, evergreen shrub, with small, olive-green leaves can easily be trimmed into simple topiary forms or a low hedge; clip as necessary in early summer to remove the white flowers.
↕1m (3ft) ↔1.2m (4ft)

TREE MEDIUM

Nyssa sinensis

This medium-sized spreading, deciduous tree makes a striking lawn specimen at maturity as the narrow, oval leaves put on a spectacular show of amber and scarlet, in autumn,especially when provided neutral to acid soil.
↕↔10m (30ft)

TREE SMALL

Juniperus scopulorum

ROCKY MOUNTAIN JUNIPER For an exclamation mark within a border or gravel area, choose one of the upright varieties of this evergreen conifer, such as 'Skyrocket' or 'Blue Arrow' AGM, both of which have fine, blue-grey foliage and are pencil thin.
↕6m (20ft) ↔75cm (30in)

SHRUB MEDIUM

Lonicera nitida

SHRUBBY HONEYSUCKLE A small-leaved, quick-growing evergreen, this is often used to shape topiary figures. It develops faster than traditional box, but need more regular trimming. 'Baggesen's Gold' AGM (above) has yellow leaves.
↕2m (6ft) ↔3m (10ft)

TREE SMALL

Parrotia persica

PERSIAN IRONWOOD As a specimen, this deciduous tree or large shrub with multiple stems has much to offer. In winter, the branches show patchwork patterns of peeling bark in greys, greens, and white, and the autumn leaf colour is striking.
↕8m (25ft) ↔10m (30ft)

PERENNIAL LARGE

Kniphofia caulescens

RED HOT POKER This perennial has a subtropical look, with its mounds of evergreen, grey-green, strap-shaped leaves topped with spikes of coral-red, opening to yellow, flowers on stout, upright stems in late summer and autumn.
↕1.2m (4ft) ↔60cm (24in)

TREE LARGE

Magnolia grandiflora

The large, glossy green leaves of this evergreen tree, often grown in the shelter of a south-facing wall, give it a tropical look, a characteristic enhanced when the giant, white, scented blooms open in midsummer.
↕18m (60ft) ↔15m (50ft)

PERENNIAL LARGE

Phormium tenax

NEW ZEALAND FLAX An evergreen, clump-forming perennial, with arching or upright, sword-shaped leaves, it creates a dramatic contrast set against less defined plants. During hot summers, tall stems of dark red flowers appear.
↕3m (10ft) ↔2m (6ft

SHRUB LARGE

Photinia glabra 'Rubens'

This evergreen shrub is grown for the colour of its new leaves, which are bright crimson, with some colour continuing through the year. Fragrant, white flower clusters appear in spring, followed by red fruits that later turn black.

↕↔3m (10ft)

TREE LARGE

Picea pungens 'Koster'

COLORADO SPRUCE This bright, silvery blue, evergreen conifer makes a handsome lawn specimen, with its relatively slow growth, symmetrical form, and stiff, needle-like leaves. Upright cones form towards the ends of the branches.

↕15m (50ft) ↔5m (15ft)

SHRUB LARGE

Pittosporum tobira 🏆

JAPANESE PITTOSPORUM In a sheltered garden, this evergreen can be grown as a hedge or used to form neat dome shapes. The oval leaves form attractive whorls and, in late spring, clusters of sweetly perfumed, creamy white flowers appear.

↕10m (30ft) ↔3m (10ft)

TREE MEDIUM

axus baccata 'Fastigiata' 🏆

SH YEW This evergreen conifer has naturally narrow, upright form, and is ed as an exclamation mark in a border to create a formal avenue of sentinels either side of a pathway. Female ants have red fruits. All parts are toxic.

0m (30ft) ↔4m (12ft)

TREE LARGE

Trachycarpus fortunei 🏆

CHUSAN PALM With an upright stem that becomes fibrous with age, and large, deeply cut, fan-like leaves, this evergreen palm makes a dramatic focal point in a border or large pot of soil-based compost. Plant it in a sheltered position.

↕up to 2m (6ft) ↔2.5m (8ft)

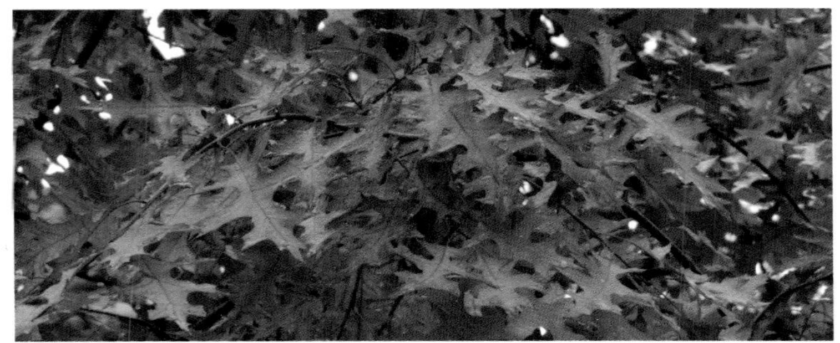

TREE LARGE

Quercus palustris 🏆

PIN OAK Planted at the end of a large lawn, this tall, pyramid-shaped oak would make a striking focal point. Apart from the attractive shape, the lobed leaves turn brilliant crimson-red and reddish brown in autumn, offering interest.

↕20m (70ft) ↔12m (40ft)

SHRUB MEDIUM

Yucca filamentosa

The sword-shaped, spine-tipped leaves of this evergreen make a useful accent in gravel gardens or large pots. 'Bright Edge' AGM (above) is particularly striking. From mid- to late summer, tall stems carry white blooms.

↕2m (6ft) ↔1.5m (5ft)

OTHER SUGGESTIONS

Perennials

Euphorbia characias 'Silver Swan' 🏆 • *Kniphofia uvaria* • *Phormium* 'Sundowner' 🏆

Shrubs

Elaeagnus angustifolia • *Euonymus fortunei* 'Emerald Gaiety' 🏆 • *Laurus nobilis* 'Angustifolia' • *Ligustrum lucidum* 🏆 • *Prunus lusitanica* • *Yucca aloifolia*

Trees

Acer palmatum 'Katsura' 🏆 • *Betula utilis* var. *jacquemontii* 'Doorenbos' 🏆 • *Cornus kousa* 'Miss Satomi' 🏆 • *Cupressus sempervirens* 'Pyramidalis' • *Cupressus sempervirens* 'Stricta' • *Ilex crenata* • *Phillyrea latifolia* 'Niwaki' • *Quercus ilex* • *Taxus cuspidata*

PLANTS for GARDEN PROBLEMS

Only the lucky few have gardens with rich, fertile soil and smooth, level beds that almost any plant will enjoy. And even those blessed with an ideal site and soil may still be plagued by rabbits or slugs determined to consume their prized plants. The good news is that most problems can be overcome by choosing plants that are adapted to your conditions. So, whether you need flowers for parched earth, foliage for soggy soils, or species that will cling to your hillside garden, there's something here for you.

Plants for sun-baked areas

Choices are limited for free-draining soils that retain few plant nutrients, yet some highly prized species have adapted to grow well in them.

Most sun-loving wild flowers abhor rich soil, and many cultivated plants, such as hollyhocks and cosmos, will also thrive in free-draining, sun-baked conditions. As well as the plants here, more may cope with poor soils once they are established, so take a look at the selection for sandy soils too (*see pp.44–51*), and provide water and some fertilizer for the first few seasons after planting. One benefit of well-drained soils is that plants that suffer in cold, wet soils over winter, such as perennial wallflowers and fleshy-leaved succulents, are more likely to survive in dry conditions.

PERENNIAL LARGE

Achillea filipendulina

YARROW Perfect for dry soil and full sun, this clump-forming perennial has finely cut, grey-green leaves and sturdy stems, punctuated with flat-topped golden-yellow flowerheads in summer. 'Gold Plate' AGM (above) is a popular variety.
↕1.2m (4ft) ↔60cm (24in)

PERENNIAL LARGE

Alcea rosea

HOLLYHOCK This tall, upright perennial may self-seed in cracks and dry, dusty soils. Its rounded leaves are joined by white, pink, red, or yellow, cup-shaped flowers on towering stems from mid- to late summer. The blooms attract bees and butterflies.
↕2m (6ft) ↔60cm (24in)

SHRUB SMALL

Caryopteris x *clandonensis*

A deciduous, bushy shrub that thrives in free-draining soils. It forms clumps of grey-green, lance-shaped leaves and dense clusters of small, purplish blue flowers from late summer to autumn. 'Worcester Gold' AGM (above) has contrasting golden foliage.
↕1m (3ft) ↔1.5m (5ft)

ANNUAL/BIENNIAL SMALL

Cosmos sulphureus

Hailing from scrublands, this upright, bushy annual produces a wealth of bowl-shaped, semi-double flowers all summer in shades of yellow, orange, or scarlet, on dry soils. The feathery, green leaves are attractive, too. It is easy to grow from seed in spring.
↕40cm (16in) ↔20cm (8in)

BULB SMALL

Crocus chrysanthus

Plant this tiny perennial bulb in autumn in a sunny spot in free-draining soil, where its cup-shaped blooms, with dark green, linear leaves, appear in early spring. Varieties include 'Gypsy Girl' (above), with purple-striped yellow flowers.
↕7cm (3in)

PERENNIAL SMALL

Delosperma nubigenum

YELLOW ICE PLANT This ground-hugging, evergreen perennial, perfect for sun-baked soils, produces mats of triangular, succulent, green leaves and daisy-like, lemon-yellow flowers in summer. Use it to edge a border; it may need frost protection in cold areas.
↕5cm (2in) ↔50cm (20in)

PERENNIAL SMALL

Dianthus alpinus

ALPINE PINK Plant this evergreen, mat-forming perennial in raised beds or at the front of a dry border. Its grey-green foliage is joined in summer by white or pink blooms, held on short stems. 'Joan's Blood' AGM (above) has solitary, deep crimson blooms.
↕8cm (3in) ↔10cm (4in)

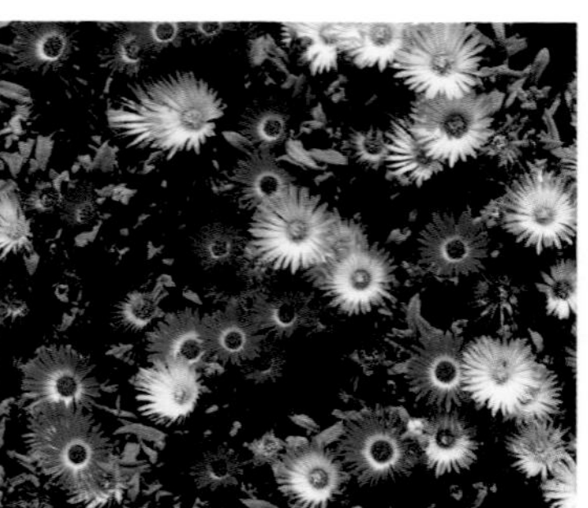

ANNUAL/BIENNIAL SMALL

Dorotheanthus bellidiformis

ICE PLANT Grow this low-growing annual to edge the front of a sun-baked border, where it will thrive. It produces small, fleshy leaves and red, pink, orange, and white, daisy-like summer flowers, which open in sun, closing again in the evenings.
↕15cm (6in) ↔30cm (12in)

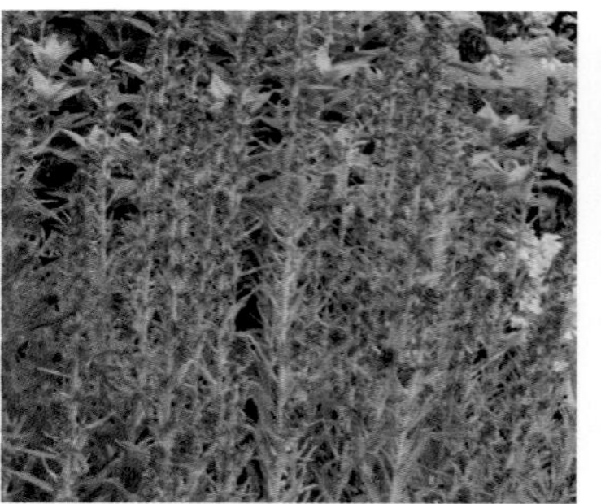

ANNUAL/BIENNIAL MEDIUM

Echium vulgare

VIPER'S BUGLOSS A bristly biennial with a basal rosette of lance-shaped leaves. In early summer, it produces slender spires of vivid violet-blue, bell-shaped blooms. This drought-tolerant plant is ideal for wildlife gardens and sun-baked borders.
↕60cm (24in) ↔30cm (12in)

PERENNIAL SMALL

Eryngium maritimum

SEA HOLLY A native of sunny coastal sites and rocky soils, this upright perennial has prickly, white-veined blue-green basal leaves. Thimble-like, metallic blue blooms with spiny collars, appear on branched stems from early summer to early autumn.
↕50cm (20in) ↔45cm (18in)

PERENNIAL MEDIUM

Erysimum 'Bowles's Mauve' ♈

PERENNIAL WALLFLOWER Ideal for a sunny border, this evergreen, drought-tolerant perennial has slim, grey-green leaves and spikes of mauve flowers from late winter to summer. Deadhead regularly.
↕60cm (24in) ↔40cm (16in)

ANNUAL/BIENNIAL SMALL

Eschscholzia californica

CALIFORNIA POPPY Easy to grow from seed in spring, this clump-forming annual bears masses of bowl-shaped, orange, red, and yellow summer flowers. Grow it in sun-baked borders, where it will offer the best show. May self-seed after its first season.
↕30cm (12in) ↔15cm (6in)

PERENNIAL SMALL

Gypsophila repens ♈

BABY'S BREATH This spreading, semi-evergreen perennial has evolved to cope with dry, stony slopes in the Mediterranean. It forms mats of narrow, bluish green leaves, above which rise sprays of round, pale pink summer flowers. Trim back after flowering.
↕20cm (8in) ↔30cm (12in) or more

PERENNIAL SMALL

Jovibarba hirta

This evergreen perennial, similar to a houseleek, produces rosettes of hairy-margined, triangular, spiny, green leaves, often red-tinted, and bell-shaped, yellow-brown flowers in summer. It tolerates very dry soils and thrives in full sun.
↕15cm (6in) ↔10cm (4in)

ANNUAL/BIENNIAL SMALL

Lobularia maritima

SWEET ALYSSUM A mat-forming annual, it is easy to grow from seed, and bears green, lance-shaped leaves and masses of white or pink, rounded, sweetly fragrant flowers all summer. Thrives in dry soils and full sun. 'Easter Bonnet' AGM (above) is popular.
↕10cm (4in) ↔30cm (12in)

PERENNIAL MEDIUM

Malva moschata

MUSK MALLOW This bushy perennial, with woody stems, will grow happily in free-draining soils and full sun. It produces deeply lobed, dark green leaves and saucer-shaped, pale pink blooms from summer to autumn. It may self-seed.
↕1m (3ft) ↔60cm (24in)

SHRUB SMALL

Phlomis fruticosa ♈

JERUSALEM SAGE This Mediterranean, evergreen shrub is adapted to rocky soils, and will thrive in full sun. It has grey-green leaves, and produces hooded, yellow flowers on upright stems in summer. The seedheads provide interest over winter.
↕1m (3ft) ↔1.5m (5ft)

PERENNIAL SMALL

Sedum erythrostictum 'Frosty Morn'

Ideal for sandy soil, this perennial's fleshy, white and grey-green variegated foliage is joined by clusters of pale pink flowers from late summer to early autumn. The brown seedheads persist through winter.
↕30cm (12in) ↔45cm (18in)

PERENNIAL SMALL

Sempervivum giuseppii ♈

HOUSELEEK Perfect for sun-baked cracks in walls or gravel beds, this diminutive, evergreen perennial produces rosettes of spiky, green leaves, with purple, pointed tips and, in summer, star-shaped, pink or red flowers on upright stems.
↕in flower 10cm (4in) ↔10cm (4in)

PERENNIAL LARGE

Solidago 'Goldkind'

GOLDENROD Tolerant of a wide range of sites and soils, this perennial thrives in sun-baked beds. It forms an upright clump of pointed, narrow, green leaves and, from midsummer to early autumn, small, golden-yellow blooms. Deadhead regularly.
↕1.2m (4ft) ↔1m (3ft)

SHRUB MEDIUM

Yucca filamentosa

Drought-tolerant and sun-loving, this evergreen shrub offers a striking display of sword-shaped, spine-tipped, dark green leaves. From mid- to late summer, tulip-like, white flowers appear. 'Bright Edge' AGM (above) is a variegated form.
↕2m (6ft) ↔1.5m (5ft)

OTHER SUGGESTIONS

Annuals and biennials

Crepis rubra • *Mesembryanthemum* 'Magic Carpet Mixed' • *Papaver rhoeas*

Perennials

Achillea millefolium • *Agave parryi* ♈ • *Alyssum montanum* • *Artemisia absinthium* • *Aubrieta* Cascade Series • *Cerastium tomentosum* • *Dianthus deltoides* ♈ • *Echeveria elegans* ♈ • *Erigeron karvinskianus* ♈ • *Eryngium agavifolium* • *Eryngium giganteum* ♈ • *Erysimum linifolium* 'Variegatum' • *Euphorbia characias* subsp. *wulfenii* • *Gypsophilia paniculata* • *Jovibarba heuffelii* • *Malva alcea* • *Sedum spathulifolium* • *Sedum spectabile* ♈ • *Sempervivum arachnoideum* ♈ • *Vinca minor*

Shrubs

Genista hispanica • *Ulex europaeus* 'Plenus' ♈

Plants for waterlogged sites

Gardens prone to waterlogging provide conditions similar to those on the margins of a natural pond where bog plants thrive (*see p.54* and *p.164*).

While most plants dislike wet feet for long periods of time, there are some, such as *Acorus*, *Gunnera*, and *Taxodium*, that thrive happily in such conditions. If your soil is intermittently waterlogged, consider choosing plants such as hostas, sedges (*Acorus* and others), kerria, and *Leycesteria*, that will tolerate a wide range of conditions. If you seek to improve your soil structure and open up air and drainage channels, dig in plenty of well-rotted manure or compost, together with horticultural grit. Also cover the surface with an organic mulch annually.

PERENNIAL LARGE

Aconitum napellus

MONKSHOOD A beautiful, upright perennial for wet soils, with deeply divided, dark green foliage and tall spires of indigo-blue blooms from mid- to late summer. The tall stems may need staking. All parts are toxic if ingested; wear gloves when handling it.
↕1.5m (5ft) ↔30cm (12in)

PERENNIAL LARGE

Actaea racemosa 🏆

BLACK COHOSH Useful for waterlogged soils, this perennial bears a combination of deeply divided leaves and bottlebrush-like spikes of white flowers in midsummer. The dried, brown seedheads extend the season of interest. Grow it at the back of a border.
↕up to 1.5m (5ft) ↔60cm (24in)

PERENNIAL SMALL

Astilbe x *arendsii*

FALSE GOAT'S BEARD This bog plant is a clump-forming perennial that thrives in waterlogged conditions. It produces deeply divided, fern-like, dark green foliage and feathery plumes of white, pink, lilac, or red flowers from early- to late summer.
↕45cm (18in) ↔30cm (12in)

SHRUB MEDIUM

Cornus sericea

This shrub is grown for its stunning, red winter stems, or bright yellow-green in varieties such as 'Flaviramea' AGM (above). The lance-shaped, dark green foliage turns orange and red in autumn. Cut old stems to the ground in late winter.
↕2m (6ft) ↔4m (12ft)

PERENNIAL LARGE

Eupatorium purpureum

JOE PYE WEED A good choice for boggy soils in wildlife or informal gardens, this upright perennial has coarse, green foliage on tall, purple-flushed stems and large, domed, fluffy, purple-pink flowerheads from late summer to early autumn.
↕2.2m (7ft) ↔1m (3ft)

PERENNIAL MEDIUM

Filipendula ulmaria

A leafy perennial, tolerant of waterlogged conditions, it produces divided, blue-green leaves that turn bright red in autumn. Upright stems bear lacy clusters of tiny, fragrant, white flowers in summer. 'Aurea' (above) has golden-yellow spring foliage.
↕90cm (36in) ↔60cm (24in)

PERENNIAL MEDIUM

Hosta sieboldiana

This leafy, medium-sized perennial is tolerant of waterlogged soils and shade. It features large, rounded, rich blue-green leaves, with a puckered texture. Spikes of lilac-tinged white flowers appear in early summer. Protect from slugs.
↕1m (3ft) ↔1.2m (4ft)

SHRUB MEDIUM

Hydrangea macrophylla

LACECAP HYDRANGEA This broad-leaved shrub grows well on clay soils and tolerates occasional waterlogging. From midsummer to early autumn, it bears tiny flowers, with petal-like florets. Other forms have white, pink, purple, or, on acid soils, blue blooms.
↕2m (6ft) ↔2.5m (8ft)

SHRUB LARGE

Hydrangea paniculata

PANICULATE HYDRANGEA Tolerant of occasional waterlogging, this upright shrub bears oval, pointed, glossy, dark green leaves and, in late summer, cone-shaped, white or pink flowerheads. Varieties include 'Phantom' AGM, with large, white blooms.
↕↔3m (10ft)

PERENNIAL MEDIUM

Iris ensata 'Rose Queen' ♀

JAPANESE FLAG This perennial adds colour to beds with waterlogged soils. It bears strap-shaped foliage and, from early- to midsummer, beardless, lilac-pink flowers, with gold marks on the lower petals. Blue, purple, and white varieties are available.

↕up to 1m (3ft) ↔indefinite

PERENNIAL MEDIUM

Iris sibirica

SIBERIAN IRIS This perennial thrives in wet soils. It has upright, sword-like, blue-green leaves and large, beardless flowers in blue, pink, white or yellow from late spring to early summer. 'Butter and Sugar' AGM (above) has yellow and white blooms.

↕1m (3ft) ↔indefinite

PERENNIAL LARGE

Ligularia przewalskii

This upright perennial thrives in boggy soils and produces large, rounded, deeply cut, dark green leaves. From mid- to late summer, it produces tall, narrow spires of spidery, daisy-like, yellow flowers. Stems may need additional support.

↕2m (6ft) ↔1m (3ft)

TREE MEDIUM

Magnolia virginiana

SWEET BAY This small tree performs best on moist soils and will tolerate occasional waterlogging. From early summer to early autumn, its large, vanilla-scented, cup-shaped, creamy white flowers are set against large, dark green leaves.

↕10m (30ft) ↔6m (20ft)

PERENNIAL MEDIUM

Mimulus cardinalis ♀

SCARLET MONKEY FLOWER This spreading perennial, grown as an annual, thrives in boggy soils. It produces a mass of small, toothed, downy, green leaves and two-lipped, small, tomato-red flowers that cover the plant from summer to early autumn.

↕90cm (36in) ↔60cm (24in)

PERENNIAL MEDIUM

Primula beesiana ♀

BEE'S PRIMROSE This primula is a moisture-loving perennial, ideal for boggy soils. It bears basal rosettes of crinkly, light green leaves and upright stems, studded with rounded clusters of purple flowers from late spring to early summer.

↕60cm (24in) ↔30cm (12in)

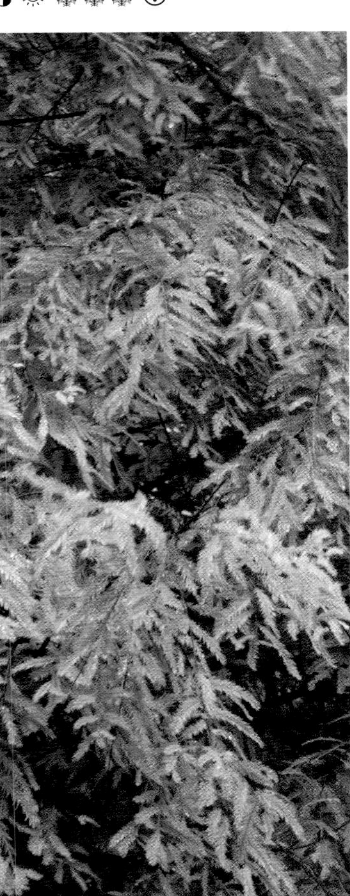

TREE LARGE

Taxodium distichum

SWAMP CYPRESS This deciduous, broadly conical conifer produces slim, green foliage that turns yellow-brown in late autumn before falling. Small, oval cones appear in summer. Grow it in damp or waterlogged soils in a large garden.

↕40m (130ft) ↔9m (28ft)

PERENNIAL LARGE

Monarda fistulosa

WILD BERGAMOT This clump-forming perennial thrives in moist soil and tolerates occasional waterlogging. In midsummer, it produces flowerheads comprised of small, two-lipped, pale lilac blooms, which attract bees and butterflies.

↕1.2m (4ft) ↔45cm (18in)

PERENNIAL LARGE

Verbena hastata

AMERICAN BLUE VERVAIN Tolerant of waterlogged conditions, this moisture-loving, upright perennial has narrow, toothed leaves and candelabra-like, violet-blue flowerheads from midsummer to autumn. 'Rosea' (above) has lilac-pink blooms.

↕1.5m (5ft) ↔60cm (24in)

PERENNIAL LARGE

Persicaria virginiana

This bog-loving perennial has oval, mid-green leaves, with dark green markings, and from late summer to early autumn, spikes of cup-shaped, green flowers that turn red. 'Painter's Palette' (above) has variegated green, yellow, and pink leaves.

↕1.2m (4ft) ↔60cm (24in)

OTHER SUGGESTIONS

Perennials and grasses

Acorus gramineus • *Astilbe chinensis* var. *pumila* ♀ • *Actaea matsumurae* • *Eupatorium cannabinum* 'Flore Pleno' • *Gunnera manicata* ♀ • *Hosta* 'Francee' ♀ • *Hosta* 'Blue Angel' ♀ • *Ligularia dentata* 'Desdemona' • *Lysimachia ephemerum* ♀ • *Lythrum salicaria* • *Matteuccia struthiopteris* ♀ • *Primula bulleyana* ♀ • *Primula pulverulenta* ♀ • *Rheum palmatum* • *Rodgersia pinnata* • *Zantedeschia aethiopica*

Shrubs and trees

Cornus alba • *Cornus florida* • *Cornus stolonifera* • *Kerria japonica* • *Leycesteria formosa* • *Nyssa sylvatica* • *Pyrus communis*

Plants for banks and slopes

Slopes and banks add interest to designs, the layers of planting producing a three-dimensional effect. For best results, use tough, drought-tolerant plants.

Slopes offer banks of colour and texture, creating stunning features, but drought and erosion can present problems on steep hills. Plants with spreading root systems, such as geraniums, junipers, heathers and heaths (*Calluna* and *Erica*), bind the soil and protect it from erosion, while a leafy canopy minimizes rain damage. Choose drought-tolerant plants that can cope with the dry soil, and if the slope is exposed, also consider their wind resistance. Most of these plants will cope with blustery conditions, but a few, including the roses and hemlock (*Tsuga*), do best on slopes sheltered from prevailing winds.

SHRUB SMALL

***Abies balsamea* Hudsonia Group** 🏆

HUDSON FIR A slow-growing, evergreen conifer, it can cope with the well-drained conditions on a slope. It forms a spreading mound of needle-like, dark grey-green leaves. Plant in neutral to acid soil.

↕↔1m (3ft)

PERENNIAL MEDIUM

Baptisia australis 🏆

FALSE INDIGO A clump-forming perennial, with grey-green, divided foliage, it thrives in well-drained soils and will decorate a slope with its spikes of small, violet-blue summer flowers. Decorative dark grey, seed pods form in autumn.

↕75cm (30in) ↔60cm (24in)

SHRUB LARGE

Berberis darwinii 🏆

DARWIN'S BARBERRY This tough and spiny evergreen shrub lends structure and height to a sloping site. Its tiny, holly-like leaves are joined in spring by pendent clusters of dark orange flowers, which attract bees and are followed by blue-black berries.

↕↔3m (10ft)

SHRUB LARGE

Berberis* x *stenophylla

BARBERRY Use this large, evergreen, prickly shrub to decorate a bank or slope with its narrow, spine-tipped, dark green leaves and small, yellow spring flowers, followed by blue-black fruits. Trim lightly in summer after the flowers have faded.

↕3m (10ft) ↔5m (15ft)

PERENNIAL SMALL

Bergenia purpurascens 🏆

PURPLE BERGENIA Tolerant of the free-draining conditions on a slope, this evergreen perennial forms clumps of large, fleshy, rounded leaves that flush purple in cold weather, and clusters of dark pink spring blooms. Remove faded flowerheads.

↕40cm (16in) ↔60cm (24in) or more

SHRUB LARGE

Buxus sempervirens

COMMON BOX This traditional, evergreen, hedging and topiary shrub tolerates a wide range of conditions and grows well on a slope. Clip its small, oval, glossy leaves to form hedges and green sculptures. Prune between early- to late summer.

↕↔5m (15ft)

SHRUB SMALL

***Calluna vulgaris* Gold Haze'**

HEATHER A ground-hugging shrub, heather has evolved to cope with life on moors and hillsides. Use it to carpet a slope with its scaly foliage and small, white flowers, which appear from late summer to autumn. Needs acid soil.

↕45cm (18in) ↔60cm (24in)

BULB MEDIUM

Chionodoxa forbesii

GLORY OF THE SNOW Use this diminutive, early spring-flowering perennial bulb on a sloping, sunny site. It has linear, mid-green foliage and star-shaped, white-eyed blue flowers. Plant the bulbs in groups in autumn.
↕15cm (6in)

SHRUB MEDIUM

Clethra alnifolia

SWEET PEPPER BUSH This deciduous shrub's late summer spires of fragrant small, white blooms add appeal to a wooded, sloping site. The green leaves turn bright yellow in autumn. It performs best on moist, acid soil at the bottom of a slope.
↕↔2.5m (8ft)

SHRUB SMALL

Cotoneaster horizontalis

The branches of this deciduous shrub are arranged in a distinctive "herringbone" pattern and will hug the contours of a slope. In summer, it produces small, white blooms, followed by vivid red berries from autumn to early winter, which are loved by birds.
↕1m (3ft) ↔1.5m (5ft)

SHRUB SMALL

Daboecia cantabrica

IRISH HEATH This drought-tolerant, low-growing shrub, ideal for a slope or bank, forms a mat of small, dark green leaves. From early summer to autumn, urn-shaped purple, white, or mauve flowers appear. 'Bicolor' (above) is a popular variety.
↕45cm (18in) ↔60cm (24in)

SHRUB LARGE

Elaeagnus x *ebbingei* 🏆

This tough, vigorous, evergreen shrub can be used as a windbreak on a sloping site. It has broadly oval, dark green leaves covered with a silvery dusting. Small, fragrant, white flowers appear in autumn. 'Gilt Edge' AGM is a colourful variegated form.
↕↔5m (15ft)

PERENNIAL SMALL

Epimedium x *youngianum* 'Niveum' 🏆

SNOWY BARRENWORT This ground-cover perennial has heart-shaped, bronze-tinted, serrated leaves that turn green, and make a decorative mat on a slope or bank. Small, white flowers appear in late spring.
↕↔30cm (12in)

SHRUB SMALL

Erica carnea

WINTER HEATH This ground-hugging, evergreen shrub will produce a mat of dark green, needle-like leaves on a sunny slope or bank. From winter to spring, it is covered with tiny pink, red, or white blooms, held on short spikes.
↕30cm (12in) ↔45cm (18in) or more

SHRUB SMALL

Gaultheria mucronata

This evergreen shrub creates a decorative addition to a slope or bank, with its dark green, prickly leaves and tiny, white late spring flowers, followed by colourful berries on female plants. 'Wintertime' AGM (above) has large, white berries.
↕↔1.2m (4ft)

PERENNIAL SMALL

Geranium endressii 🏆

Forming a clump of lobed, green leaves, this semi-evergreen perennial will clothe a slope with its foliage from spring to autumn. In summer, light pink blooms, with slightly notched petals, add to the effect. Easy to grow, it thrives in sun or part shade.
↕45cm (18in) ↔60cm (24in)

SHRUB LARGE

Itea virginica

This spreading, evergreen shrub thrives on sheltered slopes, protected from prevailing winds. It bears arching stems of oval, spiny, dark green leaves and, from midsummer to early autumn, decorative greenish white catkins. Needs acid soil.
↕3m (10ft) ↔1.5m (5ft)

PERENNIAL SMALL

Geranium macrorrhizum

CRANESBILL This easy-to-grow perennial will cover a slope or bank with a spreading carpet of aromatic, deeply lobed leaves. In early summer, it bears a mass of small, magenta blooms, and may produce a second flush if cut back in summer. May be invasive.
↕38cm (15in) ↔60cm (24in)

PERENNIAL SMALL

Hepatica nobilis 🏆

LIVERLEAF A low-growing, semi-evergreen perennial, ideal for a shaded, sloping garden, where it will form a neat dome of rounded, lobed, fleshy, green leaves and, in early spring, cup-shaped white, pink, blue, or purple blooms.
↕8cm (3in) ↔12cm (5in)

SHRUB SMALL

Juniperus procumbens

CREEPING JUNIPER This compact evergreen conifer is often used as ground cover on banks and sloping sites, where its dense, prickly green leaves offer a mat of colour. It bears small, round, brown or black fruits. Plant it in sun or light shade.
↕20cm (8in) ↔75cm (30in)

SHRUB SMALL

Lonicera pileata

BOX-LEAVED HONEYSUCKLE This spreading, dense evergreen shrub is tolerant of the dry conditions on a sloping site. It covers the ground with stems of dark green, glossy leaves. Violet-purple berries follow the funnel-shaped, white late spring flowers.
↕60cm (24in) ↔2.5m (8ft)

PERENNIAL SMALL

Luzula sylvatica

GREATER WOODRUSH Use this clump-forming, evergreen perennial as ground cover on a shady slope. It has glossy, grassy, green leaves, yellow-green from winter to spring, in 'Aurea' (above), and tiny, brown flowers from late spring to early summer.
↕40cm (16in) ↔45cm (18in)

SHRUB SMALL

Microbiota decussata 🏆

Useful as ground cover on sunny, sloping sites, this low-growing or prostrate evergreen, coniferous shrub produces green, scale-like leaves, which turn bronze-purple in winter. Yellow-brown fruits appear in autumn.
↕1m (3ft) ↔indefinite

BULB MEDIUM

Narcissus 'February Gold' 🏆

DAFFODIL This reliable daffodil copes well with free-draining, sloping sites, while the short stems resist wind damage. It bears narrow, mid-green leaves and scented yellow early spring flowers. Plant the bulbs in groups in autumn, and let them naturalize.
↕30cm (12in)

SHRUB MEDIUM

Rhododendron yakushimanum 🏆

YAKUSHIMA RHODODENDRON Native to windswept mountainsides in Japan, this evergreen shrub, with leathery, dark green leaves and bell-shaped, pink-budded, white spring flowers, is ideal for sloping gardens.
↕↔2m (6ft)

SHRUB SMALL

Rosa Flower Carpet Series

Plant this spreading, ground-cover rose on a sheltered slope to protect it from wind. Its cupped, semi-double blooms appear from summer to autumn above disease-resistant, glossy green leaves. 'Ruby' (above) is among the colours available.
↕45cm (18in) ↔1.2m (4ft)

SHRUB SMALL

Rosa KENT ♡

A spreading, ground-cover rose, ideal for sloping gardens, with disease-resistant, glossy, mid-green leaves and clusters of flat, semi-double, white flowers from summer to autumn. Apply a mulch of well-rotted manure annually in spring.
↕80cm (32in) ↔90cm (36in)

PERENNIAL MEDIUM

Stachys officinalis

Perfect for free-draining soils on hillside gardens, this upright perennial produces scalloped, oblong leaves, above which rise spikes of small, reddish purple flowers in summer and early autumn. Easy to grow, it is happy in sun or part shade.
↕60cm (24in) ↔30cm (12in)

SHRUB SMALL

Stephanandra incisa 'Crispa'

This deciduous, dwarf shrub will cover a sloping bed with mounds of crinkled foliage, which turns yellow and orange in autumn. In summer, clusters of tiny, greenish white flowers appear. Easy to grow, it tolerates most sites and soils.
↕60cm (24in) ↔3m (10ft)

SHRUB MEDIUM

Symphoricarpos albus var. *laevigatus*

COMMON SNOWBERRY Vigorous and tolerant of most sites, this deciduous, dense shrub bears small, round, dark green leaves. Round, white fruits follow the pink summer flowers. Trim to make a waist-high hedge.
↕↔2m (6ft)

PERENNIAL SMALL

Trollius pumilus

This perennial can be seen growing wild in the foothills of the Himalayas, and will brighten up late spring gardens with its yellow, buttercup-like flowers, which form above the glossy, green leaves. Happy in sun or partial shade, it prefers moist soil.
↕30cm (12in) ↔20cm (8in) or more

SHRUB LARGE

Tsuga canadensis

EASTERN HEMLOCK A slow-growing evergreen conifer, with a fountain of pendent stems of needle-like, blue-green leaves, white beneath. It also bears small, brown cones. Ideal for weeping over a bank or growing on a gentle, sheltered slope.
↕↔5m (15ft)

OTHER SUGGESTIONS

Perennials and grasses

Aegopodium podagraria 'Variegata' • *Anemone apennina* ♡ • *Anemone nemorosa* ♡ • *Festuca glauca*

Bulb

Chionodoxa siehei ♡ • *Galanthus plicatus* ♡

Shrubs, trees, and climbers

Acer rubrum • *Betula utilis* var. *jacquemontii* • *Calluna vulgaris* 'Kerstin' ♡ • *Cotoneaster dammeri* • *Diervilla* x *splendens* • *Genista aetnensis* ♡ • *Genista lydia* ♡ • *Hedera colchica* 'Sulphur Heart' ♡ • *Hypericum forrestii* ♡ • *Juniperus chinensis* • *Juniperus* 'Grey Owl' ♡ • *Mahonia* x *wagneri* 'Undulata' • *Pyracantha rogersiana* • *Rhododendron* 'Dopey' ♡ • *Rhododendron macabeanum* ♡ • *Salix caprea* 'Kilmarnock' • *Salix reticulata* ♡ • *Spiraea japonica* 'Anthony Waterer' • *Taxus baccata* 'Repens Aurea' ♡

Plants for weed-proof ground cover

Spreading plants that cover the soil with mats of weed-proof stems and leaves can convert a labour-intensive garden into an easy-care, relaxing space.

Although these plants are tough, most have specific site requirements so check that your garden conditions suit your choices. Choose sedums and heathers for sunny areas, and *Pachysandra*, ivies, and ferns for gloomy spots. As well as spreading leafy stems, many ground-cover plants also sport beautiful flowers – a succession of *Alyssum* and *Arabis* followed by hardy geraniums and roses will provide months of colourful blooms. However, these vigorous plants may spread beyond their boundaries if not kept in check, so trim occasionally to prevent them swamping more delicate specimens.

PERENNIAL SMALL

Ajuga reptans

BUGLE This evergreen perennial has a spreading habit, forming close-knit rosettes of glossy, dark green leaves. In late spring and early summer, it bears short spikes of small, deep blue flowers. 'Atropurpurea' (above) has bronze-tinted leaves.
↕15cm (6in) ↔90cm (36in)

PERENNIAL SMALL

Alchemilla mollis

LADY'S MANTLE This perennial forms a mound of rounded, pale green leaves that suppress surrounding annual weeds. In summer, it produces lax, branching sprays of small, lime-green blooms. Best planted in shade, it is ideal for the front of a border.
↕↔50cm (20in)

PERENNIAL MEDIUM

Amsonia tabernaemontana

EASTERN BLUESTAR This upright perennial forms a dense mound of weed-suppressing growth. It has star-shaped, blue spring flowers and bold autumn foliage. Plant it in groups or alongside other ground-cover plants for the best effect.
↕1m (3ft) ↔30cm (12in)

PERENNIAL LARGE

Anemone x *hybrida*

JAPANESE ANEMONE With branching stems and flat-faced, white and pink blooms that last into autumn, this perennial is ideal for late summer colour. It has a spreading habit and forms a low canopy of summer foliage. 'Elegans' AGM (above) is popular.
↕1.2m (4ft) ↔indefinite

PERENNIAL SMALL

Arabis alpina subsp. *caucasica*

ROCK CRESS This mat-forming, evergreen perennial has hairy, grey-green leaves and small, white flowers that appear from early spring to early summer. Plant it in groups at the front of a border, or on a rockery.
↕↔15cm (6in)

PERENNIAL SMALL

Armeria maritima

THRIFT A clump-forming, evergreen perennial, with grassy mounds of slender green leaves. In summer, it bears rounded heads of small, white or pink, papery blooms. Plant it at the front of a free-draining bed and allow it to naturalize.
↕10cm (4in) ↔15cm (6in)

PERENNIAL SMALL

Asarum canadense

WILD GINGER This ground-cover perennial produces ornamental, rounded, kidney-shaped leaves that form a green carpet. Small, cup-shaped, purple-brown flowers appear in summer, but are hidden. Plants are tolerant of deep shade.
↕15cm (6in) ↔30cm (12in)

PERENNIAL MEDIUM

Astrantia major

MASTERWORT An upright, clump-forming perennial, with divided green leaves and, from midsummer to early autumn, sprays of small, greenish white, pink, or red flowers. 'Sunningdale Variegated' AGM (above) has cream-accented leaves.
↕90cm (36in) ↔45cm (18in)

SHRUB SMALL

Calluna vulgaris

HEATHER Grown for its pink, red, or white flowers, which appear from midsummer to autumn, this low-growing, evergreen shrub has a spreading habit and makes excellent all-year ground cover. There are many varieties to grow, and all need acid soil.
↕60cm (24in) ↔75cm (30in)

PERENNIAL SMALL

Ceratostigma plumbaginoides 🏆

BLUE-FLOWERED LEADWORT A bushy, mound-forming perennial, with oval, green leaves and small, vivid blue late summer to autumn flowers. Plant it at the front of a border where it can spread; may be invasive.
↕45cm (18in) ↔20cm (8in)

PERENNIAL SMALL

Cornus canadensis

CREEPING DOGWOOD A low-growing, carpet-forming perennial, it bears single, white flowers in late spring, which appear in the centre of the rounded clusters of green leaves. It is slow growing but ideal for shady and woodland gardens.
↕15cm (6in) ↔30cm (12in) or more

SHRUB SMALL

Daboecia cantabrica

IRISH HEATH This spreading, evergreen shrub has small, dark green leaves, and flowers from early summer to autumn, bearing urn-shaped single or double, white, purple, or mauve flowers. 'Bicolor' AGM (above) is a popular variety.
↕45cm (18in) ↔60cm (24in)

PERENNIAL LARGE

Dryopteris filix-mas 🏆

MALE FERN A deciduous or semi-evergreen fern, with large, mid-green fronds that arch outwards from a central crown, shading the surrounding soil, and suppressing weeds. Ideal for shady and woodland gardens, plant it in groups or leave it to naturalize.
↕1.2m (4ft) ↔1m (3ft)

PERENNIAL SMALL

Epimedium grandiflorum 🏆

BARRENWORT This clump-forming perennial forms a low canopy of heart-shaped, tough green leaves. In spring, nodding white, yellow, pink, or purple flowers appear. Cut back tired foliage. 'Lilafee' (above) is a popular choice.
↕25cm (10in) ↔30cm (12in)

PERENNIAL SMALL

Euphorbia polychroma

This perennial forms mounds of small, green leaves held on upright stems. In spring, it bears small, yellow flowers surrounded by lime-yellow bracts. Perfect for the front of a border, it is best planted in groups to cover the ground effectively.
↕↔50cm (20in)

SHRUB SMALL

Gaultheria procumbens

CHECKERBERRY This dwarf, evergreen shrub forms a dense mat of oval, glossy green leaves. In summer, it produces small pink and white flowers that lead to attractive scarlet berries in autumn. It is ideal for planting beneath large shrubs.
↕15cm (6in) ↔indefinite

PERENNIAL SMALL

Geranium 'Johnson's Blue'

A clump-forming perennial, with deeply lobed, green leaves and saucer-shaped, lavender-blue flowers in summer. Plant it in groups at the front of a border and leave it to create a mat of colourful ground cover. Cut back tired foliage in summer.
↕30cm (12in) ↔60cm (24in)

PERENNIAL SMALL

Geranium macrorrhizum

CRANESBILL A resilient perennial that creates a carpet of aromatic, deeply lobed leaves. In early summer, abundant small, magenta blooms appear. Ideal for the front of a border or beneath large shrubs, where it will spread steadily; may be invasive.
↕38cm (15in) ↔60cm (24in)

Plants for weed-proof ground cover

PERENNIAL SMALL

Geranium x *oxonianum*

This vigorous, spreading evergreen perennial covers large areas with its divided, toothed, green leaves and rounded, pink flowers, which appear from late spring to midsummer. 'Wargrave Pink' AGM (above) is a popular variety.

↕45cm (18in) ↔60cm (24in)

PERENNIAL MEDIUM

Hosta sieboldiana

This perennial forms spreading clumps that shade the soil and suppress the surrounding weeds. It has large, grey-blue leaves and flowers briefly in early summer, bearing spikes of pale lilac-grey, bell-shaped blooms. Protect from slugs.

↕1m (3ft) ↔1.2m (4ft)

SHRUB SMALL

Hypericum calycinum

ROSE OF SHARON This vigorous evergreen, dwarf shrub forms a dense carpet of dark green leaves on arching stems and, from midsummer to mid-autumn, bright yellow flowers. Ideal for sun or shade, use it to cover large areas. Can be invasive.

↕60cm (24in) ↔indefinite

SHRUB SMALL

Juniperus horizontalis

CREEPING JUNIPER Providing year-round ground cover, this creeping, evergreen conifer has scaly, blue-green foliage that can suppress even perennial weeds. It is ideal for covering large areas of ground. 'Bar Harbor' has purple-tinted winter foliage.

↕50cm (20in) ↔indefinite

PERENNIAL SMALL

Lamium maculatum

Ideal for growing in shady areas and beneath large shrubs, this mat-forming perennial makes attractive ground cover. It has white-variegated green foliage and white flower spikes from late spring to summer. 'White Nancy' (above) has silver-variegated leaves.

↕15cm (6in) ↔1m (3ft)

PERENNIAL SMALL

Nepeta racemosa 🏆

DWARF CATMINT This spreading perennial forms a clump of lax stems, covered with aromatic, grey-green leaves and violet-blue flower spikes in summer. Use it at the front of a border for colourful ground cover. 'Walker's Low' AGM (above) is popular.

↕30cm (12in) ↔45cm (18in)

PERENNIAL SMALL

Persicaria affinis 'Superba' 🏆

A vigorous perennial, it forms a dense mat of oval, green leaves and, from early- to late summer, upright spikes of soft pink blooms that darken with age. Useful for the front of a border, it is quick-growing and may need restricting on smaller plots.

↕25cm (10in) ↔60cm (24in)

PERENNIAL SMALL

Phlox subulata

This evergreen, trailing perennial has needle-like, green foliage. It is covered with star-shaped, pink, white, or mauve flowers in summer. Ideal for sunny sites where low-growing ground cover is needed. 'Emerald Cushion' (above) has pink flowers.

↕15cm (6in) ↔50cm (20in)

SHRUB SMALL

Potentilla fruticosa

CINQUEFOIL A compact shrub, with divided, green leaves and small, yellow, white, red, pink, or orange flowers, depending on the variety, from summer to autumn. Plant in groups and allow them to merge. 'Goldfinger' (above) has yellow flowers.

↕1m (3ft) ↔1.5m (5ft)

PERENNIAL SMALL

Pulmonaria angustifolia 🏆

This perennial forms a clump of matt green leaves and, in spring, clusters of rich blue trumpet-shaped blooms on upright stems. Useful for spring colour, plant it at the front of border, near other plants for later interest New foliage appears in summer.

↕23cm (9in) ↔30cm (12in) or more

SHRUB SMALL

Rosa Flower Carpet Series

This ground-cover rose forms a spreading mass of disease-resistant, glossy, green foliage and semi-double flowers in shades of white, pink, red, and yellow. The leafy stems are useful for covering large areas. 'Ruby' (above) is a popular variety.

↕45cm (18in) ↔1.2m (4ft)

PERENNIAL SMALL

Sedum kamtschaticum var. *kamtschaticum*

STONECROP Use this spreading semi-evergreen perennial with orange-yellow late summer blooms to cover small areas, or plant in groups. 'Variegatum' AGM (above) has pink-tinted, cream-edged leaves.

↕8cm (3in) ↔20cm (8in)

ANNUAL/BIENNIAL SMALL

Tropaeolum majus

NASTURTIUM Useful for summer-long ground cover, this annual forms a dense canopy of round, blue-green leaves and orange or yellow flowers. Sow the seeds in a sunny site; self-seeds readily. Alaska Series AGM (above) has variegated foliage.

↕30cm (12in) ↔45cm (18in)

SHRUB SMALL

Rosa KENT

This spreading shrub rose has disease-resistant, glossy, mid-green leaves and clusters of flat, semi-double, white flowers from summer to autumn. Deadhead regularly for colourful ground cover, and mulch plants with compost in spring.

↕80cm (32in) ↔90cm (36in)

PERENNIAL SMALL

Veronica gentianoides

This perennial forms a ground-hugging mat of tightly-knit, glossy, evergreen leaves, above which rise slender spires of pale blue early summer flowers. Use it to cover small areas; grow with later-flowering plants for prolonged interest. Remove spent blooms.

↕↔45cm (18in)

PERENNIAL SMALL

Saxifraga stolonifera ♀

CREEPING SAXIFRAGE Useful for covering shady corners, this spreading perennial forms rosettes of round, hairy, silver veined dark green leaves on spreading stems. In summer, it bears branching stems of flag-like, small, white blooms.

↕15cm (6in) or more ↔30cm (12in) or more

PERENNIAL SMALL

Stachys byzantina

LAMB'S EARS Ideal for dry, sunny areas, this spreading, evergreen perennial forms a dense carpet of downy, grey-green leaves and, in summer, spikes of small, pink blooms. An attractive, front-of-border plant, it offers contrast to brightly coloured plants.

↕38cm (15in) ↔60cm (24in)

PERENNIAL SMALL

Waldsteinia ternata

This semi-evergreen perennial forms spreading mats of scallop-edged, lobed, dark green leaves. Saucer-shaped, bright yellow flowers appear in late spring and early summer. It offers good ground cover in deep shade, but may become invasive.

↕10cm (4in) ↔30cm (12in)

PERENNIAL MEDIUM

Tellima grandiflora

FRINGE CUPS Ideal for shady borders and woodland gardens, this semi-evergreen perennial forms a clump of textured green, heart-shaped leaves and, from late spring to midsummer, spikes of bell-shaped, cream flowers. It spreads gradually.

↕↔60cm (24in)

OTHER SUGGESTIONS

Perennials and grasses

Ajuga reptans 'Burgundy Glow' • *Alyssum saxatile* 'Summit' • *Dryopteris affinis* ♀ • *Geranium endressii* 'Wargrave Pink' • *Geranium* 'Orion' ♀ • *Persicaria bistorta* • *Pulmonaria officinalis* • *Sedum spathulifolium* • *Stachys macrantha* • *Viola labradorica*

Shrubs, trees, and climbers

Erica vagans • *Hebe* 'Carl Teschner' syn. *Hebe* 'Youngii' ♀ • *Hedera helix* 'Manda's Crested' • *Helianthemum nummularium* • *Lithospermum diffusum* 'Heavenly Blue' • *Pachysandra terminalis* 'Green Carpet' • *Rosa* 'Cambridgeshire' • *Rosa* 'Harlow Carr' • *Rosa* 'Smarty' • *Thymus* Coccineus Group ♀

Plants for allergy sufferers

Gardens filled with pollen-rich plants are great for attracting bees and butterflies, but they can also promote allergy attacks.

If you are a sufferer, choose plants with complex flower forms, such as fully double roses or pompon dahlias, whose intricate petals trap pollen inside the blooms. Weeds can produce large quantities of pollen, so remove them promptly before they flower, and do not plant ornamental grasses which, like their wild cousins, are responsible for the majority of allergic reactions. As well as planting low-allergen flowers, consider a boundary hedge, which may help reduce allergy attacks by trapping wind-borne pollen and preventing it from flying into your garden.

PERENNIAL LARGE

Agapanthus **Headbourne hybrids**

This perennial has strap-shaped foliage and spherical clusters of blue, late summer to early autumn flowers that produce little pollen. Grow it in a sheltered site in free-draining soil; mulch over winter.

↕1.2m (4ft) ↔60cm (24in)

PERENNIAL LARGE

Agave americana

CENTURY PLANT A low-allergen plant, this evergreen perennial has lance-shaped, sharply pointed, grey-green leaves and, in mature plants, cream flowers. Grow it in a pot; overwinter indoors. 'Variegata' AGM (above) has cream-edged leaves.

↕↔ up to 1.5m (5ft)

PERENNIAL SMALL

Ajuga reptans **'Atropurpurea'**

BUGLE This evergreen, rhizome-forming perennial, with small rosettes of glossy, deep bronze-purple leaves, spreads freely by runners and makes an excellent ground cover. Short spikes of blue flowers, which produce little pollen, appear in spring.

↕15cm (6in) ↔90cm (36in)

BULB LARGE

Allium **'Purple Sensation'**

This perennial bulb has sturdy stems topped with spherical heads of purple early summer flowers, which have a low pollen content. Plant the bulbs in groups in autumn between perennials that will disguise the leaves, which fade as the flowers appear.

↕80cm (32in)

PERENNIAL MEDIUM

Campanula trachelium

NETTLE-LEAVED BELLFLOWER This tough perennial has nettle-like, toothed-edged leaves and tall, leafy stems that hold clusters of bell-shaped, blue to lilac flowers in summer. 'Bernice' has double, lilac-blue flowers that are ideal for allergy sufferers.

↕up to 1m (3ft) ↔30cm (12in)

BULB LARGE

Dahlia **'Franz Kafka'**

This pompon dahlia produces mauve-pink blooms that contain little pollen. The flowers appear on long stems above green, divided leaves and look good in borders or large pots of gritty, soil-based compost.

↕80cm (32in)

SHRUB LARGE

Fargesia murielae

UMBRELLA BAMBOO This large, clump-forming evergreen bamboo is perfect for allergy sufferers, since it rarely produces flowers. Use the arching, yellow-green canes and lance-shaped, bright green leaves to create a screen or focal point.

↕4m (12ft) ↔indefinite

PERENNIAL MEDIUM

Geranium psilostemon

A clump-forming perennial, ideal for a low-allergen garden. This species produces lobed, toothed, mid-green leaves that turn red in autumn and, from early- to late summer, saucer-shaped magenta flowers, with black centres and veins.

↕↔60cm (24in)

ANNUAL/BIENNIAL MEDIUM

Helianthus annuus

SUNFLOWER A robust annual, with heart-shaped leaves and large, daisy-like, orange or yellow flowers in summer, followed by edible seedheads. Double-flowered varieties, such as 'Teddy Bear' (above), are good choices for allergy sufferers.

↕up to 90cm (36in) ↔60cm (24in)

PERENNIAL SMALL

Hemerocallis 'Arctic Snow' ♡

DAYLILY The strap-like, green leaves of this deciduous, low-allergen perennial form a foil for the sturdy stems of cream blooms, with yellow margins and throats. The flowers appear in succession from summer to early autumn, each lasting a day.

↕55cm (22in) ↔50cm (20in)

PERENNIAL SMALL

Hosta 'Halcyon' ♡

This leafy, compact perennial provides both colour and texture in low-allergen, shady gardens. The long, heart-shaped, blue-grey leaves have a ribbed texture, and are joined briefly by tall stems of lavender-grey flowers in summer.

↕40cm (16in) ↔70cm (28in)

PERENNIAL MEDIUM

Iris sibirica

SIBERIAN IRIS Use this moisture-loving, clump-forming perennial in a low-allergen garden, where it will decorate borders with its broad, grass-like leaves and blue, purple, or white blooms from early- to midsummer. The flowers open from tapered buds.

↕1m (3ft) ↔60cm (24in)

ANNUAL/BIENNIAL SMALL

Petunia PHANTOM

Plant this eye-catching annual in pots or baskets filled with multi-purpose compost to decorate a low-allergen patio garden. It produces a mound of trumpet-shaped, almost black flowers, with a golden central star, and oval, mid-green leaves.

↕↔30cm (12in)

SHRUB LARGE

Phyllostachys viridiglaucescens

A clump-forming, evergreen bamboo that rarely flowers and is ideal for a low-allergen garden. It has bright green, linear leaves and greenish brown canes that mature to yellow-green. Cut dead and weak canes in spring.

↕8m (25ft) ↔indefinite

PERENNIAL MEDIUM

Polemonium caeruleum

JACOB'S LADDER A clump-forming, upright perennial, with finely divided, ferny foliage. Small, lavender-blue flowers with orange-yellow stamens, which have a low pollen content, appear in early summer. Plants self-seed freely; may become invasive.

↕↔60cm (24in)

CLIMBER MEDIUM

Rosa 'Aloha' ♡

This climbing hybrid tea rose produces fragrant fully double, rose- and salmon-pink flowers, which have little pollen, in summer and again in autumn. The dark green foliage is disease resistant. Train the stems over an arch or on a wall.

↕2.5m (8ft)

PERENNIAL MEDIUM

Veronica spicata

This perennial bears spikes of star-shaped, blue, pink, white, or purple summer flowers that produce very little pollen. The blooms appear above slim, lance-shaped, green foliage. 'Romily Purple' (above) is a popular variety. Use to create tall accents in borders.

↕60cm (24in) ↔45cm (18in)

PERENNIAL SMALL

Viola cornuta ♡

HORNED VIOLET A dainty low-allergen, evergreen perennial, with oval, toothed leaves and flat-faced, purplish blue, at times white, flowers from spring to late summer. Ideal for ground cover and for containers filled with soil-based compost.

↕20cm (8in) ↔20cm (8in) or more

OTHER SUGGESTIONS

Annuals and biennials

Nigella damascena

Perennials and bulbs

Astilbe 'Fanal' ♡ • *Begonia semperflorens* • *Cerastium tomentosum* • *Dahlia* 'Natal' • *Digitalis purpurea* • *Dryopteris affinis* ♡ • *Geranium himalayense* • *Geranium macrorrhizum* • *Hosta* 'Honeybells' • *Iris unguicularis* • *Narcissus* spp. • *Pelargonium peltatum* • *Polemonium* 'Lambrook Mauve' • *Sidalcea malviflora*

Shrubs and climbers

Camellia japonica • *Clematis montana* • *Clematis alpina* • *Fuchsia magellanica* ♡ • *Hebe* 'Red Edge' ♡ • *Juniperus conferta* • *Rhododendron decorum* ♡ • *Rosa* 'Compassion' ♡ • *Viburnum tinus* • *Weigela florida*

Pest-proof plants: slugs

Slug-resistant plants include many beautiful species, and there are ranges to suit a variety of different sites and soils.

Slugs attack a wide range of plants and cause severe damage to them as they nibble at the tender, succulent leaves and stems. They are, however, less keen on woody plants and aromatic foliage. While some plants are never attacked by slugs, including cosmos and delphiniums, others are only consumed when young and tender, but become relatively mollusc-resistant once mature. To protect them during this vulnerable stage, use small quantities of slug pellets, encircle plants with copper collars – which give slugs a mild electric shock – or surround the stems with sand and grit.

PERENNIAL LARGE

Aconitum carmichaelii

MONK'S HOOD Suitable for the back of a border, this upright perennial has deeply divided, green leaves and tall spikes of hooded, lavender-blue flowers in autumn. It requires support. 'Arendsii' AGM (above) has rich blue flowers.

↕1.5m (5ft) ↔30cm (12in)

PERENNIAL SMALL

Alchemilla mollis 🏆

LADY'S MANTLE A clump-forming perennial, it has rounded, pale green leaves, with crinkled edges and, in summer, sprays of small, lime green blooms. Best planted in shade, it is ideal for the front of a border. Trim untidy leaves in summer.

↕↔50cm (20in)

BULB MEDIUM

Allium cristophii 🏆

STAR OF PERSIA This perennial bulb has short-lived, grey-green leaves that wither in early summer, when its large, rounded heads of star-shaped, violet flowers form. These dry *in situ*, giving slug-proof interest into late summer. Ideal for borders and pots.

↕40cm (16in)

PERENNIAL MEDIUM

Aquilegia vulgaris var. *stellata*

An upright perennial with tough, divided leaves and tall stems of double, pompon-like blooms from late spring to early summer. Deadhead regularly to prevent unwanted self-seeding. 'Nora Barlow' (above) has green-tipped pink blooms.

↕75cm (30in) ↔50cm (20in)

PERENNIAL MEDIUM

Astilbe 'Venus'

This perennial has coarse, heavily divided leaves, unpalatable to slugs and, in midsummer, feathery, upright plumes of tiny, pale pink flowers, which dry and provide interest into winter. It is best in damp, fertile soil in sun or light shade.

↕↔1m (3ft)

PERENNIAL MEDIUM

Crocosmia x *crocosmiiflora*

MONTBRETIA This upright perennial forms clumps of arching, sword-like, green leaves and, from summer to autumn, flared, trumpet-shaped flowers in shades of yellow, orange, and red. Varieties include 'Star of the East' AGM (above).

↕70cm (28in) ↔8cm (3in)

PERENNIAL MEDIUM

Astrantia major

MASTERWORT A clump-forming perennial, with divided, mid-green leaves. It produces sprays of small, greenish white, pink, or red flowers from midsummer to early autumn. Deadhead regularly to prolong the display. It is best in moist but free-draining soil.

↕60cm (24in) ↔45cm (18in)

PERENNIAL SMALL

Bergenia purpurascens 🏆

PURPLE BERGENIA Perfect for the front of a border, this evergreen perennial has large, fleshy, rounded leaves that spread to form a clump, and flush purple in cold weather. It bears dark pink spring flower clusters. Cut damaged leaves in summer and autumn.

↕40cm (16in) ↔60cm (24in) or more

PERENNIAL SMALL

Corydalis lutea

A mound-forming, evergreen perennial, with delicate, ferny, grey-green leaves and, from late spring to summer, heads of small, tubular, yellow blooms. It is suitable for shady and wild gardens but self-seeds freely, and may be invasive.

↕↔30cm (12in)

PERENNIAL SMALL

Dianthus 'Dad's Favourite'

GARDEN PINK Forming a low mound of resilient, grey-green, grass-like leaves, this evergreen perennial also bears scented, semi-double, white blooms, with maroon markings, all summer. All forms make good edging for a sunny border.

↕up to 45cm (18in) ↔30cm (12in)

PERENNIAL LARGE

Echinops ritro

GLOBE THISTLE This upright perennial has coarse, prickly, divided, green leaves and spiky, metallic blue, globe-shaped flowerheads in late summer. Plant it at the back of a border. Dried flowerheads are short-lived, but may last into winter.

↕1.2m (4ft) ↔75cm (30in)

PERENNIAL SMALL

Epimedium perralderianum

BARRENWORT Ideal for ground cover in shady borders and woodland gardens, this semi-evergreen perennial has green, heart-shaped foliage on branching stems. Prune in spring before new leaves and spikes of pendent, bright yellow blooms appear.

↕30cm (12in) ↔45cm (18in)

ANNUAL/BIENNIAL SMALL

Eschscholzia californica

CALIFORNIA POPPY This quick-growing annual has finely cut, grey-green leaves and large, cup-shaped flowers that form throughout summer in shades of yellow, orange, and pink. Sow seeds directly in a sunny border. Deadhead regularly.

30cm (12in) ↔15cm (6in)

PERENNIAL MEDIUM

Euphorbia griffithii

This bushy perennial has copper-flushed, green leaves and domed clusters of orange-red flowers in early summer. Suitable for light shade, it spreads to form a large clump, and is best suited to large gardens. The sap is a skin irritant.

↕75cm (30in) ↔50cm (20in)

PERENNIAL SMALL

Euphorbia myrsinites

SPURGE An evergreen perennial with creeping, woody stems and small, pointed, fleshy, grey leaves. It bears clusters of bright yellow-green flowers in spring. Grow in full sun and let it spread along the front of a border. Remove spent flowers.

↕8cm (3in) ↔20cm (8in)

SHRUB SMALL

Fuchsia 'Swingtime'

This shrub has dark green leaves and is grown for its pendent, frilly, red and white flowers that are produced throughout summer. Tolerant of partial shade, it is suitable for borders and large containers. Bring plants indoor in winter in cold regions.

↕↔1m (3ft)

BULB MEDIUM

alanthus nivalis

NOWDROP This perennial bulb forms umps of grassy, grey-green leaves and, in te winter, single, nodding, white flowers, ith green-tipped inner petals. It emerges efore most slugs are active, so escapes amage. Plant in groups and let it naturalize.

5cm (6in)

PERENNIAL SMALL

Galium odoratum

SWEET WOODRUFF This perennial has rounded clusters of lance-shaped, green leaves, which form a spreading carpet. From late spring to midsummer, it bears a mass of tiny, white, star-shaped blooms. Ideal for naturalistic and woodland gardens.

↕15cm (6in) ↔30cm (12in) or more

PERENNIAL SMALL

Geranium macrorrhizum

CRANESBILL Forming a spreading carpet, this perennial has aromatic, deeply lobed leaves and abundant small, magenta blooms in early summer. Ideal for the front of a border or beneath large shrubs, where it can be left to spread. It can be invasive.

↕38cm (15in) ↔60cm (24in)

PERENNIAL MEDIUM

Geranium phaeum

DUSKY CRANESBILL An upright, clump-forming perennial, with lobed, green leaves and nodding, maroon or white flowers, carried on lax stems, from late spring to early summer. Plant it beneath shrubs or hedges, where it will slowly spread.

↕75cm (30in) ↔45cm (18in)

PERENNIAL SMALL

Helleborus x *hybridus*

LENTEN ROSE Ideal for early-season colour, this evergreen, clump-forming perennial has divided foliage and nodding, cup-shaped, white, pink, or purple flowers from winter to early spring. Deadhead to prevent unwanted self-seeding.

↕↔60cm (24in)

PERENNIAL SMALL

Hemerocallis 'Stella de Oro'

DAYLILY This perennial has narrow, upright, arching, glossy, green foliage and, from mid- to late summer, large, yellow, trumpet blooms, each lasting a day. There are many varieties to grow, flowering in a wide range of colours.

↕30cm (12in) ↔45cm (18in)

SHRUB MEDIUM

Hydrangea macrophylla

LACECAP HYDRANGEA A bushy shrub, with broad, green leaves and flattened, pink or blue flowerheads in summer. Blue flowers will only return their colour on acid soils. 'Lanarth White' AGM (above) is a popular variety.

↕↔1.5m (5ft)

SHRUB LARGE

Hydrangea paniculata

PANICULATE HYDRANGEA This deciduous, upright shrub has oval, pointed, green, deeply veined leaves and frilly cones of pink or white flowers, depending on the variety, at the stem tips in late summer. Best in large gardens, plant at the back of a border.

↕↔3m (10ft)

PERENNIAL MEDIUM

Knautia macedonica

MACEDONIAN SCABIOUS This perennial, grown for its crimson, pincushion-like blooms, held on upright, wiry stems above deeply dissected foliage, suits naturalistic planting schemes. 'Melton Pastels' (above) has a mix of pink and dark red flowers.

↕75cm (30in) ↔60cm (24in)

SHRUB SMALL

Lavandula angustifolia

ENGLISH LAVENDER This shrub is grown for its silver-green foliage and heads of small, tubular summer blooms. Plant in full sun in free-draining soil. Also suitable for containers, it has a long season of interest. Slugs find the aromatic growth unpalatable.

↕80cm (32in) ↔60cm (24in)

BULB LARGE

Lilium henryi

This upright, perennial bulb has tall, arching stems clothed with pointed, green leaves and topped with bright orange, turks-cap flowers, with spotted faces. It is a good choice for a shady border and can be left to slowly spread.

↕1m (3ft)

PERENNIAL MEDIUM

Papaver orientale

ORIENTAL POPPY This upright perennial has divided, mid-green leaves and large, saucer-shaped, pink, white, mauve, or red blooms in summer. Its flowering season is short, but can be spectacular. 'Karine' AGM (above) has salmon-pink flowers.

↕↔up to 90cm (36in)

PERENNIAL SMALL

Pelargonium Multibloom Series

GERANIUM Often used in annual bedding displays, this perennial has green, rounded leaves, with red markings, and white, pink, or red flowers throughout summer. Water, feed, and deadhead regularly.

↕↔30cm (12in)

PERENNIAL SMALL

Penstemon 'Evelyn' ♈

Suitable for summer borders and containers, this upright perennial has slender, pointed foliage and bears spikes of trumpet-shaped, white, pink, red, and mauve blooms from summer to early autumn. Mulch in autumn.

↕↔ 45cm (18in)

PERENNIAL MEDIUM

Potentilla fruticosa

This deciduous perennial, from summer to early autumn, bears simple, rounded, red, white, yellow, and orange flowers, set against small, dissected, green leaves. It is suitable for well-drained borders. Varieties include 'Red Ace' (above).

↕ 1m (3ft) ↔ 1.5cm (5ft)

PERENNIAL MEDIUM

Rudbeckia hirta

This upright, short-lived perennial, often grown as an annual, bears daisy-like, brown-centred golden or red flowers from summer to autumn. Plant it at the front of a sunny border or in a summer container. Varieties include 'Becky Mixed' (above).

up to 90cm (36in) ↔ 45cm (18in)

PERENNIAL MEDIUM

Polemonium caeruleum

JACOB'S LADDER This clump-forming perennial bears clusters of blue or white, bell-shaped flowers in early summer, above the divided, lance-shaped leaves. Ideal for shady borders in naturalistic and woodland-style gardens.

↕↔ 60cm (24in)

PERENNIAL SMALL

Pulmonaria officinalis

LUNGWORT A semi-evergreen perennial, with coarse, bristly, green leaves, which are disliked by slugs. It bears funnel-shaped, pink, blue, or white spring blooms, depending on the variety. 'Sissinghurst White' AGM (above) has white flowers.

↕ 30cm (12in) ↔ 60cm (24in)

PERENNIAL MEDIUM

Santolina chamaecyparissus

COTTON LAVENDER This evergreen perennial forms a mound of finely cut, aromatic, silvery grey foliage. It bears yellow, pompon-like flowers in summer. Best in full sun, it is suitable for borders. Trim off the spent flowers to keep it tidy.

↕ 75cm (30in) ↔ 1m (3ft)

PERENNIAL LARGE

Polystichum setiferum ♈

SOFT SHIELD FERN This deciduous, clump-forming fern is grown for its large, arching fronds, consisting of tiny leaflets, held on lateral stems that emerge from a central crown. Best in full or partial shade, it is ideal for woodland gardens.

↕ 1.2m (4ft) ↔ 90cm (36in)

PERENNIAL MEDIUM

Verbascum chaixii

MULLEIN This perennial forms a rosette of large, coarse, green leaves, and is grown for its upright spikes of pale yellow flowers during summer. Grow it in a sunny border. Varieties include 'Album' (above), with purple-eyed, white blooms.

↕ 90cm (36in) ↔ 45cm (18in)

CLIMBER LARGE

Rosa banksiae 'Lutea' ♈

YELLOW BANKSIAN Too vigorous to be affected by slugs, this climber bears masses of small, pale yellow, double flowers that clothe the thornless stems during late spring in a single, spectacular flush. Suited to large gardens, it needs a large support.

↕ 10m (30ft)

OTHER SUGGESTIONS

Perennials and grasses

Aquilegia vulgaris var. *stellata* 'Black Barlow' • *Bergenia cordifolia* • *Calamagrostis emodensis* • *Chionochloa conspicua* • *Crocosmia masoniorum* ♈ • *Delphinium elatum* • *Dianthus* 'Doris' ♈ • *Dianthus alpinus* ♈ • *Digitalis purpurea* • *Euphorbia dulcis* 'Chameleon' • *Geranium nodosum* • *Lythrum virgatum* • *Miscanthus sinensis* • *Pelargonium tomentosum* ♈ • *Penstemon* 'Garnet' • *Sedum niveum* ♈

Bulbs

Allium 'Purple Sensation' ♈ • *Galanthus* 'Atkinsii' ♈ • *Tulipa* species

Shrubs and climbers

Cotinus coggygria • *Forsythia* x *intermedia* • *Hedera helix* 'Erecta' • *Kerria japonica* • *Leycesteria formosa* • *Parthenocissus henryana* ♈ • *Santolina rosmarinifolia*

Pest-proof plants: rabbits and deer

Deer and rabbits have an appetite for garden plants, crops, and trees and can be a menace in gardens in the countryside or close to parks and open spaces.

Both rabbits and deer tend to nibble on anything that we find tasty or ornamental. One way to keep them at bay is to install barriers. To keep out rabbits, you will need a chicken wire fence that is at least 1m (3ft) high and buried 7cm (3in) in the ground to prevent them from digging under it. Wire mesh fences 2m (6ft) tall are required to keep deer out, or plant a tall hedge around your garden using a pest-proof plant, such as *Lonicera nitida* or *Forsythia*. If you do not have the resources, or it is not feasible to erect fences, opt for plants, such as *Euphorbia griffithii*, and those described here, which these pests find unpalatable.

PERENNIAL LARGE

Achillea filipendulina

YARROW This perennial forms a clump of finely cut, aromatic, grey-green leaves and bears large, flat-topped, bright yellow flowerheads on sturdy stems in summer. The flowers can be cut and dried. Varieties include 'Gold Plate' AGM (above).

↕1.2m (4ft) ↔60cm (24in)

PERENNIAL LARGE

Aconitum napellus

MONKSHOOD This upright perennial has deeply divided, dark green foliage and bears tall spires of indigo-blue blooms from mid- to late summer. Best planted in a shady site, it needs additional support. All parts are toxic if ingested, handle it with gloves.

↕1.5m (5ft) ↔30cm (12in)

SHRUB LARGE

Aesculus parviflora

BOTTLEBRUSH BUCKEYE Best suited to large gardens, this deciduous shrub has palm-shaped leaves and bears spires of fluffy, white flowers from mid- to late summer. It has a suckering habit and spreads to form a large clump of stems.

↕3m (10ft) ↔5m (15ft)

PERENNIAL MEDIUM

Artemisia ludoviciana 'Silver Queen'

WESTERN MUGWORT This clump-forming perennial is grown for its slender, aromatic, silvery grey leaves carried on upright stems. In summer, it bears small, yellow flowers, which are best removed.

↕↔75cm (30in)

PERENNIAL MEDIUM

Asplenium scolopendrium

HART'S-TONGUE FERN This evergreen fern forms a small clump of long, tapering, tough, glossy, green fronds, which are replenished in late spring. Plant beneath trees and taller shrubs, or as ground cover in shady borders. Remove damaged leaves in spring.

↕75cm (30in) ↔45cm (18in)

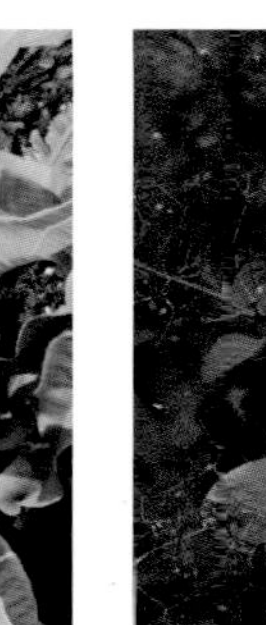

SHRUB LARGE

Cotinus coggygria

SMOKE BUSH Grown for its colourful leaves, this shrub developes fiery autumn tints and plumes of smoke-like summer flowers. 'Royal Purple' AGM (above) has rich purple leaves and flower plumes. Plant it near a border as a deer-deterrent screen.

↕↔5m (15ft)

PERENNIAL MEDIUM

Euphorbia griffithii

This bushy perennial has red stems of narrow, copper-tinged, dark green leaves, topped with clusters of orange-red flowers during early summer. Suited to larger borders, it spreads fast. Animals find the milky, toxic sap unpalatable.

↕75cm (30in) ↔50cm (20in)

SHRUB SMALL

Fuchsia 'Tom Thumb'

Suitable for borders and containers, this dwarf variety has oval, dark green leaves and flowers from summer to autumn, bearing bell-shaped, red and mauve blooms. If attacked by deer or rabbits in spring or summer, the plant recovers soon.

↕↔50cm (20in)

PERENNIAL SMALL

Geranium sanguineum

BLOODY CRANESBILL This spreading perennial forms a neat mound of deeply dissected, dark green leaves and bears single, magenta-pink flowers in summer. Tolerant of sun or shade, it can be planted in many positions in the garden.

↕25cm (10in) ↔30cm (12in) or more

PERENNIAL SMALL

Helleborus x *hybridus*

LENTEN ROSE With dense, divided foliage this evergreen, clump-forming perennial bears nodding, cup-shaped, white, pink, or purple flowers in winter and early spring. Plants hybridize readily and are highly variable. Deadhead to prevent self-seeding.

↕↔60cm (24in)

SHRUB SMALL

Hypericum 'Hidcote' ♡

ST JOHN'S WORT This evergreen or semi-evergreen shrub forms a dense bush of narrow, dark green leaves and has masses of large, golden-yellow flowers from midsummer to early autumn. Use it as a low barrier to deter deer and rabbits.
↕1.2m (4ft) ↔1.5m (5ft)

PERENNIAL SMALL

Lamium maculatum

This semi-evergreen, low-growing perennial has white-variegated green foliage, with spikes of hooded, white, pink, or purple flowers in late spring. Grow it as ground cover. 'White Nancy' (above) has pure white flowers and silver leaves.
↕15cm (6in) ↔1m (3ft)

TREE LARGE

Liquidambar styraciflua

SWEET GUM This deciduous tree is grown for its lobed, aromatic, glossy green leaves that turn brilliant shades of orange, red, and purple in autumn. Plant it near borders as a deer-resistant barrier. Compact varieties for small garden are also available.
↕25m (80ft) ↔12m (40ft)

PERENNIAL MEDIUM

Paeonia officinalis

PEONY A resilient, bushy perennial, it has divided, dark green leaves and tough, bowl-shaped, pink, red, or white flowers from early- to midsummer. Grow it mid-border among later flowering plants. 'Rubra Plena' AGM (above) has double, red blooms.
↕↔75cm (30in)

PERENNIAL MEDIUM

Physalis alkekengi ♡

CHINESE LANTERN A spreading perennial that gives good autumn colour, it has upright stems and blooms in midsummer, producing cream flowers that lead to decorative, orange "lanterns" containing a berry-like fruit. Best in a larger border, provide support.
↕75cm (30in) ↔90cm (36in)

PERENNIAL MEDIUM

Polystichum acrostichoides

CHRISTMAS FERN This evergreen fern has tough, lance-shaped, dark green fronds, divided into comb-like segments. Slow-growing, it forms a clump over time and grows in most soils. Ideal for woodland gardens, plant it beneath larger shrubs.
↕60cm (24in) ↔45cm (18in)

SHRUB LARGE

Rhododendron luteum ♡

YELLOW AZALEA This deciduous shrub is grown for its fragrant, yellow spring flowers and vivid autumnal foliage. Ideal for large and woodland-style gardens, it recovers quickly if attacked by deer. It prefers a cool site with fertile, acid soil.
↕↔4m (12ft)

PERENNIAL MEDIUM

Rudbeckia fulgida

This perennial has narrow, green leaves and bears daisy-like, golden flowers with brown centres from late summer to autumn. The flowers fade, but the centres remain, offering winter interest. Plant it mid-border and stake the stems.
↕90cm (36in) ↔45cm (18in)

PERENNIAL MEDIUM

Sedum telephium Atropurpureum Group

ORPINE This perennial is grown for its fleshy, purple stems and leaves, and domed heads of pink-white flowers, which appear from late summer to autumn, then dry and persist into winter. Provide support.
↕60cm (24in) ↔30cm (12in)

PERENNIAL MEDIUM

Trollius x *cultorum*

GLOBEFLOWER A clump-forming, upright perennial, ideal for wildlife gardens, with lobed, toothed, dark green leaves and bowl-shaped, buttercup-like flowers from late spring to early summer. 'Lemon Queen' (above) has pale yellow blooms.
↕60cm (24in) ↔45cm (18in)

OTHER SUGGESTIONS

Annuals and biennials

Digitalis purpurea • *Tagetes patula* • *Zinnia elegans*

Perennials and grasses

Acanthus mollis • *Agapanthus* 'Blue Giant' • *Ajuga reptans* • *Aquilegia vulgaris* • *Aster ericoides* • *Aster novi-angliae* • *Aster novi-belgii* • *Bergenia stracheyi* • *Echinops ritro* ♡ • *Kniphofia uvaria* • *Liriope muscari* ♡ • *Lysimachia clethroides* ♡ • *Miscanthus sinensis* • *Pulmonaria saccharata* • *Sisyrinchium striatum* • *Verbena bonariensis* ♡ • *Viola odorata*

Bulbs

Allium schoenoprasum • *Chinodoxa lucilae* • *Narcissus* 'Mount Hood' ♡

Shrubs

Ceanothus species • *Fatsia japonica* ♡ • *Hydrangea* species • *Ruscus aculeatus*

Index

A

B

C

F

G

H

I

J

K

L

M

N

O

P

Acknowledgments

Picture credits

The publisher would like to thank the following for their kind permission to reproduce their photographs:
(Key: a-above; b-below/bottom; c-centre; f-far; l-left; r-right; t-top)

Every effort has been made to trace the copyright holders. Dorling Kindersley apologizes for any unintentional omissions and would be pleased, if any such case should arise, to add an appropriate acknowledgement in future editions.

4 Dorling Kindersley: Arnie Maynardrhs/ Designer: RHS Chelsea Flower Show 2012 (br). **5 Dorling Kindersley:** Arnie Maynardrhs/ Designer: RHS Chelsea Flower Show 2012 (bl). **11 Neil Fletcher:** (clb). **14 Dorling Kindersley:** Brian North (bl). **16 Dorling Kindersley:** Glazenwood, Essex (bl). **17 Lucy Claxton:** (tr). **24 Dorling Kindersley:** Thomas Hoblyn/ Designer: RHS Chelsea Flower Show 2012 (c). **27 Dorling Kindersley:** Kari Beardsell/ Designer: RHS Hampton Court 2008 (b). **33 Dorling Kindersley:** Andy Sturgeon/ Designer: RHS Chelsea Flower Show 2012 (t). **44 Dorling Kindersley:** Waterperry Gardens (tr). **52 Dorling Kindersley:** Zia Allaway (bc). **63 Dorling Kindersley:** Claudia de Yong/ Designer: RHS Hampton Court Flower Show 2008. **96 Dorling Kindersley:** Ayletts Nurseries, St Albans (br). **100 Alan Buckingham** (c). **Dorling Kindersley:** (tl). **101 Lucy Claxton:** (c). **Alan Buckingham** (c) **Dorling Kindersley:** (bc/ Acid cherries). **102 Alan Buckingham** (c). **Dorling Kindersley:** (bl, tl, fcl, cl). **103 Alan Buckingham** (c) **Dorling Kindersley:** (bc). **119 Lucy Claxton:** (bc). **Dorling Kindersley:** Ness Botanic Gardens/Designer: RHS Tatton Park Flower Show 2008 (t). **126 Dorling Kindersley:** Waterperry Gardens (bc). **132 Dorling Kindersley:** Geoff Whiten/Designer: RHS Tatton Park Flower Show 2008 (cra). **133 Dorling Kindersley:** Dan Miller/Designer: RHS Tatton Park Flower Show 2008. **150 Dorling Kindersley:** Waterperry Gardens (tr). **157 Dorling Kindersley:** Noel Duffy/Designer: RHS Hampton Court Flower Show 2008 (crb). **183 Dorling Kindersley:** Waterperry Gardens (c). **188 Getty Images:** Photodisc/S. Solum/ PhotoLink (tc/Peppermint). **190 Dorling Kindersley:** Linda Fairman of Oasthouse Nursery/ Designer: Hampton Court 2012 (cr). **222 Lucy Claxton:** (bc). **223 Dorling Kindersley:** Jack Dunckley/Designer: RHS Hampton Court Flower Show 2012 (clb); Brian North (tr). **224 Dorling Kindersley:** OneAbode Ltd/Designer: RHS Hampton Court Flower Show 2012 (bl). **225 Dorling Kindersley:** Nillufer Danis/Designer: RHS Hampton Court Flower Show 2012 (l). **227 Dorling Kindersley:** Paul Dyer/Designer: RHS Tatton Park Flower Show 2008 (tl). **228–229 Dorling Kindersley:** Chris Beardshaw/Designer: RHS Chelsea Flower Show 2012 (b). **230 Dorling Kindersley:** Chris Beardshaw/Designer: RHS Chelsea Flower Show 2012 (cr). **232 Dorling Kindersley:** Arnie Maynardrhs/Designer: RHS Chelsea Flower Show 2012. **266 Dorling Kindersley:** Waterperry Gardens (c). **305 Justyn Willsmore:** (tc/*Pittosporum tenuifolium*). **310 Dorling Kindersley:** Arnie Maynardrhs/ Designer: RHS Chelsea Flower Show 2012. **320 Lucy Claxton:** (bc). **350 Dorling Kindersley:** Waterperry Gardens (tr). **372 Dorling Kindersley:** Waterperry Gardens (bc). Designer: RHS Chelsea Flower Show 2012 **Dorling Kindersley:** Fire and Iron and Castle Gardens/Designer: RHS Hampton Court Flower Show 2008 (-l)

All other images ©Dorling Kindersley
For further information see: www.dkimages.com

Dorling Kindersley would like to thank the following:

DK photographers Peter Anderson, Brian North

Design assistance Elaine Hewson, Neha Wahi, Pooja Verma

Editorial assistance Alexander Scammell, Arani Sinha, K Nungshithoibi Singha, Aditi Batra, Dorothy Kikon, Charis Bhagianathan, Manasvi Vohra

Indexing Jane Coulter